Essentials of Human Anatomy and Physiology

FIFTH EDITION

Elaine N. Marieb, R.N., Ph.D.

Edison Community College
Holyoke Community College, Emeritus

An imprint of Addison Wesley Longman, Inc.

Menlo Park, California • Reading, Massachusetts • New York • Harlow, England
Don Mills, Ontario • Sydney • Mexico City • Madrid • Amsterdam

Executive Editor: Johanna Schmid

Sponsoring Editor: Daryl Fox

Assistant Editor: Thor Ekstrom

Managing Editor: Wendy Earl

Production Editor: Bradley Burch

Photo Editor: Lisa Lougee

Text Designer: Gary Head, Gary Head Design

Cover Designer: Juan Vargas, Vargas/Williams/Design

Composition Supervisor: Lillian Hom

Manufacturing Supervisor: Merry Free Osborn

This edition is dedicated to the memory of
"Zeb," my husband of 38 years.
He died July 22, 1996.
This one is all for you, my love.

Library of Congress Cataloging-in-Publication Data
Marieb, Elaine Nicpon
    Essentials of human anatomy & physiology /
Elaine N. Marieb.—5th ed.
        p.   cm.
    Includes index.
    ISBN 0-8053-4185-4
    1. Human physiology.   2. Human anatomy.   I. Title.
    [DNLM: 1. Anatomy.   2. Physiology.   QS 4 M334e
    1996]
QP34.5.M455   1996
612—dc20
DNLM/DLC   96-2396
for Library of Congress   CIP

ISBN 0-8053-4185-4
1 2 3 4 5 6 7 8 9 10—VH—00 99 98 97 96

2725 Sand Hill Road
Menlo Park, California 94025

The Benjamin/Cummings Series in Human Anatomy and Physiology

Preface to the Instructor

The fifth edition of *Essentials of Human Anatomy and Physiology,* like those that preceded it, is designed to introduce students pursuing careers in the allied health fields to the structure and function of the human body. Sections dealing with timely topics, such as immunology and clinical advances, have been updated to ensure that explanations are both accurate and current. In response to reviewers' requests and suggestions, the discussion of joints now has a structural (instead of a functional) basis and an A.D.A.M. Correlations Appendix has been added. Most importantly, this edition retains the writing style and presentation that, in former editions, has successfully helped students with limited backgrounds in the sciences grasp the fundamental concepts of human anatomy and the inner workings of the body.

ORGANIZATION

Instructors often prefer different topic sequences in their presentations and lectures on body systems. Extensive reviews by anatomy and physiology instructors helped me decide on a particular presentation. My choice was to construct a sequence of topics that encourages learning. The first chapter orients students to the new world of learning in which they will soon become immersed in anatomical terminology, names of major body areas, and brief overviews of functions that must be performed to sustain life.

Chemistry, cells, tissues, and the first organ system (skin) are treated in succession. Most students find that the skin is an interesting system, and thus

the transition to the organ system discussions is easily accomplished. The subsequent chapters include systems requiring a good deal of anatomical terminology (for example, skeletal, muscular, and nervous systems). Each organ system is approached from simple to increasingly complex levels. Building-in an understanding of the concepts, rather than rote memorization, is emphasized.

ORGANIZATIONAL FLEXIBILITY

A good textbook should meet the needs of both instructors and students. Because every instructor has a personal style and philosophic approach to teaching and course organization, I have written each chapter to allow for flexibility in topic presentation. All chapters are self-contained, hence an instructor may choose a chapter sequence that is personally preferred. Topic omission may be necessary in programs in which less than one semester is devoted to anatomical and physiological concepts.

NEW TO THIS EDITION

Much thought was given to the learning process when designing and incorporating the unique features in this edition. These new features include:

- **Increased coverage of key topics.** In response to reviewers' feedback, the fifth edition includes more material on bone growth, energy sources for muscle activity, and the body's handling of cholesterol. New art has been added to support the text and aid understanding of these complicated, but important, processes.

- **New "Systems in Sync" feature.** This new feature summarizes in simple form for the student the "must know" information about the interactions that occur between the system being studied and all other body systems. This type of approach, provided for each body system, helps to emphasize that the body is a dynamic composite of interdependent parts.

- **New A.D.A.M. Correlations guide.** Provided as a new appendix (Appendix D), this handy reference guide links text figures to appropriate views and corresponding layers in A.D.A.M. Standard.® This appendix joins those present in the previous edition, i.e., Appendix A which lists important word roots, prefixes, and suffixes; Appendix B, the Periodic Table; and Appendix C which provides key information on vitamins and minerals.

SPECIAL FEATURES RETAINED FROM PREVIOUS EDITIONS

- **Presentation of anatomy, physiology, and clinically important diseases.** The content of this edition continues to present a balance of anatomic and physiologic concepts, along with discussions of diseases and dysfunctions where they enhance and reinforce an understanding of normal human structure and function. Additionally, complex topics are explained consistently by the *use of analogies* to foster, rather than discourage, student interest. In instances requiring more than the usual amounts of memorization of terminology and facts, the material is presented in both a text discussion and tabular form to reinforce learning.

- **Writing style.** The writing style is intentionally informal. New terminology is set in boldface or italic type, followed by phonetic pronunciation. Terms are defined within the text and again in the glossary.

- **Art program.** Art plays a critically important role in helping students visualize anatomical and physiological concepts. The art program is full color throughout and has been completely updated. The quality of the art, the labeling, and the text and art integration are designed to retain student attention. Each piece of art is designed to help students understand and remember structure and function, and make the body come alive.

- **Pedagogical devices.** Several pedagogical devices are used to ensure that students learn important terminology and concepts. Each chapter begins with student objectives. A summary of text contents and review questions are found at the end of each chapter. Additionally, a list of word roots, prefixes, and suffixes is included in Appendix A.

- **"At the Clinic" question sections.** Unique "At the Clinic" sections end each chapter and ask questions that help students apply new knowledge to clinical situations.

- **"Closer Look" boxes.** This edition has even more "Closer Look" boxes to capture student interest with new advances in science and intriguing topics. For example, two new boxes explore the dangers of scuba diving (p. 398) and certain diets (p. 438). "Closer Look" box subjects also include timely topics such as AIDS (p. 376), drug abuse (p. 213), and recent medical advances such as the use of light to treat some forms of depression (p. 284).

- **Clinical examples with an emphasis on homeostasis.** In recognition of the fact that the normal and most desirable condition of the body is homeostasis, or maintenance of a stable internal environment, the concept of homeostasis is introduced early (Chapter 1) and is then emphasized throughout the book. Also stressed is the understanding that loss of homeostasis leads to some kind of pathology or disease—temporary or permanent. Thus, pathological conditions are introduced and integrated with the text material as appropriate to clarify normal functioning, not as an end in themselves. The normal functional process is always explained first. The clinical examples chosen are those seen most often by people working in the allied health fields and are intended to familiarize students with the possible consequences when body structures are damaged or functional processes diverge from normal pathways. In each case, the disease conditions are indicated visually by the symbol of a red seesaw that has been tipped off balance (⚖) to remind students that disease is a loss of homeostasis.

- **Developmental aspects and aging.** Each chapter ends with a section on developmental aspects. The section presents the formation of that system in the embryo and follows the changes that occur through old age. Important health problems unique to that system are introduced, and emphasis is placed on problems that accompany the aging process. Because many health workers are intricately involved in care of elderly people, this emphasis should be valuable to today's student.

TEACHING PACKAGE

Teaching anatomy and physiology is a challenging and rewarding endeavor. To assist instructors in that task, our teaching package is designed to help them use and derive the greatest benefit from this text. Please see your local Benjamin/Cummings sales representative if you are interested in additional teaching resources.

- **Instructor's Guide for *Essentials of Human Anatomy & Physiology.*** The *Instructor's Guide,* organized by text chapter for easier instructor use, includes suggested lecture outlines and lecture hints. The classroom demonstrations and activities, the test bank, the answers to end-of-chapter questions, and the list of audiovisual aids have all been updated.

- **Transparency Acetates.** New to this edition is a set of 135 full-color transparency acetates. Available to adopters, the acetates are key illustrations selected from *Essentials,* with type enlarged for easy viewing.

- ***Anatomy & Physiology Coloring Workbook: A Complete Study Guide, Fifth Edition.*** Available for student use, this unique combination of a coloring book and study guide will provide a learning/testing tool of multiple choice, matching, diagram-labeling, and coloring exercises. These exercises complement and enhance the learning and retention of textual material.

- ***Human Anatomy and Physiology Laboratory Manual, Fourth Edition.*** For those instructors who have a lab component, a laboratory manual is available. The manual includes in-class dissection techniques for demonstration by instructor or student.

ACKNOWLEDGMENTS

Many people contributed to my efforts in the creation of the first four editions of this book. I am especially grateful to the many students who allowed me to use them as "sounding boards" to try out my analogies and ideas.

I also would like to thank the following reviewers of this work, both present and past, for their thoughtful critiques: Kay Brashear, El Centro College, Dallas, TX; Yolanda Browne, Del Mar College, Corpus Christi, TX; Ray Canham, Richland College, Dallas, TX; Elizabeth Carl, Nassau Technological Center (BOCES), Westbury, NY; Lydia Chavana, South Texas Vocational Technical College, McAllen, TX; Jean Cons, College of San Mateo, San Mateo, CA; Luci Constant, W.M.L. Dawson School of Practical Nursing, Chicago, IL; Eleanor Shorette Fowler, School of Practical Nursing, Tewksbury Hospital, Tewksbury, MA; Annette Gould, Coordinator, Broome-Delaware-Tioga BOCES, Binghamton, NY; Sue D. Hall, Rogue Community College, Grants Pass, OR; John P. Harley, Eastern Kentucky University, Richmond, KY; Gloria Hillert, Triton College, River Grove, IL; Carol Holley, San Jacinto College, Pasadena, TX; Dawn Holtzmeier, Hocking Technical College, Nelsonville, OH; Mary Hopkins, Kiamichi Voc-Tech, Wilburton, OK; Drucilla B. Jolly, Forsyth Technical College, Winston-Salem, NC; Dennis Kalichstein, Ocean County College, Toms River, NJ; Fred Klaus, East Texas State University, Commerce, TX; Bonnie Kroemmelbein, Upper Bucks County AVTS Practical Nursing Program, Perkasie, PA; John Lammert, Gustavus Adolphus College, St. Peter, MN; Jeri Lindsey, Tarrant County Jr. College, Hurst, TX; Bennie Marshal, STOP/CETA School of Practical Nursing, Norfolk, VA; William Matthai, Tarrant County Jr. College, Hurst, TX; Walter Matulis,

Mid-Michigan Community College, Harrison, MI; Richard McKeeby, Union County College, Cranford, NJ; Roxine McQuitty, Milwaukee Area Technical College, Milwaukee, WI; Lew Milner, North Central Technical College, Mansfield, OH; Lynne Osborn, Middlesex Community College, Bedford, MA; Bridget Price, Vocational Nursing School of California, Anaheim, CA; Dell Redding, Evergreen College, San Jose, CA; Joan Sandell, Gogebic Community College, Ironwood, MI; Leba Sarkis, Aims Community College, Greeley, CO; Ann Senisi, Coordinator, Nursing/Health Programs Nassau Technological Center (BOCES), Westbury, NY; Jan Strandjord, Indiana Vocational Technical College, Indianapolis, IN; Carlene R. Tonini, College of San Mateo, San Mateo, CA; Patricia Turner, Howard Community College, Columbia, MD; Jacqueline Tuttle, Forsyth Technical College, Winston-Salem, NC; Ann Welch, Indiana Vocational/Technical College, Indianapolis, IN; Eileen Williams, Nassau Technological Center (BOCES), Westbury, NY; and Iris Winkelhake, Southeast Community College, Lincoln, NE.

The staff of The Benjamin/Cummings Publishing Company contributed immensely in the form of support and guidance and deserve a hearty round of applause, one and all. I would especially like to thank Thor Ekstrom, my assistant editor, for his thoughful direction and careful guidance of the changing manuscript until it was ready to be delivered to the production team at Wendy Earl Productions. There too was a "star"—Bradley Burch—whose diligent handling and control of the many details of production made that process as harassment-free as humanly possible. And last, but not least, I want to thank Linda Weinland, my colleague at Edison Community College, who helped proofread the manuscript. Her eagle-eyed efforts were extraordinary.

Elaine N. Marieb

Edison Community College
Department of Biology
Collier County Campus
7007 Lely Cultural Parkway
Naples, Florida 34113

Preface to the Student

This book is written with you, the student, in mind. Human anatomy and physiology is more than just interesting—it is fascinating. To help get you involved in the study of this subject, a number of special features are incorporated throughout the book.

The *informal writing style* invites you to learn more about anatomy and physiology without intimidation. We want you to enjoy reading this book.

Topic boxes and *tables* are designed with you in mind. The topic boxes present scientific information that can be applied to your daily life. When reading the topic boxes, you will probably find yourself saying, "I didn't know that," or "Now I understand why. . . ." The tables are summaries of important information in the text. You should be able to use the tables when studying for an exam or reviewing an important topic.

Important terms are defined within the text as they are introduced and are listed at the ends of the chapters. An extensive *glossary* is provided at the back of the book to help you review these terms.

Phonetic spellings are provided for many of these important terms, especially those which are likely to be unfamiliar to you. To read these, you will need to remember the following rules:

1. Accent marks follow stressed syllables. The primary stress is shown by ', and the secondary stress by ".

2. Unless otherwise noted, assume that vowels at the ends of syllables are long and vowels followed by consonants are short. Exceptions to this rule are indicated by a bar (¯) over the vowel, which indicates a long vowel, or a breve sign (˘) over the vowel, indicating that the vowel is short.

For example, the phonetic spelling of "thrombophlebitis" is *throm″bo-flĕ-bi′ tis*. The next-to-the-last syllable *(bi′)* receives the greatest stress, and the first syllable *(throm″)* gets the secondary stress. The vowel in the second syllable comes at the end of the syllable and is long. The vowel that comes at the end of the third syllable is short because it has a breve sign.

Any exam causes anxiety. Exams in anatomy and physiology are no exception. To help you better prepare for an exam or comprehend the material you have just read, extensive *summaries, review questions,* and *clinical problems* ("At the Clinic") are found at the ends of the chapters. The *Systems in Sync* feature, which accompanies each system, will also be very helpful in your preparations because it lays out very simply the most important ways body systems interact with each other.

The art program is designed to help you learn the different structures and functions of the human body. All figures are referred to within the text discussion. The best way to make use of the extensive art program is to carefully study each illustration when it is discussed in the text.

I hope that you enjoy *Essentials of Human Anatomy and Physiology* and that this book makes learning A & P a fun and rewarding process. Write to me and let me know how you make out in your A & P course.

Elaine N. Marieb

Edison Community College
Department of Biology
Collier County Campus
7007 Lely Cultural Parkway
Naples, Florida 34113

Contents

1
THE HUMAN BODY: AN ORIENTATION 1

2
BASIC CHEMISTRY 21

5
THE SKELETAL SYSTEM 113

6
THE MUSCULAR SYSTEM 153

7
THE NERVOUS SYSTEM 195

8
SPECIAL SENSES 241

9
THE ENDOCRINE SYSTEM 267

10
BLOOD 295

11
THE CIRCULATORY SYSTEM 313

12
BODY DEFENSES 355

13
THE RESPIRATORY SYSTEM 383

14

THE DIGESTIVE SYSTEM AND BODY METABOLISM 407

15

THE URINARY SYSTEM 449

16
THE REPRODUCTIVE SYSTEM 471

APPENDIXES

1 The Human Body: An Orientation

YOUR GOALS

After completing this chapter, you will have mastered the objectives listed below.

AN OVERVIEW OF ANATOMY AND PHYSIOLOGY

Most of us have a natural curiosity about our bodies; we want to know what makes us tick. This curiosity is seen in infants, who can keep themselves happy for long periods of time staring at their own hands or pulling their mother's nose. Older children wonder where food goes when they swallow it, and some believe that they will grow a watermelon in their "belly" if they swallow the seeds. They scream loudly when approached by medical personnel (fearing body mutilation) but they like to play "doctor." Adults become upset when their hearts pound, when they have uncontrollable "hot flashes," or when they cannot keep their weight down.

Anatomy and physiology, subdivisions of biology, explore many of these topics as they describe how our bodies are put together and how they work.

Anatomy

Anatomy (ah-nat′o-me) is the study of the structure and shape of the body and body parts and their relationships to one another. Whenever we look at our own body or study large body structures such as the heart or bones, we are observing *gross anatomy;* that is, we are studying large, easily observable structures. Indeed, the term *anatomy,* derived from the Greek words meaning to cut *(tomy)* apart *(ana),* is related most closely to gross anatomy studies, because in such studies preserved animals or their organs are dissected (cut up) to be examined. On the other hand, if a microscope or magnifying instrument is used to see very small structures in the body, we are studying *microscopic anatomy.* The cells and tissues of the body can only be seen through a microscope.

Physiology

Physiology (fiz″e-ol′o-je) is the study of how the body and its parts work or function (*physio* = nature; *ology* = the study of). Like anatomy, physiology has many subdivisions. For example, *neurophysiology* explains the workings of the nervous system, and *cardiac physiology* studies the function of the heart, which acts as a muscular pump to keep blood flowing throughout the body.

Relationship between Anatomy and Physiology

In the real world, anatomy and physiology are always related. The parts of your body are combined and arranged to form a well-organized unit, and each of those parts has a job to do to make the body operate as a whole. Structure determines what functions can take place. For example, the lungs are not muscular chambers like the heart and cannot pump blood through the body, but because the walls of their air sacs are very thin, they *can* exchange gases and provide oxygen to the body. The intimate relationship between anatomy and physiology is stressed throughout this book to make your learning meaningful.

LEVELS OF STRUCTURAL ORGANIZATION

From Atoms to Organisms

The human body exhibits many levels of structural complexity (see Figure 1.1). The simplest level of the structural ladder is the *chemical level,* which we will study in Chapter 2. At this level, **atoms,** tiny building blocks of matter, combine to form *molecules* such as water, sugar, and proteins. Molecules, in turn, associate in specific ways to form microscopic **cells,** the smallest units of all living things. The *cellular level* is examined in Chapter 3. Individual cells vary widely in size and shape, reflecting their particular functions in the body.

The simplest living creatures are composed of single cells, but in complex organisms like human beings, the structural ladder continues on to the *tissue level.* **Tissues** consist of groups of similar cells that have a common function. As we will discuss in Chapter 3, each of the four basic tissue types plays a definite but different role in the body.

An **organ** is a structure, composed of two or more tissue types, that performs a specific function for the body. At the *organ level* of organization, ex-

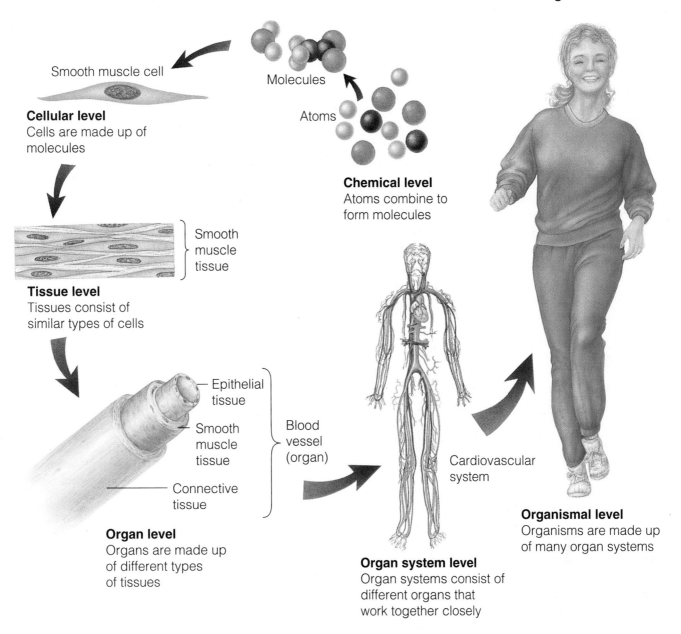

Smooth muscle cell

Cellular level
Cells are made up of
molecules

Molecules

Atoms

Chemical level
Atoms combine to
form molecules

Smooth
muscle
tissue

Tissue level
Tissues consist of
similar types of cells

Epithelial
tissue

Smooth
muscle
tissue

Connective
tissue

Blood
vessel
(organ)

Cardiovascular
system

Organismal level
Organisms are made up
of many organ systems

Organ level
Organs are made up
of different types
of tissues

Organ system level
Organ systems consist of
different organs that
work together closely

Figure 1.1   Levels of structural complexity. In this diagram, components of
the cardiovascular system are used to illustrate the various levels of complexity
in a human being.

tremely complex functions become possible. For
example, the small intestine, which digests and absorbs food, is composed of all four tissue types. All
the body's organs are grouped so that a number of
organ systems are formed. An **organ system** is a
group of organs that cooperate to accomplish a
common purpose. For example, the digestive system includes the esophagus, the stomach, and the

small and large intestines (to name a few of its organs). Each organ has its own job to do and, working together, they keep food moving through the
digestive system so that it is properly broken down
and absorbed into the blood, providing fuel for all
the body's cells. In all, 11 organ systems make up
the living body, or the **organism,** which represents the highest level of structural organization,

the *organismal level*. The major organs of each of the systems are shown in Figure 1.2. Refer to the figure as you read through the following descriptions of the organ systems.

Organ System Overview

Integumentary System

The **integumentary** (in-teg″u-men′tar-e) **system** is the external covering of the body, or the skin. It waterproofs the body and cushions and protects the deeper tissues from injury. It also excretes salts and urea in perspiration and helps regulate body temperature. Temperature, pressure, and pain receptors located in the skin alert us to what is happening at the body surface.

Skeletal System

The **skeletal system** consists of bones, cartilages, ligaments, and joints. It supports the body and provides a framework that the skeletal muscles can use to cause movement. It also has a protective function (for example, the skull encloses and protects the brain). *Hematopoiesis* (hem″ah-to-poi-e′sis), or formation of blood cells, goes on within the cavities of the skeleton. The hard substance of bones acts as a storehouse for minerals.

Muscular System

The muscles of the body have only one function—to *contract* or shorten. When this happens, movement occurs. Hence, muscles can be viewed as the "machines" of the body. The mobility of the body as a whole reflects the activity of *skeletal muscles,* the large, fleshy muscles attached to bones. When these contract, you are able to walk, leap, grasp, throw a ball, or smile. The skeletal muscles form the **muscular system.** These muscles are distinct from the muscles of the heart and of other hollow organs, which move fluids (blood, urine) or other substances (such as food) along definite pathways within the body.

Nervous System

The **nervous system** is the body's fast-acting control system. It consists of the brain, spinal cord, nerves, and sensory receptors. The body must be able to respond to irritants or stimuli coming from outside the body (such as light, sound, or changes in temperature) and from inside the body (such as decreases in oxygen or stretching of tissue). The sensory receptors detect these changes and send messages (via electrical signals called *nerve impulses*) to the central nervous system (brain and spinal cord) so that it is constantly informed about what is going on. The central nervous system then assesses this information and responds by activating the appropriate body muscles or glands.

Endocrine System

Like the nervous system, the **endocrine** (en′do-krin) **system** controls body activities, but it acts much more slowly. The endocrine glands produce chemical molecules called *hormones* and release them into the blood to travel to relatively distant target organs.

The endocrine glands include the pituitary, thyroid, parathyroids, adrenals, thymus, pancreas, pineal, ovaries (in the female), and testes (in the male). The endocrine glands are not connected anatomically in the same way that parts of the other organ systems are. What they have in common is that they all secrete hormones, which regulate other structures. The body functions controlled by hormones are many and varied, involving every cell in the body. Growth, reproduction, and food use by cells are all controlled (at least in part) by hormones.

Circulatory System

The **circulatory system** is a transport and delivery system. It consists of two distinct organ systems, the cardiovascular and the lymphatic systems, which play complementary roles. The primary organs of the **cardiovascular system** are the heart and blood vessels. Using blood as the transporting fluid, the cardiovascular system carries oxygen, nutrients, hormones, and other substances

Figure 1.2 The body's organ systems. The structural components of each organ system are illustrated in the diagrammatic views. The major functions of the organ system are listed beneath each illustration. Note that the cardiovascular system (**f**) and the lymphatic system (**g**) are considered together in this book (Chapter 11) as the circulatory system.

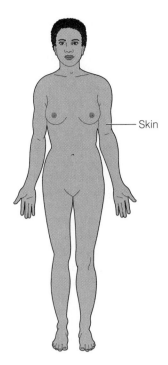

(a) Integumentary system

Forms the external body covering; protects deeper tissue from injury; synthesizes vitamin D; location of cutaneous (pain, pressure, etc.) receptors; and sweat and oil glands.

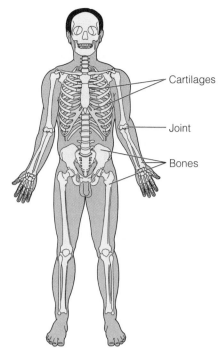

Cartilages

Joint

Bones

(b) Skeletal system

Protects and supports body organs; provides a framework the muscles use to cause movement; blood cells are formed within bones; stores minerals.

Skeletal muscles

(c) Muscular system

Allows manipulation of the environment, locomotion, and facial expression; maintains posture; produces heat.

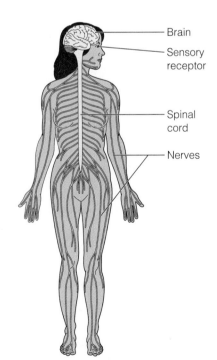

Brain

Sensory receptor

Spinal cord

Nerves

(d) Nervous system

Fast-acting control system of the body; responds to internal and external changes by activating appropriate muscles and glands.

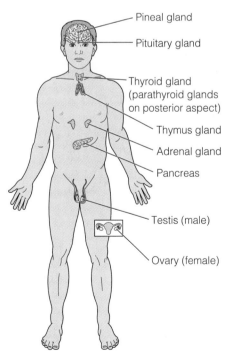

Pineal gland

Pituitary gland

Thyroid gland (parathyroid glands on posterior aspect)

Thymus gland

Adrenal gland

Pancreas

Testis (male)

Ovary (female)

(e) Endocrine system

Glands secrete hormones that regulate processes such as growth, reproduction, and nutrient use (metabolism) by body cells.

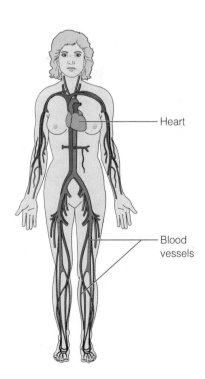

Heart

Blood vessels

(f) Cardiovascular system

Blood vessels transport blood which carries oxygen, carbon dioxide, nutrients, wastes, etc.; the heart pumps blood.

➤

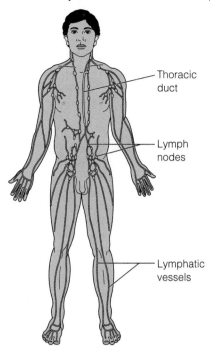

Thoracic duct

Lymph nodes

Lymphatic vessels

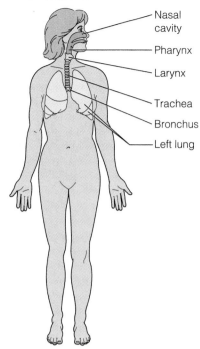

Nasal cavity

Pharynx

Larynx

Trachea

Bronchus

Left lung

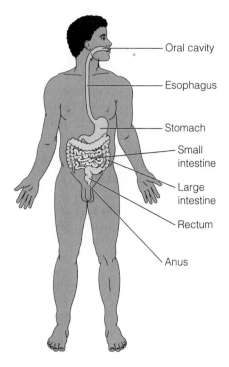

Oral cavity

Esophagus

Stomach

Small intestine

Large intestine

Rectum

Anus

(g) Lymphatic system

Picks up fluid leaked from blood vessels and returns it to blood; disposes of debris in the lymphatic stream; houses white blood cells involved in immunity.

(h) Respiratory system

Keeps blood constantly supplied with oxygen and removes carbon dioxide; the gaseous exchanges occur through the walls of the air sacs of the lungs.

(i) Digestive system

Breaks down food into absorbable units that enter the blood for distribution to body cells; indigestible food stuffs are eliminated as feces.

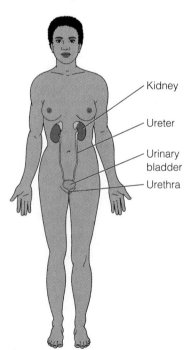

Kidney

Ureter

Urinary bladder

Urethra

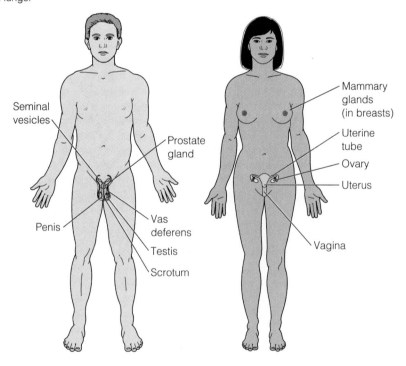

Seminal vesicles

Prostate gland

Penis

Vas deferens

Testis

Scrotum

Mammary glands (in breasts)

Uterine tube

Ovary

Uterus

Vagina

(j) Urinary system

Eliminates nitrogenous wastes from the body; regulates water, electrolyte, and acid-base balance of the blood.

(k) Male reproductive system

(l) Female reproductive system

Overall function of the reproductive system is production of offspring. Testes produce sperm and male sex hormone; ducts and glands aid in delivery of viable sperm to the female reproductive tract. Ovaries produce eggs and female sex hormones; remaining structures serve as sites for fertilization and development of the fetus. Mammary glands of female breast produce milk to nourish the newborn.

Figure 1.2 (continued)

to and from the tissue cells where exchanges are made. White blood cells and chemicals in the blood help to protect the body from such "foreign invaders" as bacteria, toxins, and tumor cells. The heart acts as the "blood pump," propelling blood through the blood vessels to all body tissues. The **lymphatic system** organs include lymphatic vessels, lymph nodes, and other lymphoid organs such as the spleen and tonsils. The lymphatic vessels return fluid leaked from the blood to the blood vessels so that blood can be kept continuously circulating through the body. The lymph nodes (and other lymphoid organs) help to cleanse the blood and house the cells involved in immunity.

Respiratory System

The jobs of the **respiratory system** are to keep the body constantly supplied with oxygen and to remove carbon dioxide. The respiratory system consists of the nasal passages, pharynx, larynx, trachea, bronchi, and lungs. Within the lungs are tiny air sacs. It is through the thin walls of these air sacs that gas exchanges are made to and from the blood.

Digestive System

The **digestive system** is basically a tube running through the body from mouth to anus. The organs of the digestive system include the mouth, oral cavity, esophagus, stomach, small and large intestines, and rectum. Their role is to break down food and deliver the products to the blood for dispersal to the body cells. The undigested food that remains in the tract leaves the body through the anus as feces. The breakdown activities that begin in the mouth are completed in the small intestine. From that point on, the major function of the digestive system is to reclaim water. The liver is considered to be a digestive organ, because the bile it produces helps to break down fats. The pancreas, which delivers digestive enzymes to the small intestine, also is functionally a digestive organ.

Urinary System

As it functions, the body produces wastes, which must be disposed of. One type of waste is nitrogen-containing waste (such as urea and uric acid),

which results from the breakdown of proteins and nucleic acids by the body cells. The **urinary system** removes the nitrogen-containing wastes from the blood and flushes them from the body in urine. This system, often called the *excretory system,* is composed of the kidneys, ureters, bladder, and urethra. Other important functions of this system include maintaining the body's water and salt balance and regulating the acid-base balance of the blood.

Reproductive System

The **reproductive system** exists primarily to produce offspring. Sperm are produced by the testes of the male. Other male reproductive system structures are the scrotum, penis, accessory glands, and the duct system, which carries sperm to the outside of the body. The ovary of the female produces the eggs, or ova; the female duct system consists of the uterine tubes, uterus, and vagina. The uterus provides the site for the development of the fetus (immature infant) once fertilization has occurred.

MAINTAINING LIFE

Necessary Life Functions

Now that we have introduced the structural levels composing the human body, the question that naturally follows is: What does this highly organized human body do? Like all complex animals, human beings maintain their boundaries, move, respond to environmental changes, take in and digest nutrients, carry out metabolism, dispose of wastes, reproduce themselves, and grow. We will discuss each of these necessary life functions briefly here and in more detail in later chapters.

Organ systems do not work in isolation; instead, they work together to promote the well-being of the entire body. Because this theme will be emphasized throughout this book, it is worthwhile to identify the most important organ systems contributing to each of the necessary life functions. As you read through this material, you may want to refer to the more detailed descriptions of the organ systems provided on pp. 4 and 7 and in Figure 1.2.

Maintaining Boundaries

Every living organism must be able to maintain its boundaries so that its "inside" remains distinct from its "outside." Every cell of the human body is surrounded by an external membrane that contains its contents and allows needed substances in while restricting the entry of potentially damaging or unnecessary substances. Additionally, the body as a whole is enclosed by the integumentary system, or skin. The integumentary system protects internal organs from drying out (which would be fatal), from bacterial invasion, and from the damaging effects of an unbelievable number of chemical substances and physical factors in the external environment.

Movement

Movement includes all the activities promoted by the muscular system, such as propelling ourselves from one place to another by walking, swimming, and so forth, and manipulating the external environment with our fingers. The muscular system is aided by the skeletal system, which provides the bones that the muscles pull on as they work. Movement also occurs when substances such as blood, foodstuffs, and urine are propelled through the internal organs of the cardiovascular, digestive, and urinary systems respectively.

Responsiveness

Responsiveness, or **irritability,** is the ability to sense changes (stimuli) in the environment and then to react to them. For example, if you cut your hand on broken glass, you involuntarily pull your hand away from the painful stimulus (the glass). It is not necessary to think about it—it just happens! Likewise, when carbon dioxide in your blood rises to dangerously high levels, the response is an increase in your breathing rate.

Because nerve cells are highly irritable and can communicate rapidly with each other by conducting electrical impulses, the nervous system bears the major responsibility for responsiveness. However, all body cells exhibit responsiveness to some extent.

Digestion

Digestion is the process of breaking down ingested foodstuffs into simple molecules that can then be absorbed into the blood for delivery to all body cells by the cardiovascular system. In a simple, one-celled organism like an amoeba, the cell itself is the "digestion factory," but in the complex, multicellular human body, the digestive system performs this function for the entire body.

Metabolism

Metabolism is a broad term that refers to all chemical reactions that occur within body cells. It includes breaking down complex substances into simpler building blocks, making larger structures from smaller ones, and using nutrients and oxygen to produce ATP molecules, energy-rich molecules that power cellular activities. Metabolism depends on the digestive and respiratory systems to make nutrients and oxygen available to the blood, and on the cardiovascular system to distribute these substances throughout the body. Metabolism is regulated chiefly by hormones secreted by the glands of the endocrine system.

Excretion

Excretion is the process of removing *excreta* (ek-skre′tah), or wastes, from the body. If the body is to continue to operate as we expect it to, it must get rid of the nonuseful substances produced during digestion and metabolism. Several organ systems participate in excretion. For example, the digestive system rids the body of indigestible food residues in feces, and the urinary system disposes of nitrogen-containing metabolic wastes in urine.

Reproduction

Reproduction, the production of offspring, can occur on the cellular or organismal level. In cellular reproduction, the original cell divides, producing two identical daughter cells that may then be used for body growth or repair. Reproduction of the human organism, or making a whole new person, is the task of the organs of the reproductive system, which produce sperm and eggs. When a sperm unites with an egg, a fertilized egg forms, which then develops into a bouncing baby within the mother's body. The function of the reproductive system is exquisitely regulated by hormones of the endocrine system.

Growth

Growth is an increase in size, usually accomplished by an increase in the number of cells. For growth to occur, cell-constructing activities must occur at a faster rate than cell-destroying activities.

Survival Needs

The goal of nearly all body systems is to maintain life. However, life is extraordinarily fragile and requires that several factors be available. These factors, which we will call *survival needs,* include nutrients (food), oxygen, water, and appropriate temperature and atmospheric pressure.

Nutrients, taken in via the diet, contain the chemicals used for energy and cell building. Carbohydrates are the major energy-providing fuel for body cells. Proteins and to a lesser extent fats are essential for building cell structures. Fats also cushion body organs and provide reserve fuel. Minerals and vitamins are required for the chemical reactions that go on in cells.

All the nutrients in the world are useless unless **oxygen** is also available, because the chemical reactions that release energy from foods require oxygen. Approximately 20 percent of the air we breathe is oxygen. It is made available to the blood and body cells by the cooperative efforts of the respiratory and cardiovascular systems.

Water accounts for 60 to 80 percent of body weight. It is the single most abundant chemical substance in the body and provides the fluid base for body secretions and excretions. Water is obtained chiefly from ingested foods or liquids and is lost from the body by evaporation from the lungs and skin and in body excretions.

For good health, **body temperature** must be maintained at around 37°C (98°F). As body temperature drops below this point, metabolic reactions become slower and slower, and finally stop. When body temperature is too high, chemical reactions proceed too rapidly, and body proteins begin to break down. At either extreme, death occurs. Most body heat is generated by the activity of the skeletal muscles.

The force exerted on the surface of the body by the weight of air is referred to as **atmospheric pressure.** Breathing and the exchange of oxygen and carbon dioxide in the lungs depend on appropriate atmospheric pressure. At high altitudes, where the air is thin and atmospheric pressure is lower, gas exchange may be inadequate.

The mere presence of these survival factors is not sufficient to maintain life. They must be present in appropriate amounts as well; excesses and deficits may be equally harmful. For example, the food ingested must be of high quality and in proper amounts; otherwise, nutritional disease, obesity, or starvation are likely outcomes.

HOMEOSTASIS

When you really think about the fact that your body contains trillions of cells in nearly constant activity, and that remarkably little usually goes wrong with it, you begin to appreciate what a marvelous machine your body really is. The word **homeostasis** (ho″me-o-sta′sis) describes the body's ability to maintain relatively stable internal conditions even though the outside world is continuously changing. Although the literal translation of homeostasis is "unchanging" (*homeo* = the same; *stasis* = standing still), the term does not really mean an unchanging state. Instead, it indicates a *dynamic* state of equilibrium, or a balance in which internal conditions change and vary, but always within relatively narrow limits.

In general, the body is in homeostasis when its needs are being adequately met and its functions are occurring smoothly. Virtually every organ system plays a role in maintaining the constancy of the internal environment. Adequate blood levels of vital nutrients must be continuously present, and heart activity and blood pressure must be constantly monitored and adjusted so that the blood is propelled with adequate force to reach all body tissues. Additionally, wastes must not be allowed to accumulate, and body temperature must be precisely controlled. Communication within the body is essential for homeostasis, and is accomplished chiefly by the nervous and endocrine systems. The

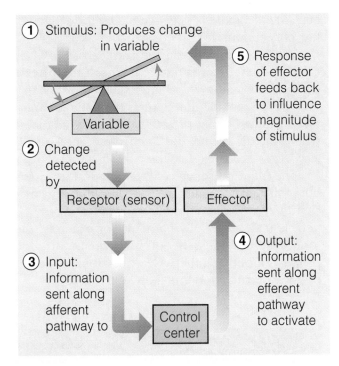

① Stimulus: Produces change in variable

⑤ Response of effector feeds back to influence magnitude of stimulus

Variable

② Change detected by

Receptor (sensor)

Effector

③ Input: Information sent along afferent pathway to

④ Output: Information sent along efferent pathway to activate

Control center

Figure 1.3 **The elements of a control system.** Communication between the receptor, control center, and effector is essential for normal operation of the system.

details of how these two great regulating systems operate are the subjects of later chapters, but the basic characteristics of control systems that promote homeostasis will be explained here.

Homeostatic Control Mechanisms

Regardless of the factor or event being regulated (this is called the *variable*), all homeostatic control mechanisms have at least three components (see Figure 1.3). The first component is a **receptor.** Essentially, it is some type of sensor that responds to changes in the environment. It responds to such changes, called *stimuli,* by sending information (input) to the second element, the *control center.* Information flows from the receptor to the control center along the *afferent pathway.*

The **control center,** which determines the level (set point) at which a variable is to be maintained, analyzes information it receives, and then determines the appropriate response.

The third component is the **effector,** which provides the means for the control center's response (output) to the stimulus. Information flows from the control center to the effector along the *ef-*

ferent pathway. The results of the response then *feed back* to influence the stimulus, either depressing it (negative feedback) so that the whole control mechanism is shut off, or enhancing it (positive feedback) so that the reaction continues at an even faster rate.

Most homeostatic control mechanisms are **negative feedback mechanisms.** In such systems, the net effect of the response to the stimulus is to shut off the original stimulus or reduce its intensity. A frequently used example of a negative feedback system is a home heating system connected to a thermostat. In this situation, the thermostat contains both the receptor and the control center. If the thermostat is set at 20°C (68°F), the heating system (effector) will be triggered ON when the house temperature drops below that setting. As the furnace produces heat, the air is warmed. When the temperature reaches 20°C or slightly higher, the thermostat sends a signal to shut off the furnace. Your body "thermostat," located in a part of your brain called the *hypothalamus,* operates in a similar fashion to regulate body temperature. Other negative feedback mechanisms regulate heart rate, blood pressure, breathing rate, and blood levels of glucose, oxygen, carbon dioxide, and minerals.

Because they tend to increase the original disturbance (stimulus) and to push the variable *farther* from its original value, positive feedback mechanisms are much more rare in the body. Typically these mechanisms control infrequent events that occur explosively and do not require continuous adjustments. Blood clotting and the birth of a baby are the most familiar examples of positive feedback mechanisms.

Homeostatic Imbalance

Homeostasis is so important that most disease is regarded as a result of its disturbance, a condition called **homeostatic imbalance.** As we age, our body organs become less efficient, and our internal conditions become less and less stable. These events place us at an ever greater risk for illness and produce the changes we associate with aging.

Examples of homeostatic imbalance will be provided throughout this book to enhance your understanding of normal physiological mechanisms. These homeostatic imbalance sections are preceded by the symbol ⚠ to alert you that an abnormal condition is being described.

THE LANGUAGE OF ANATOMY

Learning about the body is exciting, but our interest is sometimes dampened when we are confronted with the terminology of anatomy and physiology. Let's face it: You can't just pick up an anatomy and physiology book and read it as though it were a novel. Unfortunately, confusion is inevitable without specialized terminology. For example, if you are looking at a ball, "above" always means the area over the top of the ball. Other directional terms can also be used consistently because the ball is a sphere. All sides and surfaces are equal. The human body, of course, has many protrusions and bends. Thus, the question becomes: Above what? To prevent misunderstanding, anatomists have accepted a set of terms that allow body structures to be located and identified clearly with just a few words. This language of anatomy is presented and explained next.

Anatomical Position and Directional Terms

To accurately describe body parts and position, we must have an initial reference point and use directional terms. To avoid confusion, it is always assumed that the body is in a standard position called the **anatomical position.** It is important to understand this position because most body terminology used in this book refers to this body positioning *regardless* of the position the body happens to be in. The face-front diagrams in Table 1.1 and those on the top row of Figure 1.5 illustrate the anatomical position. As you can see, the body is erect with the feet together and the arms hanging at the sides with the palms facing forward.

- Stand up and assume the anatomical position. Notice that it is not very comfortable because the hands are held unnaturally forward rather than hanging cupped toward the thighs.

Directional terms used by medical personnel and anatomists allow them to explain exactly where one body structure is in relation to another. For example, we could describe the relationship between the ears and the nose informally by saying, "The ears are located on each side of the head to the right and left of the nose." Using anatomical

terminology, this condenses to, "The ears are lateral to the nose." Thus, using anatomical terminology saves a good deal of description and, once learned, is much clearer. Commonly used directional terms are defined and illustrated in Table 1.1. Although most of these terms are also used in everyday conversation, keep in mind that their anatomical meanings are very precise.

Before continuing, take a minute to check your understanding of what you have read in Table 1.1. Give the relationship between the following body parts using the correct anatomical terms.

The wrist is _____ to the hand.

The breastbone is _____ to the spine.

The brain is _____ to the spinal cord.

The lungs are _____ to the stomach.

The thumb is _____ to the fingers. (Be careful here. Remember the anatomical position.)

Regional Terms

There are many visible landmarks on the surface of the body. Once you know their proper anatomical names, you can be specific in referring to different regions of the body.

Anterior Body Landmarks

Look at Figure 1.4a to find the following body regions. Once you have identified all the anterior body landmarks, cover the labels that describe what the structures are, and again go through the list pointing out the specific areas on your own body.

- **abdominal** (ab-dom′ĭ-nal): anterior body trunk inferior to ribs
- **antecubital** (an″te-ku′bĭ-tal): anterior surface of elbow
- **axillary** (ak′sĭ-lar″e): armpit
- **brachial** (bra′ke-al): arm
- **buccal** (buk′al): cheek area
- **cervical** (ser′vĭ-kal): neck region
- **digital** (dij′ĭ-tal): fingers, toes

Table 1.1 Orientation and Directional Terms

Term	Definition	Illustration	Example
Superior (cranial or cephalad)	Toward the head end or upper part of a structure or the body; above		The forehead is superior to the nose.
Inferior (caudal)	Away from the head end or toward the lower part of a structure or the body; below		The navel is inferior to the breastbone.
Anterior (ventral)*	Toward or at the front of the body; in front of		The breastbone is anterior to the spine.
Posterior (dorsal)*	Toward or at the backside of the body; behind		The heart is posterior to the breastbone.
Medial	Toward or at the midline of the body; on the inner side of		The heart is medial to the arm.
Lateral	Away from the midline of the body; on the outer side of		The arms are lateral to the chest.
Intermediate	Between a more medial and a more lateral structure		The collarbone is intermediate between the breastbone and shoulder.
Proximal	Close to the origin of the body part or the point of attachment of a limb to the body trunk		The elbow is proximal to the wrist (meaning that the elbow is closer to the shoulder or attachment point of the arm than the wrist).
Distal	Farther from the origin of a body part or the point of attachment of a limb to the body trunk		The knee is distal to the thigh.
Superficial	Toward or at the body surface		The skin is superficial to the skeleton.
Deep	Away from the body surface; more internal		The lungs are deep to the skin.

*Ventral and *anterior* are synonymous in humans; this is not the case in four-legged animals. *Ventral* refers to the "belly" of an animal and thus is the inferior surface of four-legged animals. Likewise, although the dorsal and posterior surfaces are the same in humans, the term *dorsal* refers to an animal's back. Thus, the dorsal surface of four-legged animals is their superior surface.

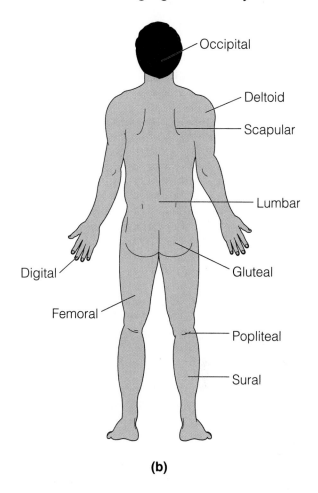

(a)   (b)

Figure 1.4   Surface anatomy: Regional terms.
(**a**) Anterior body landmarks. (**b**) Posterior body landmarks.

- **femoral** (fem′or-al): thigh
- **inguinal** (in′gwĭ-nal): area where thigh meets body trunk
- **oral** (o′ral): mouth
- **orbital** (or′bĭ-tal): eye area
- **patellar** (pah-tel′er): anterior knee
- **pubic** (pu′bik): genital region
- **thoracic** (tho-ras′ik): chest
- **umbilical** (um-bil′ĭ-kal): navel

Posterior Body Landmarks

Identify the following body regions in Figure 1.4b and then locate them on yourself without referring to this book.

- **deltoid** (del′toyd): curve of shoulder formed by large deltoid muscle
- **gluteal** (gloo′te-al): buttock

- **lumbar** (lum′bar): area of back between ribs and hips
- **occipital** (ok-sip′ĭ-tal): posterior surface of head
- **popliteal** (pop-lit′e-al): posterior knee area
- **scapular** (skap′u-lar): shoulder blade region
- **sural** (soo′ral): the posterior surface of lower leg; the calf

Body Planes and Sections

When preparing to look at the internal structures of the body, medical students find it necessary to make a **section** or cut. When the section is made through the body wall or through an organ, it is made along an imaginary line called a **plane.** Since the body is three-dimensional, we can refer to three types of planes or sections that lie at right angles to one another (see Figure 1.5).

(a) Median (midsagittal)

(b) Frontal (coronal) plane

(c) Transverse plane

Brain

Spinal cord

Nose

Tongue

Trachea

Right lung

Heart

Left lung

Liver

Stomach

Spleen

Spinal cord

Aorta

Spleen

Liver

Subcutaneous fat layer

Stomach content

Figure 1.5 The anatomical position and planes of the body. The top row of the figure illustrates three major planes of space (sagittal, frontal, and transverse) relative to humans in the anatomical position. Selected areas of the body, visualized using magnetic resonance imaging (MRI) scans taken at corresponding planes, are illustrated in the center row. Diagrams identifying body organs seen in the MRI scans are at the bottom.

A **sagittal** (saj'ĭ-tal) **section** is a cut made along the lengthwise, or longitudinal, plane of the body, dividing the body into right and left parts. If the cut is made down the median plane of the body and the right and left parts are equal in size, it is called a **midsagittal,** or **median, section.**

A **frontal section** is a cut made along a lengthwise plane that divides the body (or an organ) into anterior and posterior parts. It is also called a **coronal** (ko-ro'nal) **section.**

A **transverse section** is a cut made along a transverse, or horizontal, plane, dividing the body or organ into superior and inferior parts. It may also be called a **cross section.**

Sectioning a body or organ along different planes often results in very different views. For example, a transverse section of the body trunk at the level of the kidneys would show kidney structure in cross section very nicely; a frontal section of the body trunk would show a different view of kidney anatomy; and a midsagittal section would miss the kidneys completely. Information on body organ positioning that can be gained by taking MRI scans along different body planes is illustrated in Figure 1.5. (MRI scans are described in the "Closer Look" box on pp. 30–31 in Chapter 2.)

Body Cavities

The body has two sets of internal cavities that provide different degrees of protection to the organs within them (see Figure 1.6).

Dorsal Body Cavity

The **dorsal body cavity** has two subdivisions, which are continuous with each other. The **cranial cavity** is the space inside the bony skull. The brain is well protected because it occupies the cranial cavity. The **spinal cavity** extends from the cranial cavity nearly to the end of the vertebral column. The spinal cord, which is a continuation of the brain, is protected by the vertebrae, which surround the spinal cavity.

Ventral Body Cavity

The **ventral body cavity** is much larger than the dorsal cavity. It contains all the structures within the chest and abdomen. Like the dorsal cavity, the ventral body cavity is subdivided. The superior **thoracic cavity** is separated from the rest of the ventral cavity by a dome-shaped muscle, the

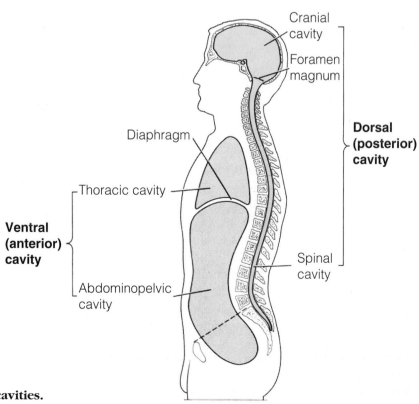

Figure 1.6   Dorsal and ventral body cavities.

Figure 1.7 Abdominopelvic cavity. (**a**) The four quadrants. (**b**) Nine regions delineated by four planes. The superior horizontal plane is just inferior to the ribs; the inferior horizontal plane is at the superior aspect of the hip bones, and the vertical planes are just medial to the nipples. (**c**) Anterior view of the abdominopelvic cavity showing superficial organs.

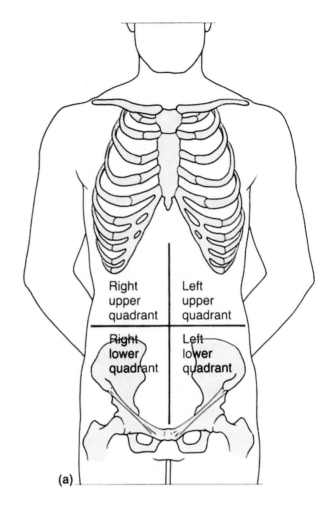

(a)

diaphragm (di'ah-fram). The organs in the thoracic cavity (lungs, heart, and others) are somewhat protected by the rib cage. The cavity inferior to the diaphragm is the **abdominopelvic** (ab-dom"ĭ-no-pel'vik) **cavity.** Some prefer to subdivide it into a superior **abdominal cavity** containing the stomach, liver, intestines, and other organs, and an inferior **pelvic cavity** with the reproductive organs, bladder, and rectum. However, there is no physical structure dividing the abdominopelvic cavity. If you look carefully at Figure 1.6, you will see that the pelvic cavity is not continuous with the abdominal cavity in a straight plane, but that it tips away from it in the posterior direction.

When the body is subjected to physical trauma (as often happens in an automobile accident, for example), the most vulnerable ab-dominopelvic organs are those within the abdominal cavity, because the cavity walls of that portion are formed only of trunk muscles and are not reinforced by bone. The pelvic organs receive a somewhat greater degree of protection from the bony pelvis in which they reside. ❑

Because the abdominopelvic cavity is quite large and contains many organs, it is helpful to divide it up into smaller areas for study. One scheme divides the abdominopelvic cavity into four more or less equal regions called *quadrants;* the quadrants are then simply named according to their relative positions—that is, right upper quadrant, right lower quadrant, left upper quadrant, and left lower quadrant (see Figure 1.7a).

Another commonly used system divides the abdominopelvic cavity into nine separate *regions*

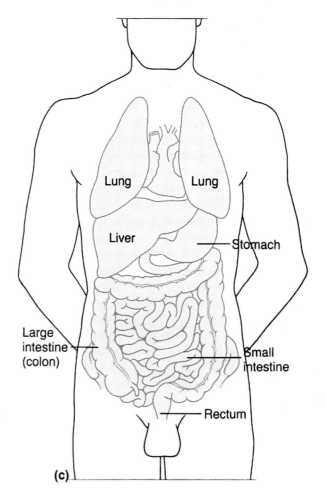

by four planes, as shown in Figure 1.7b. Although the names of the nine regions are unfamiliar to you now, with a little patience and study they will become easier to remember. As you locate these regions in the figure, notice the organs they contain by referring to Figure 1.7c.

- The **umbilical region** is the centermost region, deep to and surrounding the umbilicus (navel).

- The **epigastric** (ep″ĭ-gas′trik) **region** is located superior to the umbilical region (*epi* = upon, above; *gastric* = stomach).

- The **hypogastric (pubic) region** is located inferior to the umbilical region (*hypo* = below).

- The **right** and **left iliac**, or **inguinal**, **regions** are lateral to the hypogastric region (*iliac* = su-

perior part of the hip bone) and lie superficial to the inferior parts of the hip bones.

- The **right** and **left lumbar regions** lie lateral to the umbilical region (*lumbus* = loin) and overlie the superior parts of the hip bones.

- The **right** and **left hypochondriac** (hi″po-kon′dre-ak) **regions** flank the epigastric region laterally (*chondro* = cartilage), and overlie the lower ribs.

Important Terms

abdominopelvic (ab-dom″ĭ-no-pel′vik) **cavity**	**nervous system**
anatomy (ah-nat′o-me)	**organism**
atom	**organ system**
cell	**physiology** (fiz″e-ol′o-je)
circulatory system	**reproductive system**
cranial cavity	**respiratory system**
digestive system	**skeletal system**
endocrine (en′do-krin) **system**	**spinal cavity**
homeostasis (ho″me-o-sta′sis)	**thoracic** (tho-ras′ik) **cavity**
integumentary (in-teg″u-men′tar-e) **system**	**tissue**
muscular system	**urinary system**
negative feedback mechanism	

Summary

An Overview of Anatomy and Physiology (p. 2)

1. Anatomy is the study of structure. Observation is used to see the sizes and relationships of body parts.

2. Physiology is the study of how a structure (which may be a cell, an organ, or an organ system) functions or works.

3. Structure determines what functions can occur; therefore, if the structure changes, the function must also change.

Levels of Structural Organization (pp. 2–7)

1. There are six levels of structural organization. Atoms (at the chemical level) combine, forming the unit of life, the cell. Cells are grouped into tissues, which in turn are arranged in specific ways to form organs. A number of organs form an organ system, which performs a specific function for the body (which no other organ system can do). Together, all of the organ systems form the organism or living body.

2. For a description of organ systems naming the major organs and functions, see pp. 4–7.

Maintaining Life (pp. 7–9)

1. To sustain life, an organism must be able to maintain its boundaries, move, respond to stimuli, digest nutrients and excrete wastes, carry on metabolism, reproduce itself, and grow.

2. Survival needs include food, oxygen, water, appropriate temperature, and normal atmospheric pressure. Extremes of any of these factors can be harmful.

Homeostasis (pp. 9–10)

1. Body functions interact to maintain homeostasis, or a relatively stable internal environment within the body. Homeostasis is necessary for survival and good health; its loss results in illness or disease.

2. All homeostatic control mechanisms have a receptor that responds to environmental changes, and a control center that assesses those changes and produces a response by activating a third element, the effector.

3. Most homeostatic control systems are negative feedback systems, which act to reduce or stop the initial stimulus.

The Language of Anatomy (pp. 11–17)

1. Anatomical terminology is relative and assumes that the body is in the anatomical position (erect, palms facing forward).

2. Directional terms
   a. Superior (cranial, cephalad): above something else, toward the head.
   b. Inferior (caudal): below something else, toward the tail.
   c. Anterior (ventral): toward the front of the body or structure.
   d. Posterior (dorsal): toward the rear or back of the body or structure.
   e. Medial: toward the midline of the body.
   f. Lateral: away from the midline of the body.
   g. Intermediate: between a more medial and a more lateral structure.
   h. Proximal: closer to the point of attachment.
   i. Distal: farther from the point of attachment.
   j. Superficial (external): at or close to the body surface.
   k. Deep (internal): below or away from the body surface.

3. Body planes and sections
   a. Sagittal section: separates the body longitudinally into right and left parts.
   b. Frontal (coronal) section: separates the body on a longitudinal plane into anterior and posterior parts.
   c. Transverse (cross) section: separates the body on a horizontal plane into superior and inferior parts.

4. Regional terms. Visible landmarks on the body surface may be used to specifically refer to a body part or area. See pp. 11 and 13 for terms referring to anterior and posterior surface anatomy.

5. Body cavities
   a. Dorsal: well protected by bone; has two subdivisions.
      (1) Cranial: contains the brain.
      (2) Spinal: contains the spinal cord.
   b. Ventral: less protected than dorsal cavity; has two subdivisions.
      (1) Thoracic: The superior cavity that extends inferiorly to the diaphragm; contains the heart and lungs, which are protected by the rib cage.
      (2) Abdominopelvic: The cavity inferior to the diaphragm that contains the digestive, urinary, and reproductive organs. The abdominal portion is vulnerable because it is protected only by the trunk muscles. There is some protection of the pelvic portion by the bony pelvis. The abdominopelvic cavity is often divided into four quadrants or nine regions (see Figure 1.7 for ease of study).

Review Questions

1. Define *anatomy* and *physiology*.

2. Why would you have a hard time trying to learn and understand physiology if you did not also understand anatomy?

3. Name in order the six levels of organization, beginning with atoms and ending with the organism (the body).

4. List the 11 organ systems of the body, briefly describe the function of each, and then name two organs in each system.

5. In addition to being able to metabolize, grow, digest food, and excrete wastes, what functions must an organism perform if it is to survive?

6. List five external factors that must be present to sustain life.

7. Define *homeostasis*.

8. What is the consequence of loss of homeostasis, or homeostatic imbalance?

9. Describe the anatomical position.

10. On what body surface is each of the following located: Nose, calf of leg, ears, umbilicus, fingernails?

11. Several pairs of structures are given next. In each case, choose the one that meets the condition given first.
    a. Distal—the knee/the foot
    b. Lateral—the cheekbone/the nose
    c. Superior—the neck/the chin
    d. Anterior—the heel/the toenails
    e. External—the skin/the skeletal muscles

12. What kind of section would have to be made to cut the brain into anterior and posterior parts?

13. Which of the following organ systems—digestive, respiratory, reproductive, circulatory, urinary, or muscular—are found in *both* subdivisions of the ventral body cavity? Which are found in the thoracic cavity only? In the abdominopelvic cavity only?

14. Make a drawing of the nine abdominopelvic regions and label each region.

At the Clinic

1. A nurse informed John that she was about to take blood from his antecubital region. What part of his body was she referring to? Later, she came back and said that she was going to give him an antibiotic shot in the deltoid region. Did he take off his shirt or drop his pants to get the shot? Before John left the office, the nurse noticed that his left sural region was badly bruised. What part of his body was bruised?

2. How is the concept of homeostasis (or its loss) related to disease and aging? Provide examples to support your reasoning.

3. When we begin to become dehydrated, we usually become thirsty, which causes us to drink fluids. On the basis of what you now know about control systems, decide whether the thirst sensation is part of a negative or positive feedback control system and defend your choice.

4. Jennie Dip fell off her motorcycle and tore a nerve in her axillary region. She also tore ligaments in her cervical and scapular regions and broke the sole bone of her right brachial region. Explain where each of her injuries is located.

2

Basic Chemistry

YOUR GOALS

After completing this chapter, you will have mastered the objectives listed below.

(continued)

(Objective Checklist, continued)

BIOCHEMISTRY: THE CHEMICAL COMPOSITION OF LIVING MATTER (pp. 37–50)

- ❏ Distinguish between organic and inorganic compounds.
- ❏ Differentiate clearly between a salt, an acid, and a base.
- ❏ List several salts (or their ions) vitally important to body functioning.
- ❏ Explain the importance of water to body homeostasis and provide several examples of the various roles of water.
- ❏ Explain the concept of pH and state the pH of blood.
- ❏ Compare and contrast carbohydrates, lipids, proteins, and nucleic acids in terms of their building blocks, structures, and functions in the body.
- ❏ Differentiate between fibrous and globular proteins.
- ❏ Compare and contrast the structure and functions of DNA and RNA.
- ❏ Define *enzyme* and explain the role of enzymes.
- ❏ Explain the importance of ATP in the body.

Many short courses in anatomy and physiology lack the time to consider chemistry as a topic. So why include it here? Very simply, the food you eat and the medicines you take when you are ill are composed of chemicals. Indeed, your entire body is made up of chemicals—thousands of them—continuously interacting with one another at an incredible pace.

Although it is possible to study anatomy without much reference to chemistry, chemical reactions underlie all body processes—movement, digestion, the pumping of your heart, and even your thoughts. This chapter presents the basics of chemistry and biochemistry (the chemistry of living material), providing the background you will need to understand body functions.

CONCEPTS OF MATTER AND ENERGY

Matter

Matter is the "stuff" of the universe. With some exceptions, it can be seen, smelled, and felt. More precisely, matter is anything that occupies space and has mass (weight). Chemistry studies the nature of matter—how its building blocks are put together and how they interact.

Matter exists in solid, liquid, and gaseous states. Examples of each state are found in the human body. Solids, like bones and teeth, have a definite shape and volume. Liquids have a definite volume, but they conform to the shape of their container. Examples of body liquids are blood plasma and the interstitial fluid that bathes all body cells. Gases have neither a definite shape nor a definite volume. The air we breathe is a gas.

Matter may be changed both physically and chemically. **Physical changes** do not alter the basic nature of a substance. Examples include changes in state, such as ice melting to become water, and cutting food into smaller pieces. **Chemical changes** *do* alter the composition of the substance—often substantially. The fermentation of grapes to make wine and the digestion of food in the body are examples of chemical changes.

Energy

In contrast to matter, **energy** is massless, does not take up space, and can only be measured by its effects on matter. Energy is commonly defined as the ability to do work or to put matter into motion. When energy is actually doing work (moving objects), it is referred to as **kinetic** (kĭ-neh′tik) **energy.** When it is inactive or stored, it is called **potential energy.** All forms of energy exhibit both kinetic and potential work capacities.

Actually, energy is a physics topic, but it is difficult to separate matter and energy. All living things are built of matter, and to grow and function they require a continuous supply of energy. Thus, matter is the substance and energy is the mover of the substance. Because this is so, it is worth taking a brief detour to introduce the forms of energy used by the body as it does its work.

Forms of Energy

- **Chemical energy** is the energy stored in the bonds of chemical substances. When the bonds are broken, the stored, or potential, energy is released for use; that is, it becomes kinetic energy, or energy in action that causes an effect on matter. For example, when gasoline molecules are broken apart in your automobile engine, the energy released powers your car. In like manner, all body activities are "run" by the chemical energy harvested from the foods we eat.

- **Electrical energy** is energy that results from the movement of charged particles. In your house, electrical energy is the flow of electrons along the wiring. In your body, an electrical current is generated when charged particles (called *ions*) move across cell membranes.

- **Mechanical energy** is *directly* involved in moving matter. When you ride a bicycle, your legs provide the mechanical energy that moves the pedals. We can take this example one step further back: As the muscles in your legs shorten, they pull on your bones, causing your limbs to move (so that you can pedal the bike).

- **Radiant energy** is energy that travels in waves, that is, energy of the electromagnetic spectrum, which includes X rays and infrared, light, radio, and ultraviolet waves. Light energy, which stimulates the retinas of your eyes, is important in vision. Ultraviolet waves are responsible for that suntan we get at the beach, but also stimulate our bodies to make vitamin D.

Energy Form Conversions

With a few exceptions, energy is easily converted from one form to another. For example, an electrical current carried to a lamp socket is converted into light energy by the bulb. In the body, chemical energy of foods is trapped in the bonds of a high-energy molecule called ATP, and ATP's energy may ultimately be transformed into the electrical energy of a nerve impulse or the mechanical energy of shortening muscles.

Energy conversions are quite inefficient, and some of the initial energy supply is always "lost" to the environment as heat. (It is not really lost, because energy cannot be created or destroyed, but the part given off as heat is *unusable*.) You can easily demonstrate this principle by touching a light bulb that has been lit for an hour or so. You will soon discover that some of the electrical energy reaching the bulb is producing heat instead of light. Likewise, all energy conversions that occur in the body liberate heat. It is this heat that makes us warm-blooded animals and contributes to our relatively high body temperature, which has an important influence on body functioning. For example, when matter is heated, its particles begin to move more quickly; that is, their kinetic energy (energy of motion) increases. This is important to the chemical reactions that occur in the body because the higher the temperature, the faster those reactions occur. We will learn more about these matters later.

COMPOSITION OF MATTER

Elements and Atoms

A long time ago, it was established that all matter is composed of a limited number of fundamental substances called **elements.** The term *element* refers to a unique substance that cannot be decomposed or broken down into simpler substances by ordinary chemical methods. Examples of elements include many commonly known substances, such as oxygen, carbon, gold, copper, and iron.

So far, 112 elements have been discovered. Approximately 90 of these occur naturally; the rest are

Table 2.1   Common Elements Comprising the Human Body

Element	Atomic symbol	Percentage of body mass	Role
Oxygen	O	65.0	A major component of both organic and inorganic molecules; as a gas, essential to the oxidation of glucose and other food fuels, during which cellular energy (ATP) is produced.
Carbon	C	18.5	The primary elemental component of all organic molecules, including carbohydrates, lipids, proteins, and nucleic acids.
Hydrogen	H	9.5	A component of most organic molecules; in ionic form, influences the pH of body fluids.
Nitrogen	N	3.2	A component of proteins and nucleic acids (genetic material).
Calcium	Ca	1.5	Found as a salt in bones and teeth; in ionic form, required for muscle contraction, neural transmission, and blood clotting.
Phosphorus	P	1.0	Present as a salt, in combination with calcium, in bones and teeth; also present in nucleic acids and many proteins; forms part of the high-energy compound ATP.
Potassium	K	0.4	In its ionic form, the major intracellular cation; necessary for the conduction of nerve impulses and for muscle contraction.
Sulfur	S	0.3	A component of proteins (particularly contractile proteins of muscle).
Sodium	Na	0.2	As an ion, the major extracellular cation; important for water balance, conduction of nerve impulses, and muscle contraction.
Chlorine	Cl	0.2	In ionic form, a major extracellular anion.
Magnesium	Mg	0.1	Present in bone; also an important cofactor for enzyme activity in a number of metabolic reactions.
Iodine	I	<0.1	Needed to make functional thyroid hormones.
Iron	Fe	<0.1	A component of the functional hemoglobin molecule (which transports oxygen within red blood cells) and some enzymes.
Chromium	Cr		
Cobalt	Co		
Copper	Cu		
Fluorine	F		
Manganese	Mn		Referred to as the *trace elements* because required in very minute amounts; many found as part of enzymes or required for enzyme activation.
Molybdenum	Mo		
Selenium	Se		
Silicon	Si		
Tin	Sn		
Vanadium	V		
Zinc	Zn		

Table 2.2 Subatomic Particles

Particle	Position in atom	Mass (automic mass units, amu)	Charge
Proton (p^+)	Nucleus	1	+
Neutron (n^0)	Nucleus	1	0
Electron (e^-)	Orbitals outside the nucleus	1/1800	−

produced artificially in accelerator devices. Only four elements—carbon, oxygen, hydrogen, and nitrogen—make up about 96% of the body, but several others are present in small or trace amounts. A complete listing of the elements appears in the **periodic table** in Appendix B. The most abundant elements found in the body and their major roles are listed in Table 2.1.

The building block of an element, or the smallest particle that still retains its special properties, is called an **atom.** Because all elements are unique, the atoms of each element differ from those of all other elements. Each element is designated by a one- or two-letter chemical shorthand called an **atomic symbol.** In most cases, the atomic symbol is simply the first (or first two) letter(s) of the element's name. For example, C stands for carbon, O for oxygen, and Ca for calcium. In a few cases, the atomic symbol is taken from the Latin name for the element. For instance, sodium is indicated by Na (from the Latin word *natrium*).

Atomic Structure

The word *atom* comes from the Greek word meaning "incapable of being divided," and until this century, this idea of an atom was accepted as a scientific truth. According to this notion, you could theoretically divide a pure element, such as a block of gold, into smaller and smaller particles until you got down to the individual atoms, and then could

subdivide no further. We now know that atoms, although indescribably small, are clusters of even smaller (subatomic) particles, and that, under very special circumstances, atoms can be split into these smaller particles. Even so, the old idea of atomic indivisibility is still very appropriate, because an atom loses the unique properties of its element when it is split into its subparticles.

The atoms representing the 112 elements are composed of different numbers and proportions of three basic subatomic particles, which differ in their mass, electrical charge, and location within the atom (see Table 2.2). **Protons** (p^+) have a positive charge, whereas **neutrons** (n^0) are uncharged, or neutral. Protons and neutrons are heavy particles and have approximately the same mass (1 atomic mass unit, or 1 amu). The tiny **electrons** (e^-) bear a negative charge equal in strength to the positive charge of the protons, but their mass is so small that it is usually designated as 0 amu.

The electrical charge of a particle is a measure of its ability to attract or repel other charged particles. Particles with the same type of charge (+ to + or − to −) repel each other, but particles with unlike charges (+ to −) attract each other. Neutral particles are neither attracted nor repelled by charged particles.

Because all atoms are electrically neutral, the number of protons an atom has must be precisely balanced by its number of electrons (the + and − charges will then cancel the effect of each other).

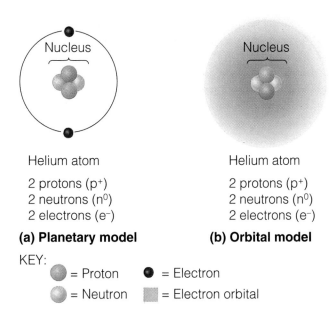

Helium atom

2 protons (p+)
2 neutrons (n⁰)
2 electrons (e⁻)

(a) Planetary model

Helium atom

2 protons (p+)
2 neutrons (n⁰)
2 electrons (e⁻)

(b) Orbital model

KEY:

= Proton   = Electron

= Neutron   = Electron orbital

Figure 2.1   The structure of an atom. The dense central nucleus contains the protons and neutrons. **(a)** In the planetary model of atomic structure, the electrons move around the nucleus in fixed orbits. **(b)** The orbital model recognizes that we never know exactly where electrons are; therefore electrons are shown as a cloud of negative charge.

Thus, hydrogen has one proton and one electron, and iron has 26 protons and 26 electrons. For any atom, the number of protons and electrons is always equal.

Planetary and Orbital Models of an Atom

The **planetary model** of an atom portrays the atom as a miniature solar system (Figure 2.1a) in which the protons and neutrons are clustered at the center of the atom in the atomic nucleus. Because the nucleus contains all the heavy particles, it is fantastically dense and positively charged. The tiny electrons orbit around the nucleus in fixed, generally circular orbits, like planets around the sun. But we can never determine the exact location of electrons at a particular time because they jump around following unknown paths. So, instead of speaking of specific orbits, chemists talk about *orbitals*—regions around the nucleus in which a given electron or electron pair is likely to be found

most of the time. This more modern model of atomic structure, called the **orbital model,** has proved to be more useful in predicting the chemical behavior of atoms. As illustrated in Figure 2.1b, the orbital model depicts the general location of electrons outside the nucleus as a haze of negative charge referred to as the *electron cloud.* Regions where electrons are most likely to be found are shown by denser shading rather than by orbit lines. Regardless of which model is used, notice that the electrons occupy nearly the entire volume of the atom and determine its chemical behavior (that is, its ability to bond with other atoms). Though now somewhat outdated, the planetary model is simple and easy to understand and use. Most of the descriptions of atomic structure in this book will use that model.

Hydrogen is the simplest atom, with just one proton and one electron. You can visualize the spatial relationships within the hydrogen atom by imagining it enlarged until its diameter equals the length of a football field. In that case, the nucleus could be represented by a lead ball the size of a gumdrop in the exact center of the sphere and the lone electron pictured as a fly buzzing about unpredictably within the sphere. This mental picture should serve to remind you that most of the volume of an atom is empty space, and nearly all of the mass is concentrated in the central nucleus.

Identifying Elements

All protons are alike, regardless of the atom being considered. The same is true of all neutrons and all electrons. So what determines the unique properties of each element? The answer is that atoms of different elements are composed of different *numbers* of protons, neutrons, and electrons.

The simplest and smallest atom, hydrogen, has one proton, one electron, and no neutrons (Figure 2.2). Next is the helium atom, with two protons, two neutrons, and two orbiting electrons. Lithium follows with three protons, four neutrons, and three electrons. If this step-by-step listing of subatomic particles were continued, all 112 atoms could be described by adding one proton and one electron at each step. The number of neutrons is not as easy to pin down, but light atoms tend to have equal numbers of protons and neutrons, whereas in larger atoms neutrons outnumber pro-

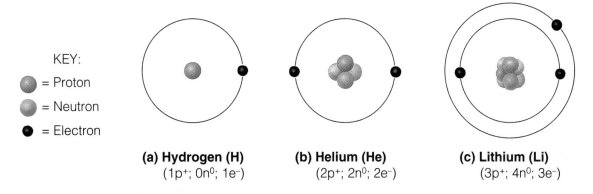

Figure 2.2   Atomic structure of the three smallest atoms.

tons. However, such an exercise would be very time-consuming, and there are easier ways of becoming informed about the various elements. All we really need to know to identify a particular element is its atomic number, mass number, and atomic weight. Taken together, these indicators provide a fairly complete picture of each element.

Atomic Number

Each element is given a number, called its **atomic number,** that is equal to the number of protons its atoms contain. Each element contains a different number of protons; hence, its atomic number is different from that of all other elements. Since the number of protons is always equal to the number of electrons, the atomic number (indirectly) also tells us the number of electrons that atom contains.

Atomic Mass Number

The **atomic mass number** of any atom is the sum of the protons and neutrons contained in its nucleus. (The mass of the electrons is so small that it is ignored.) Hydrogen has one bare proton in its nucleus; thus its atomic number and atomic mass number are the same (1). Helium, with 2 protons and 2 neutrons, has a mass number of 4. The atomic mass number is written as a superscript to the left of the atomic symbol (see the examples in Figure 2.3).

Atomic Weight and Isotopes

At first glance, it would seem that the **atomic weight** of an atom should be equal to its atomic mass. This would be so if there were only one type

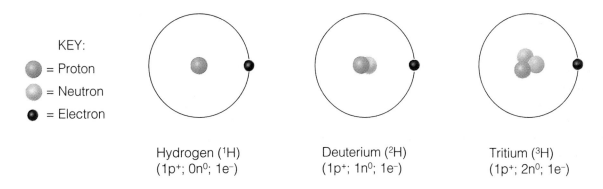

Figure 2.3   Isotopes of hydrogen.

Table 2.3 Atomic Structures of the Most Abundant Elements in the Body

Element	Symbol	Atomic number	Mass number	Atomic weight	Electrons in valence shell
Calcium	Ca	20	40	40.08	2
Carbon	C	6	12	12.011	4
Chlorine	Cl	17	35	35.453	7
Hydrogen	H	1	1	1.008	1
Iodine	I	53	127	126.905	7
Iron	Fe	26	56	55.847	2
Magnesium	Mg	12	24	24.305	2
Nitrogen	N	7	14	14.007	5
Oxygen	O	8	16	15.999	6
Phosphorus	P	15	31	30.974	5
Sodium	Na	11	23	22.99	1
Sulfur	S	16	32	32.064	6

of atom representing each element. However, the atoms of almost all elements exhibit two or more structural variations; these varieties are called **isotopes** (i′so-tōps). Isotopes have the same number of protons and electrons, but vary in the number of *neutrons* they contain. Thus, the isotopes of an element have the same atomic number, but have different atomic masses. Because all of an element's isotopes have the same number of electrons (and electrons determine bonding properties), their chemical properties are *exactly* the same. As a general rule, the atomic weight of any element is approximately equal to the mass number of its most abundant isotope. For example, hydrogen has an atomic number of 1, and has isotopes with atomic masses of 1, 2, and 3 (Figure 2.3); but its atomic weight is 1.0079, which reveals that its lightest isotope is present in much greater amounts than its heavier forms. The atomic numbers, mass numbers, and atomic weights for elements commonly found in the body are provided in Table 2.3.

The heavier isotopes of certain atoms are unstable and tend to decompose to become more stable; such isotopes are called **radioisotopes.** The why of this process is very complex, but apparently the "glue" that holds the atomic nuclei together is weaker in the heavier isotopes. This process of spontaneous atomic decay is called **radioactivity** and can be compared to a tiny explosion. All types of radioactive decay involve the ejection of particles (*alpha* or *beta*) or electromagnetic energy (*gamma rays*) from the atom's nucleus, and are damaging to living cells. Alpha emission has the least penetrating power; gamma radiation has the most.

Radioisotopes are used in minute amounts to tag biological molecules so that they can be followed, or traced, through the body, and are valuable tools for medical diagnosis and treatment. For example, PET scans, which use radioisotopes, are discussed in the box on pp. 30–31. Additionally, a radioisotope of iodine can be used to check blood circulation through the lungs or to scan the thyroid gland of a patient suspected of having a thyroid tumor. Radium, cobalt, and certain other radioisotopes are used to destroy localized cancers.

MOLECULES AND COMPOUNDS

When two or more atoms of the same or different elements chemically combine, **molecules** are formed. When two or more atoms of the same element bond together, a molecule of that element is produced. For example, when two hydrogen atoms bond, the product is a molecule of hydrogen gas:

$$H \text{ (atom)} + H \text{ (atom)} \rightarrow H_2 \text{ (molecule)}*$$

In the example given, the atoms taking part in the reaction are indicated by their atomic symbols, and the composition of the product is indicated by a *molecular formula* that shows its atomic makeup. The chemical reaction is shown by writing a *chemical equation*. When two or more *different* atoms bind together to form a molecule, the molecule is more specifically referred to as a molecule of a **compound.** For example, four hydrogen atoms and one carbon atom can interact chemically to form methane:

$$4H + C = CH_4 \text{ (methane)}$$

Thus, a molecule of methane is a compound, but a molecule of hydrogen gas is not.

It is important to understand that compounds always have properties quite different from those of the atoms making them up, and it would be next to impossible to determine the atoms making up a compound without analyzing it chemically. Notice that just as an atom is the smallest particle of an element that still retains that element's properties, a molecule is the smallest particle of a compound that still retains the properties of that compound. If you break the bonds between the atoms of the compound, properties of the atoms, rather than those of the compound, will be exhibited.

---

*Notice that when the number of atoms is written as a subscript, the subscript indicates that the atoms are joined by a chemical bond. Thus, 2H represents two unjoined atoms, but H_2 indicates that the two hydrogen atoms are bonded together to form a molecule.

CHEMICAL BONDS AND CHEMICAL REACTIONS

Chemical reactions occur whenever atoms combine with, or dissociate from, other atoms. When atoms unite chemically, chemical bonds are formed.

Bond Formation

It is important to understand that a chemical bond is not an actual physical structure, like a pair of handcuffs linking two people together. Instead, it is an energy relationship that involves interactions between the electrons of the reacting atoms. Because this is so, we will devote a few words to the role of electrons in bond formation.

Role of Electrons

As illustrated in Figure 2.2, electrons occupy generally fixed regions of space around the nucleus; these regions are called **energy levels** or **electron shells.** The maximum number of energy levels in any atom known so far is 7, and these are numbered 1 to 7 from the nucleus outward. The electrons closest to the nucleus are those most strongly attracted to its positive charge, and those farther away are less securely held. As a result, the more distant electrons are likely to interact with other atoms.

Perhaps this situation can be compared to the development of a child. During infancy and the toddler years, the child spends most of its time at home, and is shaped and molded by the ideas and demands of its parents. However, when the child goes to school, it is increasingly influenced by friends and other adults, such as teachers and coaches. Thus, just as the child is more likely to become involved with "outsiders" as it roams farther from home, electrons are more influenced by other atoms as they get farther and farther away from the positive influence of the nucleus.

There is an upper limit to the number of electrons that each electron shell can hold. Shell 1, closest to the nucleus, is small and can accommodate only 2 electrons. Shell 2 holds a maximum of 8. Shell 3 can accommodate up to 18 electrons.

A CLOSER LOOK

MEDICAL IMAGING: ILLUMINATING THE BODY

By bombarding the body with energy, new scanning techniques not only reveal the structure of our internal organs, but also wring out information about the private and, until now, secret working of their molecules. These new breeds of imaging techniques are changing the face of medical diagnosis.

Until about 40 years ago, the magical but murky X ray was the only means of peering into a living body. What X rays did and still do best was visualize hard, bony structures and locate abnormally dense structures (tumors, TB nodules) in the lungs. The 1950s saw the birth of nuclear medicine, which uses radioisotopes to scan the body, and ultrasound techniques. In the 1970s, CT, PET, and MRI scanning techniques were introduced.

The best known of new imaging devices is *computed tomography (CT)* (formerly called *computerized*

(a)

(b)

Three different methods for illuminating the body. (**a**) CT scan through the thorax showing the lungs flanking the heart. (The gurney on which the patient rests during the scan is seen at the base of the photo.) (**b**) PET scan of the brain of a patient who suffered a stroke. The area of damage and reduced brain activity is the darkened area at the upper left of the photo. (**c**) Computer-enhanced ultrasound image of a fetus in utero; the head, trunk, and limbs are clearly visible.

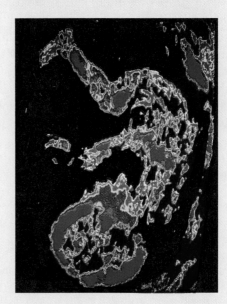

(c)

axial tomography [CAT]), a refined version of X ray. A CT scanner confines its beam to a thin slice of the body and ends the confusion resulting from images of overlapping structures seen in conventional X rays. CT's clarity has all but eliminated exploratory surgery. As the patient is slowly moved through the doughnut-shaped CT machine, its X-ray tube rotates around the body. Different tissues absorb the radiation in varying amounts. The device's computer translates this information into a detailed, cross-sectional picture of the body region scanned (see photo a). CT scans are at the forefront in evaluating most problems that affect the brain, as well as abdominal problems and certain skeletal problems, particularly those of the spine.

Special high-speed CT devices have produced a technique called *dynamic spatial reconstruction (DSR),* which provides three-dimensional images of body organs from any angle. It also allows their movements and changes in their internal volumes to be observed at normal speed, in slow motion, and at a specific moment in time. Although DSR can be used to evaluate the lungs and certain other mobile organs, its greatest value has been to visualize the heart beating and blood flowing through blood vessels. This allows heart defects, constricted blood vessels, and the status of coronary bypass grafts to be assessed.

Just as the X ray spawned a "big brother," so too did nuclear medicine in the form of *positron-emission tomography (PET).* The special advantage of PET is that its images send messages about metabolic processes. PET's greatest clinical value has been its ability to provide insights into brain activity in those affected by mental illness, Alzheimer's disease, and epilepsy. The patient is given an injection of short-lived radioisotopes that have been tagged to biological molecules (such as glucose) and then positioned in the PET scanner. As the radioisotopes are absorbed by the most active brain cells, high-energy gamma rays are produced. The computer analyzes the gamma emission and produces a picture of the brain's biochemical activity in vivid colors (see photo b).

Sonography, or *ultrasound imaging,* has distinct advantages over the approaches described so far. First, the equipment is inexpensive, and second, it employs high-frequency sound waves (ultrasound) as its energy source. Ultrasound, unlike ionizing forms of radiation, has no harmful effects on living tissues (as far as we know). The body is probed with pulses of sound waves, which are reflected and scattered to different extents by body tissues, producing echoes. The echoes are analyzed by a computer to construct visual images of body organs of interest. Because of its safety, ultrasound is the imaging technique of choice for obstetrics, i.e., for determining fetal age and position and locating the placenta (see photo c). Since sound waves have very low penetrating power and are rapidly scattered in air, sonography is of little value for looking at air-filled structures (the lungs) or those protected by bone (the brain and spinal cord).

Another technique that depends on nonionizing radiation is *magnetic resonance imaging (MRI),* a technique that uses magnetic fields 3,000 to 60,000 times stronger than the earth's to pry information from the body's tissues. The patient is inserted into a chamber within a huge magnet. Hydrogen molecules spin like tops in the magnetic field, and their energy is further enhanced by radio waves. When the radio waves are turned off, the energy is released and translated by the computer into a visual image. MRI is immensely popular because of its ability to do many things a CT scan cannot. Since dense structures do not show up at all in MRI, the bones of the skull and/or vertebral column do not impair the view of soft tissues such as the brain (see Figure 1.4a). MRI is also particularly good at detecting degenerative disease of various kinds. Multiple sclerosis plaques, for example, do not show up well in CT scans but are dazzlingly clear in MRI scans. Despite its advantages, the powerful magnetic field of MRI presents some thorny problems. For example, it can "suck" metal objects, such as implanted pacemakers and loose tooth fillings, from the body. Additionally, MRI is extremely costly and requires special quarters to insulate it from all external radio waves.

As you can see, modern medical science has many remarkable tools at its disposal. CT scans, along with nuclear medicine, now account for about 25 percent of all imaging. Ultrasonography, because of its safety and low cost, has become the most widespread of the new techniques. Even so, let's not forget conventional X rays; they remain the workhorse of diagnostic imaging techniques and still account for more than half of all imaging currently done.

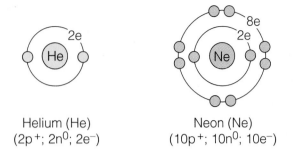

Helium (He)
($2p^+$; $2n^0$; $2e^-$)

Neon (Ne)
($10p^+$; $10n^0$; $10e^-$)

(a) Chemically inert elements (valence shell complete)

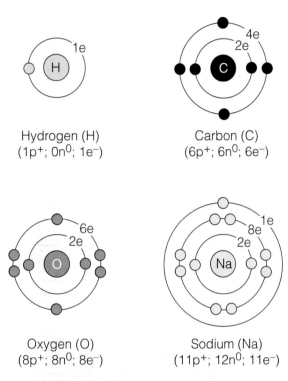

Hydrogen (H)
($1p^+$; $0n^0$; $1e^-$)

Carbon (C)
($6p^+$; $6n^0$; $6e^-$)

Oxygen (O)
($8p^+$; $8n^0$; $8e^-$)

Sodium (Na)
($11p^+$; $12n^0$; $11e^-$)

(b) Chemically active elements (valence shell incomplete)

Figure 2.4   Chemically inert and reactive elements. (a) Helium and neon are chemically inert because in each case the outermost energy level (valence shell) is fully occupied by electrons. **(b)** Elements in which the valence shell is incomplete are chemically reactive. Such atoms tend to interact with other atoms to gain, lose, or share electrons to fill their valence shells. (To simplify the diagrams, each atomic nucleus is shown as a circle with the atom's symbol in it; individual protons and neutrons are not shown.)

Subsequent shells hold larger and larger numbers of electrons. In most (but not all) cases the shells tend to be filled consecutively.

The only electrons that are important when considering bonding behavior are those in the atom's outermost shell. This shell is called the **valence shell,** and its electrons determine the chemical behavior of the atom. As a general rule, the electrons of inner shells do not take part in bonding.

When the valence shell of an atom contains 8 electrons, the atom is completely stable and is chemically inactive (inert). When the valence shell contains fewer than 8 electrons, an atom will tend to gain, lose, or share electrons with other atoms to reach the stable state. When any of these events occurs, chemical bonds are formed. Examples of chemically inert and reactive elements are shown in Figure 2.4.

The key to chemical reactivity is referred to as the *"rule of 8s";* that is, atoms interact in such a way that they will have 8 electrons in their valence shell. The first energy level represents an exception to this rule, because it is "full" when it has 2 electrons. As you might guess, atoms must approach each other very closely for their electrons to interact—in fact, their outermost energy levels must overlap.

Types of Chemical Bonds

Ionic Bonds   Ionic (i-on′ik) **bonds** form when electrons are completely transferred from one atom to another. Atoms are electrically neutral, but when they gain or lose electrons during bonding, their positive and negative charges are no longer balanced, and charged particles, called **ions,** result. When an atom gains an electron, it acquires a net negative charge because it now has more electrons than protons. Negatively charged ions are more specifically called *anions.* When an atom loses an electron, it becomes a positively charged ion, a *cation,* because it now possesses more protons than electrons. (It may help you to remember that a cation is a positively charged ion by thinking of its "t" as a plus [+] sign.) Both anions and cations result when an ionic bond is formed. Since opposite charges attract, the newly created ions tend to stay close together.

The formation of sodium chloride (NaCl), common table salt, provides a good example of ionic

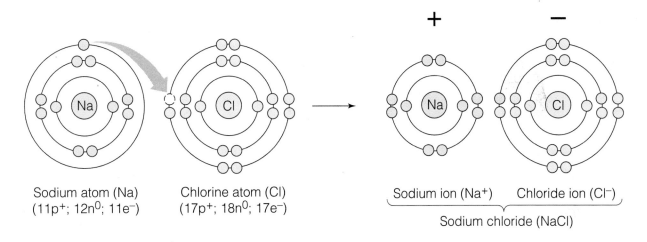

Figure 2.5   Formation of an ionic bond. Both sodium and chlorine atoms are chemically reactive because their valence shells are incompletely filled. Sodium gains stability by losing one electron, whereas chlorine becomes stable by gaining one electron. After electron transfer, sodium becomes a sodium ion (Na^+), and chlorine becomes a chloride ion (Cl^-). The oppositely charged ions attract each other.

bonding. As illustrated in Figure 2.5, sodium's valence shell contains only 1 electron and so is incomplete. However, if this single electron is "lost" to another atom, the valence shell becomes shell 2, which contains 8 electrons; thus sodium becomes a cation (Na^+) and achieves stability. Chlorine only needs 1 electron to fill its valence shell, and it is much easier to gain 1 electron (forming Cl^-) than it is to try to "give away" 7. Thus, the ideal situation is for sodium to donate its valence-shell electron to chlorine, and this is exactly what happens in the interaction between these two atoms. Sodium chloride and most other compounds formed by ionic bonding fall into the general category of chemicals called **salts.**

Covalent Bonds   Electrons do not have to be completely lost or gained for atoms to become stable. Instead, they can be shared in such a way that each atom is able to fill its valence shell at least part of the time.

Molecules in which atoms share electrons are called *covalent molecules,* and their bonds are **covalent bonds** (*co* = with; *valent* = having power). For example, hydrogen, with its single electron, can become stable if it fills its valence shell (level 1) by sharing a pair of electrons—its own and one from another atom. As shown in Figure 2.6a, a hydrogen atom can share an electron pair with another hydrogen atom to form a molecule of hydrogen gas. The shared electron pair orbits the whole molecule and satisfies the stability needs of both hydrogen atoms. Likewise, 2 oxygen atoms can share 2 pairs of electrons (form double bonds) with each other (Figure 2.6b) to form a molecule of oxygen gas (O_2).

A hydrogen atom may also share its electron with an atom of a different element. Carbon has 4 valence-shell electrons but needs 8 to achieve stability. As shown in Figure 2.6c, when methane (CH_4) is formed, carbon shares 4 electron pairs with 4 hydrogen atoms (1 pair with each hydrogen atom). Because the shared electrons orbit and "belong to" the whole molecule, each atom has a full valence shell enough of the time to satisfy its stability needs.

In the covalent molecules described thus far, electrons have been shared *equally* between the atoms of the molecule. Such molecules are called *nonpolar covalently bonded molecules.* However, electrons are not shared equally in all cases. When covalent bonds are made, the molecule formed always has a definite three-dimensional shape. A molecule's shape plays a major role in determining just what other molecules (or atoms) it can interact

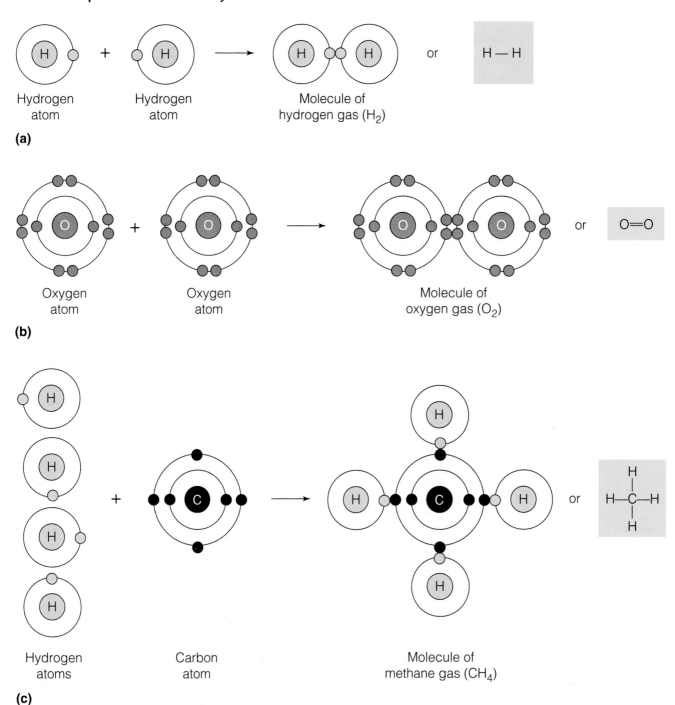

(a)

(b)

(c)

Figure 2.6   Formation of covalent bonds. (**a**) Formation of a single covalent bond between two hydrogen atoms to form a molecule of hydrogen gas.
(**b**) Formation of a molecule of oxygen gas. Each oxygen atom shares 2 electron pairs with its partner; thus a double covalent bond is formed. (**c**) Formation of a molecule of methane. A carbon atom shares 4 electron pairs with 4
hydrogen atoms. In the diagrams of molecules shown in the shaded boxes at
the far right, each pair of shared electrons is indicated by a single line between the sharing atoms.

(a) Carbon dioxide (CO_2)

(b) Water (H_2O)

Figure 2.7   Molecular models illustrating the three-dimensional structure of carbon dioxide and water molecules.

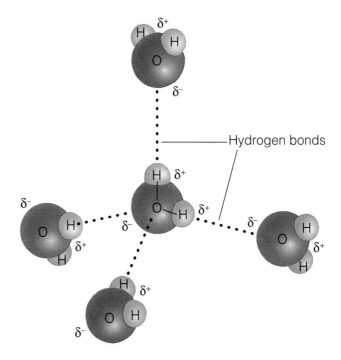

Hydrogen bonds

Figure 2.8   Hydrogen bonding between polar water molecules. The slightly positive ends (indicated by δ^+) of the water molecules become aligned with the slightly negative ends (indicated by δ^-) of other water molecules.

with; the shape may also result in unequal electron-pair sharing. The following two examples illustrate this principle (see Figure 2.7).

Carbon dioxide is formed when a carbon atom shares its 4 valence-shell electrons with 2 oxygen atoms. Oxygen is a very electron-hungry atom and attracts the shared electrons much more strongly than does carbon. However, because the carbon dioxide molecule is linear (O=C=O), the electron-pulling power of one oxygen atom is offset by that of the other, like a tug-of-war at a standoff. As a result, the electron pairs are shared equally and orbit the entire molecule, and carbon dioxide is a nonpolar molecule.

In contrast, a water molecule is formed when 2 hydrogen atoms bind covalently to a single oxygen atom. Each hydrogen atom shares an electron pair with the oxygen atom, and again the oxygen has the stronger electron-attracting ability. But in this case, the molecule formed is V-shaped (H\O/H).

The 2 hydrogen atoms are located at one end of the molecule, and the oxygen atom is at the other. Consequently, the electron pairs are not shared equally and spend more time in the vicinity of the oxygen atom, causing that end of the molecule to become slightly more negative (indicated by δ^-) and the hydrogen end to become slightly more positive (indicated by δ^+). In other words, a *polar molecule,* a molecule with two charged poles, is formed.

Polar molecules orient themselves toward other polar molecules or charged particles (ions,

proteins, and others), and they play an important role in chemical reactions that occur in body cells. Because body tissues are 60 to 80 percent water, the fact that water is a polar molecule is particularly significant, as will be described shortly.

Hydrogen Bonds **Hydrogen bonds** are extremely weak bonds formed when a hydrogen atom bound to one electron-hungry nitrogen or oxygen atom is attracted by another electron-hungry atom, and the hydrogen atom forms a "bridge" between them. Hydrogen bonding is common between water molecules (Figure 2.8), and is reflected in water's surface tension, seen in its ability to "ball up" or form spheres when it sits on a surface, and in the fact that some insects, such as water striders, can walk on water as long as they tread lightly.

Hydrogen bonds are also important *intramolecular bonds;* that is, they help to bind different parts of the *same* molecule together into a special three-dimensional shape. These rather fragile bonds are very important in helping to maintain

(a) Example of synthesis reaction: amino acids are joined to form a protein molecule

(b) Example of a decomposition reaction: breakdown of glycogen to release glucose units

Glucose + Adenosine triphosphate (ATP)

Glucose phosphate + Adenosine diphosphate (ADP)

(c) Example of an exchange reaction: ATP transfers its terminal phosphate group to glucose to form glucose phosphate

Figure 2.9   Patterns of chemical reactions. (a) In synthesis reactions, smaller particles (atoms, ions, or molecules) are bonded together to form larger, more complex molecules. **(b)** Decomposition reactions involve the breaking of bonds. **(c)** In exchange, or displacement, reactions, bonds are both made and broken.

the structure of protein molecules, which are essential functional molecules and body-building materials.

Patterns of Chemical Reactions

Chemical reactions involve the making or breaking of bonds between atoms. The total number of atoms remains the same, but the atoms appear in new combinations. Most chemical reactions have one of the three recognizable patterns described below.

Synthesis Reactions

Synthesis reactions occur when two or more atoms or molecules combine to form a larger, more complex molecule, which can be simply represented as

$$A + B \rightarrow AB$$

Synthesis reactions always involve bond formation. Because energy must be absorbed to make bonds, synthesis reactions are energy-absorbing reactions.

Synthesis reactions underlie all anabolic (constructive) activities that occur in body cells. They are particularly important for growth, and for repair of worn-out or damaged tissues. As shown in Figure 2.9a, the formation of a protein molecule by the joining of amino acids into long chains is a synthesis reaction.

Decomposition Reactions

Decomposition reactions occur when a molecule is broken down into smaller molecules, atoms, or ions, and can be indicated by

$$AB \rightarrow A + B$$

Essentially, decomposition reactions are synthesis reactions in reverse. Bonds are always broken, and the products of these reactions are smaller and simpler than the original molecules. As bonds are broken, chemical energy is released.

Decomposition reactions underlie all catabolic (destructive) processes that occur in body cells; that is, they are molecule-degrading reactions. Examples of decomposition reactions that occur in the body include the digestion of foods into their building blocks, and the breakdown of glycogen (a large carbohydrate molecule stored in the liver) to release glucose (Figure 2.9b) when blood sugar levels start to decline.

Exchange Reactions

Exchange reactions involve both synthesis and decomposition reactions: Bonds are both made and broken. During exchange reactions, a switch is made between molecule parts (changing partners, so to speak), and different molecules are made. Thus, an exchange reaction can be generally indicated as

$$AB + C \rightarrow AC + B \quad and \quad AB + CD \rightarrow AD + CB$$

An exchange reaction occurs when ATP reacts with glucose and transfers its end phosphate group to glucose, forming glucose-phosphate (Figure 2.9c). At the same time, the ATP becomes ADP. This important reaction, which occurs whenever glucose enters a body cell, effectively traps the glucose fuel molecule inside the cell.

BIOCHEMISTRY: THE CHEMICAL COMPOSITION OF LIVING MATTER

All chemicals found in the body fall into one of two major classes of molecules; they are either organic or inorganic compounds. The class of the compound is determined solely by the presence or absence of carbon. **Organic compounds** are carbon-containing compounds. The important organic compounds in the body are *carbohydrates, lipids, proteins,* and *nucleic acids.* All organic compounds are fairly (or very) large covalently bonded molecules. With a few so far unexplainable exceptions (such as carbon dioxide gas), **inorganic compounds** lack carbon and tend to be simpler, smaller molecules. Examples of inorganic compounds found in the body are *water, salts,* and many (but not all) *acids and bases.*

Organic and inorganic compounds are equally essential for life. Trying to put a price tag on which is more valuable can be compared to trying to decide whether the ignition system or the engine is more essential to the operation of a car.

Inorganic Compounds

Water

Water is the most abundant inorganic compound in the body. It accounts for about two-thirds of body weight. Among the properties that make water so vital are the following:

1. Water has a *high heat capacity;* that is, it absorbs and releases large amounts of heat before its temperature changes appreciably. Thus, it prevents the sudden changes in body temperature that might otherwise result from intense sun exposure, chilling winter winds, or internal events, such as vigorous muscle activity, that liberate large amounts of heat.

2. Because of its polarity, water is an excellent solvent; indeed, it is often called the "universal solvent." A *solvent* is a liquid or gas in which smaller amounts of other substances, called *solutes* (which may be gases, liquids, or solids), can be dissolved or suspended. The resulting *mixture* is called a *solution* when the solute particles are exceedingly minute, and a *suspension* when the solute particles are fairly large. Translucent mixtures with solute particles of intermediate size are called *colloids.*

   Small reactive chemicals such as salts, acids, and bases dissolve easily in water, and become evenly distributed. Because molecules cannot react chemically unless they are in solution, virtually all chemical reactions that occur in the body depend upon water's solvent properties. Because nutrients, respiratory gases (oxygen and carbon dioxide), and wastes can dissolve in water, water can act as a transport and exchange medium in the body. For example, all these substances are carried from one part of the body to another in blood plasma, and are exchanged between the blood and tissue cells by passing through interstitial fluid.

3. Water is an important *reactant* in some types of chemical reactions. For example, to digest foods or break down biological molecules, water molecules are added to the bonds of the larger molecules. Such reactions are called *hydrolysis reactions,* a term that specifically recognizes this role of water.

4. Water serves as the base for all body lubricants. Mucus eases feces along the large intestine;

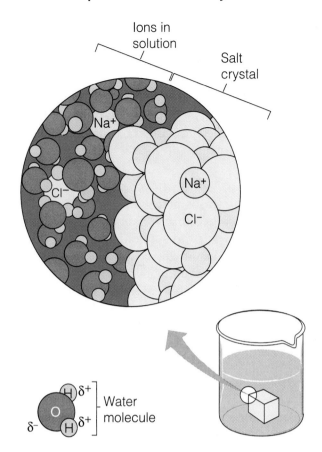

Figure 2.10   Dissociation of a salt in water. The slightly negative ends of the water molecules (δ^-) are attracted to Na^+, whereas the slightly positive ends of water molecules (δ^+) orient toward Cl^-, causing the ions of the salt crystal to be pulled apart.

saliva moistens food and prepares it for digestion. Serous fluids reduce friction between internal organs, and synovial fluids "oil" the ends of bones as they move within joint cavities.

5. Water also serves a protective function. In the form of cerebrospinal fluid, water forms a cushion around the brain that helps to protect it from physical trauma. Amniotic fluid, which surrounds a developing fetus within the mother's body, plays a similar role in protecting the fetus.

Salts

The **salts** of many metal elements are commonly found in the body, but the most plentiful salts are those containing calcium and phosphorus, found chiefly in bones and teeth. When dissolved in body fluids, salts, which are ionic compounds, easily separate into their ions. This process, called *dissociation,* occurs rather easily because the ions have already been formed. All that remains is to pull the ions apart. This is accomplished by water molecules, which orient themselves with their slightly negative ends toward the cations and their slightly positive ends toward the anions, and overcome the attraction between them (Figure 2.10).

Salts, both in their ionic forms and in combination with other elements, are vital to body functioning. For example, sodium and potassium ions are essential for nerve impulses, and iron forms part of the hemoglobin molecule that transports oxygen within red blood cells.

Because ions are charged particles, all salts are **electrolytes**—substances that conduct an electrical current in solution. When ionic (or electrolyte) balance is severely disturbed, virtually nothing in the body works. The functions of the elements found in body salts are summarized in Table 2.1 (on page 24).

Acids and Bases

Like salts, acids and bases are electrolytes. That is, they ionize and then dissociate in water and conduct an electrical current.

Characteristics of Acids   Acids have a sour taste, and can dissolve many metals or "burn" a hole in your rug; but the most useful definition of an acid is that it is a substance that can release hydrogen ions (H^+) in detectable amounts. Because a hydrogen ion is essentially a hydrogen nucleus (a "naked proton"), acids are also defined as **proton donors.**

When acids are dissolved in water, they release hydrogen ions and some anions. The anions are unimportant; it is the release of the protons that determines an acid's effects on the environment. The ionization of hydrochloric acid (an acid produced by stomach cells that aids digestion) is shown in the following equation:

$$\underset{\substack{\text{(hydrochloric} \\ \text{acid)}}}{\text{HCl}} \longrightarrow \underset{\text{(proton)}}{H^+} + \underset{\text{(anion)}}{Cl^-}$$

Other acids found or produced in the body include acetic acid (the acidic component of vinegar) and carbonic acid.

Acids that ionize completely and liberate all their protons, as does hydrochloric acid, are called *strong acids*. Acids that ionize incompletely, as do acetic and carbonic acid, are called *weak acids*. For example, when carbonic acid dissolves in water, only some of its molecules ionize to liberate H^+.

$$H_2CO_3 \rightarrow H^+ + HCO_3^- + H_2CO_3$$
(carbonic   (anion)
acid)

Characteristics of Bases **Bases** have a bitter taste, feel slippery, and are **proton acceptors.** The hydroxides are common inorganic bases. Like acids, the hydroxides ionize and dissociate in water; but in this case, the *hydroxyl ion (OH$^-$)* and some cations are released. The ionization of sodium hydroxide (NaOH), commonly known as lye, is shown as:

$$NaOH \rightarrow Na^+ + OH^-$$
(sodium   (cation)   (hydroxyl
hydroxide)   ion)

The hydroxyl ion is an avid proton (H^+) seeker, and any base containing this ion is considered a strong base. By contrast, *bicarbonate ion* (HCO_3^-), an important base in blood, is a fairly weak base.

When acids and bases are mixed, they react with each other (in an exchange reaction) to form water and a salt:

$$HCl + NaOH \rightarrow H_2O + NaCl$$
(acid)   (base)   (water)   (salt)

This type of reaction is more specifically called a **neutralization reaction.**

pH: Acid-Base Concentrations The relative concentration of hydrogen (and hydroxyl) ions in various body fluids is measured in concentration units called **pH** (pe-āch) **units.** The idea for a pH scale was devised in 1909 by a Danish biologist (and part-time beer brewer) named Sörensen, and is based on the number of protons in solution expressed in terms of moles per liter. (The *mole* is a concentration unit; its precise definition need not concern us here.) The pH scale runs from 0 to 14 (see Figure 2.11), and each successive change of 1 pH unit represents a 10-fold change in hydrogen-ion concentration.

Figure 2.11   The pH scale and pH values of representative substances. The pH scale is based on the number of hydrogen ions in solution. The actual concentration (expressed in moles per liter) of hydrogen ions [H^+] and the corresponding hydroxyl ion concentration [OH^-] are indicated for each pH value noted. At a pH of 7, the concentrations of hydrogen and hydroxyl ions are equal and the solution is neutral. A solution with a pH below 7 is acidic; above 7, basic or alkaline.

(a) Simple sugar (monosaccharide)

(b) Double sugar (disaccharide)

(c) Starch (polysaccharide)

Figure 2.12   Carbohydrates. (a) The generalized structure of a monosaccharide. **(b)** and **(c)** The basic structures of a disaccharide and a polysaccharide, respectively.

At a pH of 7, the scale midpoint, the number of hydrogen ions is exactly equal to the number of hydroxyl ions, and the solution is neutral; that is, neither acidic nor basic. Solutions with a pH lower than 7 are acidic: The hydrogen ions outnumber the hydroxyl ions. A solution with a pH of 6 has 10 times as many hydrogen ions as a solution with a pH of 7, and a pH of 3 indicates a 10,000-fold (10 × 10 × 10 × 10) increase in hydrogen-ion concentration. Solutions with a pH number higher than 7 are alkaline, or basic, and solutions with a pH of 8 and 12 (respectively) have 1/10 and 1/100,000 the number of hydrogen ions present in a solution with a pH of 7.

Living cells are extraordinarily sensitive to even slight changes in pH; and homeostasis of acid-base balance is carefully regulated by the kidneys, lungs, and a number of chemicals called **buffers,** which are present in body fluids. As described in some detail in Chapter 15, weak acids and weak bases are important components of the body's buffer systems. Because blood comes into close contact with nearly every body cell, regulation of blood pH is especially critical.

Normally, blood pH varies in a narrow range, from 7.35 to 7.45. When blood pH changes more than a few tenths of a pH unit from these limits, death becomes a distinct possibility. Although there are hundreds of examples that could be given to illustrate this point, we will provide just one very important one: When blood pH begins to dip into the acid range, the amount of life-sustaining oxygen that the hemoglobin in blood can carry to body cells begins to decline rapidly to dangerously low levels. The approximate pH values of several body fluids and of a number of commonly ingested substances appear in Figure 2.11.

Organic Compounds

Carbohydrates

Carbohydrates, which include sugars and starches, contain carbon, hydrogen, and oxygen. With slight variations, the hydrogen and oxygen atoms appear in the same ratio as in water; that is, 2 hydrogen atoms to 1 oxygen atom. This is reflected in the word *carbohydrate,* which means "hydrated carbon," and in the molecular formulas of sugars. For example, glucose is $C_6H_{12}O_6$ and ribose is $C_5H_{10}O_5$.

Carbohydrates are classified according to size as monosaccharides, disaccharides, or polysaccharides. Because monosaccharides are joined to form the molecules of the other two groups, they are the structural units, or building blocks, of carbohydrates.

Glucose   Fructose   Sucrose   Water

Figure 2.13   Dehydration synthesis and hydrolysis of a molecule of sucrose. In the reaction going to the right (the dehydration synthesis reaction), glucose and fructose are joined through a process that involves the removal of a water molecule at the bond site. The resulting disaccharide is sucrose. Sucrose is broken down to its simple sugar units when the reaction is reversed (goes to the left). In this hydrolysis reaction, a water molecule must be added to the bond to release the monosaccharides.

Monosaccharides **Monosaccharide** means one (mono) sugar (saccharide), and thus monosaccharides are also referred to as *simple sugars*. They are single-chain or single-ring structures, containing from 3 to 7 carbon atoms (see Figure 2.12a).

The most important monosaccharides in the body are glucose, fructose, galactose, ribose, and deoxyribose. *Glucose,* also called *blood sugar,* is the universal cellular fuel. *Fructose* and *galactose* are converted to glucose for use by body cells. *Ribose* and *deoxyribose* form part of the structure of nucleic acids, another group of organic molecules.

Disaccharides **Disaccharides,** or *double sugars* (Figure 2.12b), are formed when two simple sugars are joined by a synthesis reaction known as *dehydration synthesis*. In this reaction, a water molecule is lost as the bond forms (see Figure 2.13).

Some of the important disaccharides in the diet are *sucrose* (glucose-fructose), which is cane sugar; *lactose* (glucose-galactose), found in milk; and *maltose* (glucose-glucose), or malt sugar. Because the double sugars are too large to pass through cell membranes, they must be broken down (digested) to their monosaccharide units to be absorbed from the digestive tract into the blood. This is accomplished by *hydrolysis*. As a water molecule is added to each bond, the bond is broken, and the simple sugar units are released (Figure 2.13).

Polysaccharides **Polysaccharides** (literally, "many sugars,") are long, branching chains of linked simple sugars (Figure 2.12c). Because they are large

insoluble molecules, they are ideal storage products. Another consequence of their large size is that they lack the sweetness of the simple and double sugars.

Only two polysaccharides, starch and glycogen, are of major importance to the body. *Starch* is the storage polysaccharide formed by plants. We ingest it in the form of "starchy" foods, such as grain products and root vegetables (potatoes and carrots, for example). *Glycogen* is a slightly smaller, but similar, polysaccharide found in animal tissues (largely in the muscles and the liver). Like starch, it is formed of linked glucose units.

Carbohydrates provide a ready, easily used source of food energy for cells, and glucose is at the top of the "cellular menu." When glucose is oxidized (combined with oxygen) in a complex set of chemical reactions, it is broken down into carbon dioxide and water. Some of the energy released as the glucose bonds are broken is trapped in the bonds of high-energy *ATP molecules,* the energy "currency" of all body cells. If not immediately needed for ATP synthesis, dietary carbohydrates are converted to glycogen or fat and stored. Those of us who have gained weight from eating too many carbohydrate-rich snacks have a first-hand personal awareness of this conversion process!

Small amounts of carbohydrates are used for structural purposes and represent 1 to 2 percent of cell mass. Some sugars are found in our genes, and others are attached to outer surfaces of cell membranes, where they act as "road signs" to guide cellular interactions.

Table 2.4 Representative Lipids Found in the Body

Lipid type	Location/function
Neutral Fats (Triglycerides)	Found in fat deposits (subcutaneous tissue and around organs); protect and insulate the body organs; the major source of stored energy in the body.
Phospholipids (Cephalin and others)	Found in cell membranes; participate in the transport of lipids in plasma; abundant in the brain and the nervous tissue in general, where they help to form insulating white matter.
Steroids	
Cholesterol	The basis of all body steroids.
Bile salts	A breakdown product of cholesterol; released by the liver into the digestive tract, where they aid in fat digestion and absorption.
Vitamin D	Produced in the skin, on exposure to UV (ultraviolet) radiation, from a modified cholesterol molecule; necessary for normal bone growth and function.
Sex hormones	Estrogen and progesterone (female hormones) and testosterone (male sex hormone) produced from cholesterol; necessary for normal reproductive function; deficits result in sterility.
Adrenal cortical hormones	Cortisol, a glucocorticoid, is a long-term antistress hormone that is necessary for life.
	Aldosterone helps regulate salt and water balance in body fluids by targeting the kidneys.
Other Lipoid Substances	
Fat-soluble vitamins:	
A	Found in orange-pigmented vegetables (carrots) and fruits (tomatoes); part of the photoreceptor pigment involved in vision.
E	Taken in via plant products such as wheat germ and green leafy vegetables; may promote wound healing and contribute to fertility, but not proven in humans; an antioxidant; may help to neutralize free radicals, highly reactive particles believed to be involved in triggering some types of cancers.
K	Made available largely by the action of intestinal bacteria; also prevalent in a wide variety of foods; necessary for proper clotting of blood.
Prostaglandins	Derivatives of fatty acids found in cell membranes; various functions, depending on the specific class, including stimulation of uterine contractions (thus inducing labor and abortions), regulation of blood pressure, and control of stomach secretion and motility of the gastrointestinal tract; involved in inflammation.
Lipoproteins	Lipid and protein-based substances that transport fatty acids and cholesterol in the bloodstream; major varieties are high-density lipoproteins (HDLs) and low-density lipoproteins (LDLs).

(a) Formation of a triglyceride

(b) Phospholipid molecule (phosphatidyl choline)

Figure 2.14 Lipids. (a) Neutral fats, or triglycerides, are synthesized by dehydration synthesis. In this process, 3 fatty acid chains are attached to a single glycerol molecule, and a water molecule is lost at each bond site. **(b)** Structure of a typical phospholipid molecule. Two fatty acid chains and a phosphorus-containing group are attached to the glycerol backbone. **(c)** The generalized structure of cholesterol. Cholesterol is the basis for all steroids made in the body.

(c) Cholesterol

Lipids

Lipids are a large and diverse group of organic compounds (see Table 2.4). They enter the body in the form of fat-marbled meats, egg yolks, milk products, and oils. The most abundant lipids in the body are neutral fats, phospholipids, and steroids. Like carbohydrates, all lipids contain carbon, hydrogen, and oxygen atoms, but in lipids, carbon and hydrogen atoms far outnumber oxygen atoms, as illustrated by the formula for a typical fat named tristearin: $C_{57}H_{110}O_6$. Most lipids are insoluble in water, but readily dissolve in other lipids and in organic solvents such as alcohol, ether, and acetone.

Neutral Fats The **neutral fats,** or **triglycerides** (tri-glis′er-īdz), are composed of two types of building blocks, **fatty acids** and **glycerol.** Their synthesis involves the attachment of 3 fatty acids to a single glycerol molecule. The result is an E-shaped molecule that resembles the tines of a fork (Figure 2.14a). Although the glycerol backbone is the same in all neutral fats, the fatty acid chains vary; this results in different kinds of neutral fats. Neutral fats may be solid (typical of animal-product fats) or liquid (plant oils). In general, animal fats tend to be *saturated,* whereas oils are *unsaturated.* In saturated fats, all carbons have single bonds. The carbons of unsaturated fats have some double (or triple) bonds and thus can bind with more hydrogen atoms.

Neutral fats represent the body's most abundant and concentrated source of usable energy. When they are oxidized, they yield large amounts of energy. They are stored chiefly in fat deposits beneath the skin and around body organs, where they help insulate the body and protect deeper body tissues from heat loss and bumps.

Phospholipids **Phospholipids** (fos′fo-lip″idz) are very similar to the neutral fats. They differ in that a phosphorus-containing group is always part of the molecule and takes the place of one of the fatty acid chains. Thus, phospholipids have 2 instead of 3 attached fatty acids (Figure 2.14b).

Because the phosphorus-containing portion bears an electrical charge, it gives phospholipids special chemical properties and polarity. For example, the charged region attracts and interacts with water and ions, but the fatty acid chains do not. The presence of phospholipids in cellular boundaries (membranes) allows cells to be selective about what may enter or leave.

Steroids **Steroids** are basically flat molecules formed of 4 interlocking rings (Figure 2.14c); thus their structure differs quite a bit from that of fats. However, like the fats, steroids are made largely of hydrogen and carbon atoms, and are fat-soluble.

The single most important steroid molecule is *cholesterol,* which enters the body in animal products such as meat, eggs, and cheese. A certain amount is also made by the liver, regardless of dietary intake. Cholesterol is found in all cell membranes, and it is particularly abundant in the brain. Cholesterol is the raw material used to form vitamin D, some hormones (sex hormones and cortisol), and bile salts.

Saturated fats, along with cholesterol, have been implicated as substances that encourage atherosclerosis (the deposit of fatty substances in artery walls) and eventual arteriosclerosis (hard-

ening of the arteries). As a result, liquid spreads or soft margarine made from polyunsaturated fats are being promoted as products that allow us to "have our cake and eat it too"—a good-tasting spread that (unlike butter) does not damage our arteries. ❑

Proteins

Proteins account for over 50 percent of the organic matter in the body, and they have the most varied functions of the organic molecules. Some are construction materials; others play vital roles in cell function. Like carbohydrates and lipids, all proteins contain carbon, oxygen, and hydrogen. In addition, they contain nitrogen and sometimes sulfur atoms as well.

The building blocks of proteins are small molecules called **amino** (ah-me′no) **acids.** About 20 common varieties of amino acids are found in proteins. All amino acids have an *amine group,* which gives them basic properties, and an *acid group,* which allows them to act as acids. In fact, all amino acids are identical except for a single group of atoms called their *R-group* (see Figure 2.15). Hence, it is differences in the R-groups that make each amino acid chemically unique.

Amino acids are joined together in chains to form large, complex protein molecules that contain from 50 to thousands of amino acids. (Amino acid chains containing fewer than 50 amino acids are called *polypeptides.*) Because each type of amino acid has distinct properties, the sequence in which they are bound together produces proteins that vary widely both in structure and function. Perhaps this can be made more understandable if the 20 amino acids are presumed to be a "20-letter alphabet." The letters (amino acids) are then used in specific combinations to form words (a protein). Just as a change in one letter of any word can produce a word that has an entirely different meaning (flour → floor) or is nonsensical (flour → fluur), changes in kinds of amino acids (letters) or in their positions in the protein allow literally thousands of different protein molecules to be made. The structures of these proteins are specified by our genes, as will be described in Chapter 3.

Fibrous and Globular Proteins Based on their overall shape and structure, proteins are classed as

(a) Generalized structure of all amino acids

(b) Glycine (the simplest amino acid)

(c) Aspartic acid (an acidic amino acid)

(d) Lysine (a basic amino acid)

(e) Cysteine (a sulfur-containing amino acid)

Figure 2.15   Amino acid structures. (**a**) Generalized structure of amino acids. All amino acids have both an amine (—NH$_2$) group and an acid (—COOH) group; they differ only in the atomic makeup of their R-groups (green). (**b**)–(**e**) Specific structures of four amino acids. The simplest (glycine) has an R-group consisting of a single hydrogen atom. An acid in the R-group makes the amino acid (aspartic acid in this example) more acidic. An amine group in the R-group makes it more basic (as in lysine). The presence of sulfur (—SH) in the R-group of cysteine hints that this is an amino acid likely to take part in intramolecular bonding.

either fibrous or globular proteins. The strandlike **fibrous proteins**, also called **structural proteins**, appear most often in body structures. They are very important in binding structures together and for providing strength in certain body tissues. For example, *collagen* (kol′ah-jen) is found in bones, cartilage, and tendons, and is the most abundant protein in the body. *Keratin* (ker′ah-tin) is the structural protein of hair and nails, and the waterproofing material of the skin.

Globular proteins are mobile, generally spherical molecules that play crucial roles in virtually all biological processes. Because they *do things* rather than just form structures, they are also called **functional proteins.** As noted in Table 2.5, the scope of their activities is remarkable. Some (antibodies) help to provide immunity; others (hormones) help to regulate growth and development. Still others, called *enzymes* (en′zīmz), are biological catalysts that regulate essentially every chemical reaction that goes on within the body.

Fibrous proteins are exceptionally stable; the globular functional proteins are quite the opposite.

Hydrogen bonds are critically important in maintaining their structure, but hydrogen bonds are fragile and are easily broken by heat and excesses of pH. When their three-dimensional structures are destroyed, the proteins are said to be *denatured* and can no longer perform their physiological roles. This is because their function depends on their specific structure—most importantly, on the presence of specific collections of atoms called **active sites** on their surface that "fit" and interact chemically with other molecules of complementary shape and charge (see Figure 2.16). As hinted earlier, hemoglobin becomes totally unable to bind and transport oxygen when blood pH becomes too acidic, and pepsin, a protein-digesting enzyme, is inactivated by alkaline pH. In each case, the structure needed for function has been destroyed by the improper pH.

Except for enzymes, most important types of functional proteins are described with the organ system or functional process to which they are closely related. For instance, protein hormones are discussed in Chapter 9 (Endocrine System), hemo-

Table 2.5   Representative Groups of Functional Proteins

Functional group	Role(s) in the body
Antibodies (immunoglobulins)	Highly specialized proteins that recognize, bind with, and inactivate bacteria, toxins, and some viruses; function in the immune response, which helps protect the body from "invading" foreign substances.
Hormones	Help to regulate growth and development. Examples include: • Growth hormone—an anabolic hormone necessary for optimal growth during childhood and normal protein metabolism in adults. • Insulin—helps regulate blood sugar levels. • Nerve growth factor—guides the growth of neurons in the development of the nervous system.
Transport proteins	Hemoglobin transports oxygen in the blood; other transport proteins in the blood carry iron, cholesterol, or other substances.
Contractile proteins	Sliding motion of two kinds of muscle proteins (actin and myosin) accomplishes muscle contraction and body movement; on the microscopic level, activity of contractile proteins represented by cell division and sperm propulsion (toward the egg).
Catalysts (enzymes)	Essential to virtually every biochemical reaction in the body; increase the rates of chemical reactions by at least a millionfold; in their absence (or destruction), biochemical reactions cease.

globin is considered in Chapter 10 (Blood), and antibodies are described in Chapter 12 (Body Defenses). However, enzymes are important in the functioning of all body cells, and so these incredibly complex molecules are considered here.

Enzymes and Enzyme Activity   Enzymes are functional proteins that act as biological catalysts. A **catalyst** is a substance that increases the rate of a chemical reaction without becoming part of the product or being changed itself. Enzymes accomplish this feat by binding to and "holding" the reacting molecules (the substrates) in the proper position for chemical interaction. While the substrates are bound to the enzyme's active site (see Figure 2.16a), they undergo structural changes that result in a new product. Once the reaction has occurred, the enzyme releases the product. Because enzymes are not changed in doing their job, they are reusable, and only small amounts of each enzyme are needed by the cells.

Enzymes are capable of catalyzing millions of reactions each minute. However, they do more than just increase the speed of chemical reactions; they also determine just which reactions are possible at a particular time. No enzyme, no reaction! Enzymes can be compared to a bellows used to fan a sluggish fire into flaming activity. Without enzymes, biochemical reactions would occur far too slowly to sustain life.

Although there are hundreds of different kinds of enzymes in body cells, they are very specific in their activities, each controlling only one (or a small group of) chemical reaction(s) and acting only on specific molecules. Most enzymes are named according to that specific type of reaction they catalyze. There are "hydrolases," which add water; "oxidases," which cause oxidation; and so on. (In most cases, an enzyme can be recognized by the suffix -**ase** forming part of its name.)

Many enzymes are produced in an inactive form, and must be activated in some way before

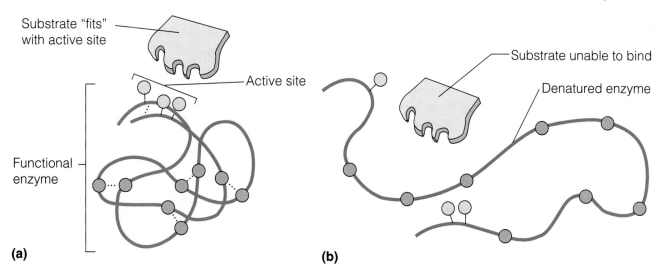

Substrate "fits" with active site

Active site

Functional enzyme

(a)

Substrate unable to bind

Denatured enzyme

(b)

Figure 2.16   Denaturation of a functional protein molecule such as an enzyme. (a) The molecule's three-dimensional structure is maintained by intramolecular bonds. Atoms composing the active site of the enzyme are shown as stalked particles. The substrate, or molecule the enzyme acts on, has a corresponding binding site, and the two sites fit together very precisely. **(b)** Breaking the intramolecular bonds that maintain the three-dimensional structure of the enzyme results in a linear molecule, with the atoms of the former active site widely separated. Enzyme-substrate binding can no longer occur.

they can function. In other cases, enzymes are inactivated immediately after they have performed their catalytic function. Both events are true of enzymes that promote blood clotting when a blood vessel has been damaged. If this were not so, large numbers of unneeded and potentially lethal blood clots would be formed.

Nucleic Acids

The role of **nucleic** (nu-kle′ik) **acids** is fundamental: They make up the genes, which provide the basic blueprint of life. Not only do they determine what type of organism you will be, but they also direct your growth and development. They do this entirely by dictating protein structure. (Remember that enzymes, which catalyze all the chemical reactions that occur in the body, are proteins.)

Nucleic acids, composed of carbon, oxygen, hydrogen, nitrogen, and phosphorus atoms, are the largest biological molecules in the body. Their building blocks, the **nucleotides** (nu′kle-o-tīdz), are quite complex. Each consists of three basic parts: (1) a nitrogen-containing base, (2) a pentose sugar, and (3) a phosphate group (Figure 2.17a).

The bases come in five varieties: *adenine* (A), *guanine* (G), *cytosine* (C), *thymine* (T), and *uracil* (U). A and G are large, two-ring bases, whereas the others are smaller, single-ring structures. The nucleotides are named according to the base they contain: A-containing bases are adenine nucleotides, C-containing bases are cytosine nucleotides, and so on.

The two major kinds of nucleic acid are **deoxyribonucleic** (de-ok″sĭ-ri″bo-nu-kle′ik) **acid (DNA)** and **ribonucleic acid (RNA).** DNA and RNA differ in many respects. DNA is the genetic material found within the cell nucleus (the control center of the cell). It has two fundamental roles: (1) It replicates itself exactly before a cell divides, thus ensuring that the genetic information in every body cell is identical. (2) It provides the instructions for building every protein in the body. RNA is located outside the nucleus and can be considered the "molecular slave" of DNA; that is, RNA carries out the orders for protein synthesis issued by DNA.

Cytosine base

Phosphate

Deoxyribose sugar

Diagrammatic representation

(a) Cytosine nucleotide

KEY:

Thymine (T)

Adenine (A)

Cytosine (C)

Guanine (G)

Deoxyribose sugar

Phosphate

— Hydrogen bond

Sugar unit

Phosphate unit

(b)

Figure 2.17 **Structure of DNA.** (**a**) The unit of DNA (deoxyribonucleic acid) is the nucleotide, composed of a linked deoxyribose sugar molecule, a phosphate group, and a nitrogen-containing base (attached to the sugar). The nucleotide illustrated contains the base cytosine. (**b**) Structure of a DNA molecule, two nucleotide chains coiled into a double helix. The "backbones" of DNA are formed by alternating sugar and phosphate molecules. The "rungs" are formed by the binding together of complementary bases (A to T, G to C) by hydrogen bonds.

Although both RNA and DNA are formed by the joining together of nucleotides, their final structures are different. As shown in Figure 2.17b, DNA is a long double chain of nucleotides. Its bases are A, G, T, and C, and its sugar is *deoxyribose*. Its two nucleotide chains are held together by hydrogen bonds between the bases, so that a ladderlike molecule is formed. Alternating sugar and phosphate molecules form the "uprights," or backbones, of the ladder, and each "rung" is formed of two joined bases (one base pair). Binding of the bases is very specific: A always binds to T, and G always binds

Adenine
NH$_2$

(a) **Adenosine triphosphate (ATP)**

Phosphates

Ribose

OH OH

Figure 2.18   ATP. (a) The structure of ATP (adenosine triphosphate). **(b)** Hydrolysis of ATP to yield ADP and inorganic phosphate.

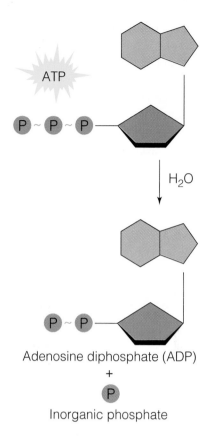

ATP

H$_2$O

Adenosine diphosphate (ADP)
+

Inorganic phosphate

(b) Hydrolysis of ATP

to C. Thus, A and T are said to be **complementary bases,** as are C and G. A base sequence of ATGA on one nucleotide chain would necessarily be bonded to the complementary base sequence TACT on the other nucleotide strand. The whole molecule is then coiled into a spiral staircaselike structure called a **double helix.**

Whereas DNA is double-stranded, RNA molecules are single nucleotide strands. RNA bases include A, G, C, and U (U replaces T found in DNA), and its sugar is ribose instead of deoxyribose. Three varieties of RNA exist—*messenger, ribosomal,* and *transfer RNA*—and each has a specific role to play in carrying out DNA's instructions for building proteins. Messenger RNA carries the information for building the protein from the DNA genes to the ribosomes, the protein-synthesizing sites. Transfer RNA ferries amino acids to the ribosomes. Ribosomal RNA forms part of the ribosomes, where it oversees the "translation" of the message and the binding together of amino acids to form the proteins. Protein synthesis will be described in greater detail in Chapter 3.

Adenosine Triphosphate (ATP)

The synthesis of **adenosine triphosphate** (ah-den′o-sēn tri-fos′fāt), or **ATP,** is all-important be-

cause it provides a form of chemical energy that is usable by all body cells. Without ATP, molecules cannot be made or broken down, cells cannot maintain their boundaries, and all life processes grind to a halt.

Although glucose is the most important "fuel" for body cells, none of the chemical energy contained in its bonds can be used directly to power cellular work. Instead, energy released as glucose is catabolized is captured and stored in the bonds of ATP molecules as small "packets" of energy.

Structurally, ATP is a modified nucleotide; it consists of an adenine base, ribose sugar, and three phosphate groups (Figure 2.18). The phosphate groups are attached by unique chemical bonds called *high-energy phosphate bonds.* When these bonds are ruptured by hydrolysis, energy that can be used immediately by the cell to power a particular activity—such as synthesizing proteins, moving substances across its membrane or, in the case of muscle cells, contracting—is liberated (Figure 2.19). ATP can be compared to a tightly coiled spring that is ready to uncoil with tremendous energy when the "catch" is released. The consequence of cleavage of its terminal phosphate bond can be represented as follows:

$$ATP \rightarrow ADP + \textcircled{P} + E$$
(adenosine   (adenosine   (inorganic   (energy)
triphosphate) diphosphate)   phosphate)

As ATP is used to provide cellular energy, ADP accumulates, and ATP supplies are replenished by oxidation of food fuels. Essentially the same amount of energy must be captured and used to reattach a phosphate group to ADP (that is, to reverse the reaction) as is liberated when the terminal phosphate is cleaved off.

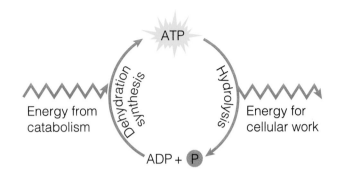

Figure 2.19   **The ATP cycle.** Energy released by oxidation of food fuels (catabolism) in the cell is used to attach a phosphate group to ADP, regenerating ATP. Energy stored in ATP drives most cellular work.

Important Terms

acid

adenosine triphosphate (ah-den′o-sēn tri-fos′fāt) **(ATP)**

atom

atomic symbol

base

carbohydrate

chemical reaction

compound

covalent bond

decomposition reaction

electrolyte

electron

element

energy

enzyme (en′zīm)

exchange reaction

hydrogen bond

inorganic compound

ionic (i-on′ik) **bond**

isotope

lipid

matter

molecule

neutron

nucleic (nu-kle′ik) **acid**

organic compound

pH (pe-āch)

protein

proton

radioactivity

radioisotope

salt

synthesis reaction

valence shell

Summary

Concepts of Matter and Energy (pp. 22–23)

1. Matter
   a. Matter is anything that occupies space and has mass.
   b. Matter exists in three states: gas, liquid, and solid.

2. Energy
   a. Energy is the capacity to do work or to put matter into motion. Energy has kinetic (active) and potential (stored) work capacities.
   b. Energy forms important in body functioning include chemical, electrical, mechanical, and radiant.
   c. Energy forms are interconvertible, but some energy is always unusable (lost as heat) in such transformations.

Composition of Matter (pp. 23–28)

1. Elements and atoms
   a. Each element is a unique substance that cannot be decomposed into simpler substances by ordinary chemical methods. A total of 110 elements exist; they differ from one another in their chemical and physical properties.
   b. Four elements (carbon, hydrogen, oxygen, and nitrogen) comprise 96 percent of living matter. Several other elements are present in small or trace amounts.
   c. The building blocks of elements are atoms. Each atom is described by an atomic symbol consisting of one or two letters.

2. Atomic structure
   a. Atoms are composed of three subatomic particles: protons, electrons, and neutrons. Because all atoms are electrically neutral, the number of protons in any atom is equal to its number of electrons.
   b. The planetary model of the atom portrays all the mass of the atom (protons and neutrons) con-

centrated in a minute central nucleus. Electrons orbit the nucleus along specific orbits. The orbital model also locates protons and electrons in a central nucleus, but depicts electrons as occupying areas of space called *orbitals* and forming an electron cloud of negative charge around the nucleus.
   c. Each atom can be identified by an atomic number, which is equal to the number of protons contained in the atom's nucleus.
   d. The atomic mass number is equal to the sum of the protons and neutrons in the atom's nucleus.
   e. Isotopes are different atomic forms of the same element; they differ only in the number of neutrons in the nucleus. Many of the heavier isotopes are unstable and decompose to a more stable form by ejecting particles or energy from the nucleus, a phenomenon called *radioactivity*. Such radioisotopes are useful in medical diagnosis and treatment, and in biochemical research.
   f. The atomic weight is approximately equal to the mass number of the most abundant isotope of any element.

Molecules and Compounds (p. 29)

1. A molecule is the smallest unit resulting from the binding of two or more atoms. If the atoms are different, a molecule of a *compound* is formed.

2. Compounds exhibit properties different from those of the atoms comprising them.

Chemical Bonds and Chemical Reactions (pp. 29–37)

1. Bond formation
   a. Chemical bonds are energy relationships. Electrons in the outermost energy level (valence shell) of the reacting atoms are active in the bonding.
   b. Atoms with a full valence shell (2 electrons in shell 1, or 8 in the subsequent shells) are chemically inactive. Those with an incomplete va-

lence shell interact by losing, gaining, or sharing electrons to achieve stability (that is, to fill the valence shell).

c. Ions are formed when valence-shell electrons are completely transferred from one atom to another. The oppositely charged ions formed attract each other, forming an ionic bond. Ionic bonds are common in salts.

d. Covalent bonds involve the sharing of electron pairs between atoms. If the electrons are shared equally, the molecule is a nonpolar covalent molecule. If the electrons are not shared equally, the molecule is a polar covalent molecule. Polar molecules orient themselves toward charged particles.

e. Hydrogen bonds are fragile bonds that bind together different parts of the same molecule (intramolecular bonds). They are common in large, complex organic molecules, such as proteins and nucleic acids.

2. Patterns of chemical reactions

a. Chemical reactions involve the formation or breaking of chemical bonds. They are indicated by the writing of a chemical equation, which provides information about the atomic composition (formula) of the reactant(s) and product(s).

b. Chemical reactions that result in larger, more complex molecules are synthesis reactions; they involve bond formation.

c. In decomposition reactions, larger molecules are broken down into simpler molecules or atoms. Bonds are broken.

d. Exchange reactions involve both the making and breaking of bonds. Atoms are replaced by other atoms.

*Biochemistry: The Chemical Composition
of Living Matter* (pp. 37–50)

1. Inorganic compounds

a. Inorganic compounds comprising living matter do not contain carbon. They include water, salts, acids, and bases.

b. Water is the single most abundant compound in the body. It acts as a universal solvent in which electrolytes (salts, acids, and bases) ionize and in which chemical reactions occur. It slowly absorbs and releases heat, thus helping to maintain homeostatic body temperature, and is the basis of transport and lubricating fluids.

c. Salts in ionic form are involved in nerve transmission, muscle contraction, blood clotting, transport of oxygen by hemoglobin, cell permeability, metabolism, and many other reactions.

Additionally, calcium salts (as bone salts) contribute to bone hardness.

d. Acids are proton donors. When dissolved in water, they release hydrogen ions. Strong acids dissociate completely; weak acids dissociate incompletely.

e. Bases are proton acceptors. The most important inorganic bases are hydroxides. Bicarbonate ions are important bases in the body. When bases and acids interact, neutralization occurs—that is, a salt and water are formed.

f. pH is a measure of the relative concentrations of hydrogen and hydroxyl ions in various body fluids. Each change of one pH unit represents a 10-fold change in hydrogen (or hydroxyl) ion concentration. A pH of 7 is neutral (that is, the concentrations of hydrogen and hydroxyl ions are equal). A pH below 7 is acidic; a pH above 7 is alkaline (basic).

g. Normal blood pH ranges from 7.35 to 7.45. Slight deviations outside this range can be fatal.

2. Organic compounds

a. Organic compounds are the carbon-containing compounds that comprise living matter. Carbohydrates, lipids, proteins, and nucleic acids are examples of organic compounds. They all contain carbon, oxygen, and hydrogen. Proteins and nucleic acids also contain substantial amounts of nitrogen.

b. Carbohydrates contain carbon, hydrogen, and oxygen in the general relationship $(CH_2O)_n$; their building blocks are monosaccharides. Monosaccharides include glucose, fructose, galactose, deoxyribose, and ribose. Disaccharides include sucrose, maltose, and lactose; and polysaccharides include starch and glycogen. Carbohydrates are ingested as sugars and starches. Carbohydrates, and in particular glucose, are the major energy source for the formation of ATP.

c. Lipids include the neutral fats (glycerol plus three fatty acid chains), phospholipids, and steroids (most importantly, cholesterol). Neutral fats are found primarily in adipose tissue, where they provide insulation and reserve body fuel. Phospholipids and cholesterol are found in all cell membranes. Cholesterol also forms the basis of certain hormones, bile salts, and vitamin D. Like carbohydrates, the lipids are degraded by hydrolysis and synthesized by dehydration synthesis.

d. Proteins are constructed from building blocks called amino acids; 20 common types of amino acids are found in the body. Amino-acid sequence determines the proteins constructed. Fibrous, or structural, proteins are the basic structural materials of the body. Globular proteins are functional molecules; examples of these include enzymes, some hormones, and hemoglobin. Disruption of the hydrogen bonds of functional proteins leads to their denaturation and inactivation.

e. Enzymes increase the rates of chemical reactions by combining specifically with the reactants and holding them in the proper position to interact. They do not become part of the product. Many enzymes are produced in an inactive form or are inactivated immediately after use.

f. Nucleic acids include deoxyribonucleic acid (DNA) and ribonucleic acid (RNA). The building

unit of nucleic acids is the nucleotide; each nucleotide consists of a nitrogenous base, a sugar (ribose or deoxyribose), and a phosphate group. DNA (the "stuff" of the genes) maintains genetic heritage by replicating itself before cell division, and contains the code-specifying protein structure. RNA acts in protein synthesis to ensure that instructions of the DNA are executed.

g. ATP (adenosine triphosphate) is the universal energy compound used by all cells of the body. When energy is liberated by the oxidation of glucose, some of that energy is captured in the high-energy phosphate bonds of ATP molecules and is stored for later use.

Review Questions

1. Why is a study of basic chemistry essential to understanding human physiology?

2. Matter and your body—how are they interrelated?

3. Matter *occupies space* and *has mass*. Describe how energy *must be* described in terms of these two factors. Then define *energy*.

4. Identify the energy *form* in use in each of the following examples:
   a. Chewing food.
   b. Vision (two types, please—think!).
   c. Bending the fingers to make a fist.
   d. Breaking the bonds of ATP molecules to energize your muscle cells to make that fist.

5. The statement has been made that "some energy is lost in every energy transformation." Explain the meaning of this statement. (Direct your response to answering the questions: Is it really lost? If not, where is it?)

6. According to Greek history, a Greek scientist went running through the streets announcing that he had transformed lead into gold. Both lead and gold are elements. On the basis of what you know about the nature of elements, explain why his rejoicing was short-lived.

7. What four elements make up the bulk of all living matter? (Give both their names and their atomic symbols.) Which of these is found primarily in proteins and nucleic acids?

8. List at least six other elements found in the body and describe one way in which each is important to body functioning.

9. What is the relationship of an atom to an element?

10. All atoms are neutral. Explain the basis of this fact.

11. Fill in the following table to fully describe an atom's subatomic particles.

Particle	Position in the atom	Charge	Mass
Proton			
Neutron			
Electron			

12. Define *isotope*.

13. If an element has three isotopes, which of them (the lightest, the one with an intermediate mass, or the heaviest) is most likely to be a radioisotope, and why? Define *radioactivity*.

14. Complete this statement: Chemical behavior results from interactions of the _____ (subatomic-particle type) found in the _____ _____ of the atom.

15. Define *molecule*. Distinguish between a molecule of an element and a molecule of a compound.

16. Explain the basis of ionic bonding. How do ionic bonds differ from covalent bonds?

17. What are the hydrogen bonds, and how are they important in the body?

18. The two oxygen atoms forming molecules of oxygen gas that you breathe are joined by a *polar* covalent bond. Explain why this statement is true or false.

19. Identify each of the following reactions as a synthesis, decomposition, or exchange reaction:

$$2 \, Hg + O_2 \rightarrow 2 \, HgO$$
$$Fe^{2+} + CuSO_4 \rightarrow FeSO_4 + Cu^{2+}$$
$$HCl + NaOH \rightarrow NaCl + H_2O$$
$$HNO_3 \rightarrow H^+ + NO_3^-$$

20. Distinguish between inorganic and organic compounds, and list the major categories of each in the body.

21. Give at least four reasons continual water intake is essential to life.

22. Salts, acids, and bases are electrolytes. What is an electrolyte?

23. Compare and contrast acids and bases.

24. Define *pH*. State which of the following would be acidic, which basic, and which neutral: pure (distilled) water, vinegar, sodium bicarbonate, and gastric juice.

25. The pH range of blood is from 7.35 to 7.45. This is slightly _____ (acidic/basic).

26. A pH of 3.3 is _____ (1, 10, 100, 1000) times more acidic than a pH of 4.3.

27. Define *monosaccharide, disaccharide,* and *polysaccharide.* Give at least two examples of each. Which of these is the building unit of carbohydrates? What is the primary function of carbohydrates in the body?

28. What are the general structures of neutral fats, phospholipids, and steroids? Give one or two important uses of each of these lipid types in the body.

29. The building block of proteins is the amino acid. Draw a diagram of the structure of a generalized amino acid. What is the importance of the R-group?

30. Name the two protein classes based on structure and function in the body, and give two examples of each.

31. Define *enzyme* and describe the mechanism of enzyme activity.

32. Virtually no chemical reaction can occur in the body in the absence of enzymes. How might excessively high body temperature or acidosis (acidic blood pH) interfere with enzyme activity?

33. What is the structural unit of nucleic acids? Name the two major classes of nucleic acid found in the body, and then compare and contrast them in terms of (a) bases and sugar content, (b) general three-dimensional structure, and (c) relative functions.

34. What is ATP, and what is its central role in the body?

At the Clinic

1. A number of antibiotics act by binding to certain essential enzymes in the target bacteria. How might these antibiotics influence the chemical reaction controlled by the enzyme? What might be the effect on the bacteria? On the person taking the antibiotic prescription?

2. Mrs. Roberts, who is in a diabetic coma, has just been admitted to Noble Hospital. Her blood pH indicates that she is in severe acidosis, and measures are quickly instituted to bring her blood pH back within normal limits. Note the normal pH of blood, and discuss why severe acidosis is a problem.

3. Evelyn is quite proud of her slender, model-like figure and boasts that she doesn't have an "ounce of excess body fat." Barbara, on the other hand, is grossly overweight. She complains of being "hot" most of the time, and on a hot day she is miserable. Evelyn generally feels chilled except on very hot days. Explain the relative sensitivity to environmental temperature of these two women on the basis of information you have been given in the Organic Compounds section of this chapter.

4. Mrs. Bigda is 3 months pregnant, and her obstetrician wants to measure the head size of the fetus in her womb to make sure development is progressing normally. Which medical imaging technique will the physician probably use and why?

3

Cells and Tissues

YOUR GOALS

After completing this chapter, you will have a working knowledge of the functions of cells and tissues and will have mastered the objectives listed below.

FUNCTION PREVIEW

- ■ Cells carry out all the chemical activities needed to sustain life.
- ■ Tissues provide for a division of labor among body cells.

OBJECTIVE CHECKLIST

CELLS (pp. 56–76)

- ❑ Name the four elements that make up the bulk of living matter and list several trace elements.
- ❑ Define *cell, organelle,* and *inclusion.*
- ❑ Identify on a cell model or diagram the three major cell regions (nucleus, cytoplasm, and plasma membrane).
- ❑ Describe the structures of the nucleus and explain the function of chromatin and nucleoli.
- ❑ Identify on a cell model or describe the organelles and discuss the major function of each.
- ❑ Define *selective permeability, diffusion* (including *dialysis* and *osmosis*), *active transport, passive transport, solute pumping, exocytosis, endocytosis, phagocytosis, pinocytosis, hypertonic, hypotonic,* and *isotonic.*
- ❑ Describe the structure of the plasma membrane, and explain how the various transport processes account for the directional movements of specific substances across the plasma membrane.
- ❑ Describe briefly the process of DNA replication and of mitosis. Explain the importance of mitotic cell division.
- ❑ In relation to protein synthesis, describe the roles of DNA and of the three varieties of RNA.

BODY TISSUES (pp. 76–87)

- ❑ Name the four major tissue types and their chief subcategories. Explain how the four major tissue types differ structurally and functionally.

(continued)

❏ Give the chief locations of the various tissue types in the body.

❏ Describe the process of tissue repair (wound healing).

DEVELOPMENTAL ASPECTS OF CELLS AND TISSUES (pp. 87–88)

❏ Define *neoplasm,* and distinguish between benign and malignant neoplasms.

❏ Explain the significance of the fact that some tissue types (muscle and nerve) are amitotic after the growth stages are over.

PART 1: CELLS

In the late 1600s, Robert Hooke was looking through a primitive microscope at some plant tissue—cork. He saw some cubelike structures that reminded him of the long rows of monk's rooms (or cells) at the monastery, so he named these structures **cells.** The living cells that had formed the cork were long since dead. However, the name stuck and is used to describe the unit, or the building block, of all living things, plants and animals alike. The human body has trillions of these microscopic building blocks.

OVERVIEW OF THE CELLULAR BASIS OF LIFE

Perhaps the most striking thing about a cell is its organization. If we chemically analyze cells, we find that they are made up primarily of four elements—carbon, oxygen, hydrogen, and nitrogen—plus trace amounts of several other elements (such as iron, sodium, and potassium). Although the four major elements build most of the cell's structure (which is largely protein), the trace elements are very important for certain cell functions. For example, calcium is needed for blood clotting (among other things), and iron is necessary to make hemoglobin, which carries oxygen in the blood. Iodine is required to make the thyroid hormone that controls metabolism. In their ionic form, many of the metals (such as calcium, sodium, and potassium) can carry an electrical charge; when

they do they are called **electrolytes** (e-lek′tro-līts). Sodium and potassium ions are essential if nerve impulses are to be transmitted and muscles are to contract. (A more detailed account of body chemistry appears in Chapter 2.)

Strange as it may seem, especially when we feel our firm muscles, living cells are about 60 percent water, which is one of the reasons water is essential for life. In addition to containing large amounts of water, all the cells of the body are constantly bathed in a dilute saltwater solution (something like seawater) called *interstitial fluid,* which is derived from the blood. All exchanges between cells and blood are made through this fluid.

Cells vary tremendously in size. The smallest known cells *(bacteria)* are about 2 micrometers (1/12,000th of an inch) in diameter. The largest may extend 10 cm (4 inches) or more across (for example, an ostrich egg), or 1 m (3 feet) or more in length (some nerve cells). Cells can also have amazingly different shapes. For example, some are disk-shaped (red blood cells), some have many threadlike extensions (nerve cells), others are like toothpicks that are pointed at each end (smooth muscle cells), and others are cubelike (some types of epithelial cells).

Cells vary dramatically in the functions, or roles, they play in the body. For example, white blood cells wander freely through the body tissues and protect the body by destroying bacteria and other foreign substances. Some cells make hormones, or chemicals that regulate other body cells. Still others take part in gas exchanges in the lungs or cleanse the blood (kidney tubule cells). A cell's structure often reflects its function; this will become clear when you read the discussion of tissues later in this chapter.

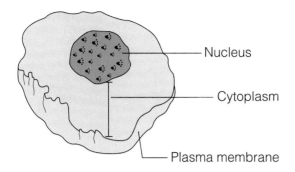

Figure 3.1   The three main regions of the generalized cell.

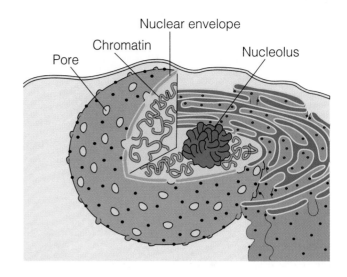

Figure 3.2   Structure of the nucleus.

ANATOMY OF A GENERALIZED CELL

Although no one cell type is exactly like all others, cells *do* have many common structural and functional features. Here we will talk about the **generalized cell,** which demonstrates the many typical structures and functions common to *all* cells.

In general, all cells have three main regions or parts—a *nucleus* (nu′kle-us), *cytoplasm* (si′to-plazm″), and a *plasma membrane* (Figure 3.1). The nucleus is usually located near the center of the cell. It is surrounded by the semifluid cytoplasm, which in turn is enclosed by the plasma membrane, which forms the outer cell boundary. (Figure 3.5 on p. 60 shows the more detailed structure of the generalized cell as revealed by the electron microscope.)

The Nucleus

Anything that works, works best when it is controlled. For cells, "headquarters" or the control center is the gene-containing **nucleus.** The genetic material, or **deoxyribonucleic acid (DNA),** is much like a blueprint that contains all the instructions needed for building the whole body; so as one might expect, human DNA differs from that of a frog. More specifically, DNA has the instructions for building **proteins.** DNA is also absolutely necessary for cell reproduction. A cell that has lost or ejected its nucleus (for whatever reason) is literally "programmed to die."

While most often oval or spherical, the shape of the nucleus usually conforms to the shape of the cell. For example, if the cell is elongated, the nucleus is usually extended as well. The nucleus has three distinct regions or structures: the nuclear membrane, nucleoli, and chromatin.

Nuclear Membrane

The nucleus is bound by a double membrane barrier called the **nuclear membrane** or **nuclear envelope** (see Figure 3.2). Between the two membranes is a fluid-filled "moat" or space. At various points, the two layers of the nuclear membrane approach each other and fuse, and *nuclear pores* penetrate through the fused regions. Like other cellular membranes, the nuclear membrane is selectively permeable, but passage of substances through it is much freer than elsewhere because of its relatively large pores. The nuclear membrane encloses a jellylike fluid called **nucleoplasm** (nu′kle-o-plazm″) in which the nucleoli and chromatin are suspended.

Nucleoli

The nucleus contains one or more small, dark-staining, essentially round bodies called **nucleoli** (nu-kle′o-li). Nucleoli are sites where ribosomes are assembled. The ribosomes, which eventually migrate into the cytoplasm, serve as the actual sites of protein synthesis, as described shortly.

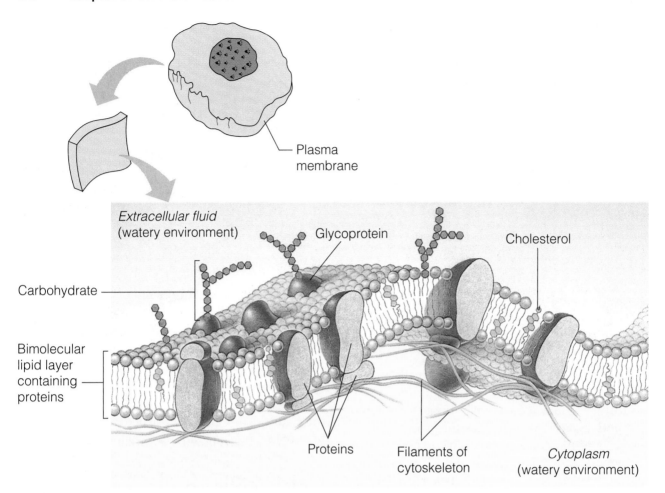

Figure 3.3 Structure of the plasma membrane.

Chromatin

When a cell is not dividing, its DNA is combined with protein and forms a loose network of bumpy threads called **chromatin** (kro′mah-tin) that is scattered throughout the nucleus. When a cell is dividing to form two daughter cells, the chromatin threads coil and condense to form dense rodlike bodies called **chromosomes**—much the way a stretched spring becomes shorter and thicker when relaxed. The functions of DNA and the mechanism of cell division are discussed in the Cell Physiology section later in this chapter.

The Plasma Membrane

The flexible **plasma membrane** is a fragile, transparent barrier that contains the cell contents and separates them from the surrounding environment. (The term *cell membrane* is often used instead, but since nearly all cellular organelles are composed of membranes, we will specifically refer to the cell's surface or outer limiting membrane as the plasma membrane.) Although the plasma membrane is important in defining the limits of the cell, it is much more than a passive envelope, or "baggie." As you will see, its unique structure allows it to play a dynamic role in many cellular activities.

The plasma membrane has a core of two lipid (fat) layers in which protein molecules float (Figure 3.3). Although the bulk of the lipid portion is phospholipids, a substantial amount of cholesterol is also found in plasma membranes. (The characteristics of these specialized lipids are described in Chapter 2, pp. 43–44.) The olive oil–like lipid bilayer forms the basic "fabric" of the membrane and is relatively impermeable to most water-soluble molecules. The cholesterol has a stabilizing effect and helps keep the membrane fluid.

The proteins scattered in the lipid bilayer are responsible for most of the specialized functions of the membrane. Some proteins are enzymes. Many of the proteins mounted on the cell exterior are receptors, or binding sites, for hormones or other chemical messengers. Most proteins that span the membrane are involved in transport functions. For example, some cluster together to form tiny *pores* through which water and small water-soluble molecules or ions can move; others act as *carriers* that bind to a substance and move it through the membrane. Branching sugar groups are attached to most of the proteins abutting the extracellular space. Such "sugar-proteins" are called *glycoproteins,* and because of their presence, the cell surface is a fuzzy, sticky, sugar-rich area. (You can think of your cells as being sugar-coated.) Among other things, these glycoproteins determine your blood type, act as receptors that certain bacteria, viruses, or toxins can bind to, and play a role in cell-to-cell interactions. Definite changes in glycoproteins occur in cells that are being transformed into cancer cells. (Cancer is discussed in the box on p. 70.)

Specializations of the Plasma Membrane

Specializations of the plasma membrane—such as microvilli and membrane junctions—are commonly displayed by the (epithelial) cells that form the linings of hollow body organs, such as the small intestine (Figure 3.4). **Microvilli** (mi″kro-vil′i) are tiny fingerlike projections that greatly increase the cell's surface area for absorption so that the process occurs more quickly.

The membrane junctions vary structurally depending on their roles. In **tight junctions,** the adjacent plasma membranes are fused together tightly like a zipper. Tight junctions bind cells together into leakproof sheets that prevent substances from passing through the extracellular space between cells. In the small intestine, for example, these junctions prevent digestive enzymes from seeping into the bloodstream. **Desmosomes** (des′mo-somz) are adhesion junctions that prevent cells subjected to mechanical stress (such as skin cells) from being pulled apart. Structurally, these junctions are buttonlike thickenings of adjacent plasma membranes, which are connected by fine protein filaments. The basic function of **gap junctions,** commonly seen in the heart and between embryonic cells, is to allow communication; chemical molecules, such as nutrients or ions, pass di-

Figure 3.4   Cell junctions. An epithelial cell is shown joined to adjacent cells by the three common types of cell junctions: tight junctions, desmosomes, and gap junctions. Also illustrated are microvilli (seen projecting from the free cell surface).

rectly from one cell to another. In gap junctions, the neighboring cells are connected by hollow cylinders (connexons) composed of proteins that span the entire width of the membrane.

The Cytoplasm

The **cytoplasm** consists of the cell material outside the nucleus and inside the plasma membrane. It is the site of most cellular activities, so it can be thought of as the "factory area" of the cell. Although early scientists believed that the cytoplasm was a structureless gel, the electron microscope has revealed that it has three major elements: the *cytosol, organelles,* and *inclusions.* The **cytosol** is semitransparent fluid which suspends the other elements. Dissolved in the cytosol, which is largely water, are nutrients and a variety of other solutes (dissolved substances).

Nucleolus

Nucleus

Chromatin

Nuclear membrane

Centriole

Vacuole

Microtubules (cytoskeleton)

Lysosome

Golgi apparatus

Secretion being released from cell by exocytosis

Cytosol

Peroxisome

Smooth endoplasmic reticulum

Rough endoplasmic reticulum

Plasma membrane

Ribosomes

Microvilli

Mitochondrion

Microfilament (cytoskeleton)

Figure 3.5   Structure of the generalized cell. No cell is exactly like this one, but this generalized cell drawing illustrates features common to many human cells.

The **organelles** (or"gah-nelz'), described in detail shortly, are the metabolic machinery of the cell. Each type of organelle is "engineered" to carry out a specific function for the cell as a whole; some synthesize proteins, others package those proteins, and so on.

Inclusions are not functioning units, but instead are chemical substances that may or may not be present, depending on the specific cell type. Most inclusions are stored nutrients or cell products. They include the fat droplets common in fat cells, glycogen granules, pigments such as melanin seen in skin and hair cells, mucus and other secretory products, and various kinds of crystals.

Cytoplasmic Organelles

The cytoplasmic organelles, literally "little organs," are specialized cellular compartments (see Figure 3.5), each performing its own job to maintain the life of the cell. Many organelles are bounded by a membrane, similar to the plasma membrane. The membrane boundaries of such organelles allow them to maintain an internal environment quite

different from that of the surrounding cytosol. This compartmentalization is crucial to their ability to perform their specialized functions for the cell. Let us consider what goes on in each of the workshops of our cellular factory.

Ribosomes   Ribosomes (ri'bo-sōmz) are tiny, round, dark bodies made of proteins and one variety of RNA called *ribosomal RNA*. Ribosomes are the actual sites of protein synthesis in the cell. Some ribosomes float free in the cytoplasm, and others attach to membranes. When ribosomes are attached to membranes, the whole ribosome-membrane combination is called the rough endoplasmic reticulum.

Endoplasmic Reticulum   The endoplasmic reticulum (en"do-plas'mik rĕ-tik'u-lum; "network within the cell") **(ER)** is a system of fluid-filled cisterns (tubules, or canals) that coil and twist through the cytoplasm. It accounts for about half of a cell's membranes. It serves as a minicirculatory system for the cell because it provides a network of channels for carrying substances (primarily proteins) from one part of the cell to another. There are two forms of ER; a particular cell may have both or only one, depending on its specific functions.

The **rough ER** is studded with ribosomes. Within its tubules, the proteins made on the ribosomes fold into their functional three-dimensional shapes, and then are dispatched to other areas of the cell. In general the amount of rough ER a cell has is a good clue to the amount of protein that cell makes. Rough ER is especially abundant in cells that export protein products—for example, pancreas cells, which produce digestive enzymes to be delivered to the small intestine.

Another function of the rough ER is to manufacture membrane lipids. Because essentially all of the building materials of cellular membranes are formed either in or on it, the rough ER can be thought of as the cell's "membrane factory."

Although the **smooth ER** is a continuation of the rough variety, it plays no role in protein synthesis. Instead it functions in cholesterol synthesis and breakdown, fat metabolism, and detoxification of drugs. Hence it is not surprising that the liver cells are chock-full of smooth ER. So too are body cells that produce steroid-based hormones—for instance, cells of the male testes that manufacture testosterone.

Golgi Apparatus The **Golgi** (gol'je) **apparatus** appears as a stack of flattened membranous sacs, associated with swarms of tiny vesicles. It is generally found close to the nucleus and is the principal "traffic director" for cellular proteins. Its major function is to modify and package proteins (sent to it by the rough ER via *transport vesicles*) in specific ways, depending on their final destination (Figure 3.6).

As proteins "tagged" for export accumulate in the Golgi apparatus, the sacs swell. Then, their swollen ends, filled with protein, pinch off and form *secretory vesicles* (ves'ĭ-kuls), which travel to the plasma membrane. When the vesicles reach the plasma membrane, they fuse with it, the membrane ruptures, and the contents of the sac are ejected to the outside of the cell. Mucus is packaged this way, as are digestive enzymes made by pancreas cells.

In addition to its packaging-for-release functions, the Golgi apparatus pinches off sacs containing proteins and phospholipids destined to become part of the plasma membrane and packages hydrolytic enzymes into membranous sacs called *lysosomes* that remain in the cell.

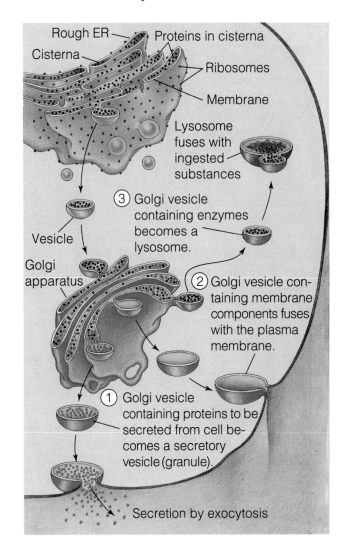

Figure 3.6   Role of the Golgi apparatus in packaging the products of the rough ER. Protein-containing transport vesicles pinch off the rough ER and migrate to fuse with the Golgi apparatus. As it passes through the Golgi apparatus, the protein product is sorted (and slightly modified). The product is then packaged within vesicles, which leave the Golgi apparatus and head for various destinations (#1–3), as shown.

Lysosomes **Lysosomes** (li'so-sōmz; literally "breakdown bodies"), which appear in different sizes, are membrane "bags" containing powerful digestive enzymes. The enzymes are capable of digesting worn-out or nonusable cell structures and foreign substances that enter the cell. Lysosomes are especially abundant in white blood cells that engulf bacteria and other potentially harmful substances because they digest and rid the body of such "foreigners." As described above, the enzymes they contain are formed by ribosomes and "packaged" by the Golgi apparatus (see Figure 3.6).

The lysosomal membrane is ordinarily quite stable, but it becomes fragile when the cell is injured or deprived of oxygen and when excessive amounts of vitamin A are present. Lysosomal rupture results in self-digestion of the cell. For this reason, lysosomes are sometimes referred to as "suicide sacs." ❑

Peroxisomes Peroxisomes (per-ok′sih-sōmz) are membranous sacs containing powerful oxidase (ok′sĭ-dāz) enzymes that use molecular oxygen (O_2) to detoxify a number of harmful or poisonous substances, including alcohol and formaldehyde. However, their most important function is to "disarm" dangerous free radicals. **Free radicals** are highly reactive chemicals with unpaired electrons that can scramble the structure of proteins and nucleic acids. Although free radicals are normal byproducts of cellular metabolism, if allowed to accumulate, they can have devastating effects on cells. Peroxisomes convert free radicals to hydrogen peroxide (H_2O_2), a function indicated in their naming (*peroxisomes* = peroxide bodies). The enzyme *catalase* (kat′ah-lās) then converts excess hydrogen peroxide to water. Peroxisomes are especially numerous in liver and kidney cells, which are very active in detoxification.

Although peroxisomes look like small lysosomes (Figure 3.5), they do not arise by budding from the Golgi apparatus. Instead, they appear to replicate themselves by simply pinching in half.

Mitochondria Mitochondria (mi″to-kon′dre-ah) are usually depicted as tiny threadlike (*mitos* = thread) or sausage-shaped organelles (Figure 3.5), but in living cells they squirm, lengthen, and change shape almost continuously. Their wall consists of a double membrane, equal to *two* plasma membranes, placed side by side. The outer membrane is smooth, but the inner membrane has shelflike protrusions called *cristae* (kris′te). Enzymes dissolved in the fluid within the mitochondria, as well as enzymes that form part of the cristae membranes, carry out the reactions in which oxygen is used to break down foods. As the foods are broken down, energy is released. Much of this energy escapes as heat, but some is captured and used to form *ATP molecules.* ATP pro-

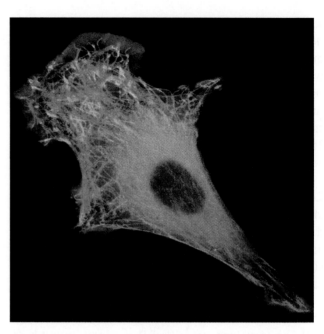

Figure 3.7   The cytoskeleton. In this light micrograph of the cytoskeleton (1400×), the microtubules appear green; the microfilaments are blue. Intermediate filaments form most of the rest of the network.

vides the energy for all cell work, and every living cell requires a constant supply of ATP for its many activities. Because the mitochondria supply most of this ATP, they are referred to as the "powerhouses" of the cell.

Metabolically "busy" cells, like liver and muscle cells, use huge amounts of ATP and have hundreds of mitochondria. By contrast, cells that are relatively inactive (an unfertilized egg, for instance) have just a few.

Cytoskeleton An elaborate network of protein structures extends throughout the cytoplasm (see Figures 3.3 and 3.5). This network, or **cytoskeleton,** acts as a cell's "bones and muscles" by furnishing an internal framework that determines cell shape, supports other organelles, and provides the machinery needed for intracellular transport and various types of cellular movements. From its largest to its smallest elements, the cytoskeleton is made up of microtubules, intermediate filaments, and microfilaments (Figure 3.7). Although there is some overlap in roles, generally speaking the strong, stable **intermediate filaments** help form desmosomes (see Figure 3.4) and provide internal guy wires to resist pulling forces on the cell, whereas **microfilaments** (such as *actin* and *myosin*) are most involved in cell motility. The tubelike **microtubules** determine the overall shape of a cell and the distribution of organelles. They are very important during cell division, as described later.

Centrioles The paired **centrioles** (sen′tre-ōlz) lie close to the nucleus. They are rod-shaped bodies that lie at right angles to each other; internally they are made up of fine *microtubules*. During cell division, the centrioles direct the formation of the *mitotic spindle* (see p. 72).

In addition to the cell structures described above, some cells have projections called **cilia** (sil′e-ah; "eyelashes"), whiplike cellular extensions that move substances along the cell surface. For example, the ciliated cells of the respiratory system lining move mucus up and away from the lungs. Where they appear, there are usually many cilia projecting from the exposed cell surface. When a cell is about to make cilia, its centrioles multiply and then line up beneath the plasma membrane at the free cell surface. Microtubules then begin to "sprout" from the centrioles and put pressure on the membrane, forming the projections. When the projections formed by the centrioles are substantially longer, they are called **flagella** (flah-jel′ah). The only example of a flagellated cell in the human body is the sperm, which has a single propulsive flagellum called its *tail. Notice that cilia propel other substances across a cell's surface, whereas a flagellum propels the cell itself.*

CELL PHYSIOLOGY

As mentioned earlier, each of the cell's internal parts is designed to perform a specific function for the cell. Most cells have the ability to *metabolize* (use nutrients to build new cell material, break down substances, and make ATP), *digest foods, dispose of wastes, reproduce, grow, move,* and *respond to a stimulus* (irritability). Most of these functions are considered in detail in later chapters.

For example, metabolism is covered in Chapter 14, and the ability to react to a stimulus is covered in Chapter 7. Here, we will consider only the functions of membrane transport (the means by which substances get through plasma membranes), protein synthesis, and cell reproduction (cell division).

Membrane Transport

The fluid environment on both sides of the plasma membrane is an example of a solution. Although the various types of mixtures were mentioned in Chapter 2 (p. 37), it is important that you really understand solutions before we dive into an explanation of membrane transport. In the most basic sense, a **solution** is a homogeneous mixture of two or more components. Examples include the air we breathe (a mixture of gases), seawater (a mixture of water and salts [solids]); and rubbing alcohol (a mixture of water and alcohol). The component or substance present in the largest amount in a solution is called the **solvent** (or dissolving medium). Water is the body's chief solvent. Components or substances present in smaller amounts are called **solutes.**

Intracellular fluid (collectively, the nucleoplasm and the cytosol) is a solution containing small amounts of gases (oxygen and carbon dioxide), nutrients, and salts, dissolved in water. So too is **interstitial fluid,** the fluid that continuously bathes the exterior of our cells. Interstitial fluid can be thought of as a rich, nutritious, and rather unusual "soup." It contains thousands of ingredients, including amino acids, sugars, fatty acids, vitamins, regulatory substances such as hormones and neurotransmitters, salts, and waste products. To remain healthy, each cell must extract from this soup the exact amounts of the substances it needs at specific times and reject the rest.

The plasma membrane is a selectively permeable barrier. **Selective permeability** means that a barrier allows some substances to pass through it while excluding others. Thus, it allows nutrients to enter the cell, but keeps many undesirable substances out. At the same time, valuable cell proteins and other substances are kept within the cell, and wastes are allowed to pass out of it.

This property of selective permeability is typical only of healthy, unharmed cells. When a

Figure 3.8 **Diffusion.** Particles in solution move continuously and collide constantly with other particles. As a result, particles tend to move away from areas where they are most highly concentrated and to become evenly distributed, as illustrated by the diffusion of sugar molecules in a cup of coffee.

cell dies or is badly damaged, its plasma membrane can no longer be selective and becomes permeable to nearly everything. This phenomenon is evident when someone has been severely burned. Precious fluids, proteins, and ions "weep" from the dead and damaged cells of the burned areas. ❏

Movement of substances through the plasma membrane happens in basically two ways—passively or actively. In **passive transport processes,** substances are transported across the membrane without any energy input from the cell. In **active transport processes,** the cell provides the metabolic energy (ATP) that drives the transport process.

Passive Transport Processes: Diffusion and Filtration

Diffusion (dĭ-fu′zhun) is an important means of passive membrane transport for every cell of the body. The other passive transport process, *filtration,* generally occurs only across capillary walls. Let us examine how these two types of passive transport differ.

Diffusion **Diffusion** is the process by which molecules (and ions) tend to scatter themselves throughout the available space. All molecules possess *kinetic energy* (energy of motion) as described in Chapter 2. As the molecules move about randomly at high speeds, they collide and change direction with each collision. Since the overall effect of this erratic movement is that molecules move away from a region where they are more concen-

trated (more numerous) to a region where they are less concentrated (fewer of them), we say that molecules move *down* their **concentration gradient.** Because the driving force (source of energy) is the kinetic energy of the molecules themselves, the speed of diffusion is affected by the size of the molecules (the smaller the faster) and temperature (the warmer the faster).

Two examples should help you understand diffusion. Picture yourself pouring a cup of coffee, and then adding a cube of sugar (but not stirring the cup). After adding the sugar, the phone rings and you are called in to work; you never do drink the coffee. Upon returning that evening, you find that the coffee tastes sweet even though it was never stirred. This is because the sugar molecules moved around all day and eventually, as a result of their activity, became equally distributed throughout the coffee (see Figure 3.8). Or try this example: You walk into a room and, after a few minutes, your coworker says, "I like your cologne." Obviously she or he did not smell the cologne until it began to diffuse out into the air of the room and away from you (the area of highest concentration).

The plasma membrane is a physical barrier to diffusion. Molecules will move *passively* through the plasma membrane by diffusion if (1) they are small enough to pass through its pores, or (2) they can dissolve in the fatty portion of the membrane (see Figure 3.9).

The unassisted diffusion of solutes through the plasma membrane (or any selectively permeable membrane) is called **simple diffusion** (Figure 3.9a). Solutes transported this way are either lipid-

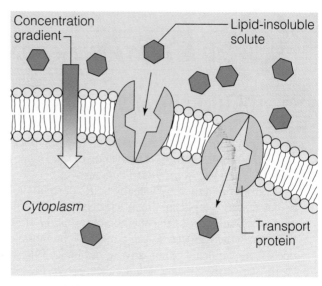

(a) Simple diffusion

(b) Facilitated diffusion

Figure 3.9   Diffusion through the plasma membrane. (a) Simple diffusion. As depicted on the left, fat-soluble molecules diffuse directly through the lipid bilayer of the plasma membrane, in which they can dissolve. On the right, small lipid-insoluble substances (water molecules or small ions) are shown diffusing through channels constructed by membrane proteins. **(b)** Facilitated diffusion moves large, lipid-insoluble molecules (e.g., glucose) across the membrane. The substance to be transported binds to a transmembrane carrier protein.

soluble (fats, fat-soluble vitamins, oxygen, carbon dioxide), or small enough to pass through the membrane pores (some small ions such as chloride ions, for example).

Diffusion of water through a selectively permeable membrane such as the plasma membrane is specifically called **osmosis** (oz-mo′sis). Because water is highly polar, it is repelled by the (nonpolar) lipid core of the plasma membrane, but it can and does pass easily through the pores created by the proteins in the membrane (Figure 3.9a). Osmosis into and out of cells is occurring all the time as water moves down its concentration gradient.

Still another example of diffusion is **facilitated diffusion** (see Figure 3.9b). Facilitated diffusion provides a means for certain needed substances, notably glucose, that are both lipid-insoluble and too large to pass through the membrane pores, to enter the cell. Although facilitated diffusion follows the "laws" of diffusion—that is, the substances move down their concentration gradient—a pro-

tein "carrier" is needed as a transport vehicle. Hence, some of the proteins in the plasma membrane act to move glucose passively across the membrane to make it available for cell use.

Substances that pass into and out of cells by diffusion save the cell a great deal of energy. When you consider how vitally important water, glucose, and oxygen are to cells, it becomes apparent just how necessary these passive transport processes really are. Glucose and oxygen continually move into the cells (where they are in lower concentration because the cells keep using them up), and carbon dioxide (a waste product) continually moves out of the cells into the blood (where it is in lower concentration).

Filtration   Filtration is the process by which water and solutes are forced through a membrane (or capillary wall) by *fluid,* or *hydrostatic, pressure.* In the body, hydrostatic pressure is usually exerted by the blood. Like diffusion, filtration is a passive

A CLOSER LOOK

IV THERAPY AND CELLULAR "TONICS"

Why is it essential that medical personnel give only the proper *intravenous (IV)*, or into-the-vein, *solutions* to patients?

Consider that there is a steady traffic of small molecules across the plasma membrane. Although diffusion of solutes across the membrane is rather slow, osmosis, which moves water across the membrane, occurs very quickly. Thus anyone administering an IV must use the correct solution to protect the patient's cells from life-threatening dehydration or rupture due to excessive water entry.

The tendency of a solution to hold water or "pull" water into it is called **osmotic pressure.** Osmotic pressure is directly related to the concentration of solutes in the solution. The higher the solute concentration, the greater the osmotic pressure and the greater the tendency of water to move into the solution. Many molecules, particularly proteins and some ions, are prevented from diffusing through the plasma membrane. Consequently, any change in their concentration on one side of the membrane forces water to move from one side of the membrane to the other, causing cells to lose or gain water. The ability of a solution to change the size and shape of cells by altering the amount of water they contain is called **tonicity** (ton-is′i-te; *ton* = strength).

Isotonic (i″so-ton′ik; "same tonicity") **solutions** (such as Ringer's lactate, 5 percent dextrose, and 0.9 percent saline) have the same solute and water concentrations as cells do. Isotonic solutions cause no visible changes in cells and when such solutions are infused into the bloodstream, red blood cells retain their normal size and disklike shape (see photo a). As you might guess, interstitial fluid and most intravenous solutions are isotonic solutions.

If red blood cells are exposed to a **hypertonic** (hi″per-ton′ik) solution—a solution that contains more solutes, or dissolved substances, than there are inside the cells—the cells will begin to shrink, or **crenate** (kre′nāt). This is because water is in higher concentration inside the cell than outside, so it follows its concentration gradient and leaves the cell (see photo b). Hypertonic solutions are sometimes given to patients who have *edema* (swollen feet and hands because of fluid retention). Such solutions draw water out of the tissue spaces into the bloodstream so that the excess fluid can be eliminated by the kidneys.

When a solution contains fewer solutes (and therefore more water) than the cell does, it is said to be **hypotonic** (hi″po-ton′ik) to the cell. Cells placed in hypotonic solutions plump up rapidly as water rushes into them (see photo c). Distilled water represents the most extreme example of a hypotonic fluid. Since it contains no solutes at all, water will enter cells until they finally burst or *lyse*. Hypotonic solutions are sometimes infused intravenously (slowly and with care) to rehydrate the tissues of extremely dehydrated patients. In less extreme cases of dehydration, drinking hypotonic fluids usually does the trick. (Many fluids that humans tend to drink regularly, such as tea, colas, apple juice, and sports drinks, are hypotonic.)

process, and a gradient is involved. In filtration, however, the gradient is a **pressure gradient** that actually pushes solute-containing fluid (*filtrate*) from the higher-pressure area to the lower-pressure area. Filtration is necessary for the kidneys to do their job properly. In the kidneys, water and small solutes filter out of the capillaries into the kidney tubules because the blood pressure in the capillaries is greater than the fluid pressure in the tubules. Part of the filtrate formed in this way eventually becomes urine. Filtration is not a very selective process. For the most part, only blood cells and protein molecules too large to pass through the membrane pores are held back.

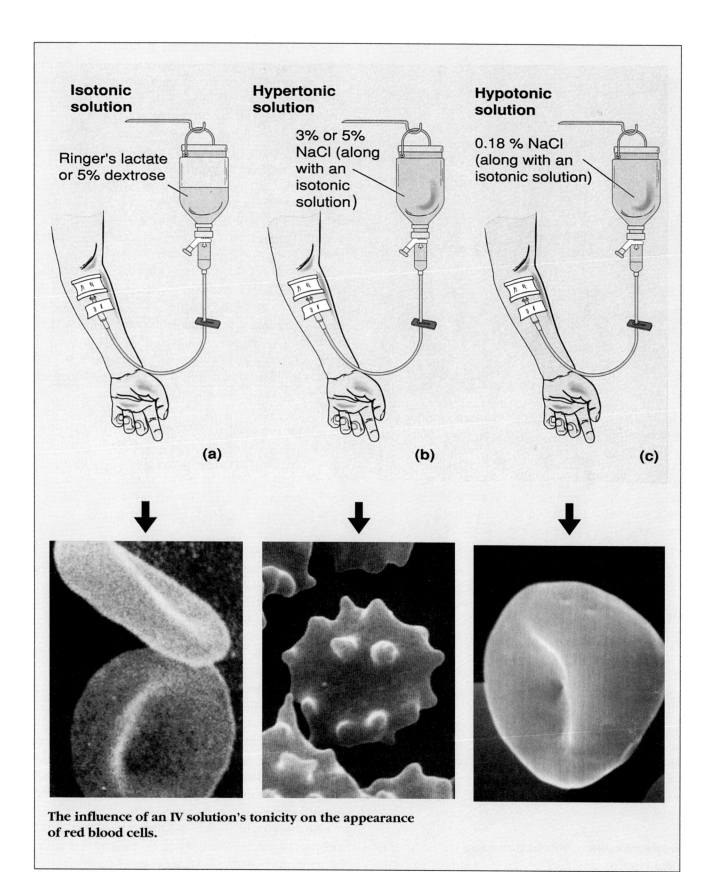

The influence of an IV solution's tonicity on the appearance of red blood cells.

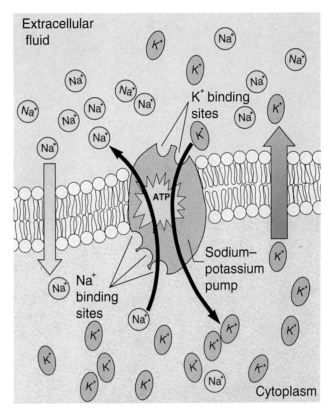

Figure 3.10 Operation of the sodium-potassium pump, a solute pump. ATP provides the energy for a "pump" protein to move three sodium ions out of the cell and two potassium ions into the cell. Both ions are moved against their concentration gradients, indicated by colored arrows moving through the membrane (yellow arrow = Na$^+$ gradient; green arrow = K$^+$ gradient).

Active Transport Processes

Whenever a cell uses energy from its ATP to move substances across the membrane, the process is referred to as *active*. Substances moved actively are usually unable to pass in the desired direction by diffusion. They may be too large to pass through the pores, they may not be able to dissolve in the fat core, or they may have to move "uphill" *against* their concentration gradients. The two most important examples of active transport mechanisms, *solute pumping* and *bulk transport,* are described next.

Solute Pumping **Solute pumping** (more simply called *active transport* by some) is similar to facilitated diffusion in that both processes require pro-

tein carriers that combine reversibly with the substances to be transported across the membrane. However, facilitated diffusion is driven by kinetic energy, whereas solute pumping uses ATP to energize its protein carriers, which are called **solute pumps.** Amino acids, some sugars, and most ions are transported by solute pumps, and in most cases these substances move *against* concentration (or electrical) gradients. This is opposite to the direction in which substances would naturally flow by diffusion, which explains the need for energy in the form of ATP. Amino acids are too large to pass through the pores and are not lipid-soluble. Sodium ions (Na$^+$) are moved out of cells by solute pumps (Figure 3.10). There are more sodium ions outside the cells than there are inside, so they tend to remain in the cell unless the cell uses ATP to force or "pump" them out. Likewise, there are relatively more potassium ions inside cells than there are in the interstitial fluid, and potassium ions that leak out of cells must be actively pumped back inside. The *sodium-potassium pump* that simultaneously carries sodium ions out of and potassium ions into the cell is absolutely necessary for normal transmission of impulses by nerve cells. Since each of the pumps in the plasma membrane transports only specific substances, solute pumping provides a way for the cell to be very selective in cases where substances cannot pass by diffusion. (No pump—no transport.)

Bulk Transport Some substances that could not get through the plasma membrane in any other way are transported with the help of ATP into or out of cells by **bulk transport.** The two types of bulk transport are *exocytosis* and *endocytosis* (Figure 3.11).

Exocytosis (ek″so-si-to′sis; "out of the cell") moves substances out of cells. It is the means by which cells actively secrete hormones, mucus, and other cell products or eject certain cellular wastes. The product to be released is first "packaged" (typically by the efforts of the Golgi apparatus) into a small membrane sac. The sac migrates to the plasma membrane and fuses with it. The fused area then ruptures, spilling the sac contents out of the cell (also see Figure 3.6).

Endocytosis (en″do-si-to′sis; "into the cell") includes those ATP-requiring processes that take up, or engulf, extracellular substances by enclosing them in a small membranous vesicle (Figure

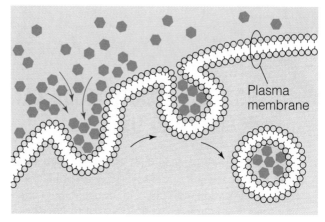

(a) Exocytosis

(b) Endocytosis

Figure 3.11   Bulk transport: Exocytosis and endocytosis. (**a**) During exocytosis, vesicles fuse with the plasma membrane and dump their contents to the outside of the cell. (**b**) During endocytosis, extracellular substances are incorporated into the cell in vesicles formed by an inward folding of the plasma membrane.

3.11b). Once the vesicle or sac is formed, it detaches from the plasma membrane and moves into the cytoplasm where it fuses with a lysosome and its contents are digested (by lysosomal enzymes). If the engulfed substances are relatively large particles, such as bacteria or dead body cells, the endocytosis process is more specifically called **phagocytosis** (fag″o-si-to′sis), a term that means "cell eating." Certain white blood cells and other "professional" phagocytes of the body act as scavenger cells that police and protect the body by ingesting bacteria and other foreign debris as well as dead body cells.

If we say that cells can eat, we can also say that they can drink, and **pinocytosis** (pi″no-si-to′sis) is cell drinking. Pinocytosis is commonly used to take in liquids that contain dissolved proteins or fats. Unlike phagocytosis, it is a routine activity of most cells. It is especially important in cells that function in absorption (for example, cells forming the lining of the small intestine and kidney tubule cells).

Cell Division

The **cell life cycle** is the series of changes a cell goes through from the time it is formed until it divides. The cycle has two major periods: **interphase,** in which the cell grows and carries on its usual metabolic activities, and **cell division,** during which it reproduces itself. Although the term *interphase* leads one to believe that it is merely a resting time between the phases of cell division, this is not the case. During interphase, which is by far the longer phase of the cell cycle, the cell is very active and is resting *only* from division. A more accurate name for interphase would be *metabolic phase*.

Preparations: DNA Replication

The function of cell division is to produce more cells for growth and repair processes. Because it is essential that all body cells have the same genetic material, an important event *always precedes* cell division: The genetic material (the DNA molecules that form part of the chromatin) is duplicated exactly. This occurs toward the end of the cell's interphase period.

You will recall from Chapter 2 that DNA is a very complex molecule (review Figure 2.17). It is composed of building blocks called *nucleotides,* each consisting of deoxyribose sugar, a phosphate group, and a nitrogen-containing base. Essentially DNA is a ladderlike molecule that is coiled into a spiral staircase shape called a *double helix*. The upright parts of the DNA "ladder" are formed of alter-

A CLOSER LOOK

NON-AGE-RELATED MODIFICATIONS IN CELLS AND TISSUES

Besides the tissue changes associated with aging, which accelerate in the later years of life, other modifications of cells and tissues may occur at any time. For example, when cells fail to honor normal controls on cell division and multiply wildly, an abnormal mass of proliferating cells called a **neoplasm** (ne′o-plazm″; "new growth") results. Neoplasms may be benign or malignant. *Benign* (be-nīn) neoplasms are strictly local affairs. They tend to be surrounded by a capsule, grow slowly, and seldom kill their hosts if they are removed before they compress vital organs. In contrast, *malignant (cancerous)* neoplasms are nonencapsulated masses that grow more relentlessly and may become killers. They invade their surroundings and tend to spread via the blood to distant parts of the body where they form new masses. This last capability is called *metastasis* (mĕ-tas′tă-sis). The left photomicrograph shows normal liver tissue; the right shows malignant liver tissue. Notice how different the cancerous cells are in appearance and how they are spreading "crablike" into adjacent tissue areas.

Various factors (such as viruses, continual irritation, or some chemicals) seem to encourage the formation of neoplasms, but .genetic changes (changes in the structure or activity of certain genes) are the final determinants of whether or not cancer will develop.

Neoplasms are diagnosed by a *biopsy* (bi′op-se), in which samples of the questionable tissue are removed surgically (or scraped off a surface) and examined under a microscope. In a cancerous growth, the cells lose the specialized look of that particular tissue or organ and come to resemble embryonic cells. The treatment of choice for either type of neoplasm is surgical removal. If surgery is not possible—as in cases where the cancer has spread widely or is inoperable—radiation and drugs (chemotherapy) are used.

Not all increases in cell number involve neoplasms. Certain body tissues (or organs) may enlarge because there is some local irritant or condition that stimulates the cells. This is called **hyperplasia** (hi″per-pla′ze-ah). For example, when one is anemic, the bone marrow undergoes hyperplasia so that red blood cells may be produced at a faster rate. When (and if) the anemia is remedied, the enlargement or overproliferation of cells in the marrow disappears. Another example is when a woman's breasts enlarge during pregnancy in response to increased hormones; this is a normal situation that doesn't have to be treated. On the other hand, **atrophy** (at′ro-fe), or decrease in size, can occur in an organ or body area if it loses its normal stimulation. For example, muscles that are not used or that have lost their nerve supply begin to atrophy and waste away rapidly.

Normal liver tissue.

Malignant liver tissue.

nating phosphate and sugar units, and the rungs of the ladder are made of pairs of nitrogen-containing bases.

The precise trigger for DNA synthesis is unknown, but once it starts, it continues until all the DNA has been replicated. The process begins as the DNA helix uncoils and gradually separates into its two nucleotide chains (see Figure 3.12). Each nucleotide strand then serves as a **template,** or set of instructions, for building a new nucleotide strand.

Remember that nucleotides join in a *complementary* way: adenine (A) always bonds to thymine (T), and guanine (G) always bonds to cytosine (C) (see p. 48). Hence, the order of the nucleotides on the template strand also determines the order on the new strand. For example, a TACTGC sequence on a template strand would bond to new nucleotides with the order ATGACG. The end result is that two DNA molecules are formed that are identical to the original DNA helix, and each consists of one old and one newly assembled nucleotide strand.

Events of Cell Division

In all cells other than bacteria and some cells of the reproductive system, cell division consists of two events. **Mitosis** (mi-to′sis), or division of the nucleus, occurs first. The second event is division of the cytoplasm, **cytokinesis** (si″to-kĭ-ne′sis), which begins when mitosis is nearly completed.

Mitosis Mitosis results in the formation of two daughter nuclei with exactly the same genes as the mother nucleus. As explained above, DNA replication precedes mitosis, so that for a short time the cell nucleus contains a "double dose" of genes. When the nucleus divides, each *daughter cell* ends up with *exactly* the same genetic information as the original mother cell and the original fertilized egg from which it came.

The stages of mitosis, diagrammed in Figure 3.13, include the following events:

- **Prophase** (pro′faz). As cell division begins, the chromatin threads coil and shorten so that visible barlike bodies called **chromosomes** (*chromo* = colored; *soma* = body) appear. Since DNA replication has already occurred,

KEY:
▷ Thymine (T)   ▮ Deoxyribose sugar
◭ Adenine (A)   ▬ Phosphate
▯ Cytosine (C) — Hydrogen bond
▮ Guanine (G)

Figure 3.12   Replication of the DNA molecule during interphase. The DNA helix unwinds (center) and its nucleotide strands are separated. Each strand then acts as a template for building a "new" complementary strand. As a result, two helixes, each identical to the original DNA helix, are formed.

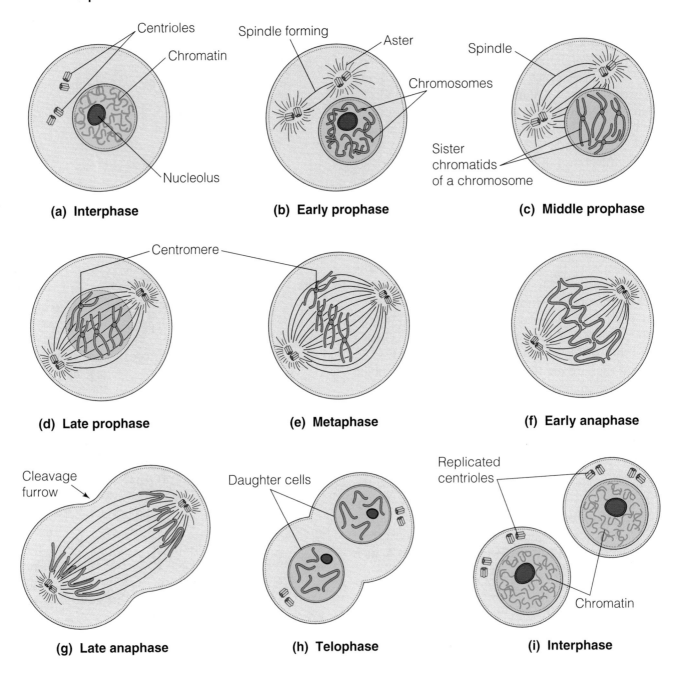

(a) Interphase

(b) Early prophase

(c) Middle prophase

(d) Late prophase

(e) Metaphase

(f) Early anaphase

(g) Late anaphase

(h) Telophase

(i) Interphase

Figure 3.13   Stages of mitosis.

each chromosome is actually made up of two strands, each called a **chromatid** (kro′mah-tid), held together by a small buttonlike body called a *centromere* (sen′tro-mer). The centrioles separate from each other and begin to move toward opposite sides of the cell, directing the assembly of the **mitotic spindle** (composed of thin microtubules) between them as

they move. The spindle provides a "scaffolding" for the attachment and movement of the chromosomes during the later mitotic stages. By the end of prophase, the nuclear membrane and the nucleoli have broken down and disappeared, and the chromosomes have become attached randomly to the spindle fibers by their centromeres.

- **Metaphase** (met′ah-faz). In this short stage, the chromosomes cluster and become aligned at the center of the spindle midway between the centrioles so that a straight line of chromosomes is seen.

- **Anaphase** (an′ah-faz). During anaphase, the centromeres that have held the chromatids together split, and the chromatids (now called chromosomes again) begin to move slowly apart, toward opposite ends of the cell. The chromosomes seem to be pulled by their half-centromeres with their "arms" dangling behind them. Anaphase is over when chromosome movement ends.

- **Telophase** (tel′o-faz). Telophase is essentially prophase in reverse. The chromosomes at opposite ends of the cell uncoil to become threadlike chromatin again. The spindle breaks down and disappears, a nuclear membrane forms around each chromatin mass, and nucleoli appear in each of the daughter nuclei.

Mitosis is basically the same in all animal cells. Depending on the type of tissue, it takes from 5 minutes to several hours to complete, but typically it lasts about 2 hours. Centriole replication is deferred until late interphase of the next cell cycle, when DNA replication begins before the onset of mitosis.

Cytokinesis Cytokinesis, or the division of the cytoplasm, usually begins during late anaphase and completes during telophase. Due to the activity of a contractile ring made of microfilaments, a **cleavage furrow** appears over the midline of the spindle and it eventually squeezes or pinches the original cytoplasmic mass into two parts.

Thus, at the end of cell division, two daughter cells exist. Each daughter cell is smaller and has less cytoplasm than the mother cell, but it is genetically identical to it. The daughter cells grow and carry out normal cell activities until it is their turn to divide.

Although mitosis and division of the cytoplasm usually go hand in hand, in some cases the cytoplasm is not divided. This condition leads to the formation of *binucleate* (two nuclei) or *multinucleate* cells. This is fairly common in the liver.

As mentioned earlier, mitosis provides the "new" cells for body growth in youth and is neces-sary to repair body tissue all through life. Mitosis "gone wild" is the basis for tumors or cancers.

Protein Synthesis

Genes: The Blueprint for Protein Structure

In addition to replicating itself for cell division, DNA serves as the master blueprint for protein syntheses. A **gene** is a DNA segment that carries the information for building one protein or polypeptide chain.

Proteins are key substances for all aspects of cell life. As described in Chapter 2, *fibrous (structural)* proteins are the major building materials for cells. Other proteins, the globular *functional* proteins, do things other than build structures. For example, all **enzymes,** biological catalysts that regulate chemical reactions in the cells, are functional proteins. The importance of enzymes cannot be overstated. Every chemical reaction that goes on in the body requires an enzyme. It follows that DNA regulates cell activities largely by specifying the structure of enzymes, which in turn control or direct the chemical reactions in which carbohydrates, fats, other proteins, and even DNA itself are made and broken down.

How does DNA bring about its miracles? It appears that DNA's information is "encoded" in the sequence of bases along each side of the ladder-like DNA molecules. Each sequence of *three* bases (a *triplet*) calls for a particular *amino acid*. (Amino acids are the building blocks of proteins that are joined during protein synthesis.) For example, a DNA base sequence of AAA specifies an amino acid called *phenylalanine,* while CCT calls for *glycine.* Just as different arrangements of notes on sheet music are played as different melodies, variations in the arrangements of A, C, T, and G in each gene allow cells to make all the different kinds of proteins needed. It has been estimated that a single gene has between 300 and 3000 base pairs in sequence.

The Role of RNA

By itself, DNA is rather like a strip of magnetic recording tape; its information is not useful until it is decoded. Furthermore, ribosomes—the manufacturing sites for proteins—are in the cytoplasm,

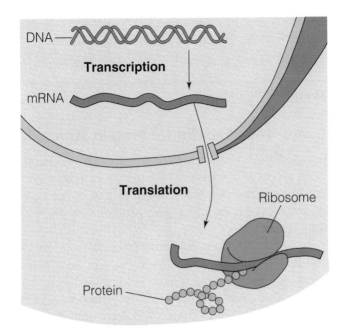

Figure 3.14 Simplified scheme of information flow from the DNA gene to protein structure. The information to be expressed in protein structure flows from DNA to messenger RNA (mRNA) (transcription) and then from mRNA to synthesis of a protein by the joining of amino acids in the designated order (translation).

but in interphase cells DNA never leaves the nucleus. Thus, DNA requires not only a decoder but also a messenger to achieve its task of specifying the structure of proteins to be built at the ribosomes. These messenger and decoder functions are carried out by the second type of nucleic acid, called **ribonucleic** (ri″bo-nu-kle′ik) **acid** or **RNA.**

As you learned in Chapter 2, RNA differs from DNA in being single-stranded and in having ribose sugar instead of deoxyribose and a uracil (U) base instead of thymine (T). There are three varieties of RNA and, as described shortly, each has a special role to play in protein synthesis. **Transfer RNA (tRNA) molecules** are small cloverleaf-shaped molecules. **Ribosomal RNA (rRNA)** helps form the ribosomes, where proteins are built. **Messenger RNA (mRNA) molecules** are long single nucleotide strands that resemble half of a DNA molecule and carry the "message" containing instructions for protein synthesis from the DNA "gene" in the nucleus to the ribosomes.

Essentially, protein synthesis involves two major phases: *transcription,* when complementary mRNA is made at the DNA gene, and *translation,*

when the information carried in mRNA molecules is "decoded" and used to assemble proteins. These steps, summarized simply in Figure 3.14, are described in more detail next.

Transcription

The word *transcription* often refers to one of the jobs done by a secretary—converting notes from one form (shorthand notes or a dictaphone recording) into another form (a typewritten letter, for example). In other words, the same information is transformed from one form or format to another. In cells, **transcription** involves the transfer of information from DNA's base sequence into the *complementary* base sequence of mRNA (Figure 3.15, step 1). Only DNA and mRNA are involved in transcription. Whereas each three-base sequence specifying a particular amino acid on the DNA gene is called a **triplet,** the corresponding three-base sequences on mRNA are called **codons.** The form is different, but the same information is being conveyed. Thus, if the (partial) sequence of DNA triplets is AAT-CGT-TCG, the related codons on mRNA would be UUA-GCA-AGC.

Translation

A translator takes words in one language and restates them in another language. In the **translation phase** of protein synthesis, the language of nucleic acids (base sequence) is "translated" into the language of proteins (amino acid sequence).

Figure 3.15 Protein synthesis. (1) Transcription. The DNA segment, or gene, specifying one polypeptide uncoils, and one strand acts as a template for the synthesis of a complementary mRNA molecule. **(2–5)** Translation. **(2)** Messenger RNA from the nucleus attaches to a ribosome in the cytoplasm. **(3)** Transfer RNA transports an amino acid to the mRNA strand and recognizes a mRNA codon calling for its amino acid by binding via its anticodon to the codon. **(4)** The ribosome moves the mRNA strand along as each codon is read sequentially. **(5)** As each amino acid is bound to the next by a peptide bond, its tRNA is released. The polypeptide chain is released when the termination (stop) codon is read.

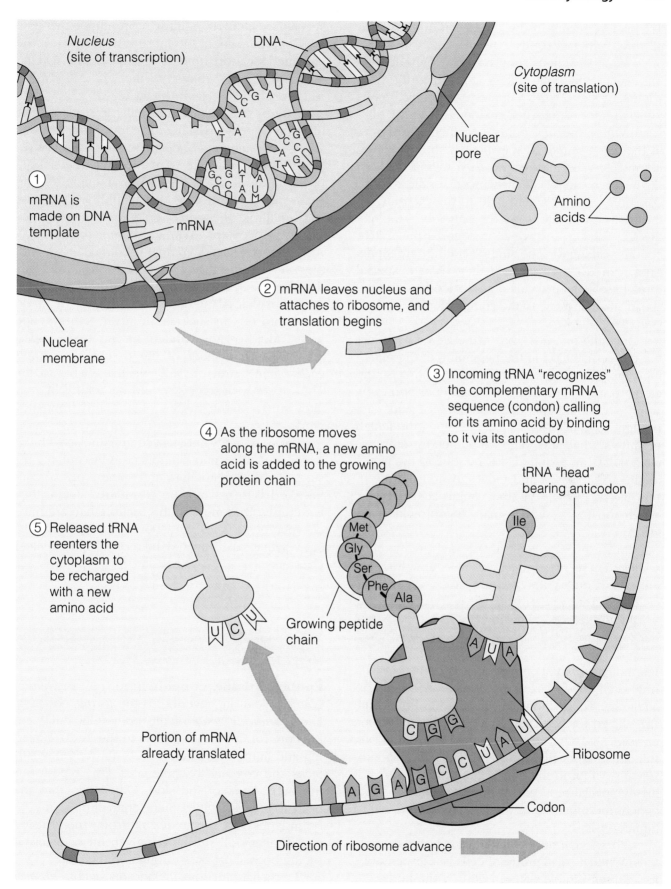

Nucleus
(site of transcription)

DNA

Cytoplasm
(site of translation)

Nuclear pore

Amino acids

① mRNA is made on DNA template

mRNA

Nuclear membrane

② mRNA leaves nucleus and attaches to ribosome, and translation begins

③ Incoming tRNA "recognizes" the complementary mRNA sequence (condon) calling for its amino acid by binding to it via its anticodon

④ As the ribosome moves along the mRNA, a new amino acid is added to the growing protein chain

tRNA "head" bearing anticodon

Ile

⑤ Released tRNA reenters the cytoplasm to be recharged with a new amino acid

Met
Gly
Ser
Phe
Ala

Growing peptide chain

Ribosome

Portion of mRNA already translated

Codon

Direction of ribosome advance

Translation occurs in the cytoplasm and involves all three varieties of RNA. As illustrated in Figure 3.15, steps 2–5, translation consists of the following events. Once the mRNA attaches to the ribosome (see step 2), tRNA comes into the picture. Its job is to ferry or "transfer" amino acids to the ribosome where they are bound together by enzymes in the exact sequence specified by the gene (and its mRNA). There are some 20 common types of tRNAs, each capable of carrying one of the 20 or so common types of amino acid to the protein synthesis sites. But that is not the only job of the tiny tRNAs. They also have to recognize the mRNA codons "calling for" the amino acid they are toting. They can do this because they have a special three-base sequence called an **anticodon** on their "head" that can bind to the complementary codons (step 3).

Once the first tRNA has maneuvered itself into the correct position at the beginning of the mRNA message, the ribosome moves the mRNA strand along, bringing the next codon into position to be "read" by another tRNA. As amino acids are brought to their proper positions along the length of mRNA, they are joined together by enzymes (step 4). As an amino acid is bonded to the chain, its tRNA is released and moves away from the ribosome to pick up another amino acid (step 5). When the last codon is read, the protein is released.

specialized cell group in the heart muscle that controls its contractions. If those particular cells are damaged or stop functioning, the heart will no longer work efficiently, and the whole body will suffer or die from lack of oxygen.

Groups of cells that are similar in structure and function are called **tissues.** The four primary tissue types—epithelium, connective tissue, nervous tissue, and muscle—interweave to form the "fabric" of the body. If we had to assign a single term to each primary tissue type that would best describe its overall role, the terms would most likely be *covering* (epithelium), *support* (connective), *movement* (muscle), and *control* (nervous). However, these terms reflect only a tiny fraction of the functions that each of these tissues performs.

As explained in Chapter 1, tissues are organized into *organs* such as the heart, kidneys, and lungs. Most organs contain several tissue types, and the arrangement of the tissues determines each organ's structure and what it is able to do. Thus, a study of tissues should be helpful in your later study of the body's organs and how they work.

For now, we want to become familiar with the major similarities and differences in the primary tissues. Because epithelium and some types of connective tissue will not be considered again, they are emphasized more in this section than are muscle, nervous tissues, and bone (a connective tissue), which are covered in more depth in later chapters.

PART 2: BODY TISSUES

The human body, complex as it is, starts out as a single cell, the fertilized egg, which divides almost endlessly. The millions of cells that result become specialized for particular functions. Some become muscle cells, others the transparent lens of the eye, still others skin cells, and so on. Thus, there is a division of labor in the body, with certain groups of highly specialized cells performing functions that benefit the organism as a whole and contribute to homeostasis.

Cell specialization carries with it certain hazards. When a small group of cells is indispensable, its loss can disable or even destroy the body. For example, the action of the heart depends on a very

EPITHELIAL TISSUE

Epithelial tissue, or **epithelium** (ep″ĭ-the′le-um; *epithe* = laid on, covering) is the *lining, covering,* and *glandular tissue* of the body. Glandular epithelium forms various glands in the body. Covering and lining epithelium covers all free body surfaces and contains versatile cells. One type forms the outer layer of the skin. Others dip into the body to line its cavities. Since epithelium forms the boundaries that separate us from the outside world, nearly all substances given off or received by the body must pass through epithelium.

Epithelial functions include *protection, absorption, filtration,* and *secretion.* For example, the epithelium of the skin protects against bacterial and chemical damage, and that lining the respiratory tract has cilia, which sweep dust and other debris away from the lungs. Epithelium specialized to absorb substances lines some digestive system organs such as the stomach and small intestine, which absorb food into the body. In the kidneys, epithelium both absorbs and filters. Secretion is a specialty of the glands, which produce such substances as perspiration, oil, digestive enzymes, and mucus.

Special Characteristics of Epithelium

Epithelium generally has the characteristics listed below:

- Epithelial cells fit closely together to form continuous sheets. Neighboring cells are bound together at many points by cell junctions, including desmosomes and tight junctions.

- The membranes always have one free (unattached) surface or edge. This so-called **apical surface** is exposed to the body's exterior or to the cavity of an internal organ. The exposed surfaces of some epithelia are slick and smooth, but others exhibit cell surface modifications, such as microvilli or cilia.

- The lower surface of an epithelium rests on a **basement membrane,** a structureless material secreted by the cells.

- Epithelial tissues have no blood supply of their own (that is, they are *avascular*) and depend on diffusion from the capillaries in the underlying connective tissue for food and oxygen.

- If well nourished, epithelial cells regenerate themselves easily.

Classification of Epithelium

The covering and lining epithelia are classified according to cell shape and arrangement (Figure 3.16). *Squamous* (skwa'mus) cells are flattened like fish scales (*squam* = scale), *cuboidal* (kuboi'dal) cells are cube-shaped like dice, and *columnar* cells are shaped like columns. The classifications by cell arrangement (layers) are **simple epithelium** (one layer of cells) and **stratified**

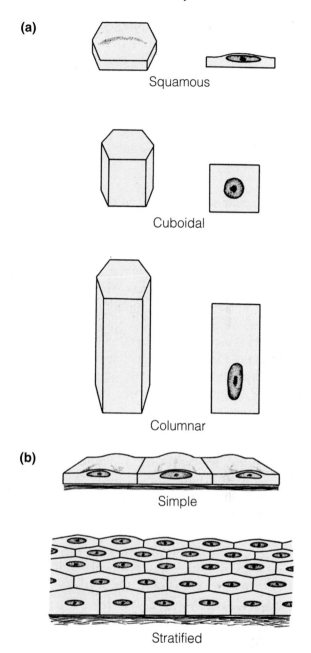

(a)

Squamous

Cuboidal

Columnar

(b)

Simple

Stratified

Figure 3.16 Classification of epithelia. (a) Classification on the basis of cell shape. For each category, a whole cell is shown on the left and a longitudinal section is shown on the right. **(b)** Classification on the basis of arrangement (layers).

epithelium (more than one cell layer). The terms describing the shape and arrangement are then combined to describe the epithelium fully. Stratified epithelia are named for the cells at the *free* *surface* of the epithelial membrane, not those resting on the basement membrane.

Simple Epithelia

The simple epithelia are most concerned with absorption, secretion, and filtration. Because they are usually very thin, protection is not one of their specialties.

Simple Squamous Epithelium   **Simple squamous epithelium** is a single layer of thin squamous cells resting on a basement membrane. The cells fit closely together, much like floor tiles. This type of epithelium usually forms membranes where filtration or exchanges of substances by rapid diffusion occur. It is in the air sacs of the lungs, where oxygen and carbon dioxide are exchanged (see Figure 3.17a), and it forms the walls of capillaries where nutrients and gases pass between the tissue cells and the blood in the capillaries. Simple squamous epithelium also forms the **serous membranes,** or **serosae** (se-ro′se), the slick membranes that line the ventral body cavity and cover the organs in that cavity. The serous membranes are described in more detail in Chapter 4.

Simple Cuboidal Epithelium   **Simple cuboidal epithelium,** which is one layer of cuboidal cells resting on a basement membrane, is common in glands and their ducts (for example, the salivary glands and pancreas). It also forms the walls of the kidney tubules, and covers the surface of the ovaries (see Figure 3.17b).

Simple Columnar Epithelium   **Simple columnar epithelium** is made up of a single layer of tall cells that fit closely together. **Goblet cells,** which produce a lubricating mucus, are often seen in this type of epithelium. Simple columnar epithelium lines the entire length of the digestive tract from the stomach to the anus (see Figure 3.17c). Epithelial membranes that line body cavities open to the body exterior are called **mucosae** (mu-ko′se) or **mucous membranes.**

Pseudostratified Columnar Epithelium   All of the cells of **pseudostratified** (soo″do-stră′tĭ-fīd) **columnar epithelium** rest on the basement membrane. However, some of its cells are shorter than others, and their nuclei appear at different heights above the basement membrane. As a result, this epithelium gives the false (pseudo) impression that it is stratified; hence its name. Like simple columnar epithelium, this variety mainly functions in absorption and secretion. A ciliated variety (more precisely called *pseudostratified ciliated columnar epithelium*) lines most of the respiratory tract (see Figure 3.17d). The mucus produced by the goblet cells in this epithelium traps dust and other debris, and the cilia propel the mucus upward and away from the lungs.

Stratified Epithelia

Stratified epithelia consist of two or more cell layers. Being considerably more durable than the simple epithelia, these epithelia function primarily to protect.

Stratified Squamous Epithelium   **Stratified squamous epithelium** is the most common *stratified* epithelium in the body. It usually consists of several layers of cells. The cells at the free edge are squamous cells, whereas those close to the basement membrane are cuboidal or columnar. Stratified squamous epithelium is found in sites that receive a good deal of abuse or friction, such as the esophagus, the mouth, and the outer portion of the skin (Figure 3.17e).

Stratified Cuboidal and Stratified Columnar Epithelia   **Stratified cuboidal epithelium** usually has just two cell layers with (at least) the surface cells being cuboidal in shape. The surface cells of **stratified columnar epithelium** are columnar cells, but its basal cells vary in size and shape. Both of these epithelia are fairly rare in the body, being found mainly in the ducts of large glands. (Because of the extremely limited distribution of these two epithelia, they are not illustrated in Figure 3.17. They are described here only to provide a complete listing of the epithelial tissues.)

Transitional Epithelium   **Transitional epithelium** is a highly modified, stratified squamous epithelium that forms the lining of only a few organs—the urinary bladder, the ureters, and part of the urethra. *All* these organs are part of the urinary system and are subject to considerable stretching (Figure 3.17f). Cells of the basal layer are cuboidal or columnar; those at the free surface

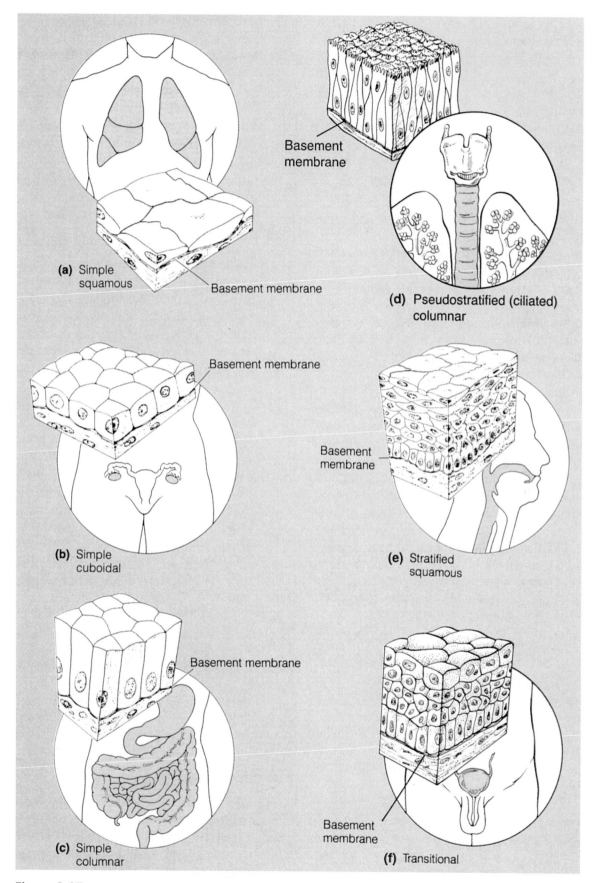

(a) Simple squamous

Basement membrane

(b) Simple cuboidal

Basement membrane

(c) Simple columnar

Basement membrane

Basement membrane

(d) Pseudostratified (ciliated) columnar

Basement membrane

(e) Stratified squamous

Basement membrane

(f) Transitional

Figure 3.17   Types of epithelia and their common locations in the body.

vary in appearance. When the organ is not stretched, the membrane is many-layered and the superficial cells are rounded and domelike. When the organ is distended with urine, the epithelium thins, and the surface cells flatten and become squamouslike. This ability of transitional cells to slide past one another and change their shape allows the ureter wall to stretch as a greater volume of urine flows through the tubelike organ. In the bladder, it allows more urine to be stored.

Glandular Epithelium

A **gland** consists of one or more cells that make and secrete a particular product. This product, called a **secretion,** typically contains protein molecules in an aqueous (water-based) fluid. Secretion is an active process in which the glandular cells obtain needed materials from the blood and use them to make their secretion, which they then discharge.

Two major types of glands develop from epithelial sheets. The **endocrine** (en′do-krin) **glands** lose their connection to the surface (duct); thus they are often called *ductless* glands. Their secretions (all hormones) diffuse directly into the blood vessels that weave through the glands. Examples of endocrine glands include the thyroid, adrenals, and pituitary. The **exocrine** (ek′so-krin) **glands** retain their ducts, and their secretions empty through the ducts to the epithelial surface. The exocrine glands, which include the sweat and oil glands, liver, and pancreas, are both internal and external. They are discussed with the organ systems to which their products are related.

CONNECTIVE TISSUE

Connective tissue, as its name suggests, connects body parts. It is found everywhere in the body. It is the most abundant and widely distributed of the tissue types.

Common Characteristics of Connective Tissue

The characteristics of connective tissue include the following:

- Variations in blood supply. Most connective tissues are well *vascularized* (that is, have a good blood supply), but there are exceptions. Tendons and ligaments have a poor blood supply, and cartilages are avascular. Consequently, all these structures heal very slowly when injured. (This is why some people say that, given a choice, they would rather have a broken bone than a torn ligament.)

- Extracellular matrix. Connective tissues are made up of many different types of cells, plus varying amounts of a nonliving substance found outside the cells, called the **extracellular matrix.**

The extracellular matrix deserves a bit more explanation because it is what makes connective tissue so different from the other tissue types. The matrix is produced by the connective tissue cells and then secreted to their exterior. Depending on the connective tissue type, the matrix may be liquid, semisolid or gel-like, or very hard. Because of its extracellular matrix, connective tissue is able to bear weight and to withstand stretching and other abuses, such as abrasion, that no other tissue could endure. But there is variation. At one extreme, fat tissue is composed mostly of cells, and the matrix is soft. At the opposite extreme, bone and cartilage have very few cells and large amounts of hard matrix, which makes them extremely strong.

Various types and amounts of fibers are deposited in and form a part of the matrix material. They include **collagen** (white) **fibers, elastic** (yellow) **fibers,** and **reticular** (fine collagen) **fibers.** Like the rest of the matrix, the fibers are made by connective tissue cells and then secreted.

Connective tissues perform many functions, but they are primarily involved in *protecting, supporting,* and *binding together* other body tissues. Find the various types of connective tissues in Figure 3.18 as you read their descriptions that follow.

Types of Connective Tissue

As noted above, all connective tissues consist of living cells surrounded by a matrix. Their major differences reflect fiber type and the amount of fibers in the matrix. From most rigid to softest, the major connective tissue classes are *bone, cartilage, dense connective tissue, loose connective tissue,* and *blood.* Each of these is described briefly next.

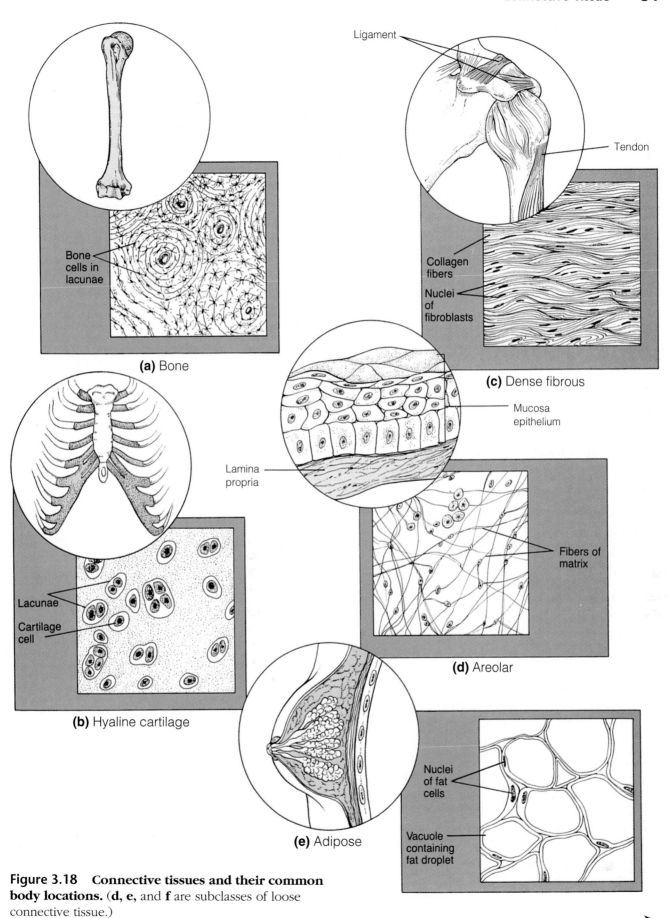

Figure 3.18   Connective tissues and their common body locations. (**d, e,** and **f** are subclasses of loose connective tissue.)

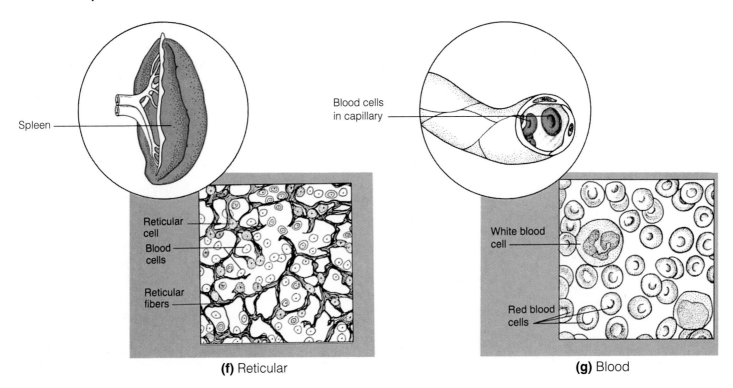

(f) Reticular

(g) Blood

Figure 3.18 (continued) **Connective tissues and their common body locations.**

Bone

Bone, sometimes called *osseous* (os′e-us) *tissue,* is composed of bone cells sitting in cavities called *lacunae* (lah-ku′ne), and surrounded by layers of a very hard matrix that contains calcium salts in addition to larger amounts of collagen fibers. Because of its rocklike hardness, bone has an exceptional ability to protect and support other body organs (for example, the skull protects the brain).

Cartilage

Cartilage is less hard and more flexible than bone. It is found in only a few places in the body. Most widespread is **hyaline** (hi′ah-lin) **cartilage,** which has abundant collagen fibers hidden by a rubbery matrix with a glassy (*hyalin* = glass) blue-white appearance. It forms the supporting structures of the larynx or voice box, attaches the ribs to the breastbone, and covers the ends of bones where they form joints. The skeleton of a fetus is made of hyaline cartilage, but by the time the baby is born most of that cartilage has been replaced by bone.

Although hyaline cartilage is the most abundant type of cartilage in the body, there are others. **Elastic cartilage** is found where a structure with elasticity is desired. For example, it supports the external ear. Highly compressible **fibrocartilage** forms the cushionlike disks between the vertebrae of the spinal column. (Elastic cartilage and fibrocartilage are not illustrated in Figure 3.18.)

Dense Connective Tissue

Dense connective tissue, also called **dense fibrous tissue,** has collagen fibers as its main matrix element. Crowded between the collagen fibers are rows of *fibroblasts* (fiber-forming cells) that manufacture the fibers. Dense connective tissue forms strong, ropelike structures such as tendons and ligaments. **Tendons** attach skeletal muscles to bones; **ligaments** connect bones to bones at joints. Ligaments are more stretchy and contain more elastic fibers than tendons. Dense connective tissue also makes up the lower layers of the skin (dermis) where it is arranged in sheets.

Loose Connective Tissue

Relatively speaking, the loose connective tissues are softer and have more cells and fewer fibers than any other connective tissue type except blood.

Areolar Tissue   Areolar (ah-re′o-lar) **tissue,** the most widely distributed connective tissue variety in the body, is a soft, pliable tissue that cushions and protects the body organs it wraps. It functions as a universal packing tissue and connective tissue "glue" because it helps to hold the internal organs together and in their proper positions. A soft layer of areolar connective tissue called the *lamina propria* (lah′mĭ-nah pro′pre-ah) underlies all mucous membranes. Its fluid matrix contains all types of fibers, which form a loose network. In fact, when viewed through a microscope, most of the matrix appears to be empty space, which explains the name of this tissue type (*areola* = small open space). Because of its loose and fluid nature, areolar connective tissue provides a reservoir of water and salts for the surrounding tissues, and essentially all body cells obtain their nutrients from and release their wastes into this "tissue fluid." When a body region is inflamed, the areolar tissue in the area soaks up the excess fluid like a sponge and the area swells and becomes puffy, a condition called **edema.** Many types of *phagocytes* wander through this tissue scavenging for bacteria, dead cells, and other "debris," which they destroy.

Adipose Tissue   Adipose (ad′ĭ-pōs) **tissue** is commonly called *fat.* Basically, it is an areolar tissue in which fat cells predominate. A glistening droplet of stored oil occupies most of a fat cell's volume and compresses the nucleus, displacing it to one side. Since the oil-containing region looks empty, and the thin rim of cytoplasm containing the bulging nucleus looks like a ring with a seal, fat cells are sometimes called *signet ring cells.*

Adipose tissue forms the subcutaneous tissue beneath the skin, where it insulates the body and protects it from extremes of both heat and cold. Adipose tissue also protects some organs individually; for example, the kidneys are surrounded by a capsule of fat, and adipose tissue cushions the eyeballs in their sockets. There are also fat "depots" in the body, such as the hips and breasts, where fat is stored and available as fuel if needed.

Reticular Connective Tissue   Reticular connective tissue consists of a delicate network of interwoven reticular fibers associated with *reticular cells,* which resemble fibroblasts. Reticular tissue is limited to certain sites: It forms the **stroma** (literally, bed or mattress), or internal supporting framework, that can support many free blood cells (largely lymphocytes) in lymphoid organs such as lymph nodes, the spleen, and bone marrow.

Blood

Blood or vascular tissue is considered a connective tissue because it consists of *blood cells,* surrounded by a nonliving fluid matrix called *blood plasma.* The "fibers" of blood are soluble protein molecules that become visible only during blood clotting. Still, we must recognize that blood is quite atypical as connective tissues go. Blood is the transport vehicle for the cardiovascular system, carrying nutrients, wastes, respiratory gases, and many other substances throughout the body. Blood is considered in detail in Chapter 10.

MUSCLE TISSUE

Muscle tissues are highly specialized to *contract* or *shorten* to produce movement. Because muscle cells are elongated to provide a long axis for contraction, they are called *muscle fibers.*

Types of Muscle Tissue

The three types of muscle tissue are illustrated in Figure 3.19. Notice their similarities and differences as you read through the descriptions below.

Skeletal Muscle

Skeletal muscle is packaged by connective tissue sheets into organs called *skeletal muscles,* which are attached to the skeleton. These muscles, which can be controlled *voluntarily* (or consciously), form the flesh of the body. When the skeletal muscles contract, they pull on bones or skin. The result of their action is gross body movements or changes in our facial expressions. The cells of skeletal muscle are long, cylindrical, and multinucleate; they have obvious **striations** (stripes).

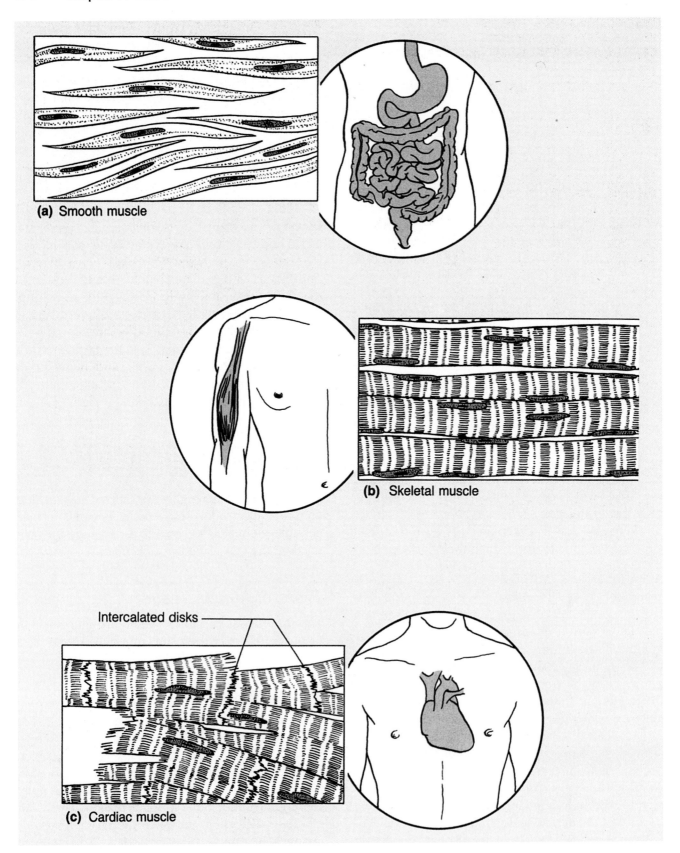

(a) Smooth muscle

(b) Skeletal muscle

Intercalated disks

(c) Cardiac muscle

Figure 3.19 **Types of muscle tissue and their common locations in the body.**

Cardiac Muscle

Cardiac muscle is found only in the heart. As it contracts, the heart acts as a pump and propels blood through the blood vessels. Like skeletal muscle, cardiac muscle has striations, but cardiac cells are uninucleate branching cells that fit tightly together (like clasped fingers) at junctions called **intercalated disks.** These intercalated disks contain gap junctions that allow ions to pass freely from cell to cell, resulting in rapid conduction of the exciting electrical impulse across the heart. Cardiac muscle is under *involuntary control,* which means that we cannot consciously control the activity of the heart. (There are, however, rare individuals who claim they have such an ability.)

Smooth Muscle

Smooth, or **visceral, muscle** is so called because no striations are visible. The individual cells have a single nucleus and are spindle-shaped (pointed at each end). Smooth muscle is found in the walls of hollow organs such as the stomach, bladder, uterus, and blood vessels. When smooth muscle contracts, the cavity of an organ alternately becomes smaller (constricts) or enlarges (dilates) so that substances are propelled through the organ along a specific pathway. Smooth muscle contracts much more slowly than the other two muscle types. *Peristalsis* (per″ĭ-stal′sis), a wavelike motion that keeps food moving through the small intestine, is typical of its activity.

NERVOUS TISSUE

When we think of **nervous tissue,** we think of cells called **neurons.** All neurons receive and conduct electrochemical impulses from one part of the body to another; thus *irritability* and *conductivity* are their two major functional characteristics. The structure of neurons is unique (see Figure 3.20). The cytoplasm is drawn out into long extensions (as much as 3 feet or more in the leg), which allows a single neuron to conduct an impulse over long distances in the body. Neurons, along with a special group of **supporting cells** that insulate, support, and protect the delicate neurons, make up the structures of the nervous system—the brain, spinal cord, and nerves.

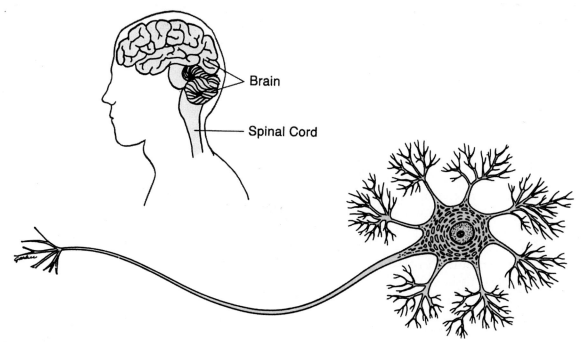

Brain

Spinal Cord

Figure 3.20   Diagrammatic view of a motor neuron. Nervous tissue (neurons and supporting cells) forms the brain, spinal cord, and nerves.

TISSUE REPAIR (WOUND HEALING)

The body has many techniques for protecting itself from "uninvited guests" or injury. Intact physical barriers such as the skin and mucous membranes, cilia, and the strong acid produced by stomach glands are just three examples of body defenses exerted at the local tissue level. When tissue injury does occur, it stimulates the body's inflammatory and immune responses and the healing process begins almost immediately. Inflammation is a non-specific body response that attempts to prevent further injury. The immune response, on the other hand, is extremely specific, and mounts a vigorous attack against recognized invaders (bacteria, viruses, toxins). These protective responses are considered in detail in Chapter 12. Here we will concentrate on the process of tissue repair itself.

Tissue repair, or wound healing, occurs in two major ways: by regeneration and by fibrosis. **Regeneration** is the replacement of destroyed tissue by the same kind of cells, whereas **fibrosis** involves repair by dense (fibrous) connective tissue; that is, by the formation of *scar tissue*. Which of these occurs depends on (1) the type of tissue damaged and (2) the severity of the injury. Generally speaking, clean cuts (incisions) heal much more successfully than ragged tears (lacerations) of the tissue.

Tissue injury sets a series of events into motion. First, the capillaries become very permeable, which allows fluid rich in clotting proteins and other substances to seep into the injured area from the bloodstream. Then the leaked clotting proteins construct a clot, which stops the loss of blood, holds the edges of the wound together, and effectively "walls off" the injured area, preventing bacteria or other harmful substances from spreading to surrounding tissues (Figure 3.21a). Where the clot is exposed to air, it quickly dries and hardens, forming a scab.

In the next phase, **granulation tissue** forms (Figure 3.21b). Granulation tissue is a delicate pink tissue composed largely of capillaries that grow into the damaged area from undamaged blood vessels nearby. These capillaries are fragile and bleed freely, as when a scab is picked away from a skin wound. Granulation tissue also contains phago-

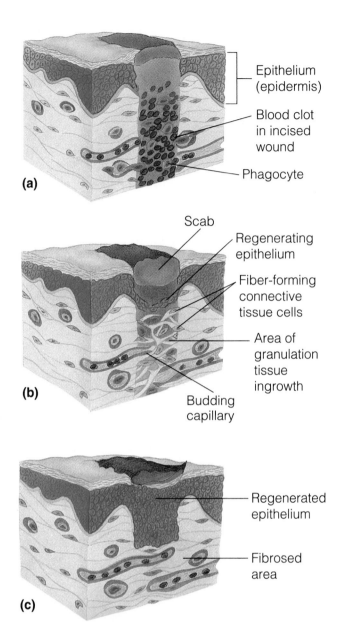

(a)

Epithelium (epidermis)

Blood clot in incised wound

Phagocyte

(b)

Scab

Regenerating epithelium

Fiber-forming connective tissue cells

Area of granulation tissue ingrowth

Budding capillary

(c)

Regenerated epithelium

Fibrosed area

Figure 3.21   Tissue repair of a simple skin wound by regeneration and fibrosis. (a) Broken blood vessels bleed, inflammatory chemicals are released, and local blood vessels dilate and become more permeable, allowing white blood cells, fluid, and plasma proteins to enter the injured area. Clotting proteins (leaked from the blood) initiate clotting; the surface dries and forms a scab. **(b)** Granulation tissue is formed. New capillaries invade the clot, restoring the blood supply. Connective tissue cells form collagen fibers that bridge the gap as phagocytes dispose of dead tissue cells and the blood clot. The surface epithelial cells divide and multiply and migrate over the granulation tissue. **(c)** Approximately one week later, the fibrosed area (scar) has contracted, and regeneration of the epithelium is progressing.

cytes that eventually dispose of the blood clot, and connective tissue cells (fibroblasts) that synthesize collagen fibers (scar tissue) to permanently bridge the gap.

While these events are going on, the surface epithelium begins to regenerate and makes its way across the granulation tissue just beneath the scab, which soon detaches. The final result is a fully regenerated surface epithelium (Figure 3.21c) that covers an underlying area of fibrosis (the scar). The scar is either invisible or visible as a thin white line, depending on the severity of the wound.

The ability of the different tissue types to regenerate varies widely. Epithelial tissues such as the skin epidermis and mucous membranes regenerate beautifully. So, too, do most of the fibrous connective tissues and bone. Skeletal muscle regenerates poorly, if at all, and cardiac muscle and nervous tissue within the brain and spinal cord are replaced only by scar tissue.

Scar tissue is strong, but it lacks the flexibility of most normal tissues. Perhaps even more important is its inability to perform the normal functions of the tissue it replaces. Thus, if scar tissue forms in the wall of the bladder, heart, or another muscular organ, it may severely hamper the functioning of that organ. ❑

DEVELOPMENTAL ASPECTS OF CELLS AND TISSUES

We all begin life as a single cell, which divides thousands of times to form our multicellular embryonic body. Very early in embryonic development, the cells begin to specialize to form the primary tissues, and by birth, most organs are well formed and functioning. The body continues to grow and enlarge by forming new tissue throughout childhood and adolescence.

Cell division is extremely important during the body's growth period. Most cells (except neurons) undergo mitosis until the end of puberty, when adult body size is reached and overall body growth ends. After this time, only certain cells routinely carry out cell division—for example, cells exposed to abrasion that continually wear away, such as skin and intestinal cells. Other cell groups, such as liver cells, stop dividing; however, they still have this ability should some of them die or become damaged and need to be replaced. Still other cell groups (for example, heart muscle and nervous tissue) completely lose their ability to divide when they are fully mature; that is, they become *amitotic* (am″ĭ-tot′ik). Amitotic tissues are severely handicapped by injury because the lost cells cannot be replaced by the same type of cells. This is why the heart of an individual who has had several severe heart attacks becomes weaker and weaker. Since the damaged cardiac muscle cannot regenerate and is replaced by scar tissue that cannot contract, the heart becomes less and less capable of acting as an efficient blood pump.

The aging process begins once maturity has been reached. (Some believe it begins at birth.) No one has been able to explain just *what* causes aging, but there have been many suggestions. Some believe it is a result of little "chemical insults," which occur continually through life—for example, the presence of toxic chemicals (such as alcohol, certain drugs, or carbon monoxide) in the blood, or the temporary absence of needed substances such as glucose or oxygen. Perhaps the effect of these chemical insults is cumulative and finally succeeds in upsetting the delicate chemical balance of the body cells. Others think that external physical factors such as radiation (X rays or ultraviolet waves) contribute to the aging process. Several believe that the aging "clock" is genetically programmed, or built into our genes. We all know of cases like the "radiant woman of 50 who looks about 35" or the "barely-out-of-adolescence man of 24 who looks 40." It appears that such traits can run in families.

There is no question that certain events are part of the aging process. For example, epithelial membranes become thinner and are more easily damaged and the skin loses its elasticity and begins to sag. The exocrine glands of the body (epithelial tissue) become much less active as we age, which is why we begin to "dry out" as less oil, mucus, and sweat are produced. Some endocrine glands produce decreasing amounts of hormones, and the body processes that they control (such as metabolism and reproduction) become less efficient or stop altogether.

Connective tissue structures also show changes with age. The amount of collagen in the body declines, bones become porous and weaken, and the repair of tissue injuries slows. Muscles and nervous tissues begin to atrophy. Although a poor diet may contribute to some of these changes, there is little doubt that decreased efficiency of the circulatory system, which reduces nutrient and oxygen delivery to body tissues, is a major factor.

Important Terms

active transport

centrioles (sen′tre-ōlz)

chromatin (kro′mah-tin)

cilia (sil′e-ah)

connective tissue

cytokinesis (si″to-kĭ-ne′sis)

cytoplasm (si′to-plazm″)

cytoskeleton

diffusion (dĭ-fu′zhun)

endoplasmic reticulum (en″do-plas′mik rĕ-tik′u-lum)

epithelial (ep″ĭ-the′le-al) **tissue**

filtration

Golgi (gol′je) **apparatus**

lysosomes (li′so-sōmz)

matrix

microvilli (mi″kro-vil′i)

mitochondria (mi″to-kon′dre-ah)

mitosis (mi-to′sis)

muscle tissue

nervous tissue

nucleoli (nu-kle′o-li)

nucleus (nu′kle-us)

organelles (or″gah-nelz′)

passive transport

phagocytosis (fag″o-si-to′sis)

pinocytosis (pi″no-si-to′sis)

ribosomes (ri′bo-sōmz)

selective permeability

solute pump

Summary

Part 1: Cells (*pp. 56-76*)

1. Overview of the Cellular Basis of Life
   a. A cell is composed primarily of four elements—carbon, hydrogen, oxygen, and nitrogen—plus many trace elements. Living matter is over 60 percent water. The major building substance of the cell is protein.
   b. Cells vary in size from microscopic to over a meter in length. Shape often reflects function. For example, muscle cells have a long axis to allow shortening.

2. Anatomy of a Generalized Cell
   a. Cells have three major regions—nucleus, cytoplasm, and plasma membrane.
      (1) The nucleus, or control center, directs cell activity and is necessary for reproduction. The nucleus contains genetic material (DNA), which carries instructions for synthesis of proteins.
      (2) The plasma membrane limits and encloses the cytoplasm and acts as a selective barrier to the movement of substances into and out of the cell. It is composed of a bilipid layer containing proteins. The water-impermeable lipid portion forms the basic fabric of the membranes. The proteins (many of which are glycoproteins) act as enzymes or carriers in membrane transport, form membrane channels or pores, provide receptor sites for hormones and other chemicals, or play a role in cellular recognition and interactions during development and immune reactions.
      Specializations of the plasma membrane include microvilli (which increase the absorptive area) and cell junctions (desmosomes, tight junctions, and gap junctions).
      (3) The cytoplasm is where most cellular activities occur. Its fluid substance, the *cytosol,* contains inclusions, stored or inactive materials in the cytoplasm (fat globules, water vacuoles, crystals, and the like) and specialized bodies called *organelles,* each with a specific function. For example, mitochondria are sites of ATP synthesis, ribosomes are sites of protein synthesis, and the Golgi apparatus packages substances for export from the cell. Lysosomes carry out intracellular digestion, and peroxisomes disarm dangerous chemicals in the cells. Cytoskeletal elements function in cellular support and motion. The centrioles play a role in cell division and form the bases of cilia and flagella.

3. Cell Physiology
   a. All cells exhibit irritability, digest foods, excrete wastes, and are able to reproduce, grow, move, and metabolize.
   b. Transport of substances through the cell membrane:
      (1) Passive transport processes include diffusion and filtration.
         (a) Diffusion is the movement of a substance from an area of its higher concentration to an area of its lower concentration. It occurs because of kinetic energy of the molecules themselves. The diffusion of dissolved solutes through the plasma membrane is called *simple diffusion*. The diffusion of water through the plasma membrane is called *osmosis*. Diffusion that requires a protein carrier is called *facilitated diffusion*.
         (b) Filtration is the movement of substances through a membrane from an area of high hydrostatic pressure to an area of lower fluid pressure. In the body, the driving force of filtration is blood pressure.
      (2) Active transport processes use energy (ATP) provided by the cell.
         (a) In solute pumping, substances are moved across the membrane by proteins called *solute pumps* against an electrical or a concentration gradient. This accounts for the transport of amino acids, some sugars, and most ions.
         (b) The two types of ATP-activated bulk transport are exocytosis and endocytosis. Exocytosis moves substances out of cells; a membrane-bound vesicle fuses with the plasma membrane, ruptures, and ejects its contents to the cell exterior. Secretions are released by exocytosis. Endocytosis, in which particles are taken up by enclosure in a plasma membrane sac, includes phagocytosis (uptake of solid particles) and pinocytosis (uptake of fluids).

c. Osmotic pressure, which reflects the solute concentration of a solution, determines whether cells gain or lose water.
    (1) Hypertonic solutions contain more solutes (and less water) than do cells. In these solutions, cells lose water by osmosis and crenate.
    (2) Hypotonic solutions contain fewer solutes (and more water) than do the cells. In these solutions, cells swell and may rupture (lysis) as water rushes in by osmosis.
    (3) Isotonic solutions, which have the same solute-to-solvent ratio as cells, cause no changes in cell size or shape.
d. Cell division has two phases, mitosis (nuclear division) and cytokinesis (division of the cytoplasm).
    (1) Mitosis begins after DNA has been replicated; it consists of four stages—prophase, metaphase, anaphase, and telophase. The result is two daughter nuclei, each identical to the mother nucleus.
    (2) Cytokinesis usually begins during anaphase and progressively pinches the cytoplasm in half.
    (3) Mitotic cell division provides an increased number of cells for growth and repair.
e. Protein synthesis involves both DNA (the genes) and RNA.
    (1) A gene is a segment of DNA that carries the instructions for building one protein. The information is in the sequence of bases in the nucleotide strands. Each three-base sequence (triplet) specifies one amino acid in the protein.
    (2) Messenger RNA carries the instructions for protein synthesis from the DNA gene to the ribosomes. Transfer RNA transports amino acids to the ribosomes. Ribosomal RNA forms part of the ribosomal structure and helps coordinate the protein building process.

Part 2: Body Tissues (pp. 76-87)

1. Epithelium is the covering, lining, and glandular tissue. Its functions include protection, absorption, and secretion. Epithelia are named according to cell shape (squamous, cuboidal, columnar) and arrangement (simple, stratified).

2. Connective tissue is the supportive, protective, and binding tissue. It is characterized by the presence of a nonliving, extracellular matrix produced and secreted by the cells; it varies in amount and consistency. Fat, ligaments and tendons, bones, and cartilage are all connective tissues or connective tissue structures.

3. Nervous tissue is composed of cells called *neurons,* which are highly specialized to receive and transmit nerve impulses and supporting cells. Neurons are important in control of body processes. Nervous tissue is located in nervous system structures—brain and spinal cord.

4. Muscle tissue is specialized to contract or shorten, which causes movement. There are three types—skeletal (attached to the skeleton), cardiac (forms the heart), and smooth (in the walls of hollow organs).

5. Tissue repair (wound healing) may involve regeneration, fibrosis, or both. In regeneration, the injured tissue is replaced by the same type of cells. In fibrosis, the wound is repaired with scar tissue. Epithelia and connective tissues regenerate well. Mature cardiac muscle and nervous tissue are repaired by fibrosis.

Developmental Aspects of Cells and Tissues (pp. 87-88)

1. Growth through cell division continues through puberty. Cell populations exposed to friction (such as epithelium) replace lost cells throughout life. Connective tissue remains mitotic and forms repair (scar) tissue. Muscle tissue becomes amitotic by the end of puberty, and nervous tissue becomes amitotic shortly after birth. Amitotic tissues are severely handicapped by injury.

2. The cause of aging is unknown, but chemical and physical insults and genetic programming are suggested.

3. Neoplasms, both benign and cancerous, represent abnormal cell masses in which normal controls on cell division are not working. Atrophy (decrease in size) of a tissue or organ occurs when the organ is no longer stimulated normally. Hyperplasia (increase in size) of tissue or organ may occur when tissue is strongly stimulated or irritated.

Review Questions

1. Name the four elements making up the bulk of living matter.

2. Define *cell* and *organelle*.

3. Although cells have differences that reflect their special functions in the body, what functional abilities do *all* cells exhibit?

4. Describe the general function of the nucleus. Describe the special function of DNA found in the nucleus. What nuclear structures contain DNA? Help to form ribosomes?

5. Describe the general structure and function of the plasma membrane.

6. Describe the general composition and function of the cytosol and inclusions of the cytoplasm.

7. Name the cellular organelles and explain the function of each.

8. What is the difference between active and passive transport processes?

9. Define *diffusion, osmosis, simple diffusion, filtration, solute pumping, exocytosis, endocytosis, phagocytosis,* and *pinocytosis.*

10. What two structural characteristics of cell membranes determine if substances can pass through them passively? What determines whether or not a substance can be actively transported through the membrane?

11. Explain the effect of the following solutions on living cells: hypertonic, hypotonic, and isotonic.

12. Briefly describe the process of DNA replication.

13. Define *mitosis*. Why is mitosis important?

14. What is the role of the spindle in mitosis?

15. Why can an organ be permanently damaged if its cells are amitotic?

16. Describe the relative roles of DNA and RNA in protein synthesis.

17. Define *tissue*. List the four major types of tissue. Which of the four major tissue types is most widely distributed in the body?

18. Describe the general characteristics of epithelial tissue. List the most important functions of epithelial tissues and give examples of each.

19. How are epithelial tissues classified?

20. Where is ciliated epithelium found, and what role does it play?

21. How do the endocrine and exocrine glands differ in structure and function?

22. What are the general structural characteristics of connective tissues? What are the functions of connective tissues? How are their functions reflected in their structures?

23. Name a connective tissue with (a) a soft fluid matrix, and (b) a stony hard matrix.

24. What is the function of muscle tissue?

25. Name the three types of muscle tissue and tell where each would be found in the body.

26. What is meant by "Smooth muscles are involuntary in action"? Which muscle type is voluntary in action?

27. What two functional characteristics are highly developed in neurons?

28. In what ways are neurons similar to other cells? In what ways are they different?

29. Define *neoplasm, atrophy,* and *hyperplasia.*

30. How are benign neoplasms different from cancerous neoplasms?

 ## At the Clinic

1. A "red-hot" bacterial infection of the intestinal tract irritates the intestinal cells and interferes with normal digestion. Such a condition is often accompanied by diarrhea, which causes loss of body water. On the basis of what you have learned about osmotic water flows, explain why diarrhea may occur.

2. Two examples of chemotherapeutic drugs (used to treat cancer) and their cellular actions are given below. Explain why each drug could be fatal to a cell.

   • Vincristine: Damages the mitotic spindle.
   • Adriamycin: Binds to DNA and blocks messenger RNA synthesis.

3. Hydrocortisone is an anti-inflammatory drug that stabilizes lysosomal membranes. Explain how this effect reduces cell damage and inflammation.

4. John has severely injured his knee during football practice. He is told that he has a torn knee cartilage and to expect that recovery and repair will take a long time. Why?

5. Three patients in an intensive care unit are examined by the resident doctor. One patient has brain damage from a stroke, another had a heart attack that severely damaged his heart muscle, and the third has a severely damaged liver (a gland) from a crushing injury in a car accident. All three patients have stabilized and will survive, but only one will have full functional recovery through regeneration. Which one and why?

6. Kareem had a nervous habit of chewing on the inner lining of his lip and the lip grew thicker and thicker. Kareem's dentist noticed his thickened lip and suggested he have it checked. A biopsy revealed hyperplasia and scattered areas of dysplasia, but no evidence of neoplasia. What do these terms mean? Did Kareem have cancer of the mouth?

4
Skin and Body Membranes

YOUR GOALS

After completing this chapter, you will have a working knowledge of the functions of the skin and body membranes system and will have mastered the objectives listed below.

FUNCTION PREVIEW

- Body membranes line or cover, protect, and lubricate body surfaces.
- As the outermost boundary of the body, the skin protects against injuries of many types.

OBJECTIVE CHECKLIST

CLASSIFICATION OF BODY MEMBRANES (pp. 94–96)

❑ List the general functions of each membrane type—cutaneous, mucous, serous, and synovial—and give its location in the body.

❑ Compare the structure (tissue makeup) of the major membrane types.

INTEGUMENTARY SYSTEM (SKIN) (pp. 96–109)

❑ List several important functions of the integumentary system and explain how these functions are accomplished.

❑ When provided with a model or diagram of the skin, recognize and name the following skin structures: epidermis, dermis (papillary and reticular layers), hair and hair follicle, sebaceous gland, and sweat gland.

❑ Name the uppermost and deepest layers of the epidermis and describe the characteristics of each.

❑ Describe the distribution and function of the epidermal derivatives—sebaceous glands, sweat glands, and hair.

❑ Name the factors that determine skin color and describe the function of melanin.

❑ Differentiate between first-, second-, and third-degree burns.

❑ Explain the importance of the "rule of nines."

❑ Summarize the characteristics of basal cell carcinoma, squamous cell carcinoma, and malignant melanoma.

(continued)

**DEVELOPMENTAL ASPECTS OF SKIN AND
BODY MEMBRANES** (p. 109)
❑ List several examples of integumentary system aging.

Body membranes, which cover surfaces, line body cavities, and form protective (and often lubricating) sheets around organs, fall into two major groups. There are (1) epithelial membranes, which include the cutaneous, mucous, and serous membranes, and (2) connective tissue membranes, represented by synovial membranes. The cutaneous membrane, generally called the skin or integumentary system, will receive most of our attention in this chapter, but first we will consider the other body membranes.

CLASSIFICATION OF BODY MEMBRANES

The two major categories of body membranes—epithelial and connective tissue—are classified in part according to their tissue makeup.

Epithelial Membranes

The **epithelial membranes** include the cutaneous membrane (skin), the mucous membranes, and the serous membranes (Figure 4.1). However, calling these membranes "epithelial" is not only misleading but also inaccurate. Although they all *do* contain an epithelial sheet, it is always combined with an underlying layer of connective tissue. Hence these membranes can actually be considered to be simple organs. Since the skin will be discussed in some detail shortly, it will not be considered here except to list it as a subcategory of the epithelial membranes.

Cutaneous Membrane

The **cutaneous** (ku-ta′ne-us) **membrane** is your skin. Its superficial epidermis is composed of a stratified squamous keratinizing epithelium. The underlying dermis is mostly dense (fibrous) connective tissue. Unlike the other epithelial membranes, the cutaneous membrane is exposed to air and is a *dry* membrane.

Mucous Membranes

A **mucous membrane (mucosa)** is composed of epithelium (the type varies with the site) resting on a loose connective tissue membrane called a *lamina propria*. This membrane type lines all body cavities that open to the exterior, such as those of the hollow organs of the respiratory, digestive, urinary and reproductive tracts (Figure 4.1b). Notice that the term *mucosa* refers only to the location of the epithelial membranes, *not* their cell makeup, which varies. However, most mucosae contain either stratified squamous epithelium (as in the mouth and esophagus) or simple columnar epithelium (as in the rest of the digestive tract). In all cases, they are "wet," or moist membranes that are almost continuously bathed in secretions, or in the case of the urinary mucosae, urine.

The epithelium of mucosae is often adapted for absorption or secretion. Although many mucosae secrete mucus, this is not a requirement. The mucosae of the respiratory and digestive tracts secrete large amounts of protective, lubricating mucus; that of the urinary tract does not.

Serous Membranes

A **serous membrane (serosa)** is composed of a layer of simple squamous epithelium resting on a thin layer of areolar connective tissue. In contrast to mucous membranes, which line open body cavities, serous membranes line body cavities that are closed to the exterior (except for the dorsal body cavity and joint cavities).

Serous membranes occur in pairs. The *parietal* (pah-ri′e-tal: *parie* = wall) *layer* lines a specific portion of the wall of the ventral body cavity. It

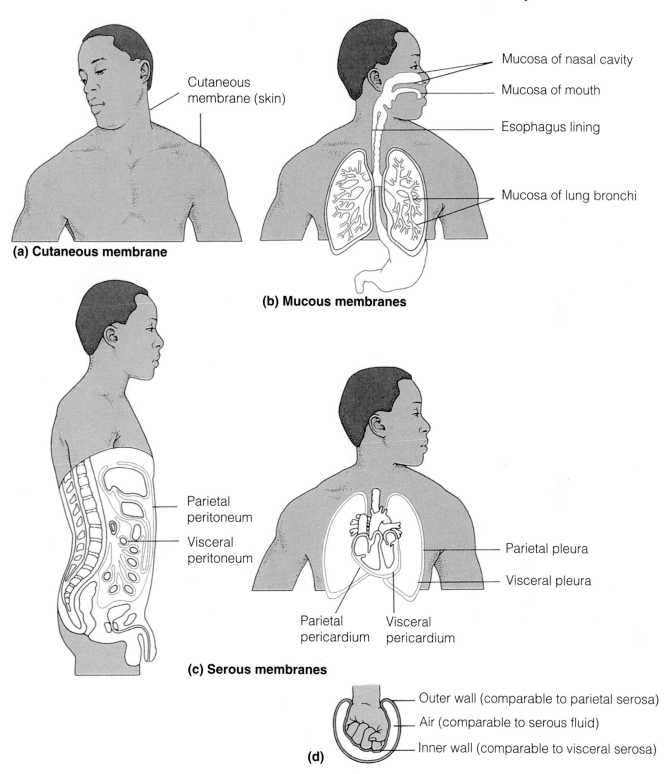

(a) Cutaneous membrane

Cutaneous membrane (skin)

(b) Mucous membranes

Mucosa of nasal cavity

Mucosa of mouth

Esophagus lining

Mucosa of lung bronchi

(c) Serous membranes

Parietal peritoneum

Visceral peritoneum

Parietal pleura

Visceral pleura

Parietal pericardium   Visceral pericardium

(d)

Outer wall (comparable to parietal serosa)

Air (comparable to serous fluid)

Inner wall (comparable to visceral serosa)

Figure 4.1   Classes of epithelial membranes. (**a**) Cutaneous membrane, or skin. (**b**) Mucous membranes (blue) line body cavities that are open to the exterior. (**c**) Serous membranes (shown in red) line ventral body cavities that are closed to the exterior. (**d**) A fist thrust into a flaccid balloon demonstrates the relationship between the parietal and visceral serous membrane layers.

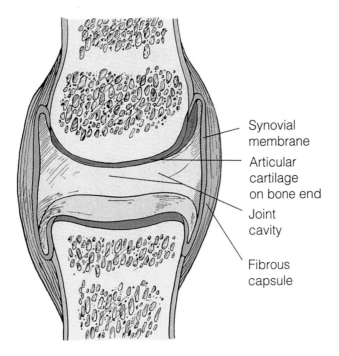

Figure 4.2   A typical synovial joint.

Synovial membrane
Articular cartilage on bone end
Joint cavity
Fibrous capsule

folds in on itself to form the *visceral* (vis'er-al) *layer,* which covers the outside of the organs in that cavity.

You can visualize the relationship between the serosal layers by pushing your fist into a limp balloon (Figure 4.1d). The part of the balloon that clings closely to your fist can be compared to the visceral serosa clinging to the organ's external surface. The outer wall of the balloon represents the parietal serosa that lines the walls of the cavity and that, unlike the balloon, is never exposed but is always fused to the cavity wall. In the body, the serous layers are separated not by air but by a thin, clear fluid, called **serous fluid,** which is secreted by both membranes. Although there is a potential space between the two membranes, they tend to lie very close to each other.

The serous fluid allows the organs to slide easily across the cavity walls and one another without friction as they carry out their routine functions. This is extremely important when mobile organs such as the pumping heart and the churning stomach are involved.

The specific names of the serous membranes depend on their locations. The serosa lining the abdominal cavity and covering its organs is the peritoneum (per"ĭ-to-ne'um), that around the lungs (Figure 4.1c) is the **pleura** (ploo'rah), and that around the heart is the **pericardium** (per"ĭ-kar'de-um).

Connective Tissue Membranes

Synovial (sĭ-no've-al) **membranes** are composed of connective tissue and contain no epithelial cells at all. These membranes line the fibrous capsules surrounding joints (see Figure 4.2), where they provide a smooth surface and secrete a lubricating fluid. They also line small sacs of connective tissue called *bursae* (ber'se) and *tendon sheaths*. Both of these structures cushion organs moving against each other during muscle activity—such as the movement of a tendon across a bone's surface.

INTEGUMENTARY SYSTEM (SKIN)

Would you be enticed by an advertisement for a coat that is waterproof, stretchable, washable, and permanent-press, that invisibly repairs small cuts, rips, and burns, and that is guaranteed to last a lifetime with reasonable care? Sounds too good to be true, but you already have such a coat—your *cutaneous membrane* or **skin.** The skin and its derivatives (sweat and oil glands, hairs, and nails) serve a number of functions, mostly protective. Together, these organs are called the **integumentary** (in-teg"u-men'ta-re) **system.**

Basic Skin Functions

Also called the **integument** (in-teg'u-ment), which simply means "covering," the skin is much more than an external body covering. It is absolutely essential because it keeps water and other precious molecules in the body. It also keeps water (and other things) out. (This is why one can swim for hours without becoming waterlogged.) Structurally, the skin is a marvel. It is pliable yet tough, which allows it to take constant punishment from external agents. Without our skin, we would quickly fall prey to bacteria and perish from water and heat loss.

Table 4.1   Functions of the Skin

Functions	How accomplished
Protects deeper tissues from:	
• Mechanical damage (bumps)	Physical barrier contains keratin, which "toughens" cells, and pressure receptors, which alert the nervous system to possible damage.
• Chemical damage (acids and bases)	Has relatively impermeable keratinized cells; contains pain receptors, which alert the nervous system to possible damage.
• Bacterial damage	Has an unbroken surface and "acid mantle" (skin secretions are acidic, and thus inhibit bacteria). Phagocytes ingest foreign substances and pathogens, preventing them from penetrating into deeper body tissues.
• Ultraviolet radiation (damaging effects of sunlight)	Melanin produced by melanocytes offers protection.
• Thermal (heat or cold) damage	Contains heat/cold/pain receptors.
• Dessication (drying out)	Contains waterproofing keratin.
Aids in body heat loss or heat retention (controlled by the nervous system)	Heat loss: By activating sweat glands and allowing blood to flush into skin capillary beds. Heat retention: By not allowing blood to flush into skin capillary beds.
Aids in excretion of urea and uric acid	Contained in perspiration produced by sweat glands.
Synthesizes vitamin D	Modified cholesterol molecules in skin converted to vitamin D by sunlight.

The skin has many functions; most, but not all, are protective (see Table 4.1). It insulates and cushions the deeper body organs and protects the entire body from mechanical damage (bumps and cuts), chemical damage (such as from acids and bases), thermal damage (heat and cold), ultraviolet radiation (in sunlight), and bacteria. The uppermost layer of the skin is full of keratin and *cornified,* or hardened, to prevent water loss from the body surface.

The skin's rich capillary network and sweat glands (both controlled by the nervous system) play an important role in regulating heat loss from the body surface. The skin acts as a mini-excretory system; urea, salts, and water are lost when we sweat. The skin also manufactures several proteins important to immunity and synthesizes vitamin D.

(Modified cholesterol molecules located in the skin are converted to vitamin D by sunlight.) Finally, the *cutaneous sensory receptors,* which are actually part of the nervous system, are located in the skin. These tiny sensors, which include touch, pressure, temperature, and pain receptors, provide us with a great deal of information about our external environment. They alert us to bumps and the presence of tissue-damaging factors as well as to the feel of wind in our hair and a caress.

Structure of the Skin

The skin is composed of two kinds of tissue. The outer **epidermis** (ep"ĭ-der'mis) is made up of stratified squamous epithelium that is capable of

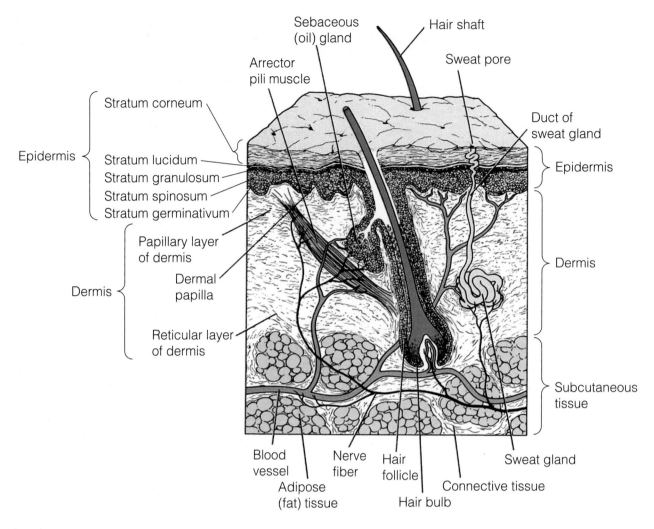

Figure 4.3   Skin structure. Three-dimensional view of the skin and under-lying subcutaneous tissue.

keratinizing (ker′ah-tin-īz-ing), or becoming hard and tough. The underlying **dermis** is made up of dense connective tissue. The epidermis and dermis are firmly cemented together. However, a burn or friction (e.g. the rubbing of a poorly fitting shoe) may cause them to separate, which results in a *blister.*

Deep to the dermis is the **subcutaneous tissue,** essentially adipose tissue. It is not considered part of the skin, but it does anchor the skin to underlying organs. Subcutaneous tissue serves as a shock absorber and insulates the deeper tissues from extreme temperature changes occurring outside the body. It is also responsible for the curves that are more a part of a woman's anatomy than a man's. The main skin areas and structures are described next. As you read, locate the described areas or structures on Figure 4.3.

Epidermis

Like all epithelial tissues, the epidermis is *avascular;* that is, it has no blood supply of its own. This explains why a man can shave daily and not bleed even though he is cutting off many cell layers each time he shaves.

The deepest cell layer of the epidermis is called the **stratum germinativum** (stra′tum jer″mĭ-na-tiv′um). It lies closest to the dermis and contains the only epidermal cells that receive adequate nourishment (through diffusion of nutrients from the dermis). These cells are constantly undergoing cell division (*germinate* = to grow), and millions of new cells are produced daily. The daughter cells are pushed upward, away from the source of nutrition, to become part of the epidermal layers closer to the skin surface. As they move away from the dermis and become part of the more superficial layers, the **stratum granulosum,** the **stratum spinosum,** and then the clear **stratum lucidum** (lu'sid-um), they become flatter, increasingly full of keratin (keratinized), and finally die. The combination of accumulating water-repellent keratin inside them and their increasing distance from the blood supply (in the dermis) effectively dooms the more superficial epidermal cells because they are unable to get adequate nutrients and oxygen.

The outermost layer, the **stratum corneum** (kor'ne-um), is 20 to 30 cell layers thick. It accounts for about three-quarters of the epidermal thickness. The shinglelike dead cell remnants, completely filled with keratin, are referred to as *cornified* or *horny cells* (*cornu* = horn). The common saying "Beauty is only skin deep" is especially interesting in light of the fact that nearly everything we see when we look at someone is dead! The water-repellent keratin is an exceptionally tough protein. Its abundance in the stratum corneum allows that layer to provide a durable "overcoat" for the body, which protects deeper cells from the hostile external environment (air) and from water loss and helps the body resist biological, chemical, and physical assaults. The stratum corneum rubs and flakes off slowly and steadily, and is replaced by cells produced by the division of the deeper stratum germinativum cells. Indeed, we have a totally "new" epidermis every 35 to 45 days.

Melanin (mel'ah-nin), a pigment which ranges in color from yellow to brown to black, is produced by special cells called **melanocytes** (mel'ah-no-sītz), found chiefly in the stratum germinativum. When the skin is exposed to sunlight, which stimulates the melanocytes to produce more of the melanin pigment, tanning occurs. The stratum germinativum cells phagocytize (eat) the pigment, and as it accumulates within them, the melanin forms a protective pigment "umbrella" over the superficial, or "sunny," side of their nuclei that shields their genetic material (DNA) from the damaging effects of ultraviolet radiation in sunlight. *Freckles* and *moles* are seen where melanin is concentrated in one spot.

Despite melanin's protective effects, excessive sun exposure eventually damages the skin. It causes the elastic fibers to clump, leading to leathery skin. It also depresses the immune system. This may help to explain why many people infected with the *herpes simplex,* or *cold sore,* virus

are more likely to have an eruption after sunbathing. Overexposure to the sun can also alter the DNA of skin cells and in this way lead to skin cancer. Black people seldom have skin cancer, attesting to melanin's amazing effectiveness as a natural sunscreen. ❏

Dermis

The dermis is your "hide." It is a strong, stretchy envelope that helps to hold the body together. When you purchase leather goods (bags, belts, shoes, and the like), you are buying the treated dermis of animals.

The dense (fibrous) connective tissue making up the dermis consists of two major regions—the *papillary* and the *reticular* areas. Like the epidermis, the dermis varies in thickness. For example, it is particularly thick on the palms of the hands and soles of the feet, but is quite thin on the eyelids.

The **papillary layer** is the upper dermal region. It is uneven and has fingerlike projections from its superior surface, called **dermal papillae** (pah-pil'e; *papill* = nipple), which indent the epidermis above. Many of the dermal papillae contain capillary loops, which furnish nutrients to the epidermis. Others house pain receptors (free nerve endings) and touch receptors called *Meissner's corpuscles* (mīs'nerz kor'puh-sulz). On the palms of the hands and soles of the feet, the papillae are arranged in definite patterns that form looped and whorled ridges on the epidermal surface that increase friction and enhance the gripping ability of the fingers and feet. Papillary patterns are genetically determined. The ridges of the fingertips are well-provided with sweat pores and leave unique, identifying films of sweat called *fingerprints* on almost anything they touch.

The **reticular layer** is the deepest skin layer. It contains blood vessels, sweat and oil glands, and deep pressure receptors called *Pacinian* (pah-sin'e-an) *corpuscles.* Many phagocytes are found here (and in fact, throughout the dermis). They act to prevent bacteria that have managed to get through the epidermis from penetrating any deeper into the body.

Both *collagen* and *elastic fibers* are found throughout the dermis. Collagen fibers are responsible for the toughness of the dermis. Collagen fibers also attract and bind water, and thus help to keep the skin hydrated. Elastic fibers give the skin

Figure 4.4   Photograph of a deep (stage III) decubitus ulcer.

its elasticity when we are young. As we age, the number of collagen and elastic fibers decreases, and the subcutaneous tissue loses fat. As a result, the skin becomes less elastic and begins to sag and wrinkle.

The dermis is abundantly supplied with blood vessels that play a role in maintaining body temperature homeostasis. When body temperature is high, the capillaries of the dermis become engorged, or swollen, with heated blood and the skin becomes reddened and warm. This allows body heat to radiate from the skin surface. If the environment is cool and body heat must be conserved, blood bypasses the dermis capillaries temporarily, allowing internal body temperature to stay high.

Any restriction of the normal blood supply to the skin results in cell death and, if severe or prolonged enough, skin ulcers. *Decubitus* (de-ku′bǐ-tus) *ulcers* (bedsores) occur in bedridden patients who are not turned regularly or who are dragged or pulled across the bed repeatedly. The weight of the body puts pressure on the skin, especially over bony projections. Because this restricts the blood supply, the skin becomes pale or blanched at pressure points. At first, the skin reddens when pressure is released, but if the situation is not corrected, the cells begin to die, and typically small cracks or breaks in the skin appear at compressed sites. Permanent damage to the super-

ficial blood vessels and tissue eventually results in degeneration and ulceration of the skin (Figure 4.4). ❑

The dermis also has a rich nerve supply. As mentioned earlier, many of the nerve endings have specialized receptor end-organs that send messages to the central nervous system for interpretation when they are stimulated by environmental factors (pressure, temperature, and the like). These cutaneous receptors are discussed in more detail in Chapter 7.

Skin Color

Three pigments contribute to skin color: (1) The amount and kind (yellow, reddish brown, or black) of melanin in the epidermis. (2) The amount of carotene deposited in the stratum corneum and subcutaneous tissue. (Carotene is an orange-yellow pigment found in abundant amounts in carrots and other orange, deep yellow, or leafy green vegetables.) (3) The amount of oxygen bound to hemoglobin (pigment in red blood cells) in the dermal blood vessels. People who produce a lot of melanin have brown-toned skin. In light-skinned (Caucasian) people, who have less melanin, the crimson color of oxygen-rich hemoglobin in the dermal blood supply flushes through the transparent cell layers above and gives the skin a rosy glow. The skin tends to take on a yellow-orange cast when large amounts of carotene-rich foods are eaten.

When hemoglobin is poorly oxygenated, both the blood and the skin of Caucasians appear blue, a condition called *cyanosis* (si″ah-no′sis). Skin often becomes cyanotic during heart failure and severe breathing disorders. In black people, the skin does not appear cyanotic because of the masking effects of melanin, but cyanosis is apparent in their mucous membranes and nail beds. ❑

Skin color is also influenced by emotional stimuli, and many alterations in skin color signal certain disease states:

- *Redness,* or *erythema* (er″ǐ-the′mah): Reddened skin may indicate embarrassment (blushing), fever, hypertension, inflammation, or allergy.

- *Pallor,* or *blanching:* Under certain types of emotional stress (fear, anger, and others), some people become pale. Pale skin may also signify anemia, low blood pressure, or impaired blood flow into the area.

- *Jaundice* (jon′dis) or a *yellow cast:* An abnormal yellow skin tone usually signifies a liver disorder in which excess bile pigments are absorbed into the blood, circulated throughout the body, and deposited in body tissues.

- *Black-and-blue marks,* or *bruises:* Black-and-blue marks reveal sites where blood has escaped from the circulation and has clotted in the tissue spaces. Such clotted blood masses are called *hematomas.* An unusual tendency to bruising may signify a deficiency of vitamin C in the diet or bleeder's disease.

Appendages of the Skin

The **skin appendages** include cutaneous glands, hair and hair follicles, and nails. Each of these appendages arises from the epidermis and plays a unique role in maintaining body homeostasis.

Cutaneous Glands

The cutaneous glands are all **exocrine glands** that release their secretions to the skin surface via ducts. They fall into two groups: *sebaceous glands* and *sweat glands.* As these glands are formed by the cells of the stratum germinativum, they push into the deeper skin regions and ultimately reside almost entirely in the dermis.

Sebaceous (Oil) Glands The **sebaceous** (seh-ba′shus) **glands,** or oil glands, are found all over the skin, except on the palms of the hands and the soles of the feet. Their ducts usually empty into a hair follicle (see Figure 4.3), but some open directly onto the skin surface.

The product of the sebaceous glands, **sebum** (se′bum; *seb* = grease), is a mixture of oily substances and fragmented cells. Sebum is a lubricant that keeps the skin soft and moist and prevents the hair from becoming brittle. Sebum also contains chemicals that *kill* bacteria, so it is important in

preventing the bacteria present on the skin surface from invading the deeper skin regions. The sebaceous glands become very active when male sex hormones are produced in increased amounts (in both sexes) during adolescence. Thus, the skin tends to become oilier during this period of life.

If a sebaceous gland's duct becomes blocked by sebum, a *whitehead* appears on the skin surface. If the accumulated material oxidizes and dries, it darkens, forming a *blackhead. Acne* is an active infection of the sebaceous glands accompanied by "pimples" on the skin. It can be mild or extremely severe, leading to permanent scarring. *Seborrhea* (seb″o-re′ah), known as "cradle cap" in infants, is caused by overactivity of the sebaceous glands. It begins on the scalp as pink, raised lesions that gradually form a yellow to brown crust that sloughs off as oily dandruff. Careful washing to remove the excessive oil often helps. ❑

Sweat Glands   **Sweat glands,** also called **sudoriferous** (su″do-rif′er-us) **glands,** are widely distributed in the skin. Their number is staggering—more than 2.5 million per person. There are two types of sweat glands, *eccrine* and *apocrine.*

The **eccrine** (ek′rin) **glands** are far more numerous and are found all over the body. They produce **sweat,** a clear secretion that is primarily water plus some salts (sodium chloride), traces of metabolic wastes (urea, uric acid), and vitamin C. Sweat is acidic (pH from 4 to 6), a characteristic that inhibits the growth of bacteria, which are always present on the skin surface. Typically, sweat reaches the skin surface via a duct that opens externally as a funnel-shaped **pore** (see Figure 4.3). Notice, however, that the facial "pores" commonly referred to when we talk about our complexion are *not* these sweat pores, but the external outlets of hair follicles.

The eccrine sweat glands are an important and highly efficient part of the body's heat-regulating equipment. They are supplied with nerve endings that cause them to secrete sweat when the external temperature or body temperature is high. When sweat evaporates off the skin surface, it carries large amounts of body heat with it. On a hot day, it is possible to lose up to 7 liters of body water in this way. The heat-regulating functions of the body are important—if internal temperature changes more than a few degrees from the normal 98.2°F,

Figure 4.5   Structure of a hair and hair follicle. (**a**) Longitudinal section of a hair within its follicle. (**b**) Enlarged longitudinal section of a hair. (**c**) Enlarged longitudinal view of the expanded hair bulb in the follicle showing the matrix, the region of actively dividing epithelial cells that produces the hair.

life-threatening changes occur in the body. Body temperature regulation is discussed in more detail in Chapter 14.

Apocrine (ap′o-krin) **sweat glands** are largely confined to the axillary and genital areas of the body. They are usually larger than eccrine glands, and their ducts empty into hair follicles. Their secretion contains fatty acids and proteins, as well as all the substances present in eccrine secretion; consequently, it may have a milky or yellowish color. The secretion is odorless, but when bacteria that live on the skin use its proteins and fats as a source of nutrients for their growth, it takes on a musky, unpleasant odor.

Apocrine glands begin to function during puberty under the influence of androgens. Although their secretion is produced almost continuously, apocrine glands play a minimal role in thermoregulation. Their precise function is not yet known, but they are activated by nerve fibers during pain and stress.

Hairs and Hair Follicles

There are millions of **hairs** scattered all over the body. But, other than serving a few minor protective functions—such as guarding the head against bumps, shielding the eyes (via eyelashes), and helping to keep foreign particles out of the respiratory tract (via nose hairs)—our body hair has lost much of its usefulness. Hairs served early humans (and still serve hairy animals) by providing insulation in cold weather, but now we have other means of keeping warm.

A hair, produced by a *hair follicle,* is a flexible epithelial structure. That part of the hair enclosed in the follicle is called the **root.** The part projecting from the surface of the scalp or skin is called the **shaft** (see Figure 4.5). A hair is formed by division of the well-nourished germinal epithelial cells in the growth zone or *matrix* of the **hair bulb** at the inferior end of the follicle. As the daughter cells are pushed farther away from the growing region, they become keratinized and die. Thus the bulk of the hair shaft, like the bulk of the epidermis, is dead material and almost entirely protein.

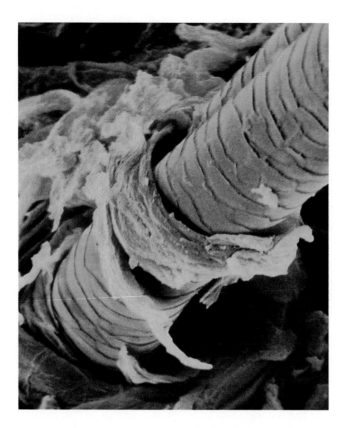

Figure 4.6   Scanning electron micrograph showing a hair shaft emerging from a follicle at the skin surface. Notice how the scalelike cells of the cuticle overlap one another (1500×).

Each hair consists of a central core called the *medulla* (me-dul′ah) surrounded by a bulky *cortex* layer. The cortex is, in turn, enclosed by an outermost *cuticle* formed by a single layer of cells that overlap one another like shingles on a roof. This arrangement of the cuticle cells helps to keep the hairs apart and keeps them from matting (see Figures 4.5b and 4.6). The cuticle is the most heavily keratinized region; it provides strength and helps keep the inner hair layers tightly compacted. Because it is most subject to abrasion, the cuticle tends to wear away at the tip of the shaft, allowing the keratin fibrils in the inner hair regions to frizz out, a phenomenon called "split ends." Hair pigment is made by melanocytes in the hair bulb and varying amounts of different types of melanin

(yellow, rust, brown, and black) combine to produce *all* varieties of hair color from pale blond to pitch black.

Hairs come in a variety of sizes and shapes. They are short and stiff in the eyebrows, long and flexible on the head, and usually nearly invisible almost everywhere else. When the hair shaft is oval, the person has wavy hair; when it is flat and ribbonlike, the hair is curly or kinky. If it is perfectly round, the hair is straight. Hairs are found all over the body surface except the palms of the hands, soles of the feet, nipples, and lips. Humans are born with as many hair follicles as they will ever have, and hairs are among the fastest growing tissues in the body. Hormones account for the development of "hairy" regions—the scalp and, in the adult, the pubic and axillary (armpit) areas.

Hair follicles are actually compound structures. The inner *sheath,* or layer, is composed of *epithelial tissue* and forms the hair. The outer sheath is actually dermal tissue. This dermal layer supplies blood vessels to the epidermal portion and reinforces it. Its nipplelike papilla provides the blood supply to the matrix, or growth region, in the hair bulb.

Look carefully at the structure of the hair follicle in Figure 4.3. Notice that it is slanted. Small bands of smooth muscle cells—**arrector pili** (ah-rek′tor pi′li)—connect each side of the hair follicle to the dermal tissue. When these muscles contract (as when we are cold or frightened), the hair is pulled upright, dimpling the skin surface with "goose bumps." This action helps keep animals warm in winter by adding a layer of insulating air to the fur. It is especially dramatic in a scared cat, whose fur actually stands on end to make it look larger to scare off its enemy. However, this "hair-raising" phenomenon is not very useful to human beings.

Nails

A **nail** is a scalelike modification of the epidermis that corresponds to the hoof or claw of other animals. Each nail has a *free edge,* a *body* (visible attached portion), and a *root* (embedded in the skin). The borders of the nail are overlapped by skin folds, called *nail folds.* The thick proximal nail fold is commonly called the *cuticle* (Figure 4.7).

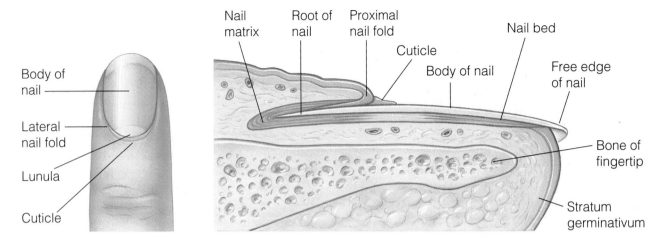

Figure 4.7 **Structure of a nail.** Surface view (left) and longitudinal section of the distal part of a finger (right), showing nail parts and the nail matrix that forms the nail.

The stratum germinativum of the epidermis extends beneath the nail as the *nail bed.* Its thickened proximal area, called the *nail matrix,* is responsible for nail growth. As the nail cells are produced by the matrix, they become heavily keratinized and die. Thus, nails, like hairs, are mostly nonliving material.

Nails are transparent and nearly colorless, but they look pink because of the rich blood supply in the underlying dermis. The exception to this is the region over the thickened nail matrix which appears as a white crescent and is called the *lunula* (loo'nyu-luh). As noted earlier, when the supply of oxygen in the blood is low, the nail beds take on a cyanotic (blue) cast.

Homeostatic Imbalances of Skin

It is difficult to scoff at anything that goes wrong with the skin because, when it rebels, it is quite a visible revolution. Loss of homeostasis in body cells and organs can reveal itself on the skin in ways that are sometimes almost unbelievable. The skin can develop more than 1000 different ailments. The most common skin disorders result from allergies or bacterial, viral, or fungal infections. Less common, but far more damaging, are burns and skin cancers. A number of the homeostatic imbalances of the skin are summarized briefly in the sections just below.

Infections and Allergies

- **Athlete's foot.** An itchy, red, peeling condition of the skin between the toes, resulting from fungus infection. Also called *tinea pedis.*

- **Boils and carbuncles** (kar'bun-kulz). Inflammation of hair follicles and sebaceous glands, common on the dorsal neck. Carbuncles are composite boils typically caused by bacterial infection (often *Staphylococcus aureus*).

- **Cold sores** (fever blisters). Small fluid-filled blisters that itch and sting, caused by a herpes simplex infection. The virus localizes in a cutaneous nerve, where it remains dormant until activated by emotional upset, fever, or UV radiation. Cold sores usually occur around the lips and in the oral mucosa of the mouth (see Figure 4.8a).

- **Contact dermatitis.** Itching, redness, and swelling of the skin, progressing to blistering. Caused by exposure of the skin to chemicals (e.g., those in poison ivy) that provoke allergic responses in sensitive individuals.

(a)   **(b)**   **(c)**

Figure 4.8   Cutaneous lesions. (a) Cold sores. (**b**) Impetigo. (**c**) Psoriasis.

- **Impetigo** (im-peh-ti′go; *impet* = an attack). Pink, water-filled, raised lesions (commonly around the mouth and nose) that develop a yellow crust and eventually rupture (Figure 4.8b). Caused by a highly contagious staphylococcus infection; common in elementary school-aged children.
- **Psoriasis** (so-ri′ah-sis). A chronic condition, characterized by reddened epidermal lesions covered with dry, silvery scales (Figure 4.8c). When severe, may be disfiguring. Its cause is unknown; may be hereditary in some cases. Attacks often triggered by trauma, infection, hormonal changes, and stress.

Burns

The skin is only about as thick as a paper towel—not too impressive as organ systems go. And yet, when it is severely damaged, nearly every body system suffers. Metabolism accelerates or may be impaired, changes in the immune system occur, and the cardiovascular system may falter. Such severe damage can be caused by **burns.** A burn is tissue damage and cell death caused by intense heat, electricity, UV radiation (sunburn), or certain chemicals (such as acids).

There are few threats to skin more serious than burns. When the skin is burned and its cells are destroyed, two life-threatening problems result. First, the body loses its precious supply of fluids containing proteins and electrolytes as these seep from the burned surfaces. Dehydration and electrolyte imbalance follow and can lead to a shutdown of the kidneys and *circulatory shock* (inadequate circulation of blood caused by low blood volume). To save the patient, the lost fluids must be replaced immediately.

The volume of fluid lost can be estimated indirectly by determining how much of the body surface is burned (extent of burns), using the **rule of nines.** This method divides the body into 11 areas, each accounting for 9 percent of the total body surface area, plus an additional area surrounding the genitals (the perineum) representing 1 percent of body surface area (see Figure 4.9). Later, infection becomes the most important threat and is the leading cause of death in burn victims. Burned skin is sterile for about 24 hours. But after that, pathogens (path′o-jenz) such as bacteria and fungi easily invade areas where the skin has been destroyed and multiply rapidly in the nutrient-rich environment of dead tissues. To make matters worse, the patient's immune system becomes depressed within one to two days after severe burn injury.

Burns are classified according to their severity (depth) as first-, second-, or third-degree burns. In **first-degree burns,** only the epidermis is damaged. The area becomes red and swollen. Except for temporary discomfort, first-degree burns are not usually serious and generally heal in two to three days without any special attention. Sunburn is usually a first-degree burn. **Second-degree burns** involve injury to the epidermis and the upper region of the dermis. The skin is red and

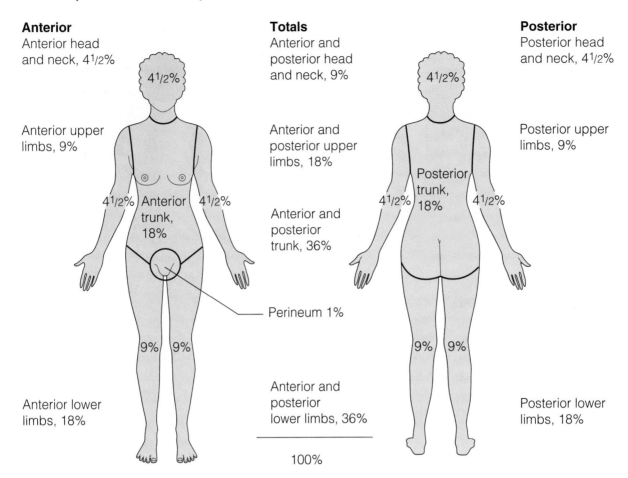

Anterior
Anterior head
and neck, 4$\frac{1}{2}$%

Anterior upper
limbs, 9%

4$\frac{1}{2}$% Anterior
trunk,
18% 4$\frac{1}{2}$%

Anterior lower
limbs, 18%

Totals
Anterior and
posterior head
and neck, 9%

Anterior and
posterior upper
limbs, 18%

Anterior and
posterior
trunk, 36%

Perineum 1%

Anterior and
posterior
lower limbs, 36%

100%

Posterior
Posterior head
and neck, 4$\frac{1}{2}$%

Posterior upper
limbs, 9%

Posterior
trunk,
18%

Posterior lower
limbs, 18%

Figure 4.9   Estimating the extent of burns by using the rule of nines.

painful and *blisters* appear. Since sufficient numbers of epithelial cells are still present, regrowth (regeneration) of the epithelium can occur. Ordinarily, no permanent scars result if care is taken to prevent infection. First- and second-degree burns are referred to as **partial-thickness burns.**

Third-degree burns destroy the entire thickness of the skin; these burns are also called **full-thickness burns.** The burned area appears blanched (gray-white) or blackened, and since the nerve endings in the area are destroyed, the burned area is not painful. In third-degree burns, regeneration is not possible, and skin grafting must be done to cover the underlying exposed tissues.

In general, burns are considered *critical* if any of the following conditions exists:

1. Over 25 percent of the body has second-degree burns,

2. Over 10 percent of the body has third-degree burns, or

3. There are third-degree burns of the face, hands, or feet.

Facial burns are dangerous because of the possibility of burned respiratory passageways, which can swell and cause suffocation. Joint injuries are troublesome because the scar tissue that eventually forms can severely limit joint mobility.

Skin Cancer

Numerous types of tumors arise in the skin. Most skin tumors are benign and do not spread (metastasize) to other body areas. (A wart, a neoplasm caused by a virus, is one such example.) However, some skin tumors are malignant, or cancerous, and they tend to invade other body areas. Indeed, skin

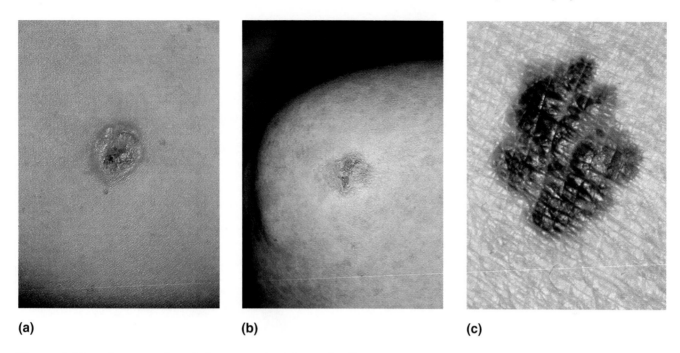

(a)   (b)   (c)

Figure 4.10 **Photographs of skin cancers.** (**a**) Basal cell carcinoma. (**b**) Squamous cell carcinoma. (**c**) Malignant melanoma typically begins as a small, irregular brown lesion.

cancer is the single most common type of cancer in humans. The cause of most skin cancers is not known, but the most important risk factor is over-exposure to ultraviolet radiation in sunlight. Frequent irritation of the skin by infections, chemicals, or physical trauma also seems to be a predisposing factor.

Basal Cell Carcinoma **Basal cell carcinoma** (kar″sĭ-no′mah) is the least malignant and most common skin cancer. Cells of the stratum germinativum, altered so that they cannot form keratin, no longer honor the boundary between epidermis and dermis. They proliferate, invading the dermis and subcutaneous tissue. The cancer lesions occur most often on sun-exposed areas of the face and appear as shiny, dome-shaped nodules (Figure 4.10a) that later develop a central ulcer with a "pearly" beaded edge. Basal cell carcinoma is relatively slow-growing, and metastasis seldom occurs before it is noticed. Full cure is the rule in 99 percent of cases where the lesion is removed surgically.

Squamous Cell Carcinoma **Squamous cell carcinoma** arises from the cells of the stratum spinosum. The lesion appears as a scaly reddened papule (small, rounded elevation) that gradually forms a shallow ulcer with a firm raised border. This variety of skin cancer appears most often on the scalp, ears, dorsum of the hands, and lower lip (Figure 4.10b). It grows rapidly and metastasizes to adjacent lymph nodes if not removed. This epidermal cancer is also believed to be sun-induced. If it is caught early and removed surgically or by radiation therapy, the chance of complete cure is good.

Malignant Melanoma **Malignant melanoma** (mel″ah-no′mah) is a cancer of melanocytes. It accounts for only about 5 percent of skin cancers, but its incidence is increasing rapidly and it is often deadly. Melanoma can begin wherever there is pigment; most such cancers appear spontaneously, but some develop from pigmented moles. It usually appears as a spreading brown to black patch (Figure 4.10c) that metastasizes rapidly to surrounding lymph and blood vessels. The chance for

A CLOSER LOOK

VAMPIRES AND TOUCH-ME-NOTS

Stories of vampires and werewolves are part of the folklore of many countries. Although many aspects of legends about vampires derive from the "behavior" of bloating corpses that shift in their graves (the "undead"), other aspects may be based on living people with rare skin conditions. One likely condition is **porphyria** (por-fēr′e-ah; "purple"), an inherited disease. People with this condition cannot make the iron-containing heme part of hemoglobin, the molecule of red blood cells that carries oxygen. Unfortunate victims who were once shunned as vampires or werewolves may actually have had this disease, which affects about 1 of every 25,000 people.

The cause of porphyria is a deficiency in some enzymes that catalyze certain reactions needed to form heme. Without these enzymes, the intermediate products of this pathway build up, spill into the circulation, and eventually cause lesions throughout the body—especially when exposed to rays of sunlight. These intermediate products are called **porphyrins,** after which porphyria is named. The porphyrias are actually a group of diseases characterized by five Ps: (1) *puberty* (the time that symptoms typically appear), (2) *psychiatric* abnormalities, (3) *pain,* (4) *polyneuropathy* (different neurological defects), and (5) *photosensitivity* (sensitivity to light) in many cases. Symptoms tend to come and go and are aggravated by exposure to many chemicals, including some in garlic.

Sunlight creates all kinds of nasty damage in some porphyria victims (a reason, perhaps, why vampires were said to hide in dark basements and coffins during daylight hours). When exposed to sunlight, the skin becomes lesioned and scarred, and the fingers, toes, and nose are often mutilated. The teeth grow prominent as the gums degenerate (the basis of large vampire "fangs"?). Rampant growth of hair causes the sufferer's face to become wolflike and the hands to resemble paws. One treatment for porphyria is to inject heme molecules extracted from healthy red blood cells. Heme injections were not available in the Middle Ages, so (perhaps) the next best thing would be to drink blood, as vampires were said to do. The claim that garlic keeps vampires away may stem from the fact that garlic severely aggravates porphyria symptoms.

Epidermolysis bullosa (literally "epidermal breakdown with blisters"), or **EB,** is a group of hereditary disorders marked by an inadequate or faulty synthesis of a specific kind of keratin found only in the basal cells of the epidermis. All of these disorders cause a lack of *cohesion* between the layers of the skin and of mucous membranes. As a result, even a simple touch can cause the layers in these membranes to separate and blister. For this reason, people suffering from EB have been called *touch-me-nots.*

As might be expected when a bodywide structural protein like collagen is involved, EB produces many symptoms. In some cases, the skin blisters and erosions are limited to the feet and hands. The ends of the digits may become fused by scar tissue, causing the hands or feet to resemble cocoons. In other cases, the mucous membranes of the digestive, respiratory, and urinary organs are affected too. In the most severe cases, widespread and fatal blistering occurs in major vital organs.

The main problem with EB occurs when the blisters break, giving pathogens (bacteria, for example) an easy route to enter the deeper body tissues. Therefore, touch-me-nots suffer frequent infections. Since there is no known cure for EB, treatments are aimed at relieving the symptoms and preventing infection with antibiotic ointments, bandages, and proper nutrition.

survival is helped by early detection. The American Cancer Society suggests that sun worshippers periodically examine their skin for new moles or pigmented spots, and apply the **ABCD rule** for recognizing melanoma. **(A) Asymmetry:** the two sides of the pigmented spot or mole do not match. **(B) Border irregularity:** the borders of the lesion are not smooth but exhibit indentations. **(C) Color:** the pigmented spot contains areas of different colors (blacks, browns, tans, and sometimes blues and reds). **(D) Diameter:** the spot is larger than 6 mm in diameter (the size of a pencil eraser). The usual therapy for malignant melanoma is wide surgical excision accompanied by chemotherapy. ❑

DEVELOPMENTAL ASPECTS OF SKIN AND BODY MEMBRANES

During the fifth and sixth months of fetal development, the soon-to-be-born infant is covered with a downy type of hair called *lanugo* (lah-noo′go), but this hairy cloak has usually been shed by birth. When a baby is born, its skin is covered with *vernix caseosa* (ver′niks kah-se-o′sah). This white, cheesy-looking substance, produced by the sebaceous glands, protects the baby's skin while it is floating in its water-filled sac inside the mother. The newborn's skin is very thin, and blood vessels can easily be seen through it. Commonly, there are accumulations in the sebaceous glands, which appear as small white spots called *milia* (mil′e-ah), on the baby's nose and forehead. These normally disappear by the third week after birth. As the baby grows, its skin becomes thicker and moist, and more subcutaneous fat is deposited.

During adolescence, the skin and hair become more oily as sebaceous glands are activated, and acne may appear. Acne usually subsides in early adulthood, and the skin reaches its optimal appearance when we are in our 20s and 30s. Then visible changes in the skin begin to appear as it is continually assaulted by abrasion, chemicals, wind, sun, and other irritants; and as its pores become clogged with air pollutants and bacteria. As a result, pimples, scales, and various kinds of *dermatitis* (der″mah-ti′tis), or skin inflammation, become more common.

During old age, the amount of subcutaneous tissue decreases, leading to the intolerance to cold so common in the elderly. The skin also becomes drier (because of decreased oil production and declining numbers of collagen fibers) and, as a result, it may become itchy and bothersome. Thinning of the skin, another result of the aging process, makes it more susceptible to bruising and other types of injuries. The decreasing elasticity of the skin, along with the loss of subcutaneous fat, allows bags to form under our eyes, and our jowls begin to sag. This loss of elasticity is speeded up by sunlight, so one of the best things you can do for your skin is to shield it from the sun by wearing sunscreens and protective clothing. In doing so, you will also be decreasing the chance of skin cancer. Although there is no way to avoid the aging of the skin, good nutrition, plenty of fluids, and cleanliness help delay the process.

Hair loses its luster as we age, and by the age of 50, the number of hair follicles has dropped by one-third and continues to decline, resulting in hair thinning and some degree of baldness or *alopecia* (al″o-pe′she-ah) in most people. Many men become obviously bald as they age, a phenomenon called *male pattern baldness.* A bald man is not really hairless—he does have hairs in the bald area. But because those hair follicles have begun to degenerate, the hairs are colorless and very tiny (and may not even emerge from the follicle). Such hairs are called *vellus* (*vell* = wool) hairs. Another phenomenon of aging is graying hair. Like balding, this is usually genetically controlled by a "delayed-action" gene. Once the gene becomes effective, the amount of melanin deposited in the hair decreases or becomes entirely absent, which results in gray-to-white hair.

Certain events can cause hair to gray or fall out prematurely. For example, many people have claimed that they turned gray nearly overnight because of some emotional crisis in their life. In addition, we know that anxiety, protein-deficient diets, therapy with certain chemicals (chemotherapy), radiation, excessive vitamin A, and certain fungal diseases (ringworm) can cause both graying and hair loss. However, when the cause of these conditions is not genetic, hair loss is usually not permanent. ❑

SYSTEMS IN SYNC

Homeostatic Relationships Between the Integumentary System and Other Body Systems

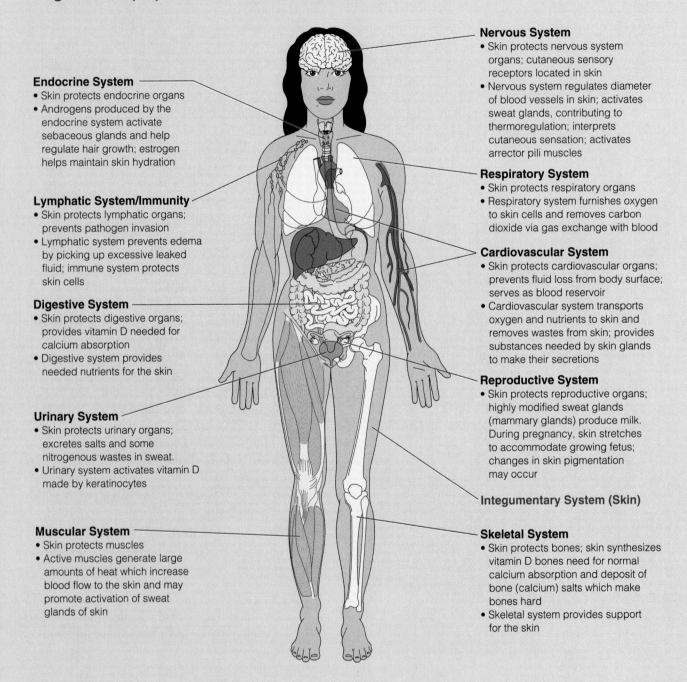

Nervous System
- Skin protects nervous system organs; cutaneous sensory receptors located in skin
- Nervous system regulates diameter of blood vessels in skin; activates sweat glands, contributing to thermoregulation; interprets cutaneous sensation; activates arrector pili muscles

Respiratory System
- Skin protects respiratory organs
- Respiratory system furnishes oxygen to skin cells and removes carbon dioxide via gas exchange with blood

Cardiovascular System
- Skin protects cardiovascular organs; prevents fluid loss from body surface; serves as blood reservoir
- Cardiovascular system transports oxygen and nutrients to skin and removes wastes from skin; provides substances needed by skin glands to make their secretions

Reproductive System
- Skin protects reproductive organs; highly modified sweat glands (mammary glands) produce milk. During pregnancy, skin stretches to accommodate growing fetus; changes in skin pigmentation may occur

Integumentary System (Skin)

Skeletal System
- Skin protects bones; skin synthesizes vitamin D bones need for normal calcium absorption and deposit of bone (calcium) salts which make bones hard
- Skeletal system provides support for the skin

Endocrine System
- Skin protects endocrine organs
- Androgens produced by the endocrine system activate sebaceous glands and help regulate hair growth; estrogen helps maintain skin hydration

Lymphatic System/Immunity
- Skin protects lymphatic organs; prevents pathogen invasion
- Lymphatic system prevents edema by picking up excessive leaked fluid; immune system protects skin cells

Digestive System
- Skin protects digestive organs; provides vitamin D needed for calcium absorption
- Digestive system provides needed nutrients for the skin

Urinary System
- Skin protects urinary organs; excretes salts and some nitrogenous wastes in sweat.
- Urinary system activates vitamin D made by keratinocytes

Muscular System
- Skin protects muscles
- Active muscles generate large amounts of heat which increase blood flow to the skin and may promote activation of sweat glands of skin

Important Terms

burns

cutaneous membrane

dermis

epidermis (ep″ĭ-der′mis)

exocrine (ek′so-krin) **glands**

follicle

integument (in-teg′u-ment)

keratinizing (ker′ah-tin-īz-ing)

melanin (mel′ah-nin)

mucous membrane

nails

sebaceous (seh-ba′shus) **glands**

serous membrane

subcutaneous tissue

sweat glands

synovial (sĭ-no′ve-al) **membrane**

Summary

Classification of Body Membranes (pp. 94–96)

1. Epithelial: Simple organs, epithelium and connective tissue components.
   a. Cutaneous (the skin): epidermis (stratified squamous epithelium) underlain by the dermis (dense connective tissue); protects body surface.
   b. Mucous: epithelial sheet underlain by a lamina propria (areolar connective tissue); lines body cavities open to the exterior.
   c. Serous: simple squamous epithelium resting on a scant connective tissue layer; lines the ventral body cavity.
2. Connective tissue: Synovial; lines joint cavities.

Integumentary System (Skin) (pp. 96–109)

1. Skin functions include protection of the deeper tissue from chemicals, bacteria, bumps, and drying; regulation of body temperature through radiation and sweating; and synthesis of defensive proteins and vitamin D. The cutaneous sensory receptors are located in the skin.
2. The epidermis, the more superficial part of the skin, is formed of stratified squamous keratinizing epithelium, and is avascular. Cells at its surface are dead and continually flake off. They are replaced by division of cells in the basal cell layer. As the cells move away from the basal layer, they accumulate keratin and die. Melanin, a pigment produced by melano-cytes, protects the nuclei of epithelial cells from the damaging rays of the sun.

3. The dermis is composed of dense connective tissue. It is the site of blood vessels, nerves, and epidermal appendages. It has two regions, the papillary and reticular layers. The papillary layer has ridges, which produce fingerprints.

4. Skin appendages are formed from the epidermis, but reside in the dermis.
   a. Sebaceous glands produce an oily product (sebum), usually ducted into a hair follicle. Sebum keeps the skin and hair soft, and contains bacteria-killing chemicals.
   b. Sweat (sudoriferous) glands, under the control of the nervous system, produce sweat, which is ducted to the epithelial surface. These glands are part of the body's heat-regulating apparatus. There are two types: eccrine (the most numerous) and apocrine (their product includes protein, which skin bacteria metabolize).
   c. A hair is primarily dead keratinized cells and is produced by the hair bulb. The root is enclosed in a sheath, the hair follicle.
   d. Nails are hornlike derivatives of the epidermis. Like hair, nails are primarily dead keratinized cells.

5. Most minor afflictions of the skin result from infections or allergic responses; more serious are burns and skin cancer. Because they interfere with skin's protective functions, burns represent a major threat to the body.

a. Burns result in loss of body fluids and invasion of bacteria. The extent of burns is assessed by the "rule of nines." The severity (depth) of burns is described as first-degree (epidermal damage only), second-degree (epidermal and some dermal injury), and third-degree (epidermis and dermis totally destroyed). Third-degree burns require skin grafting.

b. The most common cause of skin cancer is exposure to ultraviolet radiation. Cure of basal cell carcinoma and squamous cell carcinoma is complete if they are removed before metastasis. Malignant melanoma, a cancer of melanocytes, is still fairly rare, but is almost always fatal.

Developmental Aspects of the Skin and Body Membranes (p. 109)

1. The skin is thick, resilient, and well-hydrated in youth but loses its elasticity and thins as aging occurs. Skin cancer is a major threat to skin exposed to excessive sunlight.

2. Balding and/or graying occurs with aging. Both are genetically determined, but other factors (drugs, emotional stress, and so on) can result in either.

Review Questions

1. How does a mucosa differ from a serosa?

2. What is the name of the connective tissue membrane found lining joint cavities?

3. What primary tissues are destroyed when the skin is damaged?

4. From what types of damage does the skin protect the body?

5. Explain why we become tanned after sitting in the sun.

6. What is a decubitus ulcer? Why does it occur?

7. Name two different categories of skin secretions and the glands that manufacture them.

8. How does the skin help to regulate body temperature?

9. What is a blackhead?

10. What are arrector pili? What do they do?

11. What are the life-threatening consequences of severe burns?

12. Distinguish between first-, second-, and third-degree burns.

13. Why does hair turn gray?

14. Name three changes that occur in the skin as one ages.

15. Is a bald man really hairless? Explain.

At the Clinic

1. A nurse tells a doctor that a patient is *cyanotic*. What is cyanosis? What does its presence indicate?

2. Both newborn infants and aged individuals have very little subcutaneous tissue. How does this affect their sensitivity to cold environmental temperature?

3. A 40-year-old beachboy is complaining to you that his suntan made him popular when he was young, but now his face is all wrinkled, and he has several darkly pigmented moles that are growing rapidly and are as big as large coins. He shows you the moles, and immediately you think "ABCD." What does that mean, and why should he be concerned?

4. Martha, the mother of a 13-month-old infant, brings her child to the clinic because his skin has turned orange. Why does the pediatrician inquire about the child's diet?

5. If the water of a swimming pool is hypotonic to our cells and body fluids (and it is), then why do we not swell and pop when we go for a swim?

5

The Skeletal System

YOUR GOALS

After completing this chapter, you will have a working knowledge of the functions of the skeletal system and will have mastered the objectives listed below.

FUNCTION PREVIEW

■ The skeletal system provides an internal framework for the body, protects organs by enclosure, and anchors skeletal muscles so that muscle contraction can cause movement.

OBJECTIVE CHECKLIST

BONES: AN OVERVIEW (pp. 114–121)

❏ Identify the subdivisions of the skeleton as axial or appendicular.

❏ List at least three functions of the skeletal system.

❏ Name the four main kinds of bones.

❏ Identify the major anatomical areas of a long bone.

❏ Explain the role of bone salts and the organic matrix in making bone both hard and flexible.

❏ Describe briefly the process of bone formation in the fetus and summarize the events of bone remodeling throughout life.

❏ Name and describe the various types of fractures.

AXIAL SKELETON (pp. 121–133)

❏ On a skull or diagram, identify and name the bones of the skull.

❏ Describe how the skull of a newborn infant (or fetus) differs from that of an adult, and explain the function of fontanels.

❏ Name the parts of a typical vertebra and explain in general how the cervical, thoracic, and lumbar vertebrae differ from one another.

❏ Discuss the importance of the intervertebral disks and spinal curvatures.

❏ Explain how the abnormal spinal curvatures (scoliosis, lordosis, and kyphosis) differ from one another.

APPENDICULAR SKELETON (pp. 133–142)

❏ Identify on a skeleton or diagram the bones of the shoulder and pelvic girdles and their attached limbs.

❏ Describe important differences between a male and female pelvis.

(continued)

Although the word *skeleton* comes from the Greek word meaning "dried-up body," our internal framework is so beautifully designed and engineered that it puts any modern skyscraper to shame. Strong, yet light, it is perfectly adapted for its functions of body protection and motion. Shaped by an event that happened more than one million years ago—when a being first stood erect on hind legs—our skeleton is a tower of bones arranged so that we can stand upright and balance ourselves. No other animal has such relatively long legs (compared to the arms or forelimbs) or such a strange foot, and few have such remarkable grasping hands. Even though the infant's backbone is like an arch, it soon changes to the swayback or S-shaped structure that is required for the upright posture.

The skeleton is subdivided into two divisions: the **axial skeleton,** the bones that form the longitudinal axis of the body, and the **appendicular skeleton,** the bones of the limbs and girdles. In addition to bones, the **skeletal system** includes *joints, cartilages,* and *ligaments* (fibrous cords that bind the bones together at joints). The joints give the body flexibility and allow movement to occur.

BONES: AN OVERVIEW

At one time or another, all of us have heard the expressions "bone tired," "dry as a bone," or "bag of bones"—pretty unflattering and inaccurate images of some of our most phenomenal organs. Our brains, not our bones, convey feelings of fatigue, and bones are far from dry. As for "bag of bones," they are indeed more obvious in some of us, but without them to form our internal skeleton, we would creep along the ground like slugs. Let's examine how our bones contribute to overall body homeostasis.

Functions of the Bones

Besides contributing to body shape and form, our bones perform several important body functions:

1. **Support.** Bones, the "steel girders" and "reinforced concrete" of the body, form the internal framework that supports and anchors all soft organs. The bones of the legs act as pillars to support the body trunk when we stand, and the rib cage supports the thoracic wall.

2. **Protection.** Bones protect soft body organs. For example, the fused bones of the skull provide a snug enclosure for the brain. The vertebrae surround the spinal cord, and the rib cage helps protect the vital organs of the thorax.

3. **Movement.** Skeletal muscles, attached to bones by tendons, use the bones as levers to move the body and its parts. As a result, we can walk, swim, throw a ball, and breathe. Before continuing, take a moment to imagine that your bones have turned to putty. What if you were running when this change took place? Now imagine your bones forming a rigid metal framework inside your body, somewhat like a system of plumbing pipes. What problems could you envision with this arrangement? These images should help you understand how well our skeletal system provides support and protection while allowing movement.

4. **Storage.** Fat is stored in the internal cavities of bones. Bone itself serves as a storehouse for minerals, the most important being calcium and phosphorus, although others are also

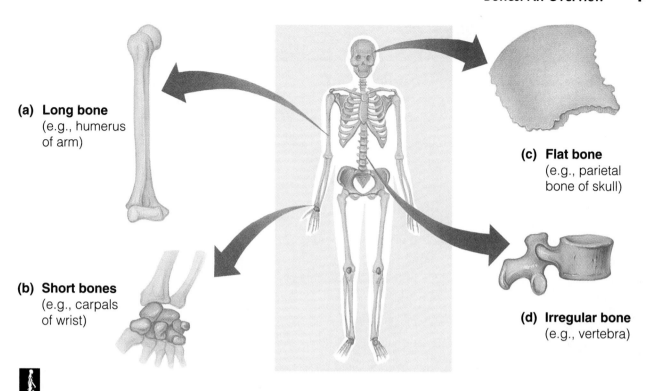

(a) Long bone
(e.g., humerus
of arm)

(c) Flat bone
(e.g., parietal
bone of skull)

(b) Short bones
(e.g., carpals
of wrist)

(d) Irregular bone
(e.g., vertebra)

Figure 5.1   Classification of bones on the basis of shape.

stored. A small amount of calcium in its ion form (Ca^{2+}) must be present in the blood at all times for the nervous system to transmit messages, for muscles to contract, and for blood to clot. Because most of the body's calcium is deposited in the bones as calcium salts, the bones are a convenient place to get more calcium ions for the blood as they are used up. Problems occur not only when there is too little calcium in the blood, but also when there is too much. Hormones control the movement of calcium to and from the bones and blood according to the needs of the body. Indeed, "deposits" and "withdrawals" of calcium (and other minerals) to and from bones go on almost all the time.

5. **Blood cell formation.** Blood cell formation, or hematopoiesis (hem″ah-to-poi-e′sis), occurs within the marrow cavities of certain bones.

Classification of Bones

The adult skeleton is composed of 206 bones. There are two basic types of osseous, or bone, tissue: **Compact bone** is dense and looks smooth and homogeneous. **Spongy bone** is composed of small needlelike pieces of bone and lots of open space.

Bones come in many sizes and shapes. For example, the tiny pisiform bone of the wrist is the size and shape of a pea, whereas the femur, or thigh bone, is nearly 2 feet long and has a large ball-shaped head. The unique shape of each bone fulfills a particular need. Bones are classified according to shape into four groups: long, short, flat, and irregular (see Figure 5.1).

As their name suggests, **long bones** are typically longer than they are wide. They have a shaft with heads at both ends. Long bones are mostly compact bone. All the bones of the limbs, except the wrist and ankle bones, are long bones.

Short bones are generally cube-shaped, and contain mostly spongy bone. The bones of the wrist and ankle are short bones.

Flat bones are thin, flattened, and usually curved. They have two thin layers of compact bone sandwiching a layer of spongy bone between them. Most bones of the skull, the ribs, and the sternum (breastbone) are flat bones.

Bones that do not fit one of the preceding categories are called **irregular bones.** The vertebrae,

Table 5.1 Bone Markings

Name of bone marking	Description	Illustration
Projections that are sites of muscle attachment		
Tuberosity	Large rounded projection; may be roughened	
Crest	Narrow ridge of bone; usually prominent	
Trochanter (tro-kan'ter)	Very large, blunt, irregularly shaped process (The only examples are on the femur.)	
Line	Narrow ridge of bone; less prominent than a crest	
Tubercle (tu'ber-kl)	Small rounded projection or process	
Epicondyle	Raised area on or above a condyle	
Spine	Sharp, slender, often pointed projection	
Projections that help to form joints		
Head	Bony expansion carried on a narrow neck	
Facet	Smooth, nearly flat articular surface	
Condyle (kon'dīl)	Rounded articular projection	
Ramus	Armlike bar of bone	
Depressions and openings allowing blood vessels and nerves to pass		
Meatus (me-a'tus)	Canal-like passageway	
Sinus	Cavity within a bone, filled with air and lined with mucous membrane	
Fossa (fos'ah)	Shallow, basinlike depression in a bone, often serving as an articular surface	
Groove	Furrow	
Fissure	Narrow, slitlike opening	
Foramen (fo-ra'men)	Round or oval opening through a bone	

Illustration labels: Line, Tuberosity, Crest, **Tibia of leg**, Trochanters, **Vertebra**, Spine, Tubercle, Epicondyle, **Femur of thigh**; Head, Facet, **Rib**, Condyle, Ramus, **Mandible**; Meatus, Sinus, Fossa, Groove, Fissure, Foramen, **Skull**

which make up the spinal column, and the hip bones fall into this group.

Even when looking casually at bones, one can see that their surfaces are not smooth but scarred with bumps, holes, and ridges. These **bone markings,** described and illustrated in Table 5.1, reveal where muscles, tendons, and ligaments were at-tached, and where blood vessels and nerves passed. There are two categories of bone markings: (a) *projections,* or *processes,* which grow out from the bone surface, and (b) *depressions,* or *cavities,* which are indentations in the bone. These terms do not have to be learned now, but they can help you remember some of the specific markings

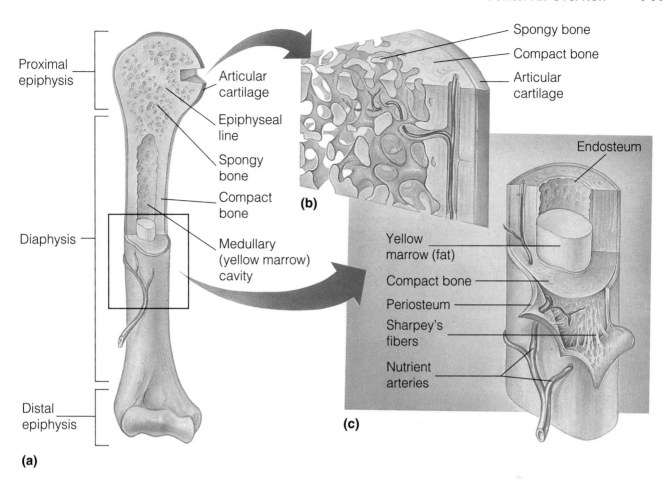

Figure 5.2   The structure of a long bone (humerus). (**a**) Anterior view with longitudinal section cut away at the proximal end. (**b**) Pie-shaped, three-dimensional view of spongy bone and compact bone of the epiphysis. (**c**) Cross section of the shaft (diaphysis). Note that the external surface of the diaphysis is covered by a periosteum, but the articular surface of the epiphysis is covered with hyaline cartilage.

on bones to which you will be introduced later in this chapter.

There is a little trick for remembering the bone markings listed in the table: All the terms beginning with *T* are projections. All the terms beginning with *F* (except *facet*) are depressions.

Structure of a Long Bone

Gross Anatomy

The gross structure of a long bone is shown in Figure 5.2. The **diaphysis** (di-af′ĭ-sis), or shaft, makes up most of the bone's length and is composed of compact bone. The diaphysis is covered and protected by a fibrous connective tissue membrane,

the **periosteum** (per-e-ŏs′te-um). Hundreds of connective tissue fibers, called *Sharpey's fibers,* secure the periosteum to the underlying bone. The **epiphyses** (ĕ-pif′ĭ-sēz) are the ends of the long bone. Each epiphysis consists of a thin layer of compact bone enclosing an area filled with spongy bone. **Articular cartilage,** instead of a periosteum, covers its external surface. Because the articular cartilage is glassy hyaline cartilage, it provides a smooth, slippery surface that decreases friction at joint surfaces.

In adult bones, there is a thin line of bony tissue spanning the epiphysis that looks a bit different from the rest of the bone in that area. This is the **epiphyseal line.** The epiphyseal line is a remnant of the **epiphyseal plate** (a flat plate of hyaline cartilage) seen in a young, growing bone. Epi-

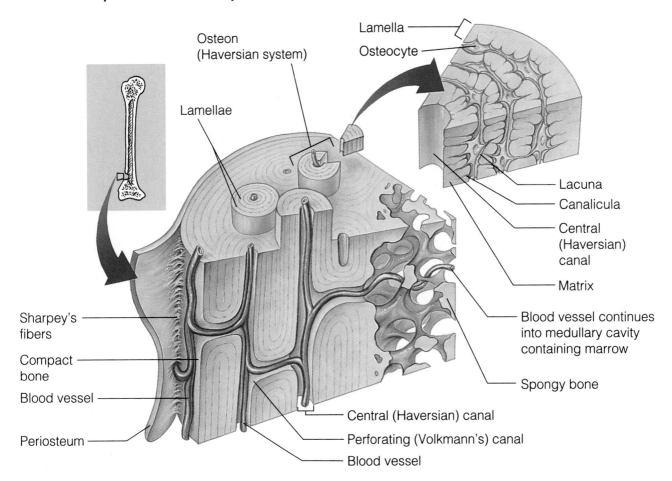

Figure 5.3 **Microscopic structure of compact bone.** Diagram of a pie-shaped segment of compact bone. (The inset shows a more highly magnified view.) Notice the position of osteocytes in lacunae (cavities in the matrix).

physeal plates cause the lengthwise growth of a long bone. By the end of puberty, when hormones stop long bone growth, epiphyseal plates have been completely replaced by bone, leaving only the epiphyseal lines to mark their previous location.

In adults the cavity of the shaft is primarily a storage area for adipose (fat) tissue. It is called the **yellow marrow,** or **medullary, cavity.** However, in infants this area forms blood cells, and **red marrow** is found there. In adult bones, red marrow is confined to the cavities of spongy bone of flat bones and the epiphyses of some long bones.

Microscopic Anatomy

To the naked eye, spongy bone has a spiky, open appearance, whereas compact bone appears to be very dense. Looking at compact bone tissue through a microscope, however, one can see that it has a complex structure (see Figure 5.3). It is rid-

dled with passageways carrying nerves, blood vessels, and the like, which provide the living bone cells with nutrients and a route for waste disposal. The mature bone cells, **osteocytes** (os′te-o-sītz″), are found in tiny cavities within the matrix called **lacunae** (lah-ku′ne). The lacunae are arranged in concentric circles called **lamellae** (lah-mel′e) around **central (Haversian) canals.** Each complex consisting of central canal and matrix rings is called an **osteon** or **Haversian system.** Central canals run lengthwise through the bony matrix, carrying blood vessels and nerves to all areas of the bone. Tiny canals, **canaliculi** (kan″ah-lik′u-li), radiate outward from the central canals to all lacunae. The canaliculi form a transportation system that connects all the bone cells to the nutrient supply through the hard bone matrix. Because of this elaborate network of canals, bone cells are well nourished in spite of the hardness of the matrix, and bone injuries heal quickly and well. The communication pathway from the outside of the bone to its interior (and the central canals) is completed by **perforating (Volkmann's) canals,** which run into the compact bone at right angles to the shaft.

Bone is one of the hardest materials in the body and, although relatively light in weight, it has a remarkable ability to resist tension and other forces acting on it. Nature has given us an extremely strong and exceptionally simple (almost crude), supporting system without giving up mobility. The calcium salts deposited in the matrix give bone its hardness, whereas the organic parts (especially the collagen fibers) provide for bone's flexibility and great tensile strength.

Bone Formation, Growth, and Remodeling

The skeleton is formed from two of the strongest and most supportive tissues in the body—cartilage and bone. In embryos, the skeleton is primarily made of hyaline cartilage, but in the young child most of the cartilage has been replaced by bone. Cartilage remains only in isolated areas such as the bridge of the nose, parts of the ribs, and the joints.

Except for flat bones, which form on fibrous membranes, most bones develop using hyaline cartilage structures as their "models." Most simply, this process of bone formation, or **ossification** (os″ĭ-fĭ-ka′shun), involves two major steps. First, the hyaline cartilage model is completely covered with bone matrix by bone-forming cells called **osteoblasts;** so, for a short period, the fetus has cartilage "bones" enclosed by "bony" bones. Then, the enclosed hyaline cartilage model is digested away, opening up a medullary cavity within the newly formed bone.

By birth or shortly after, most hyaline cartilage models have been converted to bone except for two regions—the *articular cartilages* that cover the bone ends, and the *epiphyseal plates*. The articular cartilages persist for life, reducing friction at the joint surfaces. The epiphyseal plates provide for longitudinal growth of the long bones during childhood. "New" cartilage is formed continuously on the external face of the articular cartilage and on the epiphyseal plate surface that is farther away from the medullary cavity. At the same time, the "old" cartilage abutting the internal face of the articular cartilage and the medullary cavity is broken down and replaced by bony matrix (Figure 5.4).

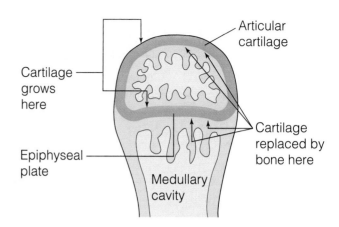

Figure 5.4   Long bone growth. "New" cartilage is formed on the external face of the articular cartilage and epiphyseal plate, as the "old" cartilage nearer the medullary cavity is broken down and replaced by bone.

This process of long-bone growth is controlled by hormones, most importantly *growth hormone* and, during puberty, the sex hormones. It ends during adolescence, when the epiphyseal plates are completely converted to bone.

Many people mistakenly think that bones are lifeless structures that never change once long-bone growth has ended. Nothing could be further from the truth; bone is a dynamic and active tissue. Bones are remodeled continually in response to changes in two factors: (1) calcium levels in the blood, and (2) the pull of gravity and muscles on the skeleton. How these factors influence bones is outlined next.

When blood calcium levels drop below homeostatic levels, the parathyroid glands (located in the throat) are stimulated to release parathyroid hormone (PTH) into the blood. PTH activates **osteoclasts,** giant bone-destroying cells in bones, to break down bone matrix and release calcium ions into the blood. On the other hand, when blood calcium levels are too high (*hypercalcemia* [hi″per-kal-se′me-ah]), calcium is deposited in bone matrix as hard calcium salts.

Bone remodeling is essential if bones are to retain normal proportions and strength during long-bone growth as the body increases in size and weight. It also accounts for the fact that bones become thicker and form large projections to increase their strength in areas where bulky muscles are attached. At such sites, osteoblasts lay down

Table 5.2 Common Types of Fractures

Fracture Type	Illustration	Description	Comment
Simple		Bone breaks cleanly, but does not penetrate the skin.	Sometimes called a "closed fracture."
Compound		Broken ends of the bone protrude through soft tissues and the skin.	An open fracture. More serious than a simple fracture; may result in a severe bone infection (osteomyelitis), requiring massive doses of antibiotics.
Comminuted		Bone breaks into many fragments.	Particularly common in the aged, whose bones are more brittle.
Compression		Bone is crushed.	Common in porous bones (i.e., osteoporotic bones).
Depressed		Broken bone portion is pressed inward.	Typical of skull fracture.
Impacted		Broken bone ends are forced into each other.	Commonly occurs when one falls and attempts to break the fall with outstretched arms; also common in hip fractures.
Spiral		Ragged break occurs when excessive twisting forces are applied to a bone.	Common sports fracture.
Greenstick		Bone breaks incompletely, much in the way a green twig breaks.	Common in children, whose bones have relatively more collagen in their matrix and are more flexible than those of adults.

new matrix and become trapped within it. (Once they are trapped, they become **osteocytes,** or mature bone cells.) On the other hand, the bones of bedridden or physically inactive people tend to lose mass and to atrophy, because they are no longer subjected to stress.

To explain the interaction between these two controlling mechanisms as simply as possible, PTH determines *when* (or *if*) bone is to be broken down or formed in response to the need for more or fewer calcium ions in the blood. On the other hand, the stresses of muscle pull and gravity acting on the skeleton determine *where* bone matrix is to be broken down or formed so that the skeleton can remain as strong and vital as possible.

Bone Fractures

 Despite their remarkable strength, bones are susceptible to **fractures,** or breaks, all

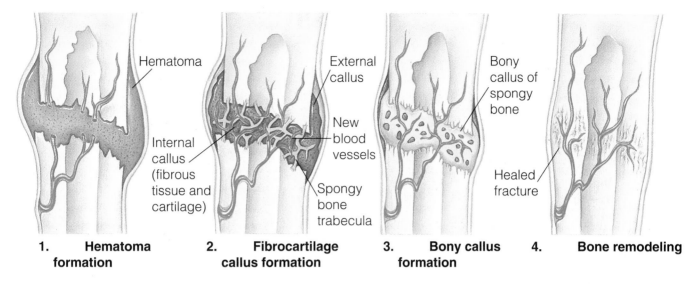

Figure 5.5   Stages in the healing of a bone fracture.

through life. During youth, most fractures result from exceptional trauma that twists or smashes the bones. Sports activities such as football, skating, and skiing jeopardize the bones, and automobile accidents certainly take their toll. In old age, bones thin and weaken, and fractures occur more frequently. Some of the many common types of fractures are illustrated and described in Table 5.2. ❑

A fracture is treated by *reduction,* which is the realignment of the broken bone ends. In **closed reduction,** the bone ends are coaxed back into their normal position by the physician's hands. In **open reductions,** surgery is performed and the bone ends are secured together with pins or wires. After the broken bone is reduced, it is immobilized by a cast or traction to allow the healing process to begin. The healing time for a simple fracture is six to eight weeks, but it is much longer for large bones and for the bones of elderly people (because of their poorer circulation).

The repair of bone fractures involves four major events (Figure 5.5):

1. A hematoma is formed. Blood vessels are ruptured when the bone breaks. As a result, a blood-filled swelling called a **hematoma** (he-mah-to′mah) forms. Bone cells deprived of nutrition die.
2. The break is splinted by a fibrocartilage callus. As described in Chapter 3, an early event of tissue repair (and bone is no exception) is the growth of new capillaries (granulation tissue) into the clotted blood at the site of the damage, and the disposal of dead tissue by phagocytes. As this goes on, connective tissue cells of various types form a mass of repair tissue, the **fibrocartilage callus** (kal′us), that contains several elements—some cartilage matrix, some bony matrix, and collagen fibers—and acts to "splint" the broken bone, closing the gap.
3. The bony callus is formed. As more osteoblasts and osteoclasts migrate into the area and multiply, fibrocartilage is gradually replaced by a callus made of spongy bone, the **bony callus.**
4. Over the next few months, the bony callus is remodeled in response to the mechanical stresses placed on it, so that it forms a strong permanent "patch" at the fracture site.

AXIAL SKELETON

As noted earlier, the skeleton is divided into two parts, the *axial* and *appendicular skeletons.* The axial skeleton, which forms the longitudinal axis of the body, is shown as the brown portion of Figure 5.6. It can be divided into three parts—the skull, the vertebral column, and the bony thorax.

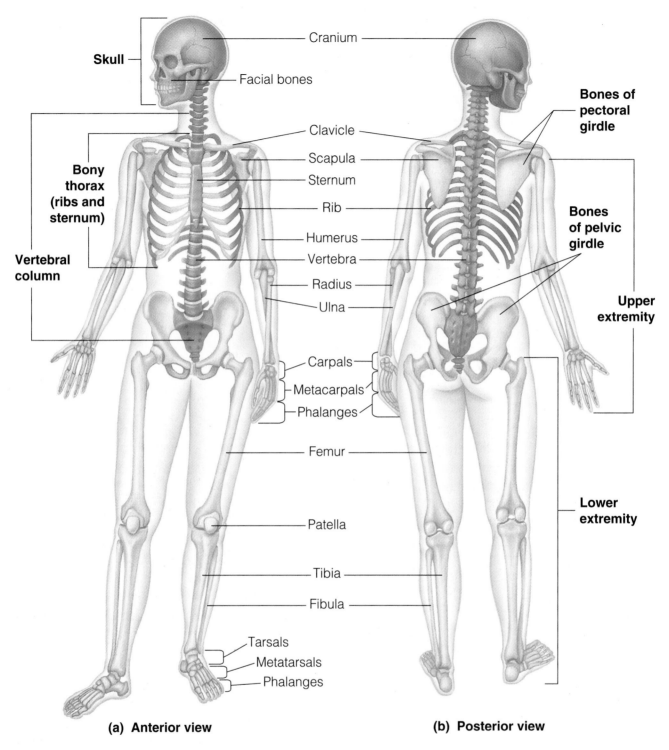

Skull

Cranium

Facial bones

Bony thorax (ribs and sternum)

Vertebral column

Clavicle

Scapula

Sternum

Rib

Humerus

Vertebra

Radius

Ulna

Carpals

Metacarpals

Phalanges

Femur

Patella

Tibia

Fibula

Tarsals

Metatarsals

Phalanges

Bones of pectoral girdle

Bones of pelvic girdle

Upper extremity

Lower extremity

(a) Anterior view

(b) Posterior view

Figure 5.6   The human skeleton. The bones of the axial skeleton are colored brown; the bones of the appendicular skeleton are shaded tan.

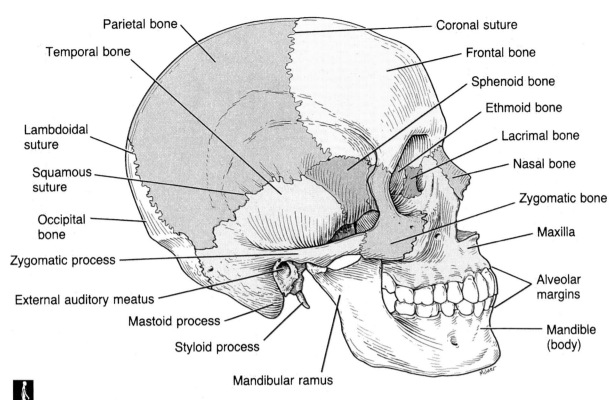

Figure 5.7   Human skull, lateral view.

Skull

The **skull** is formed by two sets of bones. The *cranium* encloses and protects the fragile brain tissue. The *facial bones* hold the eyes in an anterior position and allow the facial muscles to show our feelings through smiles or frowns. All but one of the bones of the skull are joined together by *sutures,* which are interlocking, immovable joints. The mandible (jawbone) is attached to the rest of the skull by a freely movable joint.

Cranium

The boxlike **cranium** is composed of eight large, flat bones. Except for two paired bones (the parietal and temporal), they are all single bones.

Frontal Bone   The frontal bone forms the forehead, the bony projections under the eyebrows, and the superior part of each eye's orbit (see Figure 5.7).

Parietal Bones   The paired parietal bones form most of the superior and lateral walls of the cranium (Figure 5.7). They meet in the midline of the skull at the *sagittal suture,* and form the *coronal suture* where they meet the frontal bone.

Temporal Bones   The temporal bones lie inferior to the parietal bones; they join them at the *squamous sutures.* Several important bone markings appear on the temporal bone (Figure 5.7):

1. The **external auditory meatus** is a canal that leads to the eardrum and the middle ear.
2. The **styloid process,** a sharp needlelike projection, is just inferior to the external auditory meatus. Many neck muscles use the styloid process as an attachment point.
3. The **zygomatic** (zi″go-mat′ik) **process** is a thin bridge of bone that joins with the cheekbone (zygomatic bone) anteriorly.
4. The **mastoid** (mas′toid) **process** is a rough projection posterior and inferior to the external

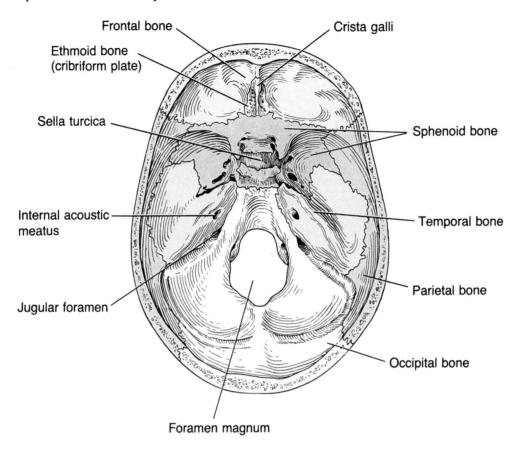

Figure 5.8   Human skull, superior view (top of cranium removed).

auditory meatus, which is full of air cavities (mastoid sinuses). It provides an attachment site for some muscles of the neck.

5.   The **jugular foramen,** at the junction of the occipital and temporal bones (see Figure 5.8), allows passage of the large *jugular vein* which drains the brain. Just anterior to it is the **carotid canal** (see Figure 5.9) through which the *internal carotid artery* runs, supplying blood to most of the brain.

The mastoid sinuses are so close to the middle ear (a high-risk spot for infections) that they may become infected, too, a condition called *mastoiditis.* Also, this area is so close to the brain, mastoiditis may spread to the brain itself. ❏

Occipital Bone   If you look at Figures 5.7, 5.8, and 5.9, you can see that the occipital (ok-sip′ĭ-tal) bone is the most posterior bone of the cranium. It forms the floor and back wall of the skull. The occipital bone joins the parietal bones anteriorly at

the *lambdoidal* (lam-doy′dal) *suture.* In the base of the occipital bone is a large opening, the **foramen magnum** (literally, "large hole"). The foramen magnum surrounds the lower part of the brain and allows the spinal cord to connect with the brain. Lateral to the foramen magnum on each side are the rockerlike **occipital condyles** (Figure 5.9), which rest on the first vertebra of the spinal column.

Sphenoid Bone   The butterfly-shaped sphenoid (sfe′noid) bone spans the width of the skull and forms part of the floor of the cranial cavity (Figure 5.8). In the midline of the sphenoid is a small depression, the **sella turcica** (sel′ah tur′sĭ-kah), or *Turk's saddle,* which holds the pituitary gland in place. Parts of the sphenoid can be seen exteriorly forming part of the eye orbits and the lateral part of the skull (see Figures 5.7 and 5.11). The central part of the sphenoid bone is riddled with air cavities, the *sphenoid sinuses* (see Figure 5.10).

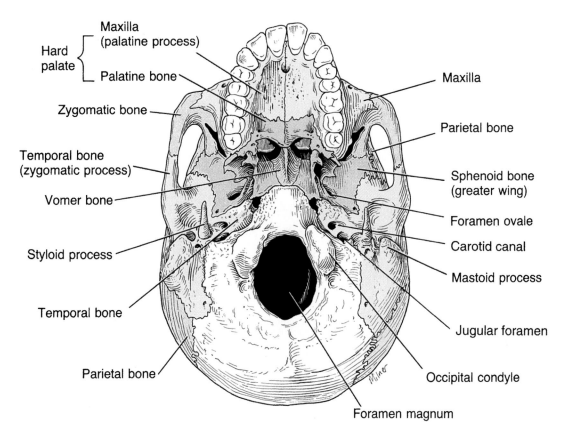

Figure 5.9   Human skull, inferior view (mandible removed).

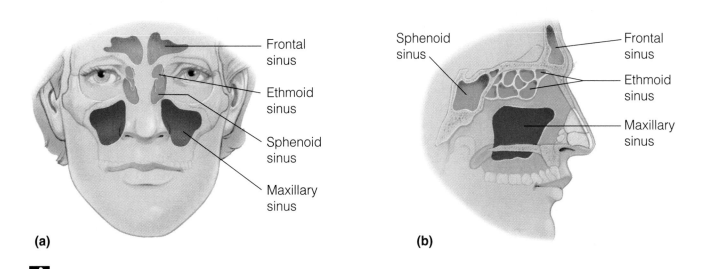

(a)   **(b)**

Figure 5.10   Paranasal sinuses. (a) Anterior view. **(b)** Medial view.

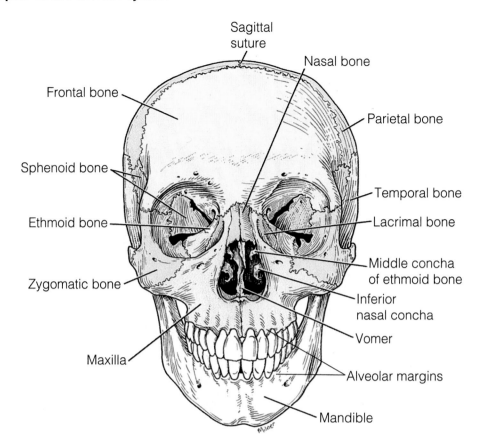

Sagittal
suture
Nasal bone
Frontal bone
Parietal bone
Sphenoid bone
Temporal bone
Ethmoid bone
Lacrimal bone
Middle concha
of ethmoid bone
Zygomatic bone
Inferior
nasal concha
Maxilla
Vomer
Alveolar margins
Mandible

Figure 5.11   Human skull, anterior view.

Ethmoid Bone   The ethmoid (eth′moid) bone is very irregularly shaped and lies anterior to the sphenoid (see Figures 5.7, 5.8, and 5.11). It forms the roof of the nasal cavity and part of the medial walls of the orbits. Projecting from its superior surface is the **crista galli** (kris′tah gah′le), literally "cock's comb" (see Figure 5.8). The outermost covering of the brain attaches to this projection. On each side of the crista galli are many small holes. These holey areas, the **cribriform** (krib′rĭ-form) **plates,** allow fibers carrying impulses from the olfactory (smell) receptors of the nose to reach the brain.

Facial Bones

Fourteen bones compose the face. Twelve are paired; only the mandible and vomer are single. Figures 5.7 and 5.11 show most of the facial bones.

Maxillae   The two maxillae (mak-si′le), or **maxillary bones,** fuse to form the upper jaw. All facial bones except the mandible join the maxillae; thus they are the main, or "keystone," bones of the face. The maxillae carry the upper teeth in the *alveolar margin.*

Extensions of the maxillae, the **palatine** (pal′ah-tīn) **processes,** form the anterior part of the hard palate of the mouth (see Figure 5.9). Like many other facial bones, the maxillae contain sinuses, which drain into the nasal passages (see Figure 5.10). These **paranasal sinuses,** whose naming reveals their position surrounding the nasal cavity, lighten the skull bones and probably act to amplify the sounds we make as we speak. They also cause many people a great deal of misery. Since the mucosa lining these sinuses is continuous with that in the nasal passages and throat, infections in these areas tend to migrate into the sinuses, causing *sinusitis*. Depending on which sinuses are infected, a headache or upper jaw pain is the usual result.

Palatine Bones
The paired palatine bones are posterior to the palatine processes of the maxillae. They form the posterior part of the hard palate (see Figure 5.9). Failure of these or the palatine processes to fuse medially results in *cleft palate.*

Zygomatic Bones   Zygomatic bones are commonly referred to as the cheekbones. They also form a good-sized portion of the lateral walls of the orbits, or eye sockets.

Lacrimal Bones   Lacrimal (lak′rĭ-mal) bones are fingernail-size bones forming part of the medial walls of each orbit. Each lacrimal bone has a groove that serves as a passageway for tears (*lacrima* = tear).

Nasal Bones   The small rectangular bones forming the bridge of the nose are the nasal bones. (The lower part of the nose is made up of cartilage.)

Vomer Bone   The single bone in the median line of the nasal cavity is the vomer. (*Vomer* means "plow," which refers to the bone's shape.) The vomer forms most of the nasal septum.

Inferior Conchae   The inferior conchae (kong′ke) are thin curved bones projecting from the lateral walls of the nasal cavity. (The superior and middle conchae are similar, but are parts of the ethmoid bone.)

Mandible   The mandible, or lower jaw, is the largest and strongest bone of the face. It joins the temporal bones on each side of the face, forming the only freely movable joints in the skull. You can find these joints on yourself by placing your fingers over your cheekbones and opening and closing your mouth. The horizontal part of the mandible (the *body*) forms the chin. Two upright bars of bone (the *rami*) extend from the body to connect the mandible with the temporal bone. The lower teeth lie in *alveoli* (sockets) in the *alveolar margin* at the superior edge of the body.

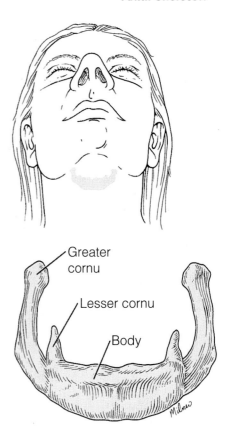

Figure 5.12   Anatomical location and structure of the hyoid bone. The hyoid bone is suspended in the midanterior neck by ligaments attached to the lesser cornua and the styloid processes of the temporal bones.

The Hyoid Bone

Though not really part of the skull, the **hyoid** (hi′oid) **bone** (Figure 5.12) is closely related to the mandible and temporal bones. The hyoid bone is unique in that it is the only bone of the body that does not articulate directly with any other bone. Instead, it is suspended in the midneck region about 2 cm (1 inch) above the larynx, where it is anchored by ligaments to the styloid processes of the temporal bones. Horseshoe-shaped, with a *body* and two pairs of *horns*, or *cornua*, the hyoid bone serves as a movable base for the tongue and an attachment point for neck muscles that raise and lower the larynx when we swallow and speak.

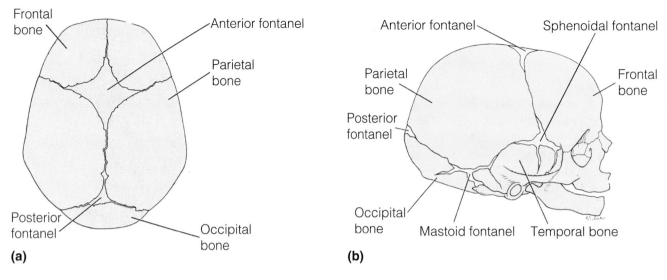

Figure 5.13   The fetal skull. (**a**) Superior view. (**b**) Lateral view.

Fetal Skull

The skull of a fetus or newborn infant is different in many ways from an adult skull. As Figure 5.13b illustrates, the infant's face is very small compared to the size of its cranium, but the skull as a whole is large compared to the infant's total body length. The adult skull represents only one-eighth of the total body length, whereas that of a newborn infant is one-fourth as long as its entire body. When a baby is born, its skeleton is still unfinished. As noted above, some areas of hyaline cartilage still remain to be ossified, or converted to bone. In the newborn, the skull also has regions that have yet to be converted to bone. These fibrous membranes connecting the cranial bones are called **fontanels** (fon"tah-nelz'). The rhythm of the baby's pulse can be felt in these "soft spots," which explains their name (*fontanel* = little fountain). The largest fontanels are the diamond-shaped *anterior fontanel* and the smaller triangular *posterior fontanel*. The fontanels allow the fetal skull to be compressed slightly during birth. In addition, because they are flexible, they allow the infant's brain to grow during the later part of pregnancy and early infancy. This would not be possible if the cranial bones were fused in sutures as in the adult skull. The fontanels are gradually converted to bone during the early part of infancy, and can no longer be felt 22 to 24 months after birth.

Vertebral Column (Spine)

Serving as the axial support of the body, the **vertebral column,** or **spine,** extends from the skull, which it supports, to the pelvis, where it transmits the weight of the body to the lower limbs. Some people think of the vertebral column as a rigid supporting rod, but that picture is inaccurate. Instead, the spine is formed from 26 irregular bones connected and reinforced by ligaments in such a way that a flexible curved structure results (see Figure 5.14). Running through the central cavity of the vertebral column is the delicate spinal cord, which it surrounds and protects.

Before birth, the spine consists of 33 separate bones called **vertebrae,** but 9 of these eventually fuse, forming the two composite bones, the *sacrum* and the *coccyx,* that construct the inferior portion of the vertebral column. Of the 24 single bones, the 7 vertebrae of the neck are cervical vertebrae, the next 12 are the thoracic vertebrae, and the remaining 5 supporting the lower back are lumbar vertebrae.

- Remembering common meal times, 7 a.m., 12 noon, and 5 p.m., may help you to recall the number of bones in these three regions of the vertebral column.

The single vertebrae are separated by pads of flexible fibrocartilage—**intervertebral disks**—which cushion the vertebrae and absorb shocks. In

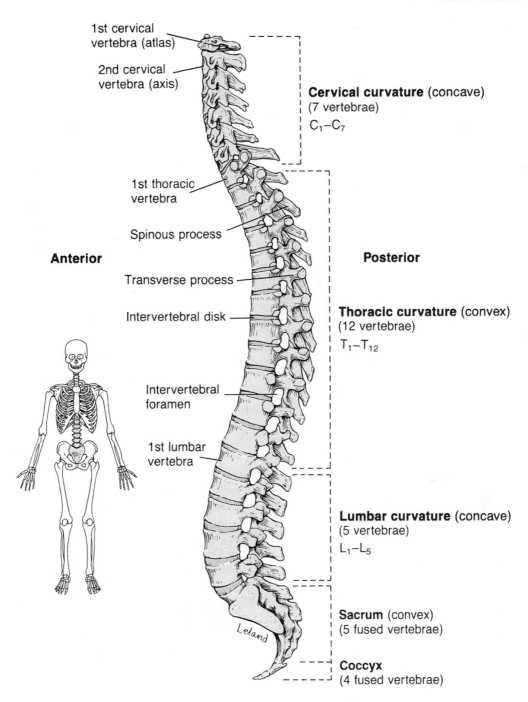

1st cervical
vertebra (atlas)

2nd cervical
vertebra (axis)

Cervical curvature (concave)
(7 vertebrae)
C_1-C_7

1st thoracic
vertebra

Spinous process

Anterior

Posterior

Transverse process

Intervertebral disk

Thoracic curvature (convex)
(12 vertebrae)
T_1-T_{12}

Intervertebral
foramen

1st lumbar
vertebra

Lumbar curvature (concave)
(5 vertebrae)
L_1-L_5

Leland

Sacrum (convex)
(5 fused vertebrae)

Coccyx
(4 fused vertebrae)

Figure 5.14   The vertebral column. Thin disks between the thoracic vertebrae allow great flexibility in the thoracic region; thick disks between the
lumbar vertebrae reduce flexibility. Note that the terms *convex* and *concave*
refer to the curvature of the posterior aspect of the vertebral column.

a young person, the disks have a high water content (about 90 percent) and are spongy and compressible. But as a person ages, the water content
of the disks decreases (as it does in other tissues
throughout the body), and the disks become
harder and less compressible.

Drying of the disks, along with a weakening
of the ligaments of the vertebral column, pre-

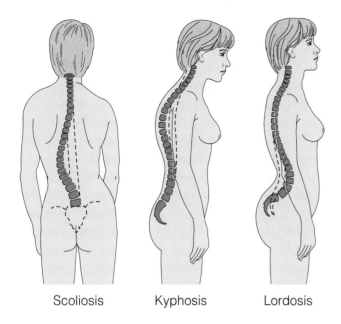

Scoliosis   Kyphosis   Lordosis

Figure 5.15   Abnormal spinal curvatures.

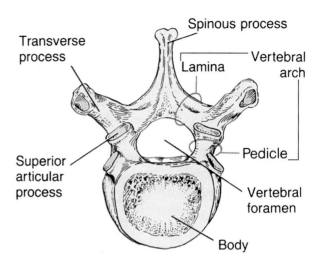

Figure 5.16   A typical vertebra, superior view. (Inferior articulating surfaces are not shown.)

disposes older people to *herniated* ("slipped") *disks*. If the protruding disk presses on the spinal cord or the spinal nerves exiting from the cord, numbness and excruciating pain can result. ❑

The disks and the S-shaped structure of the vertebral column work together to prevent shock to the head when we walk or run. They also make the body trunk flexible. The spinal curvatures in the thoracic and sacral regions are referred to as **primary curvatures** because they are present when we are born. Later, the **secondary curvatures** develop. The cervical curvature appears when a baby begins to raise its head, and the lumbar curvature develops when the baby begins to walk.

There are several types of abnormal spinal curvatures. Figure 5.15 shows three of these—*scoliosis* (sko″le-o′sis), *kyphosis* (ki-fo′sis), and *lordosis* (lor-do′sis). These abnormalities may be congenital (present at birth) or result from disease or poor posture. As you look at these diagrams, try to pinpoint how each of the conditions differs from the normal healthy spine. ❑

All vertebrae have a similar structural pattern (see Figure 5.16). The common features are listed below:

- **Body** or **centrum:** disklike, weight-bearing part of the vertebra facing anteriorly in the vertebral column.
- **Vertebral arch:** arch formed from the joining of all posterior extensions, the *laminae* and *pedicles,* from the vertebral body.
- **Vertebral foramen:** canal through which the spinal cord passes.
- **Transverse processes:** two lateral projections from the body.
- **Spinous process:** single projection arising from the posterior aspect of the vertebral arch (actually the fused laminae).
- **Superior and inferior articular processes:** paired projections lateral to the vertebral foramen, allowing a vertebra to form joints with adjacent vertebrae.

In addition to the common features just described, vertebrae in the different regions of the spine have very specific structural characteristics. These unique regional characteristics of the vertebrae are described next.

Cervical Vertebrae

The seven **cervical vertebrae** (identified as C_1 to C_7) form the neck region of the spine. The first two vertebrae *(atlas and axis)* are different because they perform functions not shared by the other cer-

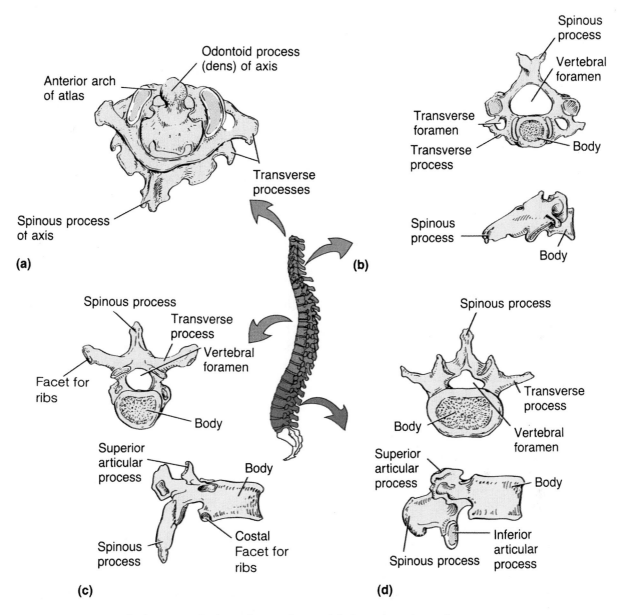

Figure 5.17 Regional characteristics of vertebrae. (a) Superior view of the articulated atlas and axis. **(b)** Cervical vertebrae; superior view above, lateral view below. **(c)** Thoracic vertebrae; superior view above, lateral view below. **(d)** Lumbar vertebrae; superior view above, lateral view below.

vical vertebrae. As you can see in Figure 5.17a, the **atlas** (C_1) has no body. The superior surfaces of its transverse processes contain large depressions that receive the occipital condyles of the skull. This joint allows you to nod "yes." The **axis** (C_2) acts as a pivot for the rotation of the atlas (and skull) above. It has a large upright process, the **odontoid** (o-don′toid) **process,** or **dens,** which acts as the pivot point. The joint between C_1 and C_2 allows

you to rotate your head from side to side to indicate "no."

The "typical" cervical vertebrae (C_3 through C_7) are shown in Figure 5.17b. They are the smallest, lightest vertebrae, and most often their spinous processes are short and divided into two branches. The transverse processes of the cervical vertebrae contain foramina (openings) through which the vertebral arteries pass on their way to the brain

Figure 5.18 **Sacrum and coccyx, posterior view.**

above. Any time you see these foramina in a vertebra, you should know immediately that it is a cervical vertebra.

Thoracic Vertebrae

The 12 **thoracic vertebrae** (T_1–T_{12}) are all typical. As seen in Figure 5.17c, they are larger than the cervical vertebrae. The body is somewhat heart-shaped and has two costal facets (articulating surfaces) on each side, which receive the heads of the ribs. The spinous process is long, and hooks sharply downward.

Lumbar Vertebrae

The five **lumbar vertebrae** (L_1–L_5) have massive, blocklike bodies and short, hatchet-shaped spinous processes (Figure 5.17d). Since most of the stress on the vertebral column occurs in the lumbar region, these are the sturdiest of the vertebrae.

Sacrum

The **sacrum** (sa′krum) is formed by the fusion of five vertebrae (Figure 5.18). Superiorly it articulates with L_5, and inferiorly it connects with the coccyx. The winglike *alae* articulate laterally with the hip bones, forming the sacroiliac joints. The sacrum forms the posterior wall of the pelvis. The vertebral canal continues inside the sacrum as the **sacral canal.**

Coccyx

The **coccyx** is formed from the fusion of three to five small, irregularly shaped vertebrae (Figure 5.18). It is the human "tailbone," a remnant of the tail that other vertebrate animals have.

Bony Thorax

The sternum, ribs, and thoracic vertebrae make up the **bony thorax.** The bony thorax is often called the *thoracic cage* because it forms a protective, cone-shaped cage of slender bones around the organs of the thoracic cavity (heart, lungs, and major blood vessels). The bony thorax is shown in Figure 5.19.

Sternum

The **sternum** (breastbone) is a typical flat bone and the result of the fusion of three bones—the **manubrium** (mah-nu′bre-um), **body,** and **xiphoid** (zif′oid) **process.** It is attached to the first seven pairs of ribs.

Because the sternum is so close to the body surface, it is easy to obtain samples of blood-forming (hematopoietic) tissue for the diagnosis of suspected blood diseases from this bone. A needle is inserted into the marrow of the sternum, and the sample is withdrawn; this procedure is called a *sternal puncture.*

Ribs

Twelve pairs of **ribs** form the walls of the thoracic cage. (Contrary to popular misconception, males do *not* have one rib less than females!) All the ribs articulate with the vertebral column posteriorly and then curve downward and toward the anterior body surface. The **true ribs,** the first seven pairs, attach directly to the sternum by costal cartilages. **False ribs,** the next five pairs, either attach indi-

Figure 5.19   Bony thorax, anterior view.

rectly to the sternum or are not attached to the sternum at all. The last two pairs of false ribs lack the sternal attachments, and so they are also called **floating ribs.**

The intercostal spaces (spaces between the ribs) are filled with the intercostal muscles that aid in breathing.

APPENDICULAR SKELETON

The *appendicular skeleton* is shaded tan in Figure 5.6. It is composed of 126 bones of the limbs (appendages) and the pectoral and pelvic girdles, which attach the limbs to the axial skeleton.

Bones of the Shoulder Girdle

Each **shoulder girdle,** or **pectoral girdle,** consists of two bones—a clavicle and a scapula (Figure 5.20).

The **clavicle** (klav′ĭ-kl), or *collarbone,* is a slender, doubly curved bone. It attaches to the

manubrium of the sternum medially and to the scapula laterally (where it helps to form the shoulder joint). The clavicle acts as a brace to hold the

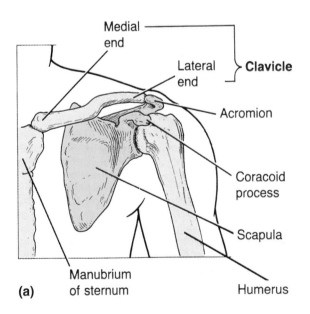

(a)

Figure 5.20a   Bones of the shoulder girdle. Relationship of the left shoulder girdle to the bones of the thorax and arm.   ➤

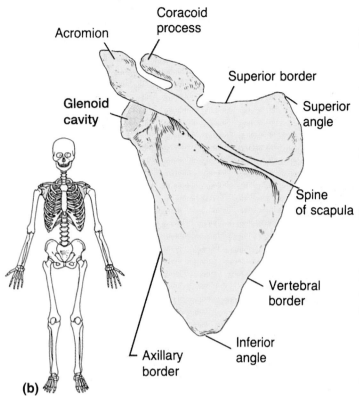

Figure 5.20b **Bones of the shoulder girdle** Left scapula, posterior view.

The shoulder girdle is very light and allows the upper limb to have exceptionally free movement. This is due to the following factors:

1. Each shoulder girdle attaches to the axial skeleton at only one point—the *sternoclavicular joint.*

2. The loose attachment of the scapula allows it to slide back and forth against the thorax as muscles act.

3. The glenoid cavity is shallow, and the shoulder joint is poorly reinforced by ligaments.

However, this exceptional flexibility also has a drawback; the shoulder girdle is very easily dislocated.

Bones of the Upper Limbs

Thirty separate bones form the skeletal framework of each upper limb (see Figures 5.21 and 5.22). They form the foundations of the arm, forearm, and hand.

Arm

The arm is formed by a single bone, the **humerus** (hu'mer-us), which is a typical long bone (Figure 5.21a and b). At its proximal end is a rounded head that fits into the shallow glenoid cavity of the scapula. Opposite the head are two bony projections—the **greater** and **lesser tubercles,** which are sites of muscle attachment. In the midpoint of the shaft is a roughened area called the **deltoid tuberosity,** where the large, fleshy deltoid muscle of the shoulder attaches. At the distal end of the humerus is the medial **trochlea** (trok'le-ah), which looks somewhat like a spool, and the lateral **capitulum** (kah-pit'u-lum). Both of these processes articulate with bones of the forearm. Above the trochlea anteriorly is a depression, the **coronoid fossa;** on the posterior surface is the **olecranon** (o-lek'rah-non) **fossa.** These two depressions allow the corresponding processes of the ulna to move freely when the elbow is bent and extended.

arm away from the top of the thorax and helps prevent shoulder dislocation. When the clavicle is broken, the whole shoulder region caves in medially, which shows how important its bracing function is.

The **scapulae** (skap'u-le), or *shoulder blades,* are triangular and are commonly called "wings" because they flare when we move our arms posteriorly. Each scapula has a flattened body and two important processes—the **acromion** (ah-kro'me-on), which is the enlarged end of the spine of the scapula, and the beaklike **coracoid** (kor'ah-koid) **process.** The acromion connects with the clavicle laterally. The coracoid process points over the top of the shoulder and anchors some of the muscles of the arm. The scapula is not directly attached to the axial skeleton; it is loosely held in place by trunk muscles. The scapula has three borders—superior, vertebral (medial), and axillary (lateral). It also has three angles—superior, inferior, and lateral. The **glenoid cavity,** a shallow socket that receives the head of the arm bone, is in the lateral angle.

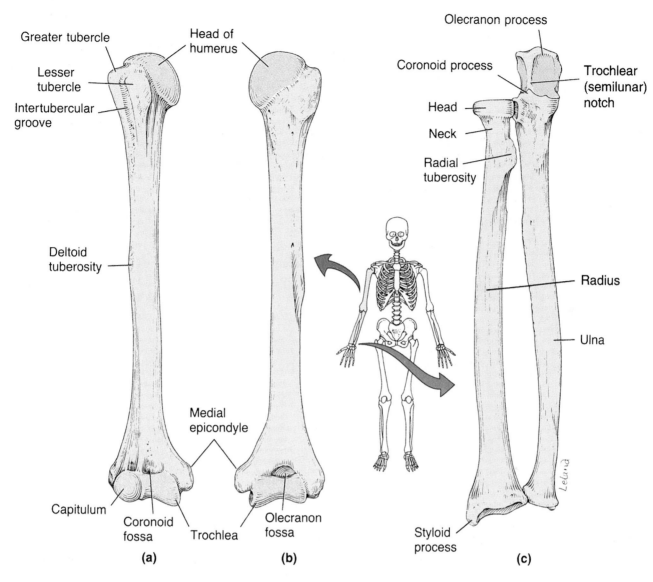

Greater tubercle

Head of humerus

Lesser tubercle

Intertubercular groove

Deltoid tuberosity

Medial epicondyle

Capitulum

Coronoid fossa

Trochlea

Olecranon fossa

(a)   **(b)**

Olecranon process

Coronoid process

Trochlear (semilunar) notch

Head

Neck

Radial tuberosity

Radius

Ulna

Styloid process

(c)

Figure 5.21   Bones of the right arm and forearm. (**a**) Humerus, anterior view. (**b**) Humerus, posterior view. (**c**) Anterior view of bones of the forearm, the radius and the ulna.

Forearm

Two bones, the radius and the ulna, form the skeleton of the forearm (Figure 5.21c). When the body is in the anatomical position, the **radius** is the lateral bone; that is, it is on the thumb side of the forearm. When the hand is rotated so that the palm faces backward, the distal end of the radius ends up medial to the ulna.

Proximally, the disk-shaped head of the radius forms a joint with the capitulum of the humerus.

Just below the head is the **radial tuberosity,** where the tendon of the biceps muscle attaches.

When the upper limb is in the anatomical position, the **ulna** is the medial bone (on the little-finger side) of the forearm. On its proximal end are the anterior **coronoid process** and the posterior **olecranon process,** which are separated by the **trochlear (semilunar) notch.** Together these two processes grip the trochlea of the humerus in a pliers-like joint.

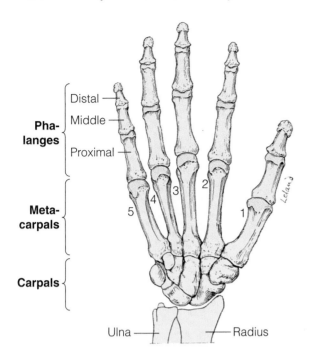

Figure 5.22 **Bones of the right hand, anterior view.**

Hand

The skeleton of the hand consists of the carpals, the metacarpals, and the phalanges (Figure 5.22). The eight **carpal bones,** arranged in two irregular rows of four bones each, form the part of the hand called the **carpus** or, more commonly, the *wrist.* The carpals are bound together by ligaments that restrict movements between them.

The palm of the hand consists of the **metacarpals.** The **phalanges** (fah-lan′jēz) are the bones of the fingers. The metacarpals are numbered 1 to 5 from the thumb side of the hand toward the little finger. When the fist is clenched, the heads of the metacarpals become obvious as the "knuckles." Each hand contains 14 phalanges. There are three in each finger (proximal, middle, and distal), except in the thumb, which has only two (proximal and distal).

Bones of the Pelvic Girdle

The **pelvic girdle** is formed by two **coxal** (kok′sal) **bones,** or **ossa coxae,** commonly called **hip bones.** Together with the sacrum and the coc-

cyx, the hip bones form the *bony pelvis* (Figure 5.23). Note that the terms "pelvic girdle" and "pelvis" have slightly different meanings. The bones of the pelvic girdle are large and heavy, and they are attached securely to the axial skeleton. The sockets, which receive the thigh bones, are deep and heavily reinforced by ligaments that attach the limbs firmly to the girdle. Bearing weight is the most important function of this girdle; the total weight of the upper body rests on the pelvis. The reproductive organs, urinary bladder, and part of the large intestine lie within and are protected by the bony pelvis.

Each hip bone is formed by the fusion of three bones: the ilium, ischium, and pubis. The **ilium** (il′e-um), which connects posteriorly with the sacrum at the **sacroiliac** (sak″ro-il′e-ac) **joint,** is a large, flaring bone that forms most of the hip bone. When you put your hands on your hips, they are resting over the ilia. The upper edge of the ilium, the **iliac crest,** is an important anatomical landmark that is always remembered by those who give injections.

The **ischium** (is′ke-um) is the "sitdown bone," since it forms the most inferior part of the coxal bone. The **ischial tuberosity** is a roughened area that receives body weight when you are sitting. The **ischial spine,** superior to the tuberosity, is another important anatomical landmark, particularly in the pregnant woman, because it narrows the outlet of the pelvis through which the baby must pass during the birth process. Another important structural feature of the ischium is the **greater sciatic notch,** which allows blood vessels and the large sciatic nerve to pass from the pelvis posteriorly into the thigh. Injections in the buttock should always be given well away from this area.

The **pubis** (pu′bis) is the most anterior part of a coxal bone. Fusion of the rami of the pubic bone anteriorly and the ischium posteriorly forms a bar of bone enclosing the **obturator** (ob′tu-ra″tor) **foramen,** an opening which allows blood vessels and nerves to pass into the anterior part of the thigh. The pubic bones of each hip bone fuse anteriorly to form a cartilaginous joint, the **pubic symphysis** (pu′bik sim′fĭ-sis).

The ilium, ischium, and pubis fuse at the deep socket called the **acetabulum** (as″ĕ-tab′u-lum), which means "vinegar cup." The acetabulum receives the head of the thigh bone.

The bony pelvis is divided into two regions. The **false pelvis** is superior to the true pelvis; it is

Figure 5.23   The pelvis. (a) Articulated male pelvis. **(b)** Articulated female pelvis. **(c)** Right coxal bone, showing the point of fusion of the ilium, ischium, and pubic bones.

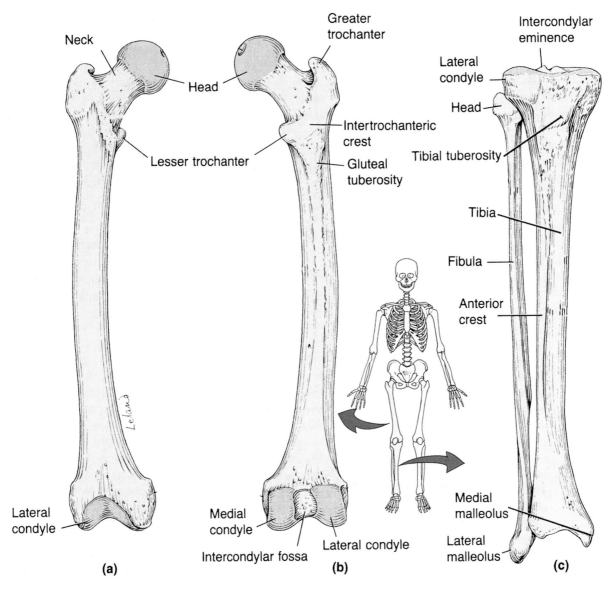

Figure 5.24 **Bones of the right thigh and leg.** (**a**) Femur (thigh bone), anterior view. (**b**) Femur, posterior view. (**c**) Tibia and fibula of the leg, anterior view.

the area medial to the flaring portions of the ilia. The **true pelvis** is surrounded by bone and lies inferior to the flaring parts of the ilia (the pelvic brim). The dimensions of the true pelvis of a woman are very important because they must be large enough to allow the infant's head (the largest part of the infant) to pass during childbirth. The dimensions of the cavity, particularly the **outlet** (the inferior opening of the pelvis), and the **inlet** (superior opening) are critical, and thus they are carefully measured by the obstetrician.

Of course, individual pelvic structures vary, but there are fairly consistent differences between a male and a female pelvis. Look at Figure 5.23 again and note the following characteristics that differ in the pelvis of the male and female.

- The female inlet is larger and more circular.
- The female pelvis as a whole is shallower, and the bones are lighter and thinner.
- The female ilia flare more laterally.
- The female sacrum is shorter and less curved.

- The female ischial spines are shorter and farther apart; thus the outlet is larger.
- The female pubic arch is more rounded because the angle of the pubic arch is greater.

Bones of the Lower Limbs

The lower limbs carry our total body weight when we are erect. Hence, it is not surprising that the bones forming the three segments of the lower limbs (thigh, leg, and foot) are much thicker and stronger than the comparable bones of the upper limb.

Thigh

The **femur** (fe'mur), or *thigh bone,* is the only bone in the thigh (Figure 5.24a and b). It is the heaviest, strongest bone in the body. Its proximal end has a ball-like head, a neck, and **greater** and **lesser trochanters** (separated posteriorly by the **intertrochanteric crest**). The trochanters, intertrochanteric crest, and the **gluteal tuberosity,** located on the shaft, all serve as sites for muscle attachment. The head of the femur articulates with the acetabulum of the hip bone in a deep, secure socket. However, the neck of the femur is a common fracture site, especially in old age.

The femur slants medially as it runs downward to join with the leg bones; this brings the knees in line with the body's center of gravity. The medial course of the femur is more noticeable in females because of the wider female pelvis. Distally on the femur are the *lateral* and *medial condyles,* which articulate with the tibia below.

Leg

Two bones, the tibia and fibula, form the leg (see Figure 5.24c). The **tibia,** or *shinbone,* is larger and more medial. At the proximal end, the *medial* and *lateral condyles* (separated by the intercondylar eminence) articulate with the distal end of the femur to form the knee joint. The patellar (kneecap) ligament attaches to the **tibial tuberosity,** a roughened area on the anterior tibial surface. Distally, the **medial malleolus** (mal-le'o-lus) process forms the inner bulge of the ankle. The anterior surface of the tibia is a sharp ridge, the **anterior**

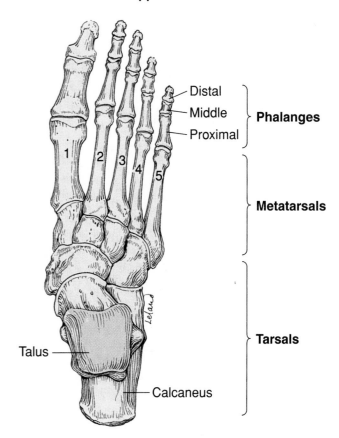

Figure 5.25   Bones of the right foot, superior view.

crest, that is unprotected by muscles; thus, it is easily felt beneath the skin.

The **fibula,** which lies alongside the tibia, is thin and sticklike. The fibula has no part in forming the knee joint. Its distal end, the **lateral malleolus,** forms the outer part of the ankle.

Foot

The foot, composed of the tarsals, metatarsals, and phalanges, has two important functions. It supports our body weight and serves as a lever that allows us to propel our bodies forward when we walk and run.

The **tarsus,** commonly called the *ankle,* is composed of seven **tarsal bones** (Figure 5.25). Body weight is mostly carried by the two largest tarsals, the **calcaneus** (kal-ka'ne-us), or heelbone, and the **talus** (ta'lus), which lies between the tibia and the calcaneus. Five **metatarsals** form the sole, and 14 **phalanges** form the toes. Like the fingers of the hand, each toe has three phalanges, except the great toe, which has two.

(Text continues on page 142)

A CLOSER LOOK

SURFACE ANATOMY—EXPLORING YOURSELF

Surface anatomy is a valuable branch of medical science. True to its name, surface anatomy is, indeed, a study of the *external surface* of the body. More importantly, it also concerns *internal* organs as they relate to external surface landmarks and as they are seen and felt through the skin. Feeling internal structures through the skin with the fingers is called **palpation** (literally, "touching").

Surface anatomy is living anatomy, better studied in live people than in cadavers. It can provide a great deal of information about the living skeleton (almost all bones can be palpated) and about the muscles and vessels that lie near the body surface. Furthermore, a skilled examiner can learn much about your heart, lungs, and other deep organs by performing a surface assessment during a standard physical examination. For those of you planning a career in the health sciences, a study of surface anatomy will show you where to take pulses, where to insert tubes and needles, where to locate broken bones and inflamed muscles, and where to listen for the sounds of the lungs, heart, and intestines. However, you must walk before you can run, so our goal here will be to conduct a very brief exploration of surface anatomy.

We will take a regional approach to surface anatomy, exploring the head first and then the girdles and limbs. You will be observing and palpating your own body as you work through this exercise, because your body is the best learning tool of all.

THE HEAD

Recall that the head is divided into the cranium and face. However, we will take a little liberty here and include the back of the neck.

1. Run your fingers over the surface of your head. Notice that the underlying cranial bones lie very near the surface. Now, feel for your *mastoid process,* the roughened area just behind your ear.

2. Grasp your auricle, the shell-like part of the external ear that surrounds the opening of the *external auditory meatus.* Insert your small finger into that canal. (See the photo.)

3. Run your hand anteriorly from your ear toward your eye, and feel the *zygomatic arch* at the high point of your cheek just deep to the skin. This bony arch is easily broken by blows to the face. Next, spread your fingers on the skin of your face and feel it bunch and stretch as you smile, frown, and make other "faces." You are now monitoring the action of your *muscles of facial expression,* which you will be studying in Chapter 6.

4. Palpate the different regions of your mandible, or lower jaw: its anterior *body* and its posterior ascending *ramus.* To feel your *temporomandibular joint* in action, place a finger directly in front of the external auditory meatus, and open and close your mouth several times. The bony structure you feel moving is the *head of the mandible.*

5. Now, turn your attention to the eye region, and trace a finger around the entire margin of an orbit. On the medial side of the orbit, feel for the fossa in the *lacrimal bone,* which contains the tear-gathering lacrimal sac.

6. Touch the most superior part of your nose, its *root,* between the eyebrows (see the photo). Just inferior to this is the *bridge* of your nose, which is formed by the *nasal bones.* Run your index finger and thumb along opposite sides of the bridge of your nose until they "slip" medially at the inferior end of the nasal bones.

7. To finish up the head and neck region, run your fingers inferiorly along the midline of the back of your neck to feel the *spinous processes* of the cervical vertebrae. The spine of C_7 is especially prominent, which is why this vertebra is sometimes called the *vertebra prominens.*

THE PECTORAL GIRDLE AND UPPER LIMB

Only selected bony areas will be palpated from this point on because of the difficulty of reaching posterior markings on yourself.

1. Palpate your *clavicle* along its entire length from the sternum to the shoulder. At the sternum-clavicle junction, identify the rigid *sternoclavicular joint.* At the high point of your shoulder, find the *acromion,* the anterior end of the scapular spine.

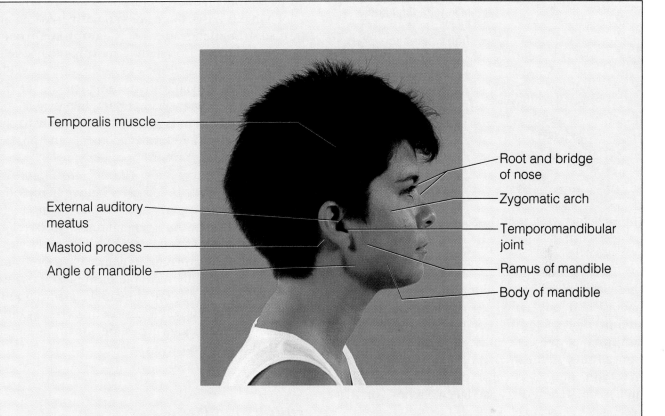

Temporalis muscle

External auditory meatus

Mastoid process

Angle of mandible

Root and bridge of nose

Zygomatic arch

Temporomandibular joint

Ramus of mandible

Body of mandible

2. Feel the medial projection at the distal end of the humerus; this is the *medial epicondyle of the humerus.*

3. Now, work your elbow—flexing and extending it—as you palpate its dorsal aspect to feel the *olecranon process* of the ulna moving in and out of the *olecranon fossa* on the backside of the humerus.

4. Clench your fist and find the first set of flexed-joint protrusions beyond your wrist. These are your *metacarpophalangeal joints,* commonly called the *knuckles.*

THE PELVIC GIRDLE AND LOWER LIMB

Find this last set of bone markings before winding up your self-exploration.

1. Rest your hands on your hips—they will be overlying the *iliac crests.* Trace an iliac crest as far anteriorly as you can. This bone marking, the *anterior superior iliac spine,* is fairly easy to feel in almost everyone, and it is clearly visible through the skin (and perhaps the clothing) of very slim people.

2. The *greater trochanter of the femur* is usually easier to locate in females than males because of the wider female pelvis and the fact that it is more likely to be covered by bulkier muscles in males. Try to locate it on yourself—it is the most lateral point of the proximal femur and it typically lies 6 to 8 inches below the iliac crest.

3. Feel your kneecap, or *patella,* and palpate the ligaments attached to its superior and inferior borders. Follow the inferior ligament to the *tibial tuberosity* to which it attaches.

4. Palpate the medial protrusion of your ankle, which is the *medial malleolus* of the distal tibia. Now feel the bulge of your lateral ankle, which is the *lateral malleolus* of your fibula.

5. Attempt to follow the extent of your *calcaneus,* or heelbone. (It helps if you are really thin!)

Figure 5.26   Arches of the foot.

The bones in the foot are arranged to form three strong arches: two longitudinal (medial and lateral) and one transverse (Figure 5.26). **Ligaments,** which bind the foot bones together, and **tendons** of the foot muscles help to hold the bones firmly in the arched position, but still allow a certain amount of give or springiness. Weak arches are referred to as "fallen arches" or "flat feet."

JOINTS

With one exception (the hyoid bone of the neck), every bone in the body forms a joint with at least one other bone. **Joints,** also called **articulations,** have two functions: They hold the bones together securely but also give the rigid skeleton mobility.

The graceful movements of a ballet dancer and the rough-and-tumble grapplings of a football player illustrate the great variety of motion allowed by joints, the sites where two or more bones meet. With fewer joints, we would move like robots. Nevertheless, the bone-binding function of joints is just as important as their role in providing mobility. The immovable joints of the skull, for instance, form a snug enclosure for our vital brain.

Joints are classified in two ways—functionally and structurally. The functional classification focuses on the amount of movement allowed by the joint. On this basis, there are **synarthroses** (sin″ar-thro′sēz) or immovable joints, **amphiarthroses** (am″fe-ar-thro′sēz) or slightly movable joints, and **diarthroses** (di″ar-thro′sēz) or freely movable joints. Freely movable joints predominate in the limbs, where mobility is important. Immovable and slightly movable joints are restricted mainly to the axial skeleton, where firm attachments and protection of internal organs are priorities.

Structurally, there are *fibrous, cartilaginous,* and *synovial joints* based on whether fibrous tissue, cartilage, or a joint cavity separates the bony regions of the joint. As a general rule, fibrous joints are immovable and synovial joints are freely movable. Although cartilaginous joints have both immovable and slightly movable examples, most are amphiarthrotic. Since the structural classification is more clear-cut, we will focus on that classification scheme here. The joint types are shown in Figure 5.27 and described next.

Fibrous Joints

In **fibrous joints,** the bones are united by fibrous tissue. The best examples of this type of joint are the *sutures* of the skull. In sutures, the irregular edges of the bones interlock and are bound tightly together by connective tissue fibers, allowing essentially no movement to occur.

Cartilaginous Joints

In **cartilaginous joints,** the bone ends are connected by cartilage. Examples of this joint type that are slightly movable (amphiarthrotic) are the *pubic symphysis* of the pelvis and *intervertebral joints* of the spinal column, where the bodies of the vertebrae are connected by pads (disks) of fibrocartilage. The hyaline-cartilage epiphyseal plates of growing long bones are immovable (synarthrotic) cartilaginous joints.

Figure 5.27 Types of joints. Joints shown to the left of the skeleton are fibrous and cartilaginous joints; joints on the right are synovial joints. (**a**) Suture connecting skull bones. (**b**) Intervertebral joints of the spinal column. (**c**) Symphysis connecting the pubic bones anteriorly. (**d–f**) Synovial joints of the shoulder, elbow, and wrist, respectively.

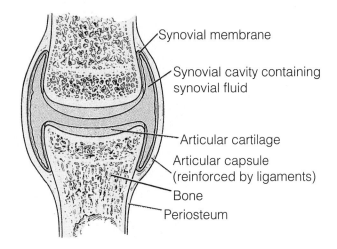

Synovial membrane

Synovial cavity containing synovial fluid

Articular cartilage

Articular capsule (reinforced by ligaments)

Bone

Periosteum

Figure 5.28   General structure of a synovial joint.

Synovial Joints

Synovial joints are those in which the articulating bone ends are separated by a joint cavity containing synovial fluid.

All synovial joints have four distinguishing features (Figure 5.28):

1. **Articular cartilage.** Articular (hyaline) cartilage covers the ends of the bones forming the joint.

2. **Fibrous articular capsule.** The joint surfaces are enclosed by a sleeve or capsule of fibrous connective tissue and the capsule is lined with a smooth *synovial membrane* (the reason these joints are called synovial joints).

3. **Joint cavity.** The articular capsule encloses a cavity, called the joint cavity, which contains lubricating synovial fluid.

4. **Reinforcing ligaments.** The fibrous capsule is usually reinforced with ligaments. Bursae (fluid-filled synovial membrane sacs) are often found cushioning tendons where they cross bone.

Synovial joints have much more freedom or flexibility than the other two joint types; however, their flexibility does vary. In the so-called *nonaxial joints,* flat or slightly curved bone surfaces allow only slipping movements to occur between the bones involved; the only examples are plane joints such as the joints of the wrist (Figure 5.29a). Some synovial joints can move in only one plane—for example, the hingelike elbow joint and the pivot joint between the radius and ulna are *uniaxial joints* (Figure 5.29b and c, respectively). Others, like the knuckles (Figure 5.29d), and the saddle joint of the thumb (Figure 5.29e), can move in two planes *(biaxial joints),* and still others are *multiaxial joints* that move in all planes, such as the ball-and-socket joints of the shoulder (Figure 5.29f) and hip. The various types of movements that occur at synovial joints are discussed in detail in the next chapter, because they relate to muscle activity.

A *dislocation* happens when a bone is forced out of its normal position in the joint cavity. The process of returning the bone to its proper position, called *reduction,* should be done only by a physician. Attempts by an untrained person to "snap the bone back into its socket" are usually more harmful than helpful. ❑

Inflammatory Disorders of Joints

Few of us pay attention to our joints unless something goes wrong with them. Joint pain and inflammation may be caused by many things. For example, falling on one's knee can cause a painful *bursitis,* called "water on the knee," due to inflammation of bursae or synovial membrane. Sprains and dislocations are other types of joint problems that result in swelling and pain. In a *sprain,* the ligaments or tendons reinforcing a joint are damaged by excessive stretching, or they are torn away from the bone. Since both tendons and ligaments are cords of dense fibrous connective tissue with a poor blood supply, sprains heal slowly and are extremely painful.

Few inflammatory joint disorders cause more pain and suffering than arthritis. The term **arthritis** (*arth* = joint; *itis* = inflammation) describes over 100 different inflammatory or degenerative diseases that damage the joints. In all its forms, arthritis is the most widespread, crippling disease in the United States. One out of seven Americans suffers its ravages. All forms of arthritis have the same initial symptoms: pain, stiffness, and swelling of the joint. Then, depending on the specific form, certain changes in the joint structure occur.

(Text continues on page 147)

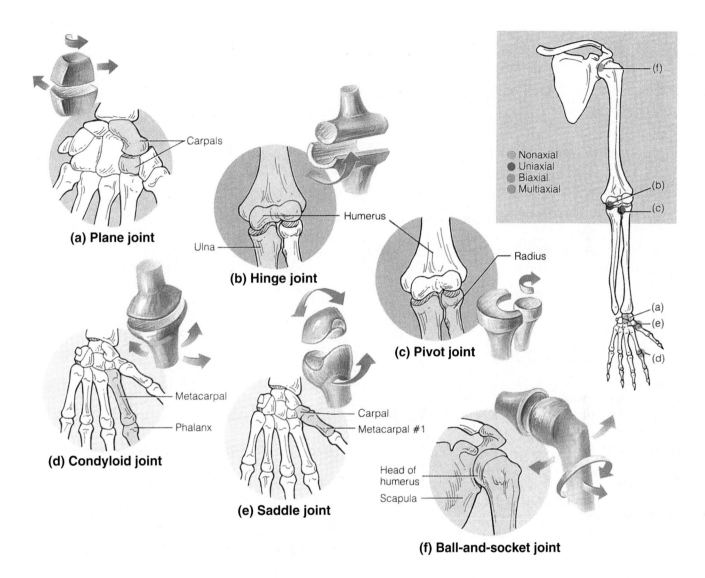

Figure 5.29   Types of synovial joints. (a) Plane joint (intercarpal and intertarsal joints). **(b)** Hinge joint (elbow and interphalangeal joints). **(c)** Pivot joint (proximal joint between the radius and the ulna). **(d)** Condyloid joint (knuckles). **(e)** Saddle joint (carpometacarpal joint of the thumb). **(f)** Ball-and-socket joint (shoulder and hip joints).

SYSTEMS IN SYNC

Homeostatic Relationships Between the Skeletal System and Other Body Systems

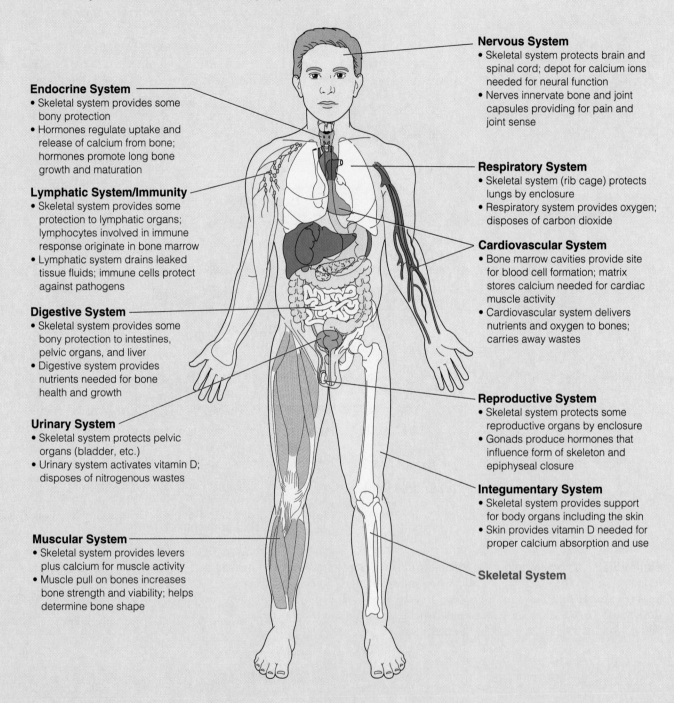

Nervous System
- Skeletal system protects brain and spinal cord; depot for calcium ions needed for neural function
- Nerves innervate bone and joint capsules providing for pain and joint sense

Respiratory System
- Skeletal system (rib cage) protects lungs by enclosure
- Respiratory system provides oxygen; disposes of carbon dioxide

Cardiovascular System
- Bone marrow cavities provide site for blood cell formation; matrix stores calcium needed for cardiac muscle activity
- Cardiovascular system delivers nutrients and oxygen to bones; carries away wastes

Reproductive System
- Skeletal system protects some reproductive organs by enclosure
- Gonads produce hormones that influence form of skeleton and epiphyseal closure

Integumentary System
- Skeletal system provides support for body organs including the skin
- Skin provides vitamin D needed for proper calcium absorption and use

Skeletal System

Endocrine System
- Skeletal system provides some bony protection
- Hormones regulate uptake and release of calcium from bone; hormones promote long bone growth and maturation

Lymphatic System/Immunity
- Skeletal system provides some protection to lymphatic organs; lymphocytes involved in immune response originate in bone marrow
- Lymphatic system drains leaked tissue fluids; immune cells protect against pathogens

Digestive System
- Skeletal system provides some bony protection to intestines, pelvic organs, and liver
- Digestive system provides nutrients needed for bone health and growth

Urinary System
- Skeletal system protects pelvic organs (bladder, etc.)
- Urinary system activates vitamin D; disposes of nitrogenous wastes

Muscular System
- Skeletal system provides levers plus calcium for muscle activity
- Muscle pull on bones increases bone strength and viability; helps determine bone shape

Figure 5.30   Photograph of a hand deformed by rheumatoid arthritis.

Acute forms of arthritis usually result from bacterial invasion and are treated with antibiotic drugs. The synovial membrane thickens and fluid production decreases, leading to increased friction and pain. Chronic forms of arthritis include osteoarthritis, rheumatoid arthritis, and gouty arthritis, which differ substantially in their later symptoms and consequences. We will focus on these forms here.

Osteoarthritis (OA), the most common form of arthritis, is a chronic degenerative condition that typically affects the aged. OA, also called "wear-and-tear arthritis," affects the articular cartilages. Over the years, there is a softening, fraying, and eventual breakdown of the cartilage. As the disease progresses, the exposed bone thickens and extra bone tissue, called *bone spurs,* grows around the margins of the eroded cartilage. The bone spurs protrude into the joint cavity, which restricts joint movement. Patients complain of stiffness on arising that lessens with activity, and the affected joints may make a crunching noise when moved. The joints most commonly affected are those of the fingers, the cervical and lumbar joints of the spine, and the large, weight-bearing joints of the lower limbs (knees and hips).

The course of osteoarthritis is usually slow and irreversible, but it is rarely crippling. In most cases, its symptoms are controllable with a mild analgesic such as aspirin, moderate activity to maintain joint mobility, and rest when the joint becomes very painful.

Rheumatoid (roo′mah-toid) **arthritis** (RA) is a chronic inflammatory disorder. Its onset is insidious and usually occurs between the ages of 30 and 40, but it may occur at any age. It affects three times as many women as men. Many joints, particularly those of the fingers, wrists, ankles, and feet, are affected at the same time and usually in a symmetrical manner. For example, if the right elbow is affected, most likely the left elbow will be affected also. The course of RA varies, and is marked by remissions and flare-ups (*rheumat* = susceptible to change or flux).

RA is an autoimmune disease—a disorder in which the body's immune system attempts to destroy its own tissues. The initial trigger for this reaction is unknown, but some suspect that it results from certain bacterial or viral infections.

RA begins with inflammation of the synovial membranes. The membranes thicken and the joints swell as synovial fluid accumulates. Inflammatory cells (white blood cells and others) enter the joint cavity from the blood and produce *pannus,* an abnormal tissue that clings to and erodes articular cartilages. As the cartilage is destroyed, scar tissue forms and connects the bone ends. The scar tissue eventually ossifies, and the bone ends become firmly fused *(ankylosis)* and often deformed (see Figure 5.30). Not all cases of RA progress to the severely crippling ankylosis stage, but all cases involve restricted joint movement and extreme pain.

Current therapy for RA involves many different kinds of drugs. However, drug therapy is usually begun with aspirin, which in large doses is an effective anti-inflammatory agent. Exercise is recommended to maintain as much joint mobility as possible. Cold packs are used to relieve the swelling and pain, and heat helps to relieve morning stiffness. Replacement joints are the last resort for severely crippled RA patients.

Gouty (gow′te) **arthritis,** or **gout,** is a disease in which uric acid (a normal waste product of nucleic acid metabolism) accumulates in the blood and may be deposited as crystals in the soft tissues of joints. This leads to an agonizingly painful attack that typically affects a single joint, often in the great toe. Gout is most common in males, and rarely appears before the age of 30. It tends to run in families, so genetic factors are definitely implicated.

Untreated gout can be very destructive; the bone ends fuse and the joint becomes immobilized. Fortunately, several drugs (colchicine and others) are successful in preventing acute gout attacks. Patients are advised to lose weight if obese, to avoid foods such as liver, kidneys, and sardines, which are high in nucleic acids, and to avoid alcohol and excessive vitamin C, which inhibit excretion of uric acid by the kidneys. ❏

DEVELOPMENTAL ASPECTS OF THE SKELETON

As described earlier, the first "long bones" in the very young fetus are formed of hyaline cartilage, and the earliest "flat bones" of the skull are actually fibrous membranes. An X ray of a fetal skeleton showing fontanels, cartilages, and ossified areas is shown in Figure 5.31. As the fetus continues to develop and grow, both the flat and the long bone models are converted to bone. At birth, some fontanels still remain in the skull to allow for brain growth, but these areas are usually fully ossified by 2 years of age. By the end of adolescence, the epiphyseal plates of long bones that provide for longitudinal growth in childhood have become fully ossified, and long-bone growth ends.

Most cases of abnormal spinal curvatures, such as scoliosis and lordosis (see p. 130), are congenital, but some can result from injuries. The abnormal curvatures are usually treated by surgery, braces, or casts when diagnosed. Generally speaking, young healthy people have no skeletal problems assuming that the diet is nutritious and they stay reasonably active.

Rickets is a disease of children in which the bones fail to calcify. As a result, the bones soften and a definite bowing of the weight-bearing bones of the legs occurs. Rickets is usually due to a lack of calcium in the diet or lack of vitamin D, which is needed to absorb calcium into the bloodstream. Rickets is not seen very often in the United States, where great stress is put on good nutrition. Milk, bread, and other foods are fortified with vitamin D, and most children drink enough calcium-

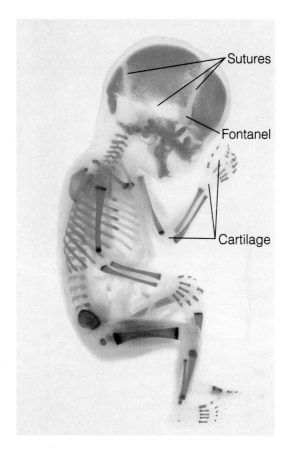

Figure 5.31 Skeleton of a 12-week-old fetus.
Darker areas of the skeleton represent ossified, or bony, areas, whereas the lighter areas indicate cartilage that has not yet been replaced by bone. The fontanels, soft membranous areas between the sutures, are still clearly visible.

rich milk. However, it remains a problem in some other parts of the world. ❏

It cannot be emphasized too strongly that bones have to be physically stressed to remain healthy. The more you use them, the stronger they get. When we remain active physically and muscles and gravity pull on the skeleton, the bones respond by becoming stronger. On the other hand, if we are totally inactive, they become thin and fragile. **Osteoporosis,** a loss in bone mass leading to thin, fragile bones (Figure 5.32) especially in the spine and the neck of the femur, is a common consequence of aging, particularly in women. Estrogen helps to maintain the health and normal density of a woman's skeleton, and the estrogen

Figure 5.32   Osteoporosis. The architecture of osteoporotic bone (left) is contrasted with that of normal bone (right).

deficiency that occurs after a woman goes through menopause ("change of life") is strongly implicated as a cause of osteoporosis. Other factors that may contribute to osteoporosis are a diet poor in calcium and protein, lack of vitamin D, smoking, and insufficient weight-bearing exercise to stress the bones. Sadly, many elderly people feel that they are helping themselves by "saving their strength" and not doing anything too physical. Their reward for this is *pathologic fractures* (spontaneous breaks

without apparent injury), which increase dramatically with age and are the single most common skeletal problem for this age group.

Advancing years also take their toll on joints. Weight-bearing joints in particular begin to degenerate and *osteoarthritis* is common. Such degenerative joint changes lead to the complaint often heard from the aging person: "My joints are getting so stiff. . . ."

Important Terms

amphiarthroses (am″fe-ar-thro′sēz)

appendicular skeleton

articular cartilage

articulations

axial skeleton

bony thorax

cartilaginous joint

compact bone

diaphysis (di-af′ĭ-sis)

diarthroses (di″ar-thro′sēz)

epiphyseal plate

epiphysis (ĕ-pif′ĭ-sis)

fibrous joint

flat bones

intervertebral disks

irregular bones

ligaments

long bones

pelvic girdle	**spongy bone**
periosteum (per-e-ŏs′te-um)	**synarthroses** (sin″ar-thro′sēz)
red marrow	**synovial joint**
short bones	**vertebral column**
shoulder girdle	**yellow marrow**
skull	

Summary

Bones: An Overview (pp. 114–121)

1. Bones support and protect body organs; serve as levers for the muscles to pull on to cause movement at joints; store calcium, fats, and other substances for the body; and contain red marrow, the site of blood cell production.

2. Bones are classified into four groups—long, short, flat, and irregular—on the basis of their shape and the amount of compact or spongy bone they contain. Bone markings are important anatomical landmarks that reveal where muscles attach and where blood vessels and nerves pass.

3. A long bone is composed of a shaft (diaphysis) with two ends (epiphyses). The shaft is compact bone; its cavity contains yellow marrow. The epiphyses are covered with hyaline cartilage; they contain spongy bone (where red marrow is found).

4. The organic parts of the matrix make bone flexible; calcium salts deposited in the matrix make bone hard.

5. Bones form on hyaline cartilage "models" or fibrous membranes. Eventually these initial supporting structures are replaced by bone tissue. Epiphyseal plates persist to provide for longitudinal growth of long bones during childhood and become inactive when adolescence ends.

6. Bones change in shape throughout life. This remodeling occurs in response to hormones (i.e., PTH, which regulates blood calcium levels) and mechanical stresses acting on the skeleton.

7. A fracture is a break in a bone. Common types of fractures include simple, compound, compression, comminuted, and greenstick. Bone fractures must be reduced to heal properly.

Axial Skeleton (pp. 121–133)

1. The skull is formed by cranial and facial bones. Eight cranial bones protect the brain: frontal, occipital, ethmoid, and sphenoid bones, and the pairs of parietal and temporal bones. The 14 facial bones are all paired (maxillae, zygomatics, palatines, nasals, lacrimals, and inferior conchae), except for the vomer and mandible. The hyoid bone, not really a skull bone, is supported in the neck by ligaments.

2. Skulls of newborns contain fontanels (membranous areas), which allow brain growth. The infant's facial bones are very small compared to the size of the cranium.

3. The vertebral column is formed from 24 vertebrae, the sacrum, and the coccyx. There are 7 cervical vertebrae, 12 thoracic vertebrae, and 5 lumbar vertebrae, which have common as well as unique features. The vertebrae are separated by fibrocartilage disks that allow the vertebral column to be flexible. The vertebral column is S-shaped to allow for upright posture. Spinal curvatures present at birth are the thoracic and sacral curvatures; secondary curvatures (cervical and lumbar) develop after birth.

4. The bony thorax is formed from the sternum and 12 pairs of ribs. All ribs attach posteriorly to thoracic vertebrae. Anteriorly, the first 7 pairs attach directly to the sternum (true ribs); the last 5 pairs attach indirectly or not at all (false ribs). The bony thorax encloses the lungs, heart, and other organs of the thoracic cavity.

Appendicular Skeleton (pp. 133–142)

1. The shoulder girdle, composed of two bones—the scapula and the clavicle—attaches the upper limb to the axial skeleton. It is a light, poorly reinforced girdle that allows the upper limb a great deal of freedom.

2. The bones of the upper limb include the humerus of the arm, the radius and ulna of the forearm, and the carpals, metacarpals, and phalanges of the hand.

3. The pelvic girdle is formed by the two coxal bones, or hip bones. Each hip bone is the result of fusion of the ilium, ischium, and pubis bones. The pelvic girdle is securely attached to the sacrum of the axial skeleton, and the socket for the thigh bone is deep and heavily reinforced. This girdle receives the weight of the upper body and transfers it to the lower limbs. The female pelvis is lighter and broader than the male's; its inlet and outlet are larger, reflecting the childbearing function of the female.

4. The bones of the lower limb include the femur of the thigh, the tibia and fibula of the leg, and the tarsals, metatarsals, and phalanges of the foot.

Joints (pp. 142-148)

1. Joints hold bones together and allow movement of the skeleton.

2. Joints fall into three functional categories: synarthroses (immovable), amphiarthroses (slightly movable), and diarthroses (freely movable).

3. Joints also can be classified structurally as fibrous, cartilaginous, or synovial joints depending on the substance separating the articulating bones.

4. Most joints of the body are synovial joints. In synovial joints, the articulating bone surfaces are covered with articular cartilage and enclosed within the joint cavity by a fibrous capsule lined with a synovial membrane. All synovial joints are diarthroses. Most fibrous joints are synarthrotic and most cartilaginous joints are amphiarthrotic.

5. The most common joint problem is arthritis, or inflammation of the joints. Osteoarthritis, or degenerative arthritis, is a result of the "wear and tear" on joints over many years and is a common affliction of the aged. Rheumatoid arthritis occurs in both young and older adults; it is believed to be an autoimmune disease. Gouty arthritis, caused by the deposit of uric acid crystals in joints, typically affects a single joint.

Developmental Aspects of the Skeleton (pp. 148-149)

1. The fetal skeleton is largely made of cartilage, most of which is replaced by bone by the time of birth. Unossified areas persist for a brief period in the skull (fontanels) and until the end of adolescence in the long bones (epiphyseal disks).

2. Abnormal spinal curvatures are usually congenital and are treated by surgery, braces, or casts. During youth, most other bone problems are due to trauma (fractures) or poor nutrition (rickets). Bones must be stressed to remain healthy.

3. Fractures are the most common bone problem in elderly persons. Osteoporosis, a condition of bone wasting that results mainly from hormone deficit or inactivity, is also common in elderly individuals.

Review Questions

1. Name three functions of the skeletal system.

2. Name the four major classifications of bones, and give two examples of each type.

3. What is the anatomical name for the shaft of a long bone? for its ends? What is yellow marrow? How do spongy and compact bone look different?

4. Why do bone injuries heal much more rapidly than injuries to cartilage?

5. What is the function of the organic part of bone matrix? of the inorganic part (bone salts)?

6. Compare and contrast the role of PTH hormone and mechanical forces acting on the skeleton in bone remodeling.

7. Define *fracture*. Which fracture types are most common in the elderly? Why are greenstick fractures more common in children?

8. Name the three major parts of the axial skeleton.

9. Name the eight bones of the cranium.

10. What bones are connected by the frontal suture? by the sagittal suture?

11. With one exception, all skull bones are joined by sutures. What is the exception?

12. Name the four bones containing the paranasal sinuses.

13. What facial bone forms the chin? the cheekbone? the upper jaw? the bony eyebrow ridges?

14. Name two ways in which the fetal skull differs from the adult skull.

15. Name the five major regions of the vertebral column.

16. Diagram the normal spinal curvatures, and then the curvatures seen in scoliosis and lordosis.

17. What is the function of the intervertebral disks?

18. Name the major components of the thorax.

19. What is a true rib? a false rib? Is a floating rib a true or a false rib? Why are floating ribs easily broken?

20. What is the general shape of the thoracic cage?

21. Name the bones of the shoulder girdle.

22. Name all the bones with which the ulna articulates.

23. The major function of the shoulder girdle is to provide flexibility. What is the major function of the pelvic girdle?

24. What bones make up each hip bone or coxal bone? Which of these is the largest? Which has tuberosities that we sit on? Which is the most anterior?

25. List three differences between the male and the female pelvis.

26. Name the bones of the lower limb superiorly to inferiorly.

27. What is the function of joints?

28. Compare the amount of movement possible in synarthrotic, amphiarthrotic, and diarthrotic joints. Relate these terms to the structural classification of joints; that is, to fibrous, cartilaginous, and synovial joints.

29. Describe the structure of a synovial joint.

30. Define *arthritis*. What type of arthritis is most common in the elderly? What type is believed to result from the immune system's attack on one's own joint tissues?

31. List two factors that keep bones healthy. List two factors that can cause bones to become soft or to atrophy.

At the Clinic

1. A 75-year-old woman and her 9-year-old granddaughter were in a train crash in which both sustained trauma to the chest. X rays showed that the grandmother had several fractured ribs but her granddaughter had none. Explain these surprisingly (?) different findings.

2. The pediatrician at the clinic explains to parents of a newborn that their child suffers from cleft palate. She tells them that the normal palate fuses in an anterior-to-posterior pattern. The child's palatine processes have not fused. Have his palatine bones fused normally?

3. After having a severe cold accompanied by nasal congestion, Helen complained that she had a frontal headache and the right side of her face ached. What bony structures probably became infected by the bacteria or viruses causing the cold?

4. Bernice, a 75-year-old woman, stumbled slightly while walking, then felt a terrible pain in her left hip. At the hospital, X rays revealed that the hip was broken. Also, the compact and spongy bone throughout her spine were very thin. What was her probable condition?

5. At work, a box fell from a shelf onto Bertha's acromial region. In the emergency room, the physician felt that the head of her humerus had moved into the axilla. What had happened to Bertha?

6
The Muscular System

YOUR GOALS

After completing this chapter, you will have a working knowledge of the functions of the muscular system and will have mastered the objectives listed below.

FUNCTION PREVIEW

■ The muscular system provides for movement of the body and its parts (as muscles shorten), maintains posture, generates heat, and stabilizes joints.

OBJECTIVE CHECKLIST

OVERVIEW OF MUSCLE TISSUES (pp. 154–158)

❏ Describe similarities and differences in the structure and function of the three types of muscle tissue and note where they are found in the body.

❏ Define and explain the role of the following: *endomysium, perimysium, epimysium, tendon,* and *aponeurosis.*

MICROSCOPIC ANATOMY OF SKELETAL MUSCLE (pp. 158–160)

❏ Define *muscular system.*

❏ Describe the microscopic structure of skeletal muscle.

SKELETAL MUSCLE ACTIVITY (pp. 160–168)

❏ Describe how an action potential is initiated in a muscle cell.

❏ Describe the events of muscle cell contraction.

❏ Define *graded response, tetanus, muscle fatigue, isotonic* and *isometric contractions,* and *muscle tone* as these terms apply to a skeletal muscle.

❏ Describe three ways in which ATP is regenerated during muscle activity.

❏ Define *oxygen debt* and *muscle fatigue* and list possible causes of muscle fatigue.

❏ Describe the effects of aerobic and resistance exercise on skeletal muscles and other body organs.

MUSCLE MOVEMENTS, TYPES, AND NAMES (pp. 170–174)

❏ Define *origin, insertion, prime mover, antagonist, synergist,* and *fixator* as they relate to muscles.

(continued)

Because flexing muscles look like mice scurrying beneath the skin, some scientist long ago dubbed them *muscles,* from the Latin word *mus* meaning "little mouse." Indeed, an image of the rippling muscles of professional boxers or weight lifters is often the first thing that comes to mind when one hears the word *muscle.* But muscle is also the dominant tissue in the heart and in the walls of other hollow organs of the body. In all its forms, it makes up nearly half the body's mass.

The essential function of muscle is *contraction* or *shortening*—a unique characteristic that sets it apart from any other body tissue. As a result of this ability, muscles are responsible for essentially all body movement, and can be viewed as the "machines" of the body.

OVERVIEW OF MUSCLE TISSUES

Muscle Types

There are three types of muscle tissue—skeletal, cardiac, and smooth. As summarized in Table 6.1, these differ in their cell structure, body location, and how they are stimulated to contract. But, before we explore their differences, let's look at some of the ways they are all the same.

First, *all* muscle cells are elongated. For this reason all muscle cells, regardless of type, are called *muscle fibers.* Second, the ability of muscle to shorten or contract depends on two types of *myofilaments,* the muscle cell equivalents of the microfilaments of the cytoskeleton studied in Chapter 3. A third similarity has to do with terminology. Whenever you see the prefixes *myo* and *mys* ("muscle") and *sarco* (meaning "flesh"), you will know that muscle is being referred to. For example, in muscle cells the cytoplasm is called *sarcoplasm* (sar′ko-plaz″um.)

Skeletal Muscle

Skeletal muscle fibers are packaged into the organs called *skeletal muscles* that attach to the body's skeleton. As the skeletal muscles cover our bony "underpinnings," they help form the much smoother contours of the body. Skeletal muscle fibers are cigar-shaped, multinucleate cells, and the largest of the muscle fiber types—some ranging up to 30 cm (1 foot) in length. Indeed, the fibers of large, hardworking muscles, such as the antigravity muscles of the hip, are so big and coarse that they can be seen with the naked eye.

Skeletal muscle is also known as *striated muscle* because its fibers appear to be striped, and as *voluntary muscle* because it is the only muscle type subject to conscious control. However, it is important to recognize that skeletal muscles are often activated by reflexes (without our "willed command") as well. When you think of skeletal muscle tissue, the key words to keep in mind are *skeletal,*

Table 6.1   Comparison of Skeletal, Cardiac, and Smooth Muscles

Characteristic	Skeletal	Cardiac	Smooth
Body location	Attached to bones or (some facial muscles) to skin	Walls of the heart	Typically in walls of hollow visceral organs (other than the heart)
Cell shape and appearance	Single, very long, cylindrical, multinucleate cells with very obvious striations	Branching chains of cells; uninucleate; striations	Single, fusiform, uninucleate; no striations
Regulation of contraction	Voluntary; via nervous system controls	Involuntary; the heart has a pacemaker; also nervous system controls; hormones	Involuntary; nervous system controls; hormones, chemicals, stretch
Speed of contraction	Slow to fast	Slow	Very slow
Rhythmic contraction	No	Yes	Yes in some

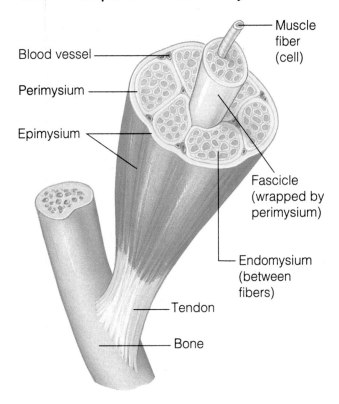

Figure 6.1   Connective tissue wrappings of skeletal muscle.

striated, and voluntary. Skeletal muscle tissue can contract rapidly and with great force, but it tires easily and must rest after short periods of activity.

Skeletal muscle fibers, like most living cells, are soft and surprisingly fragile. Yet skeletal muscles can exert tremendous power—how so? The reason they are not ripped apart as they exert force is that thousands of their fibers are bundled together by connective tissue, which provides strength and support to the muscle as a whole (Figure 6.1). Each muscle fiber is enclosed in a delicate connective tissue sheath called an **endomysium** (en″do-mis′e-um). Several sheathed muscle fibers are then wrapped by a coarser fibrous membrane called a **perimysium** to form a bundle of fibers called a **fascicle** (fas′ĭ-kul). Many fascicles are bound together by an even tougher "overcoat" of connective tissue called an **epimysium,** which covers the entire muscle. The epimysia blend into the strong, cordlike **tendons,** or into sheetlike **aponeuroses** (ap″o-nu-ro′sēz), which attach muscles indirectly to bones, cartilages, or connective tissue coverings of each other.

Besides simply acting to anchor muscles, tendons perform several functions. The most important are providing durability and conserving space. Since tendons are mostly tough collagenic fibers, they can cross rough bony projections, which would tear the more delicate muscle tissues. Because of their relatively small size, more tendons than fleshy muscles can pass over a joint.

Many people think of muscles as always having an enlarged "belly" that tapers down to a tendon at each end. However, muscles vary considerably in the way their fibers are arranged. Many are spindle-shaped as just described, but in others, the fibers are arranged in a fan-shape or a circle.

Smooth Muscle

Smooth muscle has no striations and is involuntary, which means that we cannot consciously control it. Found in the walls of hollow visceral organs such as the stomach, urinary bladder, and respiratory passages, smooth muscle acts to propel substances along a definite tract, or pathway, within the body. We can best describe smooth muscle using the terms visceral, nonstriated, and involuntary.

As described in Chapter 3, smooth muscle cells are spindle-shaped and have a single nucleus (see also Table 6.1 on p. 155). They are arranged in sheets or layers. Most often there are two such layers, one running circularly and the other longitudinally, as shown in Figure 6.2a. As the two layers alternately contract and relax, they change the size and shape of the organ. Movement of food through the digestive tract and emptying the bowels and bladder are examples of "housekeeping" activities normally handled by smooth muscles. Contractions of smooth muscle fibers are slow and sustained. If skeletal muscle is like a speedy windup car that quickly runs down, then smooth muscle is like a steady, heavy-duty engine that lumbers along tirelessly.

Cardiac Muscle

Cardiac muscle is found in only one place in the body—the heart. The heart serves as a pump, propelling blood into the blood vessels and to all tis-

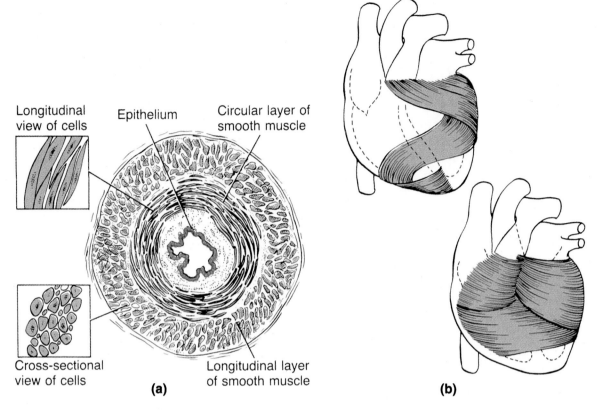

Figure 6.2   Arrangement of smooth and cardiac muscle cells. (a) Diagrammatic view of a cross section of a sperm duct; smooth muscle cells are shown in both longitudinal and cross-sectional views. **(b)** Longitudinal view of the heart, showing the spiral arrangement of the cardiac muscle cells in its walls.

sues of the body. Cardiac muscle is like skeletal muscle in that it is striated, and like smooth muscle in that it is involuntary and cannot be consciously controlled by most of us. Important key words to jog your memory for this muscle type are *cardiac, striated,* and *involuntary.*

The cardiac fibers are cushioned by small amounts of soft connective tissue and arranged in spiral or figure **8**-shaped bundles, as shown in Figure 6.2b. When the heart contracts, its internal chambers become smaller, forcing the blood into the large arteries leaving the heart. Recall that cardiac muscle fibers are branching cells joined by special junctions called *intercalated disks* (see Table 6.1 on p. 155, and Figure 3.19 on p. 84). These two structural features and the spiral arrangement of the muscle bundles in the heart allow heart activity to be closely coordinated. Cardiac muscle usually contracts at a fairly steady rate

set by the heart's "in-house" pacemaker, but the heart can also be stimulated by the nervous system to shift into "high gear" for short periods, as when you race to catch the bus.

Muscle Functions

Muscle plays four important roles in the body: it *produces movement, maintains posture, stabilizes joints,* and *generates heat.*

Producing Movement

Just about all movements of the human body are a result of muscle contraction. Mobility of the body as a whole reflects the activity of skeletal muscles, which are responsible for all locomotion (walking, swimming, and cross-country skiing, for example) and manipulation. They enable us to respond quickly to changes in the external environment; for example, their speed and power enable us to jump out of the way of a runaway car and then follow its flight with our eyes. They also allow us to express our emotions with the silent language of smiles and frowns.

These are distinct from the smooth muscle of blood vessel walls and cardiac muscle of the heart which work together to circulate blood and maintain blood pressure, and the smooth muscle of other hollow organs which forces fluids (urine, bile) and other substances (food, a baby) through internal body channels.

Maintaining Posture

We are rarely aware of the workings of the skeletal muscles that maintain body posture. Yet, they function almost continuously, making one tiny adjustment after another so that we can maintain an erect or seated posture despite the never-ending downward pull of gravity.

Stabilizing Joints

As the skeletal muscles pull on bones to cause movements, they also stabilize the joints of the skeleton. Indeed, muscle tendons are extremely important in reinforcing and stabilizing joints that have poorly fitting articulating surfaces (the shoulder joint, for example).

Generating Heat

The fourth function of muscle, generation of body heat, is a by-product of muscle activity. As ATP is used to power muscle contraction, nearly three-quarters of its energy escapes as heat. This heat is vital in maintaining normal body temperature. Since skeletal muscle accounts for at least 40 percent of body mass, it is the muscle type most responsible for heat generation.

As you can see, each of the three muscle types has a structure and function well suited for its job in the body. But since the term **muscular system** applies specifically to skeletal muscle, we will be concentrating on this muscle type in this chapter. The most important structural and functional aspects of the three muscle types are outlined in Table 6.1.

MICROSCOPIC ANATOMY OF SKELETAL MUSCLE

As mentioned above and illustrated in Figure 6.3a, skeletal muscle cells are multinucleate. Many oval nuclei can be seen just beneath the plasma membrane, which is called the **sarcolemma** (sar″ko-lem′ah; "muscle husk") in muscle cells. The nuclei are pushed aside by long ribbonlike organelles, the **myofibrils** (mi″o-fi′brilz), which nearly fill the cytoplasm. Alternating **light (I)** and **dark (A) bands** along the length of the perfectly aligned myofibrils give the muscle cell as a whole its striped appearance. A closer look at the banding pattern reveals that the light I band has a midline interruption, a darker area called the *Z line,* and the dark A band has a lighter central area called the *H zone* (Figure 6.3b). So why are we bothering with all these terms—dark this and light that? Because the banding pattern reveals the working structure of the myofibrils. First, we find that the myofibrils are actually chains of tiny contractile units called **sarcomeres** (sar′ko-merz), which are aligned end-to-end like boxcars in a train along the length of the myofibrils. Second, it is the arrangement of even smaller structures (myofilaments) *within* sarcomeres that actually produces the banding pattern.

Let's examine how the arrangement of the myofilaments leads to the banding pattern. There are two types of threadlike protein **myofilaments** within each of our "boxcar" sarcomeres (Figure 6.3c). The larger **thick filaments,** also called *myosin filaments,* are made mostly of bundled molecules of the protein **myosin,** but they also contain ATPase enzymes which split ATP to generate the power for muscle contraction. Notice that the thick filaments extend the entire length of the dark A band. Also, notice that the midparts of the thick filaments are smooth, but their ends are studded with small projections (Figure 6.3d). These projections or myosin *heads* are sometimes called **cross bridges** because they link the thick and thin filaments together during contraction. The **thin filaments** are composed of the contractile protein called **actin,** plus some regulatory proteins which play a role in allowing (or preventing) myosin head binding to actin. The thin filaments, also called *actin filaments,* are anchored to the Z line (which is actually a membrane). Notice that the light I band is an area that includes parts of two ad-

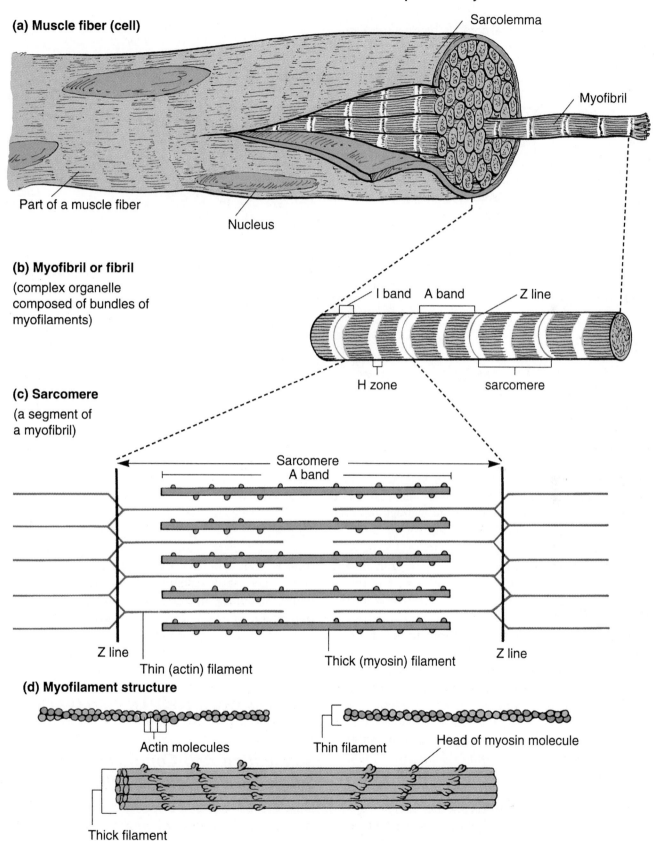

(a) Muscle fiber (cell)

Sarcolemma

Myofibril

Part of a muscle fiber

Nucleus

(b) Myofibril or fibril

(complex organelle
composed of bundles of
myofilaments)

I band   A band   Z line

H zone   sarcomere

(c) Sarcomere

(a segment of
a myofibril)

Sarcomere
A band

Z line

Thin (actin) filament

Thick (myosin) filament

Z line

(d) Myofilament structure

Actin molecules   Thin filament   Head of myosin molecule

Thick filament

Figure 6.3   Anatomy of a skeletal muscle cell (fiber). (a) A muscle fiber.
One myofibril has been extended. **(b)** Enlarged view of a myofibril showing its
banding pattern. **(c)** Enlarged view of one sarcomere (contractile unit) of a myofibril. **(d)** Structure of the thick and thin myofilaments found in the sarcomeres.

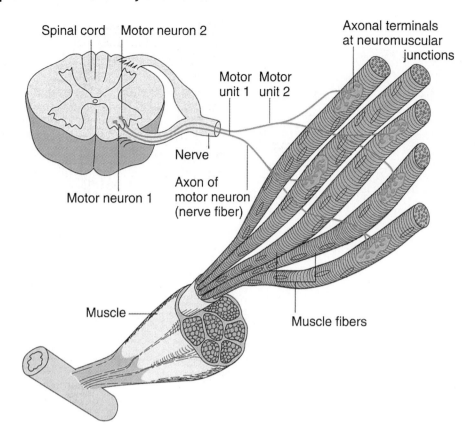

Figure 6.4   **Motor units.** Each motor unit consists of a motor neuron and all the muscle fibers it activates. Portions of two motor units are shown. The motor neurons reside in the spinal cord, and their axons extend to the muscle. Within the muscle, each axon divides into a number of axonal terminals, distributed to muscle fibers scattered throughout the muscle.

jacent sarcomeres and contains *only* the thin filaments. Although the thin filaments overlap the ends of the thick filaments, they do not extend into the middle of a relaxed sarcomere, and thus the central region, called the *H zone,* looks a bit lighter. When contraction occurs, and the actin-containing filaments slide toward each other into the center of the sarcomeres, these light zones disappear because the actin and myosin filaments are completely overlapped. For now, however, just recognize that it is the precise arrangement of the myofilaments in the myofibrils that produces the banding pattern, or striations, in skeletal muscle cells.

Not shown in Figure 6.3 is another very important muscle fiber organelle—the **sarcoplasmic reticulum (SR),** a specialized smooth endoplasmic reticulum. The interconnecting tubules and sacs of the SR surround each and every myofibril like the sleeve of a loosely crocheted sweater sur-

rounds your arm. The major role of this elaborate system is to store calcium and to release it "on demand" when the muscle fiber is stimulated to contract. As you will see, calcium provides the final "go" signal for contraction.

SKELETAL MUSCLE ACTIVITY

Stimulation and Contraction of Single Skeletal Muscle Cells

Muscle cells have some special functional properties that enable them to perform their duties. The first of these is *irritability,* the ability to receive and respond to a stimulus. The second, *contractility,* is the ability to shorten (forcibly) when an adequate stimulus is received.

Figure 6.5   The neuromuscular junction. (a) Axonal terminal of a motor neuron forming a neuromuscular junction with a muscle fiber. (b) The axonal terminal contains vesicles filled with the neurotransmitter acetylcholine (ACh), which is released when the nerve impulse reaches the axonal terminal. Acetylcholine diffuses across the synaptic cleft and attaches to ACh receptors on the sarcolemma, initiating changes in the electrical condition of the sarcolemma. The sarcolemma is highly invaginated (folded) adjacent to the synaptic cleft, allowing it to have many ACh receptors in this region.

The Nerve Stimulus and the Action Potential

Skeletal muscle cells must be stimulated by nerve impulses to contract. One motor neuron (nerve cell) may stimulate a few muscle cells or hundreds of them, depending on the particular muscle and the work it does. One neuron and all the skeletal muscle cells it stimulates are a **motor unit** (Figure 6.4). When a long threadlike extension of the neuron, called the *nerve fiber* or *axon,* reaches the muscle, it branches into a number of *axonal terminals,* each of which forms junctions with the sarcolemma of a different muscle cell (Figure 6.5).

These junctions are called **neuromuscular** (literally, "nerve-muscle") **junctions.** Although the nerve endings and the muscle cells' membranes are very close, they never touch. The gap between them, the **synaptic cleft,** is filled with tissue (interstitial) fluid.

Now that we have described the structure of the neuromuscular junction, we are ready to examine what happens there. When the nerve impulse reaches the axonal terminals, a chemical referred to as a **neurotransmitter** is released. The specific neurotransmitter that stimulates skeletal muscle cells is **acetylcholine** (as″e-til-ko′len) or

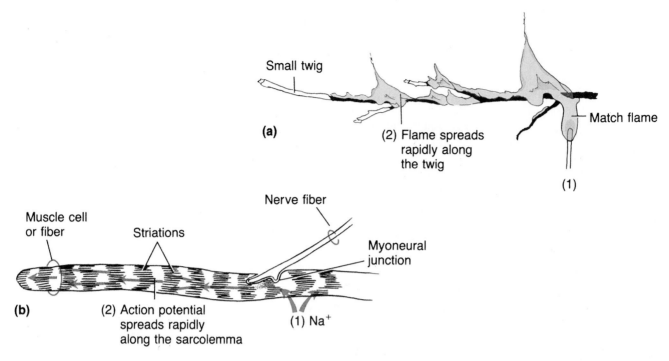

Figure 6.6   Comparison of the action potential to a flame consuming a dry twig. (a) The first event in igniting a dry twig is holding the match flame under one area of the twig. The second event is the twig's bursting into flame when it has been heated enough and the flame's spreading to burn the entire twig. **(b)** The first event in exciting a muscle cell is the rapid diffusion of sodium ions (Na^+) into the cell when the permeability of the sarcolemma changes. The second event is the spreading of the action potential along the sarcolemma when enough sodium ions have entered to upset the electrical conditions in the cell.

ACh. Acetylcholine diffuses across the synaptic cleft and attaches to receptors (membrane proteins), which are part of the sarcolemma. If enough acetylcholine is released, the sarcolemma at that point becomes *temporarily* permeable to sodium ions (Na^+), which rush into the muscle cell. This sudden inward rush of sodium ions gives the cell interior an excess of positive ions, which upsets and changes the electrical conditions of the sarcolemma. This "upset" generates an electrical current called an **action potential.** Once begun, the action potential is unstoppable; it travels over the entire surface of the sarcolemma, conducting the electrical impulse from one end of the cell to the other. The result is contraction of the muscle cell.

This series of events is explained more fully on p. 204 in the discussion of nerve physiology, but perhaps it would be helpful to compare this to some common event, such as lighting a match under a small dry twig (Figure 6.6). The flame, which soon begins to char the twig, can be compared to the change in membrane permeability that allows sodium ions into the cell. When that part of the twig becomes hot enough (when enough sodium ions have entered the cell), the twig will suddenly burst into flame and the flame will consume the twig (the action potential will be conducted along the entire length of the sarcolemma). The events that return the cell to its resting state include (1) diffusion of potassium ions (K^+) out of the cell, and (2) activation of the sodium-potassium pump, the active transport mechanism that moves the sodium and potassium ions back to their initial positions.

Figure 6.7   Diagrammatic views of a sarcomere. (**a**) Relaxed; (**b**) contracted. Notice that in the contracted sarcomere, the light H zone in the center of the A band has disappeared, the Z lines are closer to the thick filaments, and the I bands have nearly disappeared. The A bands move closer together but do not change in length.

Mechanism of Muscle Contraction: The Sliding Filament Theory

What causes the filaments to slide? This question brings us back to the cross bridges (myosin heads) that protrude all around the ends of the thick filaments. When muscle fibers are activated by the nervous system as just described, the cross bridges attach to myosin binding sites on the thin filaments, and the sliding begins. Energized by ATP, each cross bridge attaches and detaches several times during a contraction, acting much like a tiny oar to generate tension and pull the thin filaments toward the center of the sarcomere. As this event occurs simultaneously in sarcomeres throughout the cell, the muscle cell shortens (Figure 6.7). The attachment of the myosin cross bridges to actin requires calcium ions, and the action potential lead-

ing to contraction causes the sarcoplasmic reticulum to release stored calcium ions into the sarcoplasm. Calcium's precise role in contraction is described in more detail in Figure 6.8. When the action potential ends, calcium ions are immediately reabsorbed into the SR storage areas, and the muscle cell relaxes and settles back to its original length. This whole series of events takes just a few thousandths of a second.

It should be mentioned that while the action potential is occurring, acetylcholine, which began the process, is broken down by enzymes present on the sarcolemma. For this reason, a single nerve impulse produces only one contraction. This prevents continued contraction of the muscle cell in the absence of additional nerve impulses. The muscle cell relaxes until stimulated by the next round of acetylcholine release.

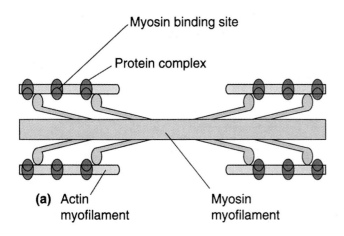

(a) Actin myofilament

Myosin binding site

Protein complex

Myosin myofilament

Figure 6.8 **Mechanism of contraction: The sliding filament theory.** In a relaxed muscle cell, the regulatory proteins forming part of the actin myofilaments prevent myosin binding (see **a**). When an action potential sweeps along its sarcolemma and a muscle cell is excited, calcium ions (Ca^{2+}) are released from intracellular storage areas (the sacs of the sarcoplasmic reticulum).

(b)   Upper part of thick filament

The flood of calcium acts as the final trigger for contraction, because as calcium binds to the regulatory proteins on the actin filaments, they change both their shape and their position on the thin filaments. This action exposes myosin binding sites on the actin, to which the myosin heads can attach (see **b**), and the myosin heads immediately begin seeking out binding sites.

(c)

Free myosin has a unique property: Its heads are "cocked," much like a set mousetrap. The physical attachment of myosin to actin "springs the trap," causing the myosin heads to snap (pivot) toward the center of the sarcomere. Since the actin and myosin are still firmly bound to each other when this happens, the thin filaments are slightly pulled toward the center of the sarcomere (see **c**). ATP provides the energy needed to release and recock each myosin head so that it is ready to take another "step" and attach to a binding site farther along the thin filament. This "walking" of the myosin cross bridges or heads along the thin filaments during muscle shortening is much like a centipede's gait. Some myosin heads ("legs") are always in contact with actin ("the ground"), so that the thin filaments cannot slide backward as this cycle is repeated again and again during contraction. Notice that the myofilaments themselves do not shorten during contraction; they simply slide past each other. When the action potential ends and calcium ions are reabsorbed into the SR storage areas, the regulatory proteins resume their original shape and position, and again block myosin binding to the thin filaments. Since myosin now has nothing to attach to, the muscle cell relaxes and settles back to its original length.

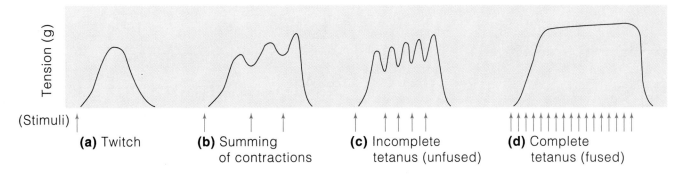

(Stimuli)

(a) Twitch   **(b)** Summing of contractions   **(c)** Incomplete tetanus (unfused)   **(d)** Complete tetanus (fused)

Figure 6.9   A whole muscle's response to different rates of stimulation. In (**a**), a single stimulus is delivered, and the muscle contracts and relaxes (a twitch contraction). In (**b**), stimuli are delivered more frequently, so that the muscle does not have time to completely relax; contraction force increases because the effects of the individual twitches are summed. In (**c**), more complete fusion of the twitches (incomplete tetanus) occurs as stimuli are delivered at a still faster rate. In (**d**), complete tetanus, a smooth continuous contraction without any evidence of relaxation, results from a very rapid rate of stimulation. (Points at which stimuli are delivered are indicated by red arrows. Tension [measured in grams] on the vertical axis refers to the relative force of muscle contraction.)

Contraction of a Skeletal Muscle as a Whole

Graded Responses

In skeletal muscles, the "all-or-none" law of muscle physiology applies to the *muscle cell*, not to the whole muscle. It states that a muscle cell will contract to its fullest extent when it is stimulated adequately; it never partially contracts. However, skeletal muscles are organs that consist of thousands of muscle cells, and they react to stimuli with **graded responses,** or different degrees of shortening. In general, graded muscle contractions can be produced two ways: (1) by changing the *speed* of muscle stimulation, and (2) by changing the *number* of muscle cells being stimulated. A muscle's response to each of these is briefly described next.

Muscle Response to Increasingly Rapid Stimulation Although **muscle twitches** (single, brief, jerky contractions) sometimes occur as a result of certain nervous system problems, this is *not* the way our muscles normally operate. In most types of muscle activity, nerve impulses are delivered to the muscle at a very rapid rate—so rapid that the cells do not

get a chance to relax completely between stimuli (see Figure 6.9). As a result, the effect of the successive contractions are "summed" (added) together and the contractions of the muscle get stronger and smoother. When the muscle is stimulated so rapidly that no evidence of relaxation is seen and the contractions are completely smooth and sustained, the muscle is said to be in **tetanus** (tet'ah-nus),* or in tetanic contraction.

Muscle Response to Stronger Stimuli Although tetanus also produces stronger muscle contractions, its primary role is to produce smooth and prolonged muscle contractions. How forcefully a muscle contracts depends to a large extent on how many of its cells are stimulated. When only a few cells are stimulated, the contraction of the muscle as a whole will be slight. In the strongest contractions, when all the motor units are active and all

---

*Tetanic contraction is normal and desirable, and is quite different from the pathologic condition of tetanus (commonly called *lockjaw*) caused by a toxin made by a bacterium. Lockjaw causes muscles to go into uncontrollable spasms, finally causing respiratory arrest.

Figure 6.10 **Methods of regenerating ATP during muscle activity.** The fastest mechanism is direct phosphorylation; the slowest is aerobic respiration.

the muscle cells are being stimulated, the muscle contraction is as strong as it can get. Thus, muscle contractions can be slight or vigorous depending on what work has to be done. (The same hand that soothes can also deliver a stinging slap!)

Providing Energy for Muscle Contraction

As a muscle contracts, the bonds of ATP molecules are hydrolyzed to release the needed energy. Surprisingly, muscles store very limited supplies of ATP—only 4 to 6 seconds' worth, just enough to get you going. Since ATP is the *only* energy source that can be used directly to power muscle activity, ATP must be regenerated continuously if contraction is to continue.

Essentially, contracting muscles use three pathways for ATP regeneration:

1. **Direct phosphorylation of ADP by creatine phosphate** (Figure 6.10a). The unique high-energy molecule **creatine phosphate (CP)** is found in muscle fibers but not other cell types. As ATP is being depleted, interactions between CP and ADP result in transfers of a high-energy phosphate group from CP to ADP, thus regenerating more ATP in a fraction of a second. Although muscle cells store perhaps five times as much CP as ATP, the CP supplies are also soon exhausted (in about 20 seconds).

2. **Aerobic respiration** (Figure 6.10c). At rest and during light exercise, ATP is regenerated almost entirely by metabolic pathways that use oxygen. These are collectively referred to as **aerobic respiration.** During aerobic respiration, glucose is broken down completely to carbon dioxide and water, and some of the energy released as the bonds are broken is captured in the bonds of ATP molecules. Although aerobic respiration provides a rich ATP harvest (36 ATP per 1 glucose), it is fairly slow and requires continuous delivery of oxygen and nutrient fuels to the muscle to keep it going.

3. **Anaerobic glycolysis and lactic acid formation** (Figure 6.10b). The initial steps of glucose breakdown occur via a pathway called *glycolysis,* which does not use oxygen, and hence is an *anaerobic* (literally "without oxygen") part of the metabolic pathway. During glycolysis, glucose is broken down to pyruvic acid and small amounts of energy are captured in ATP bonds (2 ATP per 1 glucose molecule). As long as enough oxygen is present, the pyruvic acid then enters the oxygen-requiring aerobic pathways to produce more ATP as described above. However, when muscle activity is intense, or oxygen and glucose delivery is temporarily inadequate to meet the needs of the working muscles, the sluggish aerobic mechanisms cannot keep up with the demands for ATP. Under

these conditions, the pyruvic acid generated during glycolysis is converted to **lactic acid** and the overall process is referred to as **anaerobic glycolysis.**

Anaerobic glycolysis produces only about 5 percent as much ATP from each glucose molecule as aerobic respiration. However, it is some $2\frac{1}{2}$ times faster and it can provide most of the ATP needed for 30 to 40 seconds of strenuous muscle activity. The main shortcomings of anaerobic glycolysis are that it uses huge amounts of glucose for a small ATP harvest, and accumulating lactic acid promotes muscle fatigue and muscle soreness.

Muscle Fatigue and Oxygen Debt

If we exercise our muscles strenuously for a long time, **muscle fatigue** occurs. A muscle is fatigued when it is unable to contract, even though it is still being stimulated. Without rest, an active or working muscle begins to tire and contracts more weakly until it finally ceases reacting and stops contracting. Muscle fatigue is believed to result from the **oxygen debt** that occurs during prolonged muscle activity: A person is not able to take in oxygen fast enough to keep the muscles supplied with all the oxygen they need when they are working vigorously. Obviously, then, the work that a muscle can do and how long it can work without becoming fatigued depends on how good its blood supply is. When muscles lack oxygen, lactic acid begins to accumulate in the muscle via the anaerobic mechanism described above. In addition, the muscle's ATP supply starts to run low. The increasing acidity in the muscle and the lack of ATP cause the muscle to contract less and less effectively and finally to stop contracting altogether.

True muscle fatigue, in which the muscle quits entirely, rarely occurs in most of us because we feel fatigued long before it happens and we simply reduce or stop our activity. It *does* happen commonly in marathon runners. Many of them have literally collapsed when their muscles became fatigued and could no longer work.

Oxygen debt, which always occurs to some extent during vigorous muscle activity, must be "paid back" whether or not fatigue occurs. During the recovery period after activity, the individual breathes rapidly and deeply. This continues until the muscles have received the amount of oxygen needed to get rid of the accumulated lactic acid and make ATP and creatine phosphate reserves.

Types of Muscle Contractions— Isotonic and Isometric

Until now, we have been discussing contraction in terms of shortening behavior, but muscles do not always shorten when they contract. (I can hear you saying, "What kind of double-talk is that?"—but pay attention.) The event that is common to all muscle contractions is that *tension* develops in the muscle as the actin and myosin myofilaments interact and attempt to slide past each other within the muscle fibers.

Isotonic contractions (literally, "same tone" or tension) are more familiar to most of us. In isotonic contractions, the myofilaments are successful in their sliding movements, the muscle shortens, and movement occurs. Bending the knee, rotating the arms, and smiling are all examples of isotonic contractions.

Contractions in which the muscles do not shorten are called **isometric contractions** (literally, "same measurement" or length). In isometric contractions, the myofilaments are "skidding their wheels" and the tension in the muscle keeps increasing. They are trying to slide, but the muscle is pitted against some more or less immovable object. For example, muscles are contracting isometrically when you try to lift a 400-pound dresser alone. When you straighten a bent elbow, the triceps muscle is contracting isotonically. But when you push against a wall with bent elbows, the wall doesn't move, and the triceps muscles, which cannot shorten to straighten the elbows, are contracting isometrically.

Muscle Tone

One aspect of skeletal muscle activity cannot be consciously controlled. Even when a muscle is voluntarily relaxed, some of its fibers are contracting—first one group and then another. Their contraction is not visible, but as a result of it the muscle remains firm, healthy, and constantly ready for action. This state of continuous partial contractions is called **muscle tone.** Muscle tone is the result of different motor units, which are scattered through the muscle, being stimulated by the nervous system in a systematic way.

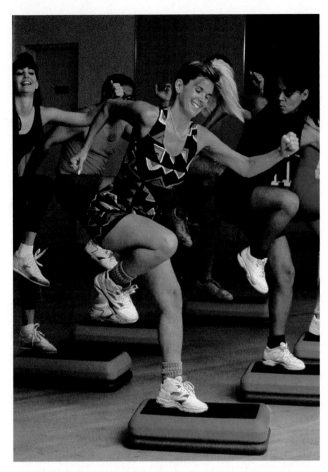

Figure 6.11 **Aerobics class.** By combining an ongoing musical beat with kicking, bending, and jumping, aerobic workouts deliver the same cardiovascular benefits as running or bicycling.

If the nerve supply to a muscle is destroyed (as in an accident), the muscle is no longer stimulated in this manner, and it loses tone and becomes paralyzed. Soon after, it becomes *flaccid* (flak′sid), or soft and flabby, and begins to atrophy (waste away). ❑

Effect of Exercise on Muscles

The amount of work done by a muscle is reflected in changes in the muscle itself. Muscle inactivity (due to a loss of nerve supply, immobilization, or whatever the cause) always leads to muscle weakness and wasting. Muscles are no exception to the saying, "Use it or lose it!"

Conversely, regular exercise increases muscle size, strength, and endurance. However, not all types of exercise produce these effects—in fact, there are important differences in the benefits of exercise.

Aerobic, or **endurance,** types of exercise, such as jogging, biking, or participating in an aerobics class (Figure 6.11), result in stronger, more flexible muscles with greater resistance to fatigue. These changes come about, at least partly, because the blood supply to the muscles increases, and the individual muscle cells form more mitochondria and store more oxygen. However, aerobic exercise benefits much more than the skeletal muscles. It makes overall body metabolism more efficient, improves digestion (and elimination), enhances neuromuscular coordination, and makes the skeleton stronger. The heart enlarges (hypertrophies) so that more blood is pumped out with each beat, fat deposits are cleared from the blood vessel walls, and gas exchanges in the lungs become more efficient. These benefits may be permanent or temporary, depending on how often and how vigorously one exercises.

Aerobic exercise does *not* cause the muscles to increase much in size, even though the exercise may go on for hours. The bulging muscles of a body builder or professional weight lifter result mainly from **resistance,** or isometric, exercise in which the muscles are pitted against some immovable object (or nearly so). Such resistance exercises require very little time and little or no special equipment. A few minutes every other day is usually sufficient. A wall can be pushed against, and buttock muscles can be strongly contracted even while standing in line at the grocery store. The key is forcing the muscles to contract with as much force as possible. The increased muscle size and strength that results is due to enlargement of individual muscle cells (they make more contractile filaments), rather than an increase in their number. The amount of connective tissue that reinforces the muscle also increases.

Because endurance and resistance exercises produce different patterns of muscle response, it is important to know what your exercise goals are. Lifting weights will not improve your endurance for a marathon. By the same token, jogging will do little to improve your muscle definition for competing in the Mr. or Ms. Muscle contest, nor will it make you stronger for moving furniture. Obviously, the best exercise program is one that includes both types of exercise.

A CLOSER LOOK

ATHLETES LOOKING GOOD—DOING BETTER WITH ANABOLIC STEROIDS?

Everyone loves a winner, and top athletes make lots of money. It is not surprising that some will grasp at anything to increase their performance—including anabolic steroids. These hormones, engineered by pharmaceutical companies, were introduced in the 1950s to treat victims of certain muscle-wasting diseases and anemia, and to prevent muscle atrophy in patients immobilized after surgery. Testosterone, a natural anabolic steroid hormone made by the body, is responsible for the increase in muscle and bone mass and other physical changes that occur during puberty and convert boys into men. Convinced that huge doses of the steroids could produce similar and enhanced masculinizing effects in grown men, many athletes were using the steroids by the early 1960s and the practice continues today.

The use of these drugs has been banned by most international athletic competitions, and users (and prescribing physicians or drug dealers) are naturally reluctant to talk about it. Nonetheless, there is little question that many professional body builders and athletes competing in events that require great muscle strength (e.g., discus throwing and weight lifting) are heavy users. Sports figures such as football players have also admitted to using steroids to help them prepare for games. Advantages of anabolic steroids cited by athletes include increased muscle mass and strength, increased oxygen-carrying capacity of the blood (because of greater red blood cell volume), and an increase in aggressive behavior (the urge to "steamroller the other guy").

But do the drugs do all that is claimed for them? Research studies have reported increases in isometric strength and body weight in steroid users. Although these are results weight lifters dream about, there is a hot dispute over whether the drugs also enhance the fine muscle coordination and endurance needed by runners and others.

Do the seemingly slight advantages conferred by steroid use outweigh the risks? This, too, is doubtful. Physicians say they cause bloated faces (a sign of steroid excess), shriveled testes, and infertility; damage the liver and promote liver cancer; and cause changes in blood cholesterol levels (which may place long-term users at risk for coronary heart disease). Additionally, about one-third of anabolic steroid users have serious mental problems. Manic behavior in which the users undergo Jekyll-Hyde personality swings and become extremely violent is common; so, too, are depression and delusions.

The question of why athletes use these drugs is easy to answer. Some say they are willing to do almost anything to win, short of killing themselves. Are they unwittingly doing this as well?

MUSCLE MOVEMENTS, TYPES, AND NAMES

This section is a bit of a hodge-podge. It includes some topics that don't really fit together, but they don't fit anywhere else any better. For example, there are five very basic understandings about gross muscle activity incorporated as the cornerstones of certain sections. I call these the *5 golden rules* of skeletal muscle activity because until you understand them, comprehending muscle movements and appreciating muscle interactions is nearly impossible. These golden rules are summarized for your quick review in Table 6.2.

Table 6.2 The 5 Golden Rules of Skeletal Muscle Activity
1. All muscles cross at least one joint.
2. Typically, the bulk of the muscle lies proximal to the joint crossed.
3. All muscles have at least two attachments: the origin and the insertion.
4. Muscles can only pull, they never push.
5. During contraction, the muscle insertion moves toward the origin.

Types of Body Movements

Every one of our 600-odd skeletal muscles is attached to bone, or to other connective tissue structures, at no less than two points. One of these points, the **origin,** is attached to the immovable or less movable bone (Figure 6.12). The **insertion** is attached to the movable bone, and when the muscle contracts, the insertion moves toward the origin.

Body movement occurs when muscles contract across joints. The type of movement depends on the mobility of the joint and on where the muscle is located in relation to the joint. The most obvious examples of the action of muscles on bones are the movements that occur at the joints of the limbs. However, less freely movable bones are also tugged into motion by the muscles, such as the

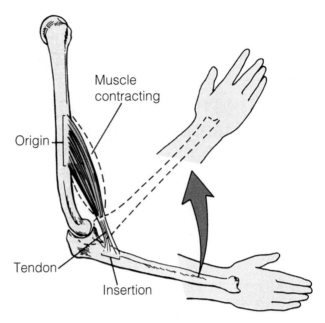

Figure 6.12   Muscle attachments (origin and insertion). When a skeletal muscle contracts, its insertion moves toward its origin.

vertebrae's movements when the torso is bent to the side.

The most common types of body movements are described next and shown in Figure 6.13. Try to demonstrate each movement as you read the following descriptions:

Flexion. A movement, generally in the sagittal plane, that decreases the angle of the joint and brings two bones closer together (Figure 6.13a and b). Flexion is typical of hinge joints (bending the knee or elbow), but it is also common at ball-and-socket joints (bending forward at the hips).

Extension. Extension is the opposite of flexion, so it is a movement that increases the angle, or the distance, between two bones or parts of the body (straightening the knee or elbow). If extension is greater than 180° (as when you tip your head posteriorly so that your chin points toward the ceiling), it is hyperextension (Figure 6.13b).

Abduction. Abduction is moving a limb away (generally on the frontal plane) from the midline, or median plane, of the body (Figure 6.13e). The terminology also applies to the fanning movement of the fingers or toes when they are spread apart (Figure 6.13c).

(b) Flexion and extension of the head

(a) Flexion and extension of the shoulder and knee

(c) Abduction and adduction of the arm

(d) Rotation of the head

Figure 6.13   Body movements.

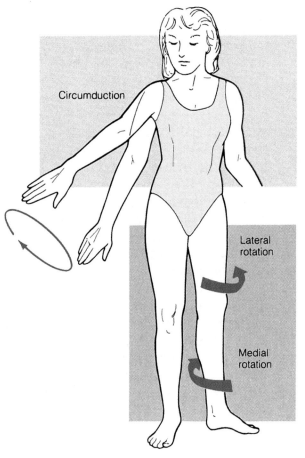

Circumduction

Lateral rotation

Medial rotation

(e) Circumduction of the arm. The distal end of the limb describes a cone. Rotation of the lower limb around its longitudinal axis is also illustrated

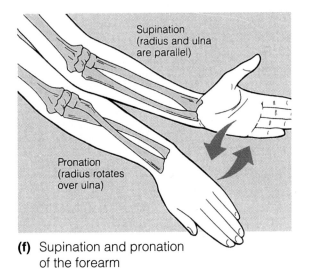

Supination (radius and ulna are parallel)

Pronation (radius rotates over ulna)

(f) Supination and pronation of the forearm

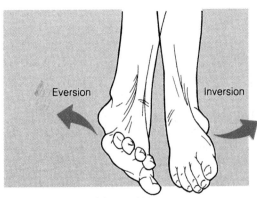

Eversion   Inversion

(g) Eversion and inversion of the foot

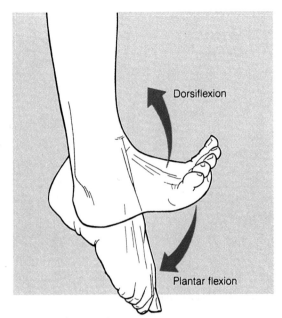

Dorsiflexion

Plantar flexion

(h) Dorsiflexion and plantar flexion of the foot

Figure 6.13 (continued)   Body movements.

Adduction. Adduction is the opposite of abduction, so it is the movement of a limb toward the body midline.

Rotation. Rotation is movement of a bone around its longitudinal axis (Figure 6.13d and e). Rotation is a common movement of ball-and-socket joints and describes the movement of the atlas around the dens of the axis (as in shaking your head "no").

Circumduction. Circumduction is a combination of flexion, extension, abduction, and adduction commonly seen in ball-and-socket joints such as the shoulder. The proximal end of the limb is stationary, and its distal end moves in a circle. The limb as a whole outlines a cone (Figure 6.13e).

Pronation. Pronation is moving the palm of the hand from an anterior, or upward-facing, position to a posterior, or downward-facing, position (Figure 6.13f). This action moves the distal end of the radius across the ulna.

Supination. Supination is moving the palm from a posterior position to an anterior position (the anatomical position). Supination is the opposite of pronation. When the forearm is supinated, the radius and the ulna are parallel.

The following four terms refer to movements of the foot (Figure 6.13g and h):

Inversion. Inversion is turning the sole of the foot so that it faces medially.

Eversion. Eversion is turning the sole of the foot laterally. Eversion is the opposite of inversion.

Dorsiflexion. Dorsiflexion is movement at the ankle that moves the instep of the foot up and dorsally toward the shin; that is, standing on your heels.

Plantar flexion. Plantar flexion is the movement that straightens the ankle joint, causing the toes to point downward; standing on your toes.

together or against each other. Muscles are arranged on the skeleton in such a way that whatever one muscle (or group of muscles) can do, another group of muscles can reverse. Because of this, muscles are able to bring about an immense variety of movements.

When several muscles are contracting at the same time, the muscle that has the major responsibility for causing a particular movement is called the **prime mover.** (This physiologic term has been borrowed by the business world to label a person who gets things done.) Muscles that oppose or reverse a movement are **antagonists** (an-tag'o-nists). When a prime mover is active, its antagonist is stretched and relaxed. Antagonists can be prime movers in their own right. For example, the biceps of the arm (prime mover of elbow flexion) is antagonized by the triceps (a prime mover of elbow extension).

Synergists (sin'er-jists; *syn* = together, *erg* = work) are muscles that help prime movers by producing the same movement, or by reducing undesirable or unnecessary movement. When a muscle crosses two or more joints, its contraction will cause movement in all the joints crossed unless synergists are there to stabilize them. For example, the finger-flexor muscles cross both the wrist and the finger joints. You can make a fist without bending your wrist because synergist muscles stabilize the wrist joints and allow the prime mover to act on the finger joints.

Fixators are specialized synergists. They act to hold a bone still or to stabilize the origin of a prime mover so all the tension can be used to move the insertion bone. The postural muscles that stabilize the vertebral column are fixators, as are the muscles that anchor the scapula to the thorax.

In summary, although prime movers seem to "get all the credit" for causing certain movements, the actions of antagonistic and synergistic muscles are also important in effecting smooth, coordinated, and precise movements.

Types of Muscles

Muscles can't push, they can only pull as they contract, so most often body movements are the result of the activity of pairs or teams of muscles acting

Naming Skeletal Muscles

Like bones, muscles come in many shapes and sizes to suit their particular tasks in the body. Muscles are named on the basis of several criteria, each of which focuses on a particular structural or functional characteristic. Paying close attention to these cues can greatly simplify your task of learning muscle names and actions:

Direction of the muscle fibers. Some muscles are named in reference to some imaginary line, usually the midline of the body or the long axis of a limb bone. When a muscle's name includes the term *rectus* (straight), its fibers run parallel to that imaginary line. For example, the rectus femoris is the straight muscle of the thigh, or femur. Similarly, the term *oblique* as part of a muscle's name tells you that the muscle fibers run obliquely (slanted) to the imaginary line.

Relative size of the muscle. Such terms as *maximus* (largest), *minimus* (smallest), and *longus* (long) are often used in the names of muscles—for example, the gluteus maximus is the largest muscle of the gluteus muscle group.

Location of the muscle. Some muscles are named for the bone with which they are associated. For example, the temporalis and frontalis muscles overlie the temporal and frontal bones of the skull.

Number of origins. When the term *biceps, triceps,* or *quadriceps* forms part of a muscle name, one can assume that the muscle has two, three, or four origins, respectively. For example, the biceps muscle of the arm has two heads or origins, and the triceps muscle has three.

Location of the muscle's origin and insertion. Occasionally, muscles are named for their attachment sites. For example, the sternocleidomastoid muscle has its origin on the sternum *(sterno)* and clavicle *(cleido)* and inserts on the *mastoid* process of the temporal bone.

Shape of the muscle. Some muscles have a distinctive shape that helps to identify them. For example, the deltoid muscle is roughly triangular (deltoid means "triangular").

Action of the muscle. When muscles are named for their actions, terms such as *flexor, extensor,* and *adductor* appear in their names. For example, the adductor muscles of the thigh all bring about its adduction, and the extensor muscles of the wrist all extend the wrist.

GROSS ANATOMY OF SKELETAL MUSCLES

It is beyond the scope of this book to describe the hundreds of skeletal muscles of the human body. Only the most important muscles are described here. In addition, all the superficial muscles considered are summarized in Tables 6.3 and 6.4 and are illustrated in overall body views in Figures 6.20 and 6.21, which accompany the tables (pp. 184–187).

Head Muscles

The head muscles (Figure 6.14) are an interesting group. They have many specific functions but are usually grouped into two large categories—facial muscles and chewing muscles. Facial muscles are unique because they are inserted into soft tissues such as other muscles or skin. When they pull on the skin of the face, they permit us to smile faintly, grin widely, frown, pout, deliver a kiss, and so forth. The chewing muscles begin the breakdown of food for the body.

Facial Muscles

Frontalis   The frontalis covers the frontal bone as it runs from the cranial aponeurosis to the skin of the eyebrows where it inserts. This muscle allows you to raise your eyebrows, as in surprise, and to wrinkle your forehead.

Orbicularis Oculi   The orbicularis oculi (or-bik″u-la′ris ok′u-li) has fibers that run in circles around the eyes. It allows you to close your eyes, squint, blink, and wink.

Orbicularis Oris   The orbicularis oris is the circular muscle of the lips. Because it closes the mouth and protrudes the lips, it is often called the "kissing" muscle.

Buccinator   The fleshy buccinator (bu′sĭ-na″tor) muscle runs horizontally across the cheek and inserts into the orbicularis oris. It flattens the cheek (as in whistling or blowing a trumpet). It is also listed as a chewing muscle, because it compresses

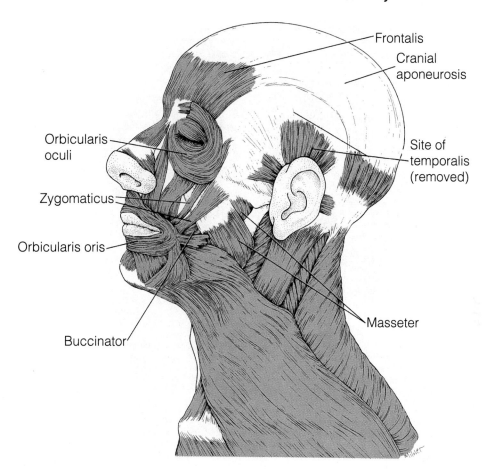

Figure 6.14   Superficial muscles of the face and neck.

the cheek to hold the food between the teeth during chewing.

Zygomaticus   The zygomaticus (zi″go-mat′i-kus) extends from the corner of the mouth to the cheekbone. It is often referred to as the "smiling" muscle, because it raises the corners of the mouth upward.

Chewing Muscles

The buccinator muscle, which is a member of this group, is described with the facial muscles.

Masseter   The masseter (mă-se′ter) covers the angle of the lower jaw as it runs from the zygomatic process of the temporal bone to the mandible. This muscle closes the jaw by elevating the mandible.

Temporalis   The temporalis is a fan-shaped muscle overlying the temporal bone (see Figure 6.20, p. 184). It inserts into the mandible and acts as a synergist of the masseter in closing the jaw.

Trunk and Neck Muscles

The neck muscles, which move the head or shoulder girdle, are small and straplike. The trunk muscles include (1) those that move the vertebral column (most of which are posterior antigravity muscles); (2) anterior thorax muscles that move the ribs, head, and arms; and (3) muscles of the abdominal wall that help to move the vertebral column but, most importantly, form the muscular "natural girdle" of the abdominal body wall.

Anterior Muscles   (Figure 6.15)

Sternocleidomastoid   The paired sternocleidomastoid (ster″no-kli″do-mas′toid) muscles are two-headed muscles, one found on each side of the neck. Of the two heads of each muscle, one arises from the sternum and the other arises from the clavicle. The heads fuse before inserting into the mastoid process of the temporal bone. When both sternocleidomastoid muscles contract together,

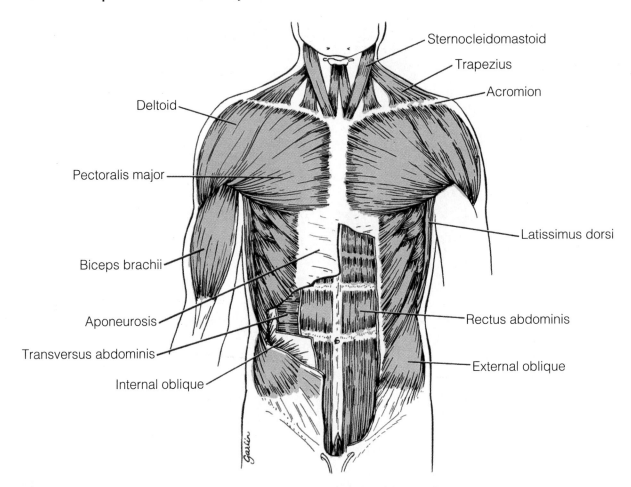

Sternocleidomastoid

Trapezius

Acromion

Deltoid

Pectoralis major

Latissimus dorsi

Biceps brachii

Aponeurosis

Rectus abdominis

Transversus abdominis

Internal oblique

External oblique

Figure 6.15   Muscles of the anterior neck, trunk, shoulder, and arm. Portions of the superficial muscles of the right side of the abdomen are cut away to reveal the deeper muscles.

they flex your neck. (It is this action of bowing the head that has led some people to call these muscles the "prayer" muscles.) If just one muscle contracts, the head is rotated toward the opposite side. In some difficult births, one of these muscles may be injured and develop spasms. A baby injured in this way has *torticollis* (tor″ti-kol′is), or wryneck. ❑

Pectoralis Major   The pectoralis (pek″to-ra′lis) major is a large fan-shaped muscle covering the upper part of the chest. Its origin is from the shoulder girdle and the first six ribs. It inserts on the proximal end of the humerus. This muscle forms the anterior wall of the axilla and acts to adduct and flex the arm.

Intercostal Muscles   The intercostal muscles are deep muscles found between the ribs. (Although they are not shown in Figure 6.15, which only

shows superficial muscles, they are illustrated in Figure 6.20.) The external intercostals are important in breathing because they help to raise the rib cage for breathing air in. The internal intercostals depress the rib cage, which helps to move air out of the lungs when you exhale forcibly.

Muscles of the Abdominal Girdle   The anterior abdominal muscles (rectus abdominis, external and internal obliques, and transversus abdominis) form a natural "girdle" that reinforces the body trunk. Taken together, they resemble the structure of plywood, because the fibers of each muscle or muscle pair run in a different direction. Just as plywood is exceptionally strong for its thickness, the abdominal muscles form a muscular wall that is well suited for its job of containing and protecting the abdominal contents.

• **Rectus abdominis.** The paired straplike rectus abdominis muscles are the most superficial

Sternocleidomastoid

Occipital bone

C₇

T₁

Trapezius

Acromion

Deltoid

Triceps brachii

Humerus

Olecranon
process
of ulna (deep
to tendon)

Latissimus dorsi

Iliac crest

Figure 6.16   Muscles of the posterior neck, trunk, and arm.

muscles of the abdomen. They run from the pubis to the rib cage, enclosed in an aponeurosis. Their main function is to flex the vertebral column. They also compress the abdominal contents during defecation and childbirth, and are involved in forced breathing.

- **External oblique.** The external oblique muscles are paired superficial muscles that make up the lateral walls of the abdomen. Their fibers run downward and medially from the last eight ribs and insert into the ilium. Like the rectus abdominis, they flex the vertebral column but they also rotate the trunk and bend it laterally.

- **Internal oblique.** The internal oblique muscles are paired muscles deep to the external obliques. Their fibers run at right angles to those of the external obliques. They arise from

the iliac crest and insert into the last three ribs. Their functions are the same as those of the external obliques.

- **Transversus abdominis.** The transversus abdominis is the deepest muscle of the abdominal wall and has fibers that run horizontally across the abdomen. It arises from the lower ribs and iliac crest and inserts into the pubis. This muscle compresses the abdominal contents.

Posterior Muscles   (Figure 6.16)

Trapezius   The trapezius (trah-pe′ze-us) muscles are the most superficial muscles of the posterior neck and upper trunk. When seen together, they form a diamond- or kite-shaped muscle mass. Their origin is very broad; each runs from the occipital bone of the skull down the vertebral column

Deltoid
muscle

Humerus

Figure 6.17   The fleshy deltoid muscle is a favored site for administering intramuscular injections.

to the end of the thoracic vertebrae. They then flare laterally to insert on the scapular spine and clavicle. The trapezius muscles extend the head (thus are antagonists of the sternocleidomastoids). They also can elevate, depress, adduct, and stabilize the scapula.

Latissimus Dorsi   The latissimus (lah-tis′ĭ-mus) dorsi is the large, flat muscle pair that covers the lower back. It originates on the lower spine and ilium and then sweeps superiorly to insert into the proximal end of the humerus. The latissimus dorsi extends and adducts the humerus. These are very important muscles when the arm must be brought down in a power stroke, as when swimming or striking a blow.

Erector Spinae   The erector spinae (e-rek′tor spi′ne) is a prime mover of back extension. Because these paired muscles are deep muscles of the back, they are not shown in Figure 6.16. Each erector spinae is a composite muscle consisting of three muscle columns that collectively span the entire length of the vertebral column. These muscles not only act as powerful back extensors ("erectors"), but they provide resistance that helps control the action of bending over at the waist. Following injury to back structures, these muscles go into spasms, a common source of lower back pain.

Deltoid   The deltoids are fleshy, triangle-shaped muscles that form the rounded shape of your shoulders. Because they are so bulky, they are a favorite injection site (see Figure 6.17) when rela-

tively small amounts of medication (less than 5 ml) must be given intramuscularly (into muscle). The origin of each deltoid winds across the shoulder girdle from the spine of the scapula to the clavicle. It inserts into the proximal humerus. The deltoids are the prime movers of arm abduction.

Muscles of the Upper Limb

The upper limb muscles fall into three groups. The first group includes muscles that arise from the shoulder girdle and cross the shoulder joint to insert into the humerus. These muscles, which move the arm, have already been considered—the pectoralis major, latissimus dorsi, and deltoid.

The second group causes movement at the elbow joint. These muscles enclose the humerus and insert on the forearm bones. Only the muscles of this second group will be described in this section. The third group includes the muscles of the forearm, which insert on the hand bones and cause their movement. The muscles of this last group are thin and spindle-shaped, and there are many of them. They will not be considered here except to mention their general naming and function. In general, the forearm muscles have names that reflect their activities. For example, the flexor carpi and flexor digitorum muscles, found on the anterior aspect of the forearm, cause flexion of the wrist and fingers, respectively. The extensor carpi and extensor digitorum muscles, found on the lateral and posterior aspect of the forearm, extend the same structures. (Some of these muscles are described briefly in Table 6.4 and illustrated in Figure 6.21.)

Muscles of the Humerus That Act on the Forearm

Biceps Brachii   The biceps brachii (bra′ke-i) is the most familiar muscle of the forearm because it bulges when the elbow is flexed (see Figure 6.15). It originates by two heads from the shoulder girdle and inserts into the radial tuberosity. This muscle is the powerful prime mover for flexion of the forearm and acts to supinate the forearm. The best way to remember its action is that "it turns the corkscrew *and* pulls the cork."

Triceps Brachii   The triceps muscle is the only muscle fleshing out the posterior humerus (see

Figure 6.16). Its three heads arise from the shoulder girdle and proximal humerus, and it inserts into the olecranon process of the ulna. Being the powerful prime mover of elbow extension, it is the antagonist of the biceps brachii. This muscle is often called the "boxer's" muscle, because it can deliver a straight-arm knockout punch.

Muscles of the Lower Limb

Muscles that act on the lower limb or extremity cause movement at the hip, knee, and foot joints. They are among the largest, strongest muscles in the body and are highly specialized for walking and balancing the body. Because the pelvic girdle is composed of heavy, fused bones that allow little movement, no special group of muscles is necessary to stabilize it. This is very different from the shoulder girdle, which requires several fixator muscles.

Many muscles of the lower limb span two joints and can cause movement at both of them. Therefore, the terms *origin* and *insertion* are often interchangeable in referring to these muscles.

Muscles acting on the thigh are massive muscles that help hold the body upright against the pull of gravity and cause various movements at the hip joint. Muscles acting on the leg form the flesh of the thigh. (Recall that in common usage the term *leg* refers to the whole lower limb, but anatomically the term refers only to that part between the knee and the ankle.) The thigh muscles cross the knee and cause its flexion or extension. Because many of the thigh muscles also have attachments on the pelvic girdle, they can cause movement at the hip joint as well. Muscles originating on the leg cause assorted movements of the ankle and foot. Only three muscles of this group will be considered, but there are many others that act to extend and flex the ankle and toe joints.

Muscles Causing Movement at the Hip Joint   (Figure 6.18)

Iliopsoas   The iliopsoas (il″e-o-so′as; the *p* is silent) is a fused muscle composed of two muscles. It runs from the iliac bone and lower vertebrae deep inside the pelvis to insert on the lesser trochanter of the femur. It is a prime mover of hip flexion. It also acts as a postural muscle to keep the upper body from falling backward when standing erect.

Adductor Muscles   The muscles of the adductor group form the muscle mass at the medial side of each thigh. As their name indicates, they adduct or press the thighs together. However, since gravity does most of the work for them, they tend to become flabby very easily. Special exercises are usually needed to keep them toned. The adductors have their origin on the pelvis and insert on the proximal aspect of the femur.

Gluteus Maximus   The gluteus maximus (gloo′te-us max′ĭ-mus) is a superficial muscle of the hip that forms most of the flesh of the buttock (Figure 6.18c). It is a powerful hip extensor that acts to bring the thigh in a straight line with the pelvis. Although not very important in walking, it is probably the most important muscle for extending the hip when power is needed, as when climbing stairs and when jumping. It originates from the sacrum and iliac bones and runs to insert on the gluteal tuberosity of the femur.

Gluteus Medius   The gluteus medius runs from the ilium to the femur, beneath the gluteus maximus for most of its length. The gluteus medius is a hip abductor and is important in steadying the pelvis during walking. The gluteus medius is an important site for giving intramuscular injections, particularly when more than 5 ml is administered (see Figure 6.18d). Although it might appear that the large, fleshy gluteus maximus that forms the bulk of the buttock mass would be a better choice, notice that the medial part of each buttock overlies the large sciatic nerve; hence this area must be carefully avoided. This can be accomplished by *mentally* dividing the buttock into four equal quadrants (shown by the dotted division lines on Figure 6.18d). The upper outer quadrant then overlies the gluteus medius muscle, which is usually a very safe site for an intramuscular injection.

Muscles Causing Movement at the Knee Joint   (Figure 6.18)

Sartorius   Compared to other thigh muscles described here, the thin, straplike sartorius (sar-to′re-us) muscle is not too important; but since it is the most superficial muscle of the thigh it is rather hard

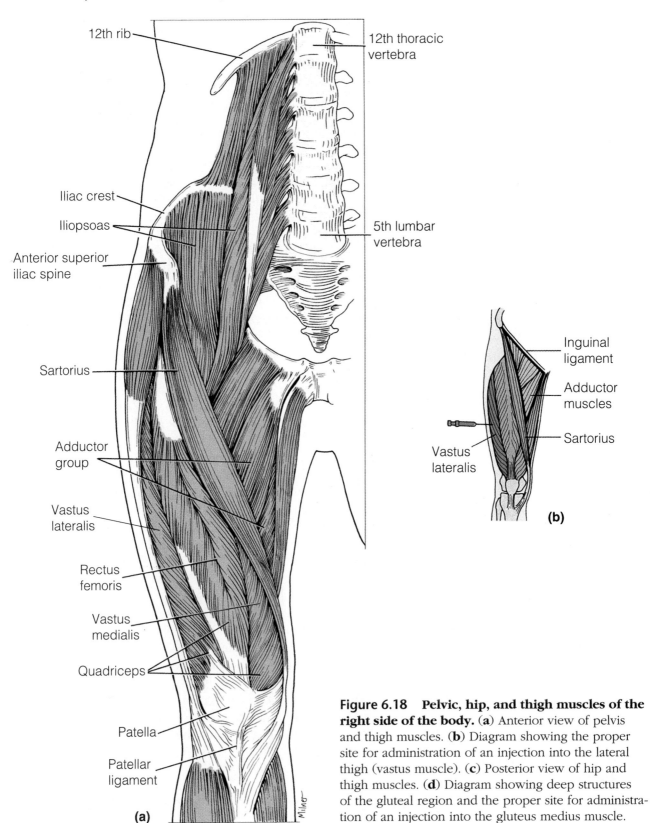

(a)

(b)

Figure 6.18 Pelvic, hip, and thigh muscles of the right side of the body. (a) Anterior view of pelvis and thigh muscles. **(b)** Diagram showing the proper site for administration of an injection into the lateral thigh (vastus muscle). **(c)** Posterior view of hip and thigh muscles. **(d)** Diagram showing deep structures of the gluteal region and the proper site for administration of an injection into the gluteus medius muscle.

Gluteus medius

Gluteus maximus

Adductor magnus

Semitendinosus

Biceps femoris

Hamstring group

Semimembranosus

Gastrocnemius

(c)

Posterior superior iliac spine

Iliac crest

Gluteus medius

Gluteus maximus

Sciatic nerve

Greater trochanter

(d)

Figure 6.19 Superficial muscles of the right leg. Anterior view is on the left; posterior view is on the right.

to miss. It runs obliquely across the thigh from the anterior iliac crest to the medial side of the tibia. It is a weak thigh flexor. The sartorius is commonly referred to as the "tailor's" muscle because it acts as a synergist to bring about the cross-legged position in which old-time tailors are often shown.

Quadriceps Group The quadriceps (kwod'rĭ-seps) group consists of four muscles—the **rectus femoris** and three **vastus muscles**—which flesh out the anterior thigh. The vastus muscles originate from the femur; the rectus femoris originates on the pelvis. All four muscles insert into the tibial tuberosity by the patellar tendon. The group as a whole acts to extend the knee powerfully, as when kicking a football. Because the rectus femoris crosses two joints, the hip and the knee, it can also help to flex the hip. The lateral vastus and rectus femoris are sometimes used as intramuscular injec-

tion sites (Figure 6.18b), particularly in infants, who have poorly developed gluteus muscles.

Hamstring Group The muscles forming the muscle mass of the posterior thigh are the hamstrings (Figure 6.18c). The group consists of three muscles, the **biceps femoris, semimembranosus,** and **semitendinosus,** which originate on the ischial tuberosity and run down the thigh to insert on both sides of the proximal tibia. Their name comes from the fact that butchers use their tendons to hang "hams" (consisting of thigh and hip muscles) for smoking. These tendons can be felt at the back of the knee.

Muscles Causing Movement at the Ankle and Foot (Figure 6.19)

Tibialis Anterior The tibialis anterior is a superficial muscle on the anterior leg. It arises from the upper tibia and then parallels the anterior crest as it runs to the tarsal bones where it inserts by a long tendon. It acts to dorsiflex and invert the foot.

Peroneus Muscles The three peroneus (per″o-ne′us) muscles of this group are found on the lateral part of the leg. They arise from the fibula and insert into the metatarsal bones of the foot. The group as a whole plantar flexes and everts the foot.

Gastrocnemius The gastrocnemius (gas″trokne′me-us) muscle is a two-bellied muscle that forms the curved calf of the posterior leg. It arises by two heads, one from each side of the distal femur, and inserts through the large Achilles (calcaneal) tendon into the heel of the foot. It is a prime mover for plantar flexion of the foot; for this reason it is often called the "toe dancer's" muscle. If its insertion tendon is cut, walking is very difficult. The foot drags because the heel cannot be lifted.

Remember that most of the superficial muscles previously described are shown in anterior and posterior views of the body as a whole in Figures 6.20 and 6.21. In addition, these muscles are summarized in Tables 6.3 and 6.4. Take the time to review these muscles again before continuing with this chapter.

DEVELOPMENTAL ASPECTS OF THE MUSCULAR SYSTEM

In the developing embryo, the muscular system is laid down in segments (much like the structural plan of an earthworm), and then each segment is invaded by nerves. The muscles of the thoracic and lumbar regions become very extensive, since they must cover and move the bones of the limbs. Development of the muscles and their control by the nervous system occur rather early in pregnancy. The expectant mother is often astonished by the first movements (called the *quickening*) of the fetus, which usually occur by the sixteenth week of pregnancy.

Very few congenital muscular problems occur. The exception is *muscular dystrophy*—a group of inherited muscle-destroying diseases that affect specific muscle groups. The muscles enlarge due to fat and connective tissue deposit, but the muscle fibers degenerate and atrophy.

The most common and serious form is *Duchenne muscular dystrophy,* which is expressed almost exclusively in males. This tragic disease is usually diagnosed between the ages of 2 and 6 years. Active, normal-appearing children become clumsy and begin to fall frequently as their muscles weaken. The disease progresses relentlessly from the extremities upward, finally affecting the head and chest muscles. Most victims must use wheelchairs by the age of 12, and affected individuals generally do not live beyond young adulthood. Although the cause of muscular dystrophy has been pinned down (the diseased muscle fibers lack a protein that helps maintain the sarcolemma), a cure is still elusive. ❑

Initially after birth, a baby's movements are all gross reflex types of movements. Since the nervous system must mature before the baby can control muscles, we can trace the increasing efficiency of the nervous system by observing a baby's development of muscle control. This development proceeds in a cephalic/caudal direction, and gross muscular movements precede fine ones. Babies can raise their heads before they can sit up, and can sit up before they can walk. Muscular control also proceeds in a proximal/distal direction; that is, babies can wave "bye-bye" and pull objects to themselves before using the pincher grasp to pick up a pin. All through childhood, the control of the skeletal muscles by the nervous system becomes more and more precise. By midadolescence, we have reached the peak level of development of this natural control and can simply accept it or bring it to a fine edge by athletic training.

Because of its rich blood supply, skeletal muscle is amazingly resistant to infection throughout life, and given good nutrition, relatively few problems afflict skeletal muscles. It should be repeated, however, that muscles, like bones, *will* atrophy, even with normal tone, if they are not used continually. On the other hand, a lifelong program of regular exercise keeps the whole body operating at its best possible level.

One rare disease that can affect muscles during adulthood is *myasthenia gravis* (mi″as-the′ne-ah gra′vis; *asthen* = weakness; *gravi* = heavy), a disease characterized by drooping of the upper eyelids, difficulty in swallowing and talking, and generalized muscle weakness and fatigability. The disease involves a shortage of acetylcholine receptors at the neuromuscular junction. The blood of many of these patients contains antibodies to acetylcholine receptors, which suggests that myasthenia gravis is an autoimmune disease. Although the receptors may initially be present in normal numbers, they appear to be destroyed as the disease progresses. Whatever the case, the muscle cells are not stimulated properly and get progressively weaker. Death usually occurs as a result of the inability of the respiratory muscles to function. This is called *respiratory failure.* ❑

As we age, the amount of connective tissue in the muscles increases and the amount of muscle tissue decreases; thus the muscles become stringier, or more sinewy. Since the skeletal muscles represent so much of the body weight, body weight begins to decline in the elderly person, as this natural loss in muscle mass occurs. Another result of the loss in muscle mass is a decrease in muscle strength; strength decreases by about 50 percent by the age of 80. Regular exercise can help offset the effects of aging on the muscular system.

Temporalis

Frontalis

Zygomaticus

Orbicularis oculi

Masseter

Orbicularis oris

Sternocleidomastoid

Deltoid

Pectoralis major

External intercostals

Biceps brachii

Rectus abdominis

External oblique

Flexor carpi radialis

Adductor longus

Sartorius

Rectus femoris — Quadriceps group

Vastus muscles

Peroneus longus

Tibialis anterior

Figure 6.20   Major superficial muscles of the anterior surface of the body.

Table 6.3   Superficial Anterior Muscles of the Body (See Figure 6.20)

Name	Origin	Insertion	Action
Head/Neck Muscles			
Frontalis	Cranial aponeurosis	Skin of eyebrows	Raises eyebrows
Orbicularis oculi	Frontal bone and maxilla	Tissue around eyes	Blinks and closes eyes
Orbicularis oris	Mandible and maxilla	Skin and muscle around mouth	Closes and protrudes lips
Temporalis	Temporal bone	Mandible	Closes jaw
Zygomaticus	Zygomatic bone	Skin and muscle at corner of lips	Raises corner of mouth
Masseter	Temporal bone	Mandible	Closes jaw
Sternocleidomastoid	Sternum and clavicle	Temporal bone (mastoid process)	Flexes neck; rotates head
Trunk Muscles			
Pectoralis major	Sternum, clavicle, and first to sixth ribs	Proximal humerus	Adducts and flexes humerus
Rectus abdominis	Pubis	Sternum and fifth to seventh ribs	Flexes vertebral column
External oblique	Lower eight ribs	Iliac crest	Flexes and rotates vertebral column
Arm/Shoulder Muscles			
Biceps brachii	Scapula of shoulder girdle	Proximal radius	Flexes elbow and supinates forearm
Deltoid	See Table 6.4		Abducts arm
Hip/Thigh/Leg Muscles			
Iliopsoas*	Ilium and lumbar vertebrae	Femur	Flexes hip
Adductor muscles	Pelvis	Proximal femur	Adducts thigh
Sartorius	Ilium	Proximal tibia	Flexes thigh on hip
Quadriceps group (vastus medialis, intermedius, and lateralis, and the rectus femoris)	Vasti: Femur Rectus femoris: Pelvis	Tibial tuberosity via patellar tendon Tibial tuberosity via patellar tendon	All extend knee; rectus femoris also flexes hip on thigh
Tibialis anterior	Proximal tibia	First cuneiform (tarsal) and first metatarsal of foot	Dorsiflexes and inverts foot
Peroneus muscles	Fibula	Metatarsals of foot	Plantar flex and evert foot

*Iliopsoas is a deep muscle and is not shown in Figure 6.20.

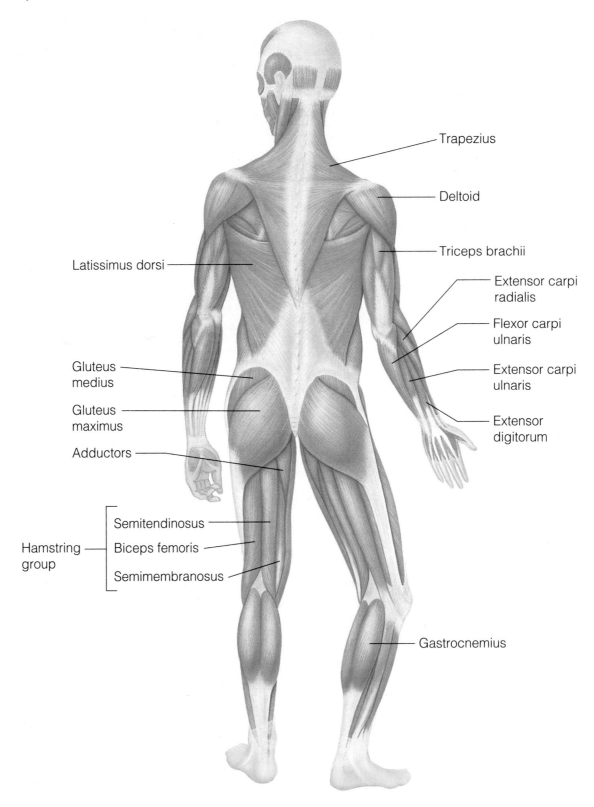

Trapezius

Deltoid

Triceps brachii

Extensor carpi radialis

Flexor carpi ulnaris

Extensor carpi ulnaris

Extensor digitorum

Latissimus dorsi

Gluteus medius

Gluteus maximus

Adductors

Hamstring group

Semitendinosus

Biceps femoris

Semimembranosus

Gastrocnemius

Figure 6.21 **Major superficial muscles of the posterior surface of the body.**

Table 6.4 Superficial Posterior Muscles of the Body (Some Forearm Muscles Also Shown) (See Figure 6.21)

Name	Origin	Insertion	Action
Neck/Trunk/ Shoulder Muscles			
Trapezius	Occipital bone and all cervical and thoracic vertebrae	Scapular spine and clavicle	Extends neck and adducts scapula
Latissimus dorsi	Lower spine and iliac crest	Proximal humerus	Extends and adducts humerus
Erector spinae*	Iliac crests, ribs 3–12, and vertebrae	Ribs, thoracic and cervical vertebrae	Extends back
Deltoid	Scapular spine and clavicle	Humerus (deltoid tuberosity)	Abducts humerus
Arm/Forearm Muscles			
Triceps brachii	Shoulder girdle and proximal humerus	Olecranon process of ulna	Extends elbow
Flexor carpi radialis	Distal humerus	Second and third metacarpals	Flexes wrist and abducts hand (see Figure 6.20)
Flexor carpi ulnaris	Distal humerus and posterior ulna	Carpals of wrist and fifth metacarpal	Flexes wrist and adducts hand
Flexor digitorum superficialis[†]	Distal humerus, ulna and radius	Middle phalanges of second to fifth fingers	Flexes wrists and fingers
Extensor carpi radialis	Humerus	Base of second and third metacarpals	Extends wrists and abducts hand
Extensor digitorum	Distal humerus	Distal phalanges of second to fifth fingers	Extends fingers and wrist
Hip/Thigh/Leg Muscles			
Gluteus maximus	Sacrum and ilium	Proximal femur	Extends hip (when forceful extension is required)
Gluteus medius	Ilium	Proximal femur	Adducts thigh
Hamstring muscles (semitendinosus, semimembranosus, biceps femoris)	Ischial tuberosity	Proximal tibia (head of fibula in case of biceps femoris)	Flex knee and extend hip
Gastrocnemius	Distal femur	Calcaneus (heel via Achilles tendon)	Plantar flexes foot and flexes knee

*Erector spinae is a deep muscle and is not shown in Figure 6.21.

[†]Although its name indicates that it is a superficial muscle, the flexor digitorum superficialis lies deep to the flexor carpi radialis and is not visible in a superficial view.

A CLOSER LOOK

PROTECT YOUR BACK—IT'S THE ONLY ONE YOU'VE GOT!

Regular exercise is vital to keep the spine strong and protect its beautifully proportioned structure. Unfortunately, most of us neglect our backs until aches and pains demand our attention. Next to sore throats and the common cold, backaches are the most common reason for visits to the physician's office. Millions of Americans suffer from back pain, and back injuries account for a quarter of all disability insurance payments.

Although backache may be symptomatic of many ailments, from spinal tumor to kidney disease, over 80 percent of back problems result from weak muscles or are anxiety-related. This is not surprising since strong abdominal muscles and flexible back and hip muscles play a major role in reinforcing the spine's delicate architecture, and stress causes muscles to "knot." Tense muscles can aggravate a back problem even if they do not cause it. Obese individuals and those who rarely exercise are prime candidates for lower back pain. Indeed, some orthopedists (bone specialists) look upon back pain as an index of affluence. The more gadgets we accumulate to do our work for us, the more we can sit or lie down—and the more our muscles deteriorate. On the other side of the coin, people who work at jobs that require heavy lifting are also "sitting ducks" for lower back pain if they do not lift properly.

Preventing back problems is better than attempting to treat agonizing back pain. Perhaps 70 to 80 percent of all cases of lower back pain are preventable with only 10 minutes of daily exercise, if proper body alignment and certain precautions are observed. Here are a few well-documented guidelines for protecting your back.

1. Keep your weight down. Just a few extra pounds of weight, particularly the unbalanced weight of a protruding potbelly, exert a strain on the spine that is much greater than that of the weight itself. The protruding abdomen has the effect of pulling the body forward, forcing the back muscles to contract more strongly to counterbalance the weight. For example, 10 extra pounds located 10 inches anterior to the spine in the abdominal region force the back muscles to exert 50 extra pounds of force to counteract the 10-pound pulling force.

2. Wear high-heel and negative-heel shoes or boots infrequently, and avoid them altogether if you have back problems. Both change the alignment of the spine. For example, high-heel shoes tilt the pelvis forward and increase the stress on the abdominal and back muscles. Negative-heel shoes tilt the pelvis backward, making weight transfer during walking very difficult.

3. Maintain good posture. Your head and back should be aligned, with abdomen and buttocks tucked inward.

4. Lift heavy objects using proper body mechanics. Every object, including a human being, has a center of gravity around which its mass is equally distributed. In a standing adult, the center of gravity is inside the pelvis, slightly posterior to the anterior border of the joint between the sacrum and the fifth lumbar vertebra. A basic

Center of gravity

Line of gravity

Base of support

Lowering the center of gravity by squatting.

principle of body mechanics is the broader your base of support and the lower your center of gravity, the more stable you are. Applying this principle to lifting demands that you take a wider than normal stance (spread your feet apart slightly), and then bend your knees (rather than your back) to reach for and pick up the object (see the diagram below). The weight is then transmitted to your stronger legs, sparing your more delicate spine. If you must move large objects, push rather than pull them.

5. Avoid sitting for prolonged periods. Sitting puts much more stress on the spine than standing:

(a)

(b)

(c)

Truck drivers have five times the normal risk for lower back problems and herniated disks. If you must sit for an extended time, resting your feet on a small stool (or the lowest desk drawer) will reduce the stress on your vertebral column.

6. Take 10 minutes a day to stretch your lower back extensor muscles and hip flexor muscles and to strengthen your abdominal muscles as illustrated in the figure. If you have any history of back problems, avoid any exercise that strains your lower back, including (1) sit-ups with straight legs, (2) double leg raises; and (3) lying on your abdomen and then raising your head, arms, and legs.

Dare to be different! Be good to your spine and you may well be saved from uttering that nearly universal complaint, "Oh, my aching back!"

(**a**) *Stretching the extensor muscles of the lower back.* Lie on back with knees bent and feet flat on the floor. Keeping your arms at your sides, raise one knee to the chest. Lower the foot to the floor with knee bent, and then slide the foot along the floor until the leg is fully extended. Rotate the leg gently from side to side.

Resume the starting position and repeat the exercise with the alternate leg. Repeat the entire exercise 5 to 6 times.

(**b**) *Stretching the flexor muscles of the hip.* Lie on back. Exhale as you bring both knees toward the chest. Then, while holding one knee to the chest, slide the opposite leg along the floor until it is fully extended. Attempt to touch the back of the knee (popliteal region) of the extended leg to the floor. Hold for a count of 6.

Resume the starting position and repeat the exercise with the alternate leg. Repeat the entire exercise 5 to 6 times.

(**c**) *Strengthening the abdominal muscles.* Lie on back with knees bent and feet flat on floor. Keeping your arms at your sides, raise one knee toward your chest. Exhale as you lift your head, attempting to touch your raised knee to your forehead. Count to 6. Resume the starting position and roll your head gently from side to side. Inhale, and repeat the exercise with the alternate leg.

Repeat the entire exercise 8 to 10 times at the beginning; work up to 25 repetitions per day.

SYSTEMS IN SYNC

Homeostatic Relationships Between the Muscular System and Other Body Systems

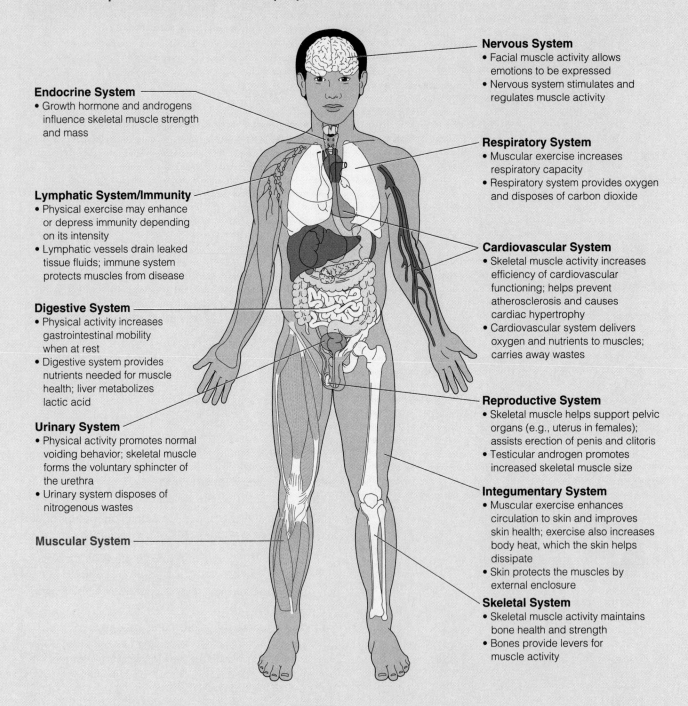

Nervous System
- Facial muscle activity allows emotions to be expressed
- Nervous system stimulates and regulates muscle activity

Endocrine System
- Growth hormone and androgens influence skeletal muscle strength and mass

Respiratory System
- Muscular exercise increases respiratory capacity
- Respiratory system provides oxygen and disposes of carbon dioxide

Lymphatic System/Immunity
- Physical exercise may enhance or depress immunity depending on its intensity
- Lymphatic vessels drain leaked tissue fluids; immune system protects muscles from disease

Cardiovascular System
- Skeletal muscle activity increases efficiency of cardiovascular functioning; helps prevent atherosclerosis and causes cardiac hypertrophy
- Cardiovascular system delivers oxygen and nutrients to muscles; carries away wastes

Digestive System
- Physical activity increases gastrointestinal mobility when at rest
- Digestive system provides nutrients needed for muscle health; liver metabolizes lactic acid

Reproductive System
- Skeletal muscle helps support pelvic organs (e.g., uterus in females); assists erection of penis and clitoris
- Testicular androgen promotes increased skeletal muscle size

Urinary System
- Physical activity promotes normal voiding behavior; skeletal muscle forms the voluntary sphincter of the urethra
- Urinary system disposes of nitrogenous wastes

Integumentary System
- Muscular exercise enhances circulation to skin and improves skin health; exercise also increases body heat, which the skin helps dissipate
- Skin protects the muscles by external enclosure

Muscular System

Skeletal System
- Skeletal muscle activity maintains bone health and strength
- Bones provide levers for muscle activity

Important Terms

abduction	**lactic acid**
acetylcholine (as″e-til-ko′len)	**motor unit**
actin	**myofibrils** (mi″o-fi′brilz)
action potential	**myosin**
adduction	**neuromuscular junction**
aerobic respiration	**neurotransmitter**
anaerobic glycolysis	**origin**
antagonists (an-tag′o-nists)	**oxygen debt**
aponeuroses (ap″o-nu-ro′sēz)	**plantar flexion**
creatine phosphate	**prime mover**
dorsiflexion	**pronation**
extension	**supination**
fixators	**synaptic cleft**
flexion	**synergists** (sin′er-jists)
insertion	**tendons**

Summary

Overview of Muscle Tissues (pp. 154–158)

1. Skeletal muscle forms the muscles attached to the skeleton, which move the limbs and other body parts. Its cells are long, striated, and multinucleate. Skeletal muscle fibers are subject to voluntary control. Connective tissue coverings (endomysium, perimysium, and epimysium) enclose and protect the muscle fibers and increase the strength of skeletal muscles.

2. Smooth muscle cells are uninucleate, spindle-shaped, and arranged in opposing layers in the walls of hollow organs. When they contract, substances (food, urine, a baby) are moved along internal pathways. Smooth muscle control is involuntary.

3. Cardiac muscle cells are striated, branching cells that fit closely together and are arranged in spiral bundles in the heart. Their contraction pumps blood through the blood vessels. Control is involuntary.

4. The sole function of muscle tissue is to contract or shorten. As it contracts, it causes movement, maintains posture, stabilizes joints, and generates heat.

Microscopic Anatomy of Skeletal Muscle (pp. 158–160)

1. The multinucleate cylindrical skeletal muscle fibers are packed with unique organelles called myofibrils. The banding pattern (striations) of the myofibrils and the cell as a whole reflects the regular arrangement of thin (actin-containing) and thick (myosin) filaments within the sarcomeres, the contractile units composing the myofibrils.

2. Each myofibril is loosely enclosed by a specialized ER, called the sarcoplasmic reticulum (SR), which

plays an important role in storing and releasing calcium ions. Calcium ions are the final "trigger" for muscle fiber contraction.

Skeletal Muscle Activity (pp. 160-168)

1. All skeletal muscle cells are stimulated by motor neurons. When the neuron releases a neurotransmitter (acetylcholine), the permeability of the sarcolemma changes, allowing sodium ions to enter the muscle cell. This produces an electrical current (action potential), which flows across the entire sarcolemma, resulting in release of calcium ions from the SR.

2. Calcium binds to regulatory proteins on the thin filaments and exposes myosin binding sites, allowing the myosin heads on the thick filaments to attach. The attached heads pivot, sliding the thin filaments toward the center of the sarcomere, and contraction occurs. ATP provides the energy for the sliding process, which continues as long as ionic calcium is present.

3. Although individual muscle cells contract completely when adequately stimulated, a muscle (an organ) responds to stimuli to different degrees, i.e., it exhibits graded responses.

4. Most skeletal muscle contractions are tetanic (smooth and sustained) because rapid nerve impulses are reaching the muscle, and the muscle cannot relax completely between contractions. The strength of muscle contraction reflects the relative number of muscle cells contracting (more = stronger).

5. ATP, the immediate source of energy for muscle contraction, is stored in small amounts in muscle fibers and is quickly used up. ATP is regenerated via three routes. From the fastest to the slowest, these are via a coupled reaction of creatine phosphate with ADP, via anaerobic glycolysis and lactic acid formation, and via aerobic respiration. Only aerobic respiration requires oxygen.

6. If muscle activity is strenuous and prolonged, muscle fatigue occurs due to an accumulation of lactic acid in the muscle and a decrease in its energy (ATP) supply. After exercise, the oxygen debt is repaid by rapid deep breathing.

7. Muscle contractions are isotonic (the muscle shortens and movement occurs) or isometric (the muscle does not shorten, but its tension increases).

8. Muscle tone keeps muscles healthy and ready to react. It is a result of a staggered series of nerve impulses delivered to different cells within the muscle.

If the nerve supply is destroyed, the muscle loses tone, becomes paralyzed, and atrophies.

9. Inactive muscles atrophy. Muscles challenged by resistance exercise to respond (almost) beyond their ability increase in size and strength. Muscles subjected to regular aerobic exercise become more efficient and stronger, and can work longer without tiring. Aerobic exercise also benefits other body organ systems.

Muscle Movements, Types, and Names (pp. 170-174)

1. All muscles are attached to bones at two points. The origin is the immovable attachment; the insertion is the movable bony attachment. When contraction occurs, the insertion moves toward the origin.

2. Body movements include flexion, extension, abduction, adduction, circumduction, rotation, pronation, supination, inversion, eversion, dorsiflexion, and plantar flexion.

3. On the basis of their general functions in the body, muscles are classified as prime movers, antagonists, synergists, and fixators.

4. Muscles are named according to several criteria, including muscle size, shape, number and location of origins, bones associated with, and action of, the muscle.

Gross Anatomy of Skeletal Muscles (pp. 174-182)

1. Muscles of the head fall into two groups. The muscles of facial expression include the frontalis, orbicularis oris and oculi, and zygomaticus. The chewing muscles are the masseter, temporalis, and buccinator, which is also a muscle of facial expression.

2. Muscles of the trunk and neck move the head, shoulder girdle, and trunk, and form the abdominal girdle. Anterior neck and trunk muscles include the sternocleidomastoid, pectoralis major, intercostals, rectus abdominis, external and internal obliques, and transversus abdominis. Posterior trunk and neck muscles include the trapezius, latissimus dorsi, and deltoid.

3. Muscles of the upper limb include muscles that cause movement at the shoulder joint, elbow, and hand. Muscles causing movement at the elbow include the biceps brachii and triceps brachii.

4. Muscles of the lower extremity cause movement at the hip, knee, and foot. They include the iliopsoas, gluteus maximus and medius, adductors, quadriceps, and hamstring groups, gastrocnemius, anterior tibialis, and peroneus muscles.

Developmental Aspects of the Muscular System
(pp. 183–187)

1. Increasing muscular control reflects the maturation of the nervous system. Muscle control is achieved in a cephalic/caudal and proximal/distal direction.

2. To remain healthy, muscles must be regularly exercised. Without exercise, they atrophy; with extremely vigorous exercise, they hypertrophy.

3. As we age, muscle mass decreases and the muscles become more sinewy.

Review Questions

1. What is the major function of muscle?

2. Compare skeletal, smooth, and cardiac muscles as to their microscopic anatomy, location and arrangement in body organs, and function in the body.

3. Specifically, what is responsible for the banding pattern seen in skeletal muscle cells?

4. Why are the connective tissue wrappings of skeletal muscles important? Name these connective tissue coverings beginning with the finest and ending with the most coarse.

5. What is the function of tendons? How is a tendon different from an aponeurosis? How is it similar?

6. Define *myoneural junction, motor unit, tetanus, graded response, aerobic respiration, anaerobic glycolysis, muscle fatigue,* and *neurotransmitter.*

7. If blue litmus paper is pressed against the cut surface of a fatigued muscle, it changes color to red, indicating a low pH or acid condition. No such change happens when blue litmus paper is pressed to the cut surface of a rested or nonfatigued muscle. Explain.

8. Describe the events that occur from the time a motor neuron releases acetylcholine at the myoneural junction until muscle cell contraction occurs.

9. How do isotonic and isometric contractions differ?

10. Muscle tone keeps muscles healthy. What is muscle tone, and what causes it? What happens to a muscle that loses its tone?

11. A skeletal muscle is attached to bones at two points. Name each of these attachment points and indicate which is movable and which is nonmovable.

12. List the 12 body movements studied in this chapter and demonstrate each.

13. How is a prime mover different from a synergist muscle? How can a prime mover also be considered an antagonist?

14. Name at least four criteria used as a basis for naming skeletal muscles and give an example of each.

15. Name the prime mover for chewing. Name three other muscles of the face and give the location and function of each.

16. The sternocleidomastoid muscles help to flex the neck. What are their antagonists?

17. Name two muscles that reverse the movement of the deltoid muscle.

18. Name the prime mover of elbow flexion. Name its antagonist.

19. Other than acting to flex the spine and compress the abdominal contents, the abdominal muscles are extremely important in protecting and containing the abdominal viscera. What is it about the arrangement of these muscles that makes them so well suited for their job?

20. The hamstring and quadriceps muscle groups are antagonists of each other, and each group is a prime mover in its own right. What action does each muscle group perform?

21. What two-bellied muscle makes up the calf region of the leg? What is its function?

22. What happens to muscles when they are exercised regularly? Exercised vigorously as in weight lifting? Not used?

23. What is the effect of aging on skeletal muscles?

At the Clinic

1. Name three muscles or muscle groups used as the site for intramuscular injections. Which is most often used in babies?

2. What are the possible harmful effects of using anabolic steroids to increase muscle mass and strength?

3. Mr. Ahmadi was advised by his physician to lose

weight and start jogging. He began to jog daily. On the sixth day, he was forced to jump out of the way of a speeding car. He heard a snapping sound that was immediately followed by pain in his right lower calf. A gap was visible between his swollen calf and his heel, and he was unable to plantar flex that foot. What do you think happened?

4. While painting her house, Sally fell off the ladder and fractured her right clavicle. Treatment prescribed by the emergency room physician included use of a sling for her right arm to immobilize the clavicle and speed its healing. What muscles are temporarily "put out of business" by the sling?

5. When Eric returned from jogging, he was breathing heavily, sweating profusely, and complained that his legs ached and felt weak. His wife poured him a sports drink and urged him to take it easy until he could "catch his breath." On the basis of what you have learned about muscle energy metabolism, respond to the following questions:
   - Why is Eric breathing heavily?
   - What ATP harvesting pathway have his working muscles been using that leads to such a breathing pattern?
   - What metabolic product(s) might account for his sore muscles and his feeling of muscle weakness?

7

The Nervous System

YOUR GOALS

After completing this chapter, you will have working knowledge of the functions of the nervous system and will have mastered the objectives listed below.

FUNCTION PREVIEW

■ The nervous system maintains body homeostasis with electrical signals; provides for sensation, higher mental functioning, and emotional response; and activates muscles and glands.

(continued)

❏ Discuss the formation and function of cerebrospinal fluid and the blood-brain barrier.

❏ Compare the signs of a CVA with Alzheimer's disease; of a contusion with a concussion.

❏ Define *EEG* and explain how it evaluates neural functioning.

❏ List two important functions of the spinal cord.

❏ Describe the structure of the spinal cord.

PERIPHERAL NERVOUS SYSTEM (pp. 222–232)

❏ Describe the general structure of a nerve.

❏ Identify the cranial nerves by number and by name, and list the major functions of each.

❏ Describe the origin and fiber composition of (a) ventral and dorsal roots, (b) the spinal nerve proper, and (c) ventral and dorsal rami.

❏ Discuss the distribution of the dorsal and ventral rami of spinal nerves.

❏ Name the four major nerve plexuses, the major nerves of each, and describe their distribution.

❏ Identify the site of origin and explain the function of the sympathetic and parasympathetic divisions of the autonomic nervous system.

❏ Contrast the effect of the parasympathetic and sympathetic divisions on the following organs: heart, lungs, digestive system, blood vessels.

DEVELOPMENTAL ASPECTS OF THE NERVOUS SYSTEM (pp. 232–234)

❏ List several factors that may have harmful effects on brain development.

❏ Briefly describe the cause, signs, and consequences of the following congenital disorders: spina bifida, anencephaly, cerebral palsy.

❏ Explain the decline in brain size and weight that occurs with age.

❏ Define *senility* and note some possible causes.

You are driving down the freeway, and a horn blares on your right. You swerve to your left. Charlie leaves a note on the kitchen table: "See you later—have the stuff ready at 6." You know that the "stuff" is chili with taco chips. You are dozing, and your infant son makes a soft cry. Instantly, you awaken. What do all these events have in common? They are all everyday examples of the functioning of your nervous system, which has your body cells humming with activity nearly all the time.

The **nervous system** is the master controlling and communicating system of the body. Every thought, action, and emotion reflects its activity. Its signaling device, or means of communicating with body cells, is electrical impulses, which are rapid, specific, and cause almost immediate responses.

To carry out its normal role, the nervous system has three overlapping functions (Figure 7.1): (1) Much like a sentry, it uses its millions of sensory receptors to *monitor changes* occurring both inside and outside the body. These changes are called *stimuli,* and the gathered information is called **sensory input.** (2) It processes and interprets the sensory input and makes decisions about what should be done at each moment—a process

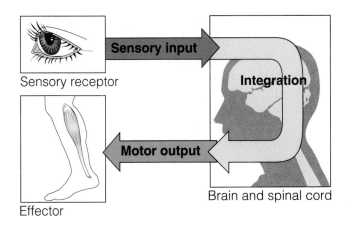

Figure 7.1   The nervous system's functions.

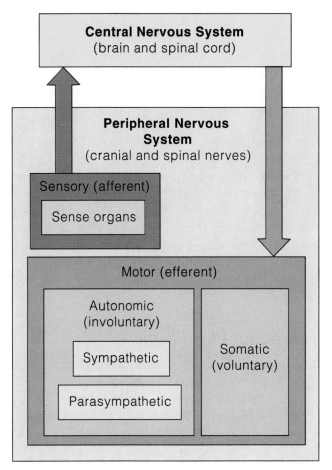

Figure 7.2   Organization of the nervous system. Organizational flowchart showing that the central nervous system receives input via sensory fibers and issues commands via motor fibers. The sensory and motor fibers together form the nerves that comprise the peripheral nervous system.

called **integration**. (3) It then *effects a response* by activating muscles or glands; the response is called **motor output**. An example will illustrate how these functions work together. When you are driving and see a red light just ahead (sensory input), your nervous system integrates this information (red light means "stop"), and your foot goes for the brake pedal (motor output).

The nervous system does not work alone to regulate and maintain body homeostasis; the endocrine system is a second important regulating system. While the nervous system controls with rapid electrical nerve impulses, the endocrine system organs produce hormones that are released into the blood. Thus, the endocrine system typically brings about its effects in a more leisurely way.

ORGANIZATION OF THE NERVOUS SYSTEM

We have only one nervous system, but because of its complexity it is difficult to consider all its parts at the same time. So, to simplify its study, we divide it in terms of its structures (structural classification) or in terms of its activities (functional classification). Each of these classification schemes is described briefly below and their relationships are illustrated in Figure 7.2. It is not necessary to memorize this whole scheme now, but as you are reading the descriptions, try to get a "feel" for the ma-

jor parts and how they fit together. This will make your learning task easier as you make your way through this chapter. Later you will meet all these terms and concepts again and in more detail.

Structural Classification

The structural classification, which includes all nervous system organs, has two subdivisions—the central nervous system and the peripheral nervous system (Figure 7.2).

The **central nervous system (CNS)** consists of the brain and spinal cord, which occupy the dorsal body cavity and act as the integrating and command centers of the nervous system. They interpret incoming sensory information and issue instructions based on past experience and current conditions.

The **peripheral** (pĕ-rif′er-al) **nervous system (PNS),** the part of the nervous system outside the CNS, consists mainly of the nerves that extend from the brain and spinal cord. *Spinal nerves* carry impulses to and from the spinal cord. *Cranial* (kra′ne-al) *nerves* carry impulses to and from the brain. These nerves serve as communication lines. They link all parts of the body by carrying impulses from the sensory receptors to the CNS and from the CNS to the appropriate glands or muscles.

The organs comprising the CNS and PNS are discussed at length later in this chapter.

Functional Classification

The functional classification scheme is concerned only with PNS structures. It subdivides them into two principal divisions (see Figure 7.2).

The **sensory,** or **afferent** (af′er-ent), **division** consists of nerve fibers that *convey impulses to the central nervous system* from sensory receptors located in various parts of the body. Sensory fibers delivering impulses from the skin, skeletal muscles, and joints are called *somatic (soma = body) sensory* (afferent) *fibers,* whereas those transmitting impulses from the visceral organs are called *visceral sensory fibers,* or *visceral afferents.* The sensory division keeps the CNS constantly informed of events going on both inside and outside the body.

The **motor,** or **efferent** (ef′er-rent), **division** *carries impulses from the CNS* to effector organs, the muscles and glands. These impulses activate muscles and glands; that is, they *effect* (bring about) a motor response.

The motor division in turn has two subdivisions (see Figure 7.2):

1. The **somatic** (so-mat′ik) **nervous system** allows us to consciously, or voluntarily, control our skeletal muscles. Hence, this subdivision is often referred to as the **voluntary nervous system.**

2. The **autonomic** (aw″to-nom′ik) **nervous system (ANS)** regulates events that are automatic, or involuntary, such as the activity of smooth and cardiac muscles and glands. This subdivision, commonly called the **involuntary nervous system,** itself has two parts, the *sympathetic* and *parasympathetic.* These will be described later.

Although it is simpler to study the nervous system in terms of its subdivisions, you should recognize that these subdivisions are made for the sake of convenience only. Remember that the nervous system acts as a coordinated unit, both structurally and functionally.

NERVOUS TISSUE: STRUCTURE AND FUNCTION

Even though it is complex, nervous tissue is made up of just two principal types of cells—*supporting cells* and *neurons.*

Supporting Cells

Supporting cells in the CNS are "lumped together" as **neuroglia** (nu-rog′le-ah), literally "nerve glue." Neuroglia includes many types of cells that generally support, insulate, and protect the delicate neurons (see Figure 7.3). In addition, each of the different types of neuroglia, also simply called **glia** (gle′ah), has special functions.

The star-shaped **astrocytes** are particularly abundant and account for nearly half of the neural tissue. They have numerous projections with swollen ends that cling to neurons, bracing them and anchoring them to their nutrient supply lines, the blood capillaries (Figure 7.3a). The astrocytes form a living barrier between capillaries and neurons, and they play a role in making exchanges between the two. In this way, they help protect the neurons from harmful substances that might be in the blood. Astrocytes also help control the chemical environment in the brain by picking up excess ions and recapturing released neurotransmitters.

The spiderlike **microglia** (mi-krog′le-ah) are phagocytes that dispose of debris—dead brain cells, bacteria, and the like (Figure 7.3b). **Ependymal** (ĕ-pen′dĭ-mal) **cells** line the cavities of the

Figure 7.3 Supporting cells of nervous tissues. (a–e) Types of glial cells found in the central nervous system. Astrocytes (a) form a living barrier between neurons and capillaries. Microglia (b) are phagocytes, whereas ependymal cells (c) line the fluid-filled cavities of the CNS. The oligodendrocytes (d) form myelin sheaths around nerve fibers in the CNS. (e) The relationship of Schwann cells (myelinating cells) and satellite cells to a neuron in the peripheral nervous system.

brain and spinal cord (Figure 7.3c). The beating of their cilia helps to circulate the cerebrospinal fluid that fills those cavities and forms a protective cushion around the CNS. The **oligodendrocytes** (ol″ĭ-go-den′dro-sītz) wrap their flat extensions tightly around the nerve fibers, producing fatty insulating coverings called *myelin sheaths* (Figure 7.3d).

Although they resemble neurons structurally, neuroglia are not able to transmit nerve impulses, a function that is highly developed in neurons. Another important difference is that neuroglia never lose their ability to divide, whereas neurons do. Consequently, most brain tumors are *gliomas*, or tumors formed by glial cells (neuroglia).

Supporting cells in the PNS come in two major varieties—Schwann cells and satellite cells (Figure 7.3e). **Schwann cells** form the myelin sheaths around nerve fibers that extend from the CNS. **Satellite cells** act as generally protective, cushioning cells.

Neurons

Anatomy

Neurons, also called **nerve cells,** are highly specialized to transmit messages (nerve impulses)

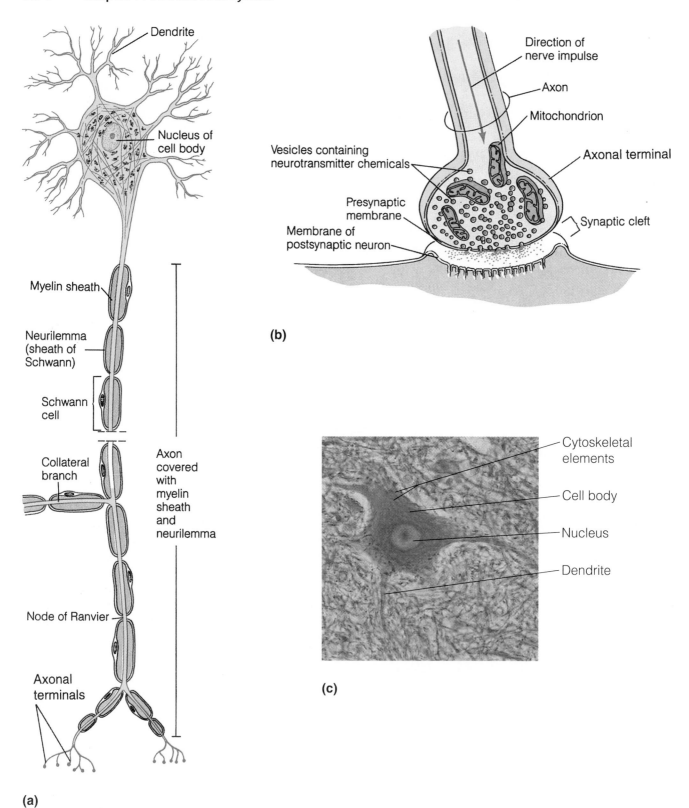

(a)

(b)

(c)

Figure 7.4 **Structure of a typical motor neuron.** (**a**) Diagrammatic view. (**b**) An enlarged synapse. (**c**) Photomicrograph (265×)

from one part of the body to another. Although neurons differ structurally, they have many common features (see Figure 7.4). All have (a) a **cell body,** which contains the nucleus and is the metabolic center of the cell, and (b) one or more slender **processes,** or **fibers,** extending from the cell body. The processes vary in length from microscopic in size to a length of 3–4 feet. The longest ones in humans reach from the lumbar region of the spine to the big toe. Neuron processes that conduct electrical currents *toward* the cell body are **dendrites** (den′drītz), whereas those that generate nerve impulses and typically conduct them *away* from the cell body are **axons** (ak′sonz). Neurons may have hundreds of the branching dendrites (*dendr* = tree), depending on the neuron type, but each neuron has only one axon.

An occasional axon gives off a *collateral branch* along its length, but all axons branch profusely at their terminal end, forming hundreds to thousands of **axonal terminals.** As we said, axons transmit nerve impulses away from the cell body. When these impulses reach the axonal terminals, they stimulate the release of chemicals called **neurotransmitters** into the extracellular space.

An enlarged view of an axonal terminal in Figure 7.4a shows that it is separated from the next neuron by a tiny gap, the **synaptic** (sĭ-nap′tik) **cleft.** Such a functional junction is called a **synapse** (*syn* = to clasp or join). Although they are close, neurons never actually touch other neurons. The figure also illustrates the tiny vesicles, or sacs, that store the neurotransmitter within the axonal terminal. We will learn more about the events that occur at the synapse a bit later.*

Most long nerve fibers are covered with a whitish, fatty material, called **myelin** (mi′ĕ-lin), that has a waxy appearance. Myelin protects and insulates the fibers and increases the transmission rate of nerve impulses. Axons outside the CNS are myelinated by **Schwann cells,** specialized supporting cells that wrap themselves tightly around the axon jelly-roll fashion (Figure 7.5). When the wrapping process is done, a tight coil of wrapped

---

*Although most neurons communicate via the *chemical* type of synapse described above, there are some examples of *electrical* synapses in which the neurons are physically joined by gap junctions and electrical currents actually flow from one neuron to the next.

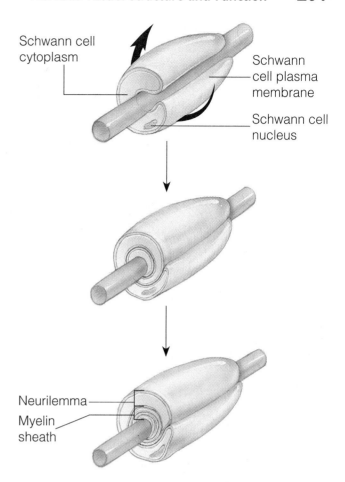

Figure 7.5 Relationship of Schwann cells to axons in the peripheral nervous system. As illustrated (top to bottom), a Schwann cell envelops part of an axon in a trough and then rotates around the axon. Most of the Schwann cell cytoplasm comes to lie just beneath the exposed part of its plasma membrane. The tight coil of plasma membrane material surrounding the axon is the myelin sheath; the Schwann cell cytoplasm and exposed membrane is referred to as the *neurilemma.*

membranes, the **myelin sheath,** encloses the axon. Most of the Schwann cell cytoplasm ends up just beneath the outermost part of its plasma membrane. This part of the Schwann cell, external to the myelin sheath, is called the **neurilemma** (nu″rĭ-lem′mah). Since the myelin sheath is formed by many individual Schwann cells, it has gaps or indentations, called **nodes of Ranvier** (rahn-vēr), at regular intervals (see Figure 7.4).

Myelinated fibers are also found in the central nervous system. However, there it is oligodendrocytes that form CNS myelin sheaths (see Figure 7.3d). In contrast to Schwann cells, each of which

Figure 7.6 Neurons classified by function. Sensory (afferent) neurons conduct impulses from sensory receptors (in the skin, viscera, muscles) to the central nervous system; most cell bodies are in ganglia in the PNS. Motor (efferent) neurons transmit impulses from the CNS (brain or spinal cord) to effectors in the body periphery. Association neurons (interneurons) complete the communication pathway between sensory and motor neurons; their cell bodies reside in the CNS.

deposits myelin around a small segment of one nerve fiber, the oligodendrocytes with their many flat extensions can coil around as many as 60 different fibers at the same time.

The importance of the myelin insulation to nerve transmission is best illustrated by observing what happens when it is not there. In people with a demyelinating disease called *multiple sclerosis (MS),* the myelin sheaths around the fibers are gradually destroyed, converted to hardened sheaths called *scleroses.* As this happens, the current is short-circuited and the affected person loses the ability to control his or her muscles and becomes increasingly disabled. Multiple sclerosis is an autoimmune disease in which a protein component of the sheath is attacked. As yet there is no cure, but injections of interferon (a hormonelike substance released by some immune cells) and oral doses of bovine myelin appear to provide some relief to MS victims. ❑

Clusters of neuron cell bodies and collections of nerve fibers are named differently when they are in the CNS than when they are part of the PNS. For the most part, cell bodies are found in the CNS in clusters called **nuclei.** This well-protected location within the bony skull or vertebral column is essential to the well-being of the nervous system—remember that neurons do not undergo cell division after birth. The cell body carries out most of

the metabolic functions of a neuron, so if it is damaged the cell dies and is not replaced. Small collections of cell bodies called **ganglia** (gang′le-ah) are found in a few sites outside the CNS in the PNS. Bundles of nerve fibers (neuron processes) running through the CNS are called **tracts,** whereas in the PNS they are called **nerves.** The terms *white matter* and *gray matter* refer respectively to myelinated versus unmyelinated regions of the CNS. As a general rule, the **white matter** consists of dense collections of myelinated fibers (tracts) and **gray matter** contains mostly unmyelinated fibers and cell bodies.

Classification

Neurons may be classified, or grouped, either according to their structure or according to how they function.

Functional Classification Functional classification groups neurons according to the direction the nerve impulse is traveling relative to the CNS. On this basis, there are sensory, motor, and association neurons (Figure 7.6). Neurons carrying impulses from sensory receptors (in the internal organs or the skin) to the CNS are **sensory,** or **afferent, neurons.** (*Afferent* literally means "to go toward.") The cell bodies of sensory neurons are always

Figure 7.7   Types of sensory receptors. (**a**) Naked nerve endings (pain and temperature receptors). (**b**) Meissner's corpuscle (touch receptor). (**c**) Pacinian corpuscle (deep pressure receptor). (**d**) Golgi tendon organ (proprioceptor). (**e**) Muscle spindle (proprioceptor). (Note: The proprioceptors respond to the degree of stretch, or tension, in the muscles and tendons.)

found in a *ganglion* outside the CNS. Sensory neurons keep us informed about what is happening both inside and outside the body.

The dendrite endings of the sensory neurons are usually associated with specialized **receptors** that are activated by specific changes occurring nearby. The very complex receptors of the special sense organs (vision, hearing, equilibrium, taste, and smell) are covered separately in Chapter 8. The simpler types of sensory receptors seen in the skin (**cutaneous sense organs**) and in the muscles and tendons (**proprioceptors** [pro″pre-o-sep′torz]) are shown in Figure 7.7. The pain receptors (actually bare dendrite endings) are the least specialized of the cutaneous receptors. They are also the most numerous, because pain warns us that some type of body damage is occurring or is about to occur. However, strong stimulation of any of the cutaneous receptors (for example, by searing heat, extreme cold, or excessive pressure) is also interpreted as pain.

The proprioceptors detect the amount of stretch, or tension, in skeletal muscles, their tendons, and joints. They send this information to the brain so that the proper adjustments can be made to maintain balance and normal posture. *Propria* comes from the Latin meaning "one's own," and the proprioceptors constantly advise our brain of our own movements.

Neurons carrying impulses from the CNS to the viscera and/or muscles and glands are **motor,** or **efferent, neurons.** The cell bodies of motor neurons are always located in the CNS.

The third category of neurons is the **association neurons,** or **interneurons.** They connect the motor and sensory neurons in neural pathways. Like the motor neurons, their cell bodies are always located in the CNS.

(a) Multipolar neuron

(b) Bipolar neuron

(c) Unipolar neuron

Figure 7.8   Classification of neurons on the basis of structure. (**a**) Multipolar. (**b**) Bipolar. (**c**) Unipolar.

Structural Classification   Structural classification is based on the number of processes extending from the cell body (Figure 7.8). If there are several, the neuron is a **multipolar neuron.** Since all motor and association neurons are multipolar, this is the most common structural type. Neurons with two processes—an axon and a dendrite—are called **bipolar neurons.** Bipolar neurons are rare in adults, found only in some special sense organs (eye, ear) where they act as sensory receptor cells. **Unipolar neurons** have a single process emerging from the cell body. However, it is very short and divides almost immediately into proximal (central) and distal (peripheral) fibers. Unipolar neurons are unique in that only the small branches at the end of the peripheral process are dendrites. The remainder of the peripheral process and the central process function as axons; thus, in this case, the axon conducts nerve impulses both toward *and* away from the cell body. Sensory neurons found in PNS ganglia are unipolar.

Physiology

Nerve Impulses   Neurons have two major functional properties: *irritability,* the ability to respond to a stimulus and convert it into a nerve impulse, and *conductivity,* the ability to transmit the impulse to other neurons, muscles, or glands. We will consider irritability first.

The plasma membrane of a resting, or inactive, neuron is **polarized,** which means that there are fewer positive ions inside the neuron than there are in the tissue fluid that surrounds it (Figure 7.9). The major positive ions inside the cell are potassium (K^+), whereas the major positive ions outside the cell are sodium (Na^+). As long as the inside remains negative and the outside remains positive, the neuron will stay inactive.

Many different types of stimuli excite neurons to become active and generate an impulse. For example, light excites the eye receptors, sound excites some of the ear receptors, and pressure excites some cutaneous receptors of the skin. However, *most* neurons in the body are excited by neurotransmitters released by other neurons, as will be described shortly. Regardless of what the stimulus is, the result is always the same—the permeability properties of the cell membrane change for a very brief period. *Normally,* sodium ions cannot diffuse through the cell membrane to any great extent; but when the neuron is adequately stimulated, the "sodium gates" in the membrane open. Because sodium is in much higher concentration outside the cell, it will then diffuse quickly into the neuron. (Remember the laws of diffusion?) This inward rush of sodium ions changes the polarity of the neuron's membrane; the inside is now more positive and the outside is less positive. This event, called **depolarization,** activates the neuron to transmit an **action potential,** also called a **nerve impulse** in neurons. The nerve impulse is an *all-or-none response.* It is either propagated (conducted) over the entire axon, or it doesn't happen at all. The nerve impulse never goes partway along an axon's length.

Almost immediately after the sodium ions rush into the neuron, the membrane permeability changes again, becoming impermeable to sodium ions. But now potassium ions are allowed to diffuse out of the neuron into the tissue fluid, and they do so very rapidly. This outflow of positive ions from the cell restores the electrical conditions at the membrane to the polarized, or resting, state,

(a) Resting membrane

(b) Depolarization and generation of the action potential

(c) Propagation of the action potential

(d) Repolarization

Figure 7.9 **The nerve impulse. (a)** Electrical conditions of a resting (polarized) membrane. The external face of the membrane is slightly positive; its internal face is slightly negative. The chief extracellular ion is sodium (Na⁺), whereas the chief intracellular ion is potassium (K⁺). The membrane is relatively impermeable to both ions. **(b)** Depolarization and generation of an action potential. A stimulus changes the permeability of a "patch" of the membrane and sodium ions diffuse rapidly into the cell. This changes the polarity of the membrane (the inside becomes more positive, the outside becomes more negative). If the stimulus is strong enough, an action potential is initiated. **(c)** Propagation of the action potential. Depolarization of the first membrane patch causes permeability changes in the adjacent membrane, and the events described in (b) are repeated. Thus, the action potential propagates rapidly along the entire length of the membrane. **(d)** Repolarization. Potassium ions diffuse out of the cell as membrane permeability changes again, restoring the negative charge on the inside of the membrane and the positive charge on the outside surface. Repolarization occurs in the same direction as depolarization. The ionic conditions of the resting state are restored later by the activity of the sodium-potassium pump.

an event called **repolarization.** *Until repolarization occurs, a neuron cannot conduct another impulse.* After repolarization occurs, the initial concentrations of the sodium and potassium ions inside and outside the neuron are restored by activation of the sodium-potassium pump. This pump uses ATP (cellular energy) to pump excess sodium ions out of the cell and to bring potassium ions back into it.

The events just described explain propagation of a nerve impulse along unmyelinated fibers. Fibers that have myelin sheaths conduct impulses much faster because the nerve impulse literally jumps, or leaps, from node to node along the length of the fiber. This occurs because no current can flow across the axonal membrane where there is fatty myelin insulation. This faster type of impulse propagation is called *saltatory* (sal′tah-to″re) *conduction* (*saltare* = to dance or leap).

A number of factors can impair the conduction of impulses. For example, alcohol, sedatives, and anesthetics all block nerve impulses by reducing membrane permeability to sodium ions. As we have seen, no sodium entry = no action potential.

Cold and continuous pressure hinder impulse conduction because they interrupt blood circulation (and hence the delivery of oxygen and nutrients) to the neurons. For example, your fingers get numb when you hold an ice cube for more than a few seconds. Likewise, when you sit on your foot, it "goes to sleep." When you warm the fingers, or remove the pressure from your foot, the impulses begin to be transmitted once again, leading to an unpleasant prickly feeling. ❑

So far we have only explained the irritability aspect of neuronal functioning. What about conductivity—how does the electrical impulse traveling along one neuron get to the next neuron (or effector cell) to influence its activity? The answer is that *it* doesn't! When the action potential reaches the axonal endings, the axonal terminals release the neurotransmitter chemical, which diffuses across the synapse and binds to receptors on the membrane of the next neuron (Figure 7.10). If enough neurotransmitter is released, the whole series of events described above (sodium entry, depolarization, etc.) will occur, leading to generation of a nerve impulse in the neuron beyond the synapse. The electrical changes prompted by neurotransmitter binding are very brief because the

Transmitting neuron

Receiving neuron

Figure 7.10   How neurons communicate at synapses. The events occurring at the synapse are numbered in order.

neurotransmitter is quickly removed from the synapse, either by re-uptake into the axonal terminal or by enzymatic breakdown. This limits the effect of each nerve impulse to a period shorter than the blink of an eye. Notice that the transmission of an impulse is an *electrochemical event.* Transmission down the length of the neuron's membrane is basically *electrical,* but the next neuron is stimulated by a neurotransmitter, which is a *chemical.* Since each neuron both receives signals from and sends signals to scores of other neurons, it carries on "conversations" with many different neurons at the same time.

Reflex Arc   Although there are many types of communication between neurons, much of what the body *must* do every day is programmed as **reflexes.** Reflexes are *rapid, predictable, and involuntary responses* to stimuli. They are much like one-way streets—once a reflex begins, it always goes in the same direction. Reflexes occur over neural pathways called **reflex arcs.**

The types of reflexes that occur in the body are classed either as autonomic or somatic reflexes. **Autonomic reflexes** regulate the activity of smooth muscles, the heart, and glands. Secretion of saliva (salivary reflex) and changes in the size of the eye pupils (pupillary reflex) are two such reflexes. Autonomic reflexes regulate such body functions as digestion, elimination, blood pressure, and sweating. **Somatic reflexes** include all reflexes that stimulate the skeletal muscles. When you quickly pull your hand away from a hot object, a somatic reflex is working.

All reflex arcs have a minimum of four elements: (1) a sensory receptor (which reacts to a stimulus), (2) an effector organ (the muscle or gland eventually stimulated), and (3 and 4) afferent and efferent neurons to connect the two. The simple *patellar* (pah-tel′ar) or *knee-jerk reflex* shown in Figure 7.11a is an example of a two-neuron reflex arc, the most simple type in humans. The patellar reflex is familiar to most of us. It is usually tested during a physical exam to determine the general health of the motor portion of the nervous system. Most reflexes are much more complex than the two-neuron reflex, involving one or more interneurons in the pathway. A three-neuron reflex arc, the *flexor reflex,* is diagrammed in Figure 7.11b. The three-neuron reflex arc consists of five elements—receptor, afferent neuron, interneuron,

Figure 7.11   Simple reflex arcs. (a) Two-neuron reflex arc.
(b) Three-neuron reflex arc.

efferent neuron, and effector. Since there is always a delay at synapses (it takes time for the neurotransmitter to diffuse through the synaptic cleft), the more synapses there are in a reflex pathway, the longer the reflex takes to happen.

Many spinal reflexes involve only spinal cord neurons and occur without brain involvement. As long as the spinal cord is functional, spinal reflexes such as the flexor reflex will work. On the other hand, some reflexes require that the brain become involved, because many different types of information have to be evaluated to arrive at the "right" response. The response of the pupils of the eyes to light is a reflex of this type.

As noted earlier, reflex testing is an important tool in evaluating the condition of the nervous system. Whenever reflexes are exaggerated, distorted, or absent, nervous system disorders are indicated. Reflex changes often can occur before the pathologic condition has become obvious in other ways.

CENTRAL NERVOUS SYSTEM

During embryonic development, the CNS first appears as a simple tube, the **neural tube,** which extends down the dorsal median plane of the developing embryo's body. By the fourth week, the anterior end of the neural tube begins to expand, and brain formation begins. The rest of the neural tube posterior to the forming brain becomes the spinal cord. The central canal of the neural tube, which is continuous between the brain and spinal cord, becomes enlarged in four regions of the brain to form chambers called **ventricles** (see p. 215).

Functional Anatomy of the Brain

The adult brain's unimpressive appearance gives few hints of its remarkable abilities. It is about two good fistfuls of pinkish gray tissue, wrinkled like a walnut, and with the texture of cold oatmeal. It weighs a little over three pounds. Because the brain is the largest and most complex mass of nervous tissue in the body, it is commonly discussed in terms of its four major regions—cerebral hemispheres, diencephalon (di"en-sef'ah-lon), brain stem, and cerebellum.

Cerebral Hemispheres

The paired **cerebral** (ser'e-bral) **hemispheres** are the most superior part of the brain, and together are a good deal larger than the other three brain regions combined. In fact, the cerebral hemispheres enclose and obscure most of the brain stem so that

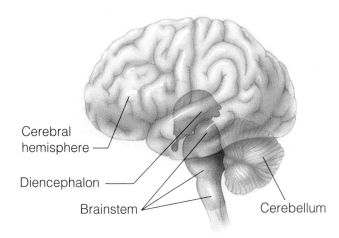

Cerebral hemisphere —

Diencephalon —

Brainstem —

Cerebellum

Figure 7.12   Regions of the human brain.
The brain can be considered in terms of four main parts: cerebral hemispheres, diencephalon, brain stem, and cerebellum. The left cerebral hemisphere is drawn so that it looks transparent, to reveal the location of the deeply situated diencephalon and superior part of the brain stem.

many brain stem structures cannot normally be seen unless a sagittal section is made. Picture how a mushroom cap covers the top of its stalk, and you have a fairly good idea of how the cerebral hemispheres cover the diencephalon and the superior part of the brain stem (Figure 7.12).

The entire surface of the cerebral hemispheres exhibits elevated ridges of tissue called **gyri** (ji′re; "twisters"), separated by shallow grooves called **sulci** (sul′ki; "furrows"). Less numerous are the deeper grooves called **fissures** (Figure 7.13a), which separate large regions of the brain. Many of the fissures and gyri are important anatomical landmarks. The cerebral hemispheres are separated by a single deep fissure, the *longitudinal fissure.* Other fissures divide each cerebral hemisphere into a number of **lobes,** named for the cranial bones that lie over them.

Speech, memory, logical and emotional response, as well as consciousness, interpretation of sensation, and voluntary movement are all functions of cerebral cortex neurons, and many of the functional areas of the cerebral hemispheres have been identified (Figure 7.13b). The **somatic sensory area** is located posterior to the *central fissure* in the **parietal lobe.** Impulses traveling from the body's sensory receptors (except for the special senses) are localized and interpreted in this area of

the brain. The somatic sensory area allows you to recognize pain, coldness, or a light touch. As illustrated in Figure 7.14 on p. 208, the body is represented in an upside-down manner in the sensory area. Body regions with the most sensory receptors—the lips and fingertips—send impulses to neurons that make up a large part of the sensory cortex. Furthermore, the sensory pathways are crossed pathways—meaning that the left side of the sensory cortex receives impulses from the right side of the body, and vice versa.

Impulses from the special sense organs are interpreted in other cortical areas (see Figure 7.13b). For example, the visual area is located in the posterior part of the **occipital lobe,** the auditory area is in the **temporal lobe** bordering the *lateral fissure,* and the olfactory area is found deep inside the temporal lobe.

The **primary motor area** that allows us to consciously move our skeletal muscles is anterior to the central fissure in the **frontal lobe.** The axons of these motor neurons form the major voluntary motor tract—the **pyramidal,** or **corticospinal** (kor″tĭ-ko-spi′nal), **tract,** which descends to the cord. As in the somatic sensory cortex, the body is represented upside-down and the pathways are crossed. Most of the neurons in this primary motor area control body areas having the finest motor control; that is, the face, mouth, and hands (Figure 7.14).

A specialized area that is very involved in our ability to speak, **Broca's** (bro′kahz) **area** (Figure 7.13b), is found at the base of the precentral gyrus (the gyrus anterior to the central fissure). Damage to this area, which is located in only one cerebral hemisphere (usually the left), causes inability to say words properly. You know what you want to say, but can't vocalize the words.

Areas involved in higher intellectual reasoning are believed to be in the anterior part of the frontal lobes. Complex memories appear to be stored in the temporal and frontal lobes. The **speech area** is located at the junction of the temporal, parietal, and occipital lobes. The speech area allows one to sound out words. This area (like Broca's area) is usually in only one cerebral hemisphere. The frontal lobes house areas involved with language comprehension (word meanings).

The cell bodies of neurons involved in the cerebral hemisphere functions named above are

(a)

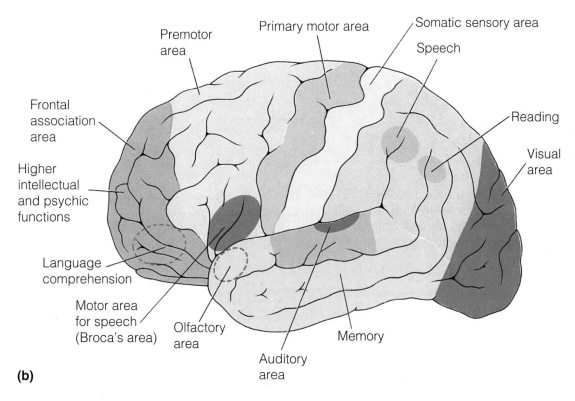

(b)

Figure 7.13   Left lateral view of the brain. Major (**a**) structural and (**b**) functional areas of the cerebral hemispheres.

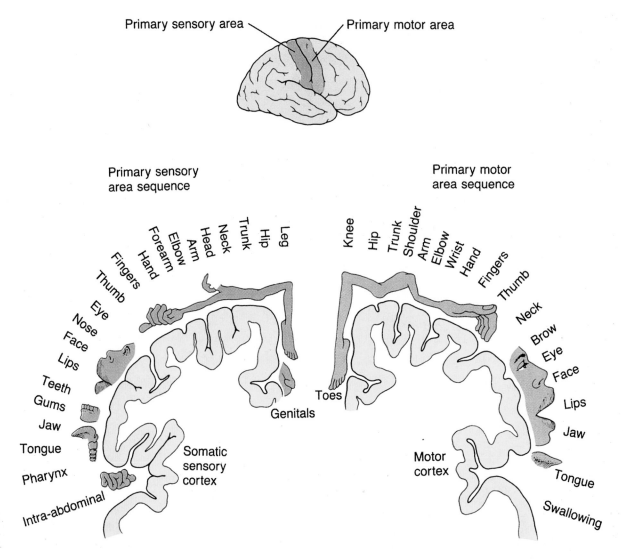

Figure 7.14   Sensory and motor areas of the cerebral cortex. The relative amount of cortical tissue devoted to each function is indicated by the amount of the gyrus occupied by the body area diagrams. The primary motor cortex is shown on the right; the somatic sensory cortex on the left.

found only in the outermost **gray matter** of the cerebrum, the **cerebral cortex.** As noted earlier, the cortical region is highly ridged and convoluted, providing more room for the thousands of neurons found there.

Most of the remaining cerebral hemisphere tissue—the deeper **cerebral white matter**—is composed of fiber tracts (bundles of nerve fibers) carrying impulses to or from the cortex. One very large fiber tract, the **corpus callosum** (kah-lo′sum) connects the cerebral hemispheres (Figure 7.15). The corpus callosum arches above the struc-

tures of the brain stem and allows the cerebral hemispheres to communicate with one another. This is important because, as already noted, some of the cortical functional areas are in only one hemisphere.

Although most of the gray matter is in the cerebral cortex, there are several "islands" of gray matter, called the **basal nuclei,** buried deep within the white matter of the cerebral hemispheres. The basal nuclei help regulate voluntary motor activities by modifying instructions sent to the skeletal muscles by the primary motor cortex.

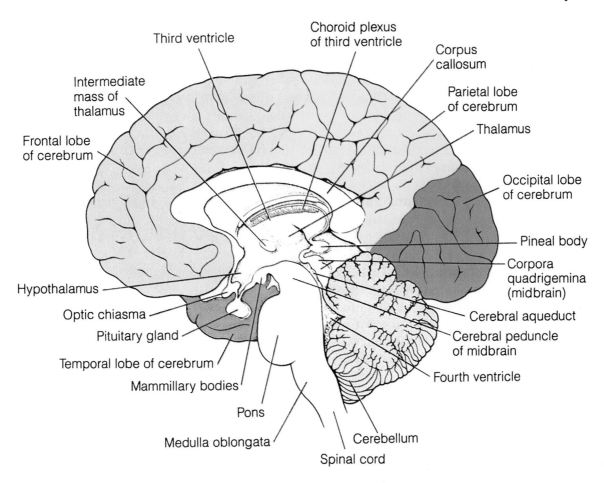

Figure 7.15   **Midsagittal section of the brain.**

Individuals who have problems with their basal nuclei are often unable to walk normally or carry out other voluntary movements in a normal way. **Huntington's chorea** (ko-re'ah), a genetic disease in which the individual is unable to control muscles and exhibits abrupt, jerky, and almost continuous movements, is one example of basal nuclei problems; **Parkinson's disease** is another. People with Parkinson's disease have trouble initiating movement or getting their muscles going. They also have a persistent hand tremor in which their thumb and index finger make continuous circles with one another in what is called the "pill-rolling" movement. Parkinson's disease is due to a deficit of the neurotransmitter *dopamine,* and drugs that help to replace it benefit some. In contrast, symptoms of chorea sufferers are helped by drugs that *block* dopamine's effects. As you can see, neurotransmitters, which are the "vocabulary" of neurons, can cause garbled neural language when things go wrong. ❏

Diencephalon

The **diencephalon,** or **interbrain,** sits atop the brain stem and is enclosed by the cerebral hemispheres (see Figure 7.12). The major structures of the diencephalon are the thalamus, hypothalamus, and epithalamus (see Figure 7.15). The **thalamus,** which encloses the shallow *third ventricle* of the brain, is a relay station for sensory impulses passing upward to the sensory cortex. As impulses surge through the thalamus, we have a crude recognition of whether the sensation we are about to have is pleasant or unpleasant. The actual localization and interpretation of the sensation is done by the neurons of the sensory cortex.

The **hypothalamus** (literally, "under the thalamus") makes up the floor of the diencephalon. It is an important autonomic nervous system center because it plays a role in the regulation of body temperature, water balance, and metabolism. The hypothalamus is also the center for many drives and emotions, and as such it is an important part of the so-called **limbic system,** or "emotional-visceral brain." For example, thirst, appetite, sex, pain, and pleasure centers are in the hypothalamus. (The central role of the pleasure center in cocaine addiction is examined in the "Closer Look" box on the opposite page.) Additionally, the hypothalamus regulates the pituitary gland (an endocrine organ) and produces two hormones of its own. The **pituitary gland** hangs from the anterior floor of the hypothalamus by a slender stalk. (Its function is discussed in Chapter 9.) The **mammillary bodies,** reflex centers involved in olfaction (the sense of smell), bulge from the floor of the hypothalamus posterior to the pituitary gland.

The **epithalamus** (ep″ĭ-thal′ah-mus) forms the roof of the third ventricle. Important parts of the epithalamus are the **pineal body** (part of the endocrine system) and the **choroid** (ko′roid) **plexus** of the third ventricle. The choroid plexuses, knots of capillaries within each ventricle, form the cerebrospinal fluid.

Brain Stem

The **brain stem** is about the size of a thumb in diameter and approximately 3 inches long. Its structures are the *midbrain, pons,* and *medulla oblongata.* In addition to providing a pathway for ascending and descending tracts, the brain stem has many small gray matter areas. These nuclei form the cranial nerves and control vital activities such as breathing and blood pressure. Extending the entire length of the brain stem is a diffuse mass of gray matter, the **reticular formation.** The neurons of the reticular formation are involved in motor control of the visceral organs. A special group of reticular formation neurons, the **reticular activating system (RAS)** plays a role in consciousness and the awake/sleep cycles. Damage to this area can result in permanent unconsciousness (coma). Identify the brain stem areas in Figure 7.15 as you read their descriptions that follow.

Midbrain The **midbrain** is a relatively small part of the brain stem. It extends from the mammillary bodies to the pons inferiorly. The **cerebral aqueduct** is a tiny canal that travels through the midbrain and connects the third ventricle of the diencephalon to the fourth ventricle below. Anteriorly, the midbrain is composed primarily of two bulging fiber tracts, the **cerebral peduncles** (literally, "little feet of the cerebrum"), which convey ascending and descending impulses. Dorsally located are four rounded protrusions called the **corpora quadrigemina** (kor′por-ah kwah″drĭ-jem′ĭ-nah) because they reminded some anatomist of two pairs of twins *(gemini).* These bulging nuclei are reflex centers involved with vision and hearing.

Pons The **pons** (ponz) is the rounded structure that protrudes just below the midbrain. Pons means "bridge," and this area of the brain stem is mostly fiber tracts. However, it does have important nuclei involved in the control of breathing.

Medulla Oblongata The **medulla oblongata** (mĕ-dul′ah ob″long-gă′tah) is the most inferior part of the brain stem. It merges into the spinal cord below without any obvious change in structure. Like the pons, the medulla is an important fiber tract area. The medulla also contains many nuclei that regulate vital visceral activities. It contains centers that control heart rate, blood pressure, breathing, swallowing, and vomiting, among others. The *fourth ventricle* lies posterior to the pons and medulla and anterior to the cerebellum.

Cerebellum

The large, cauliflowerlike **cerebellum** (ser″e-bel′um) projects dorsally from under the occipital lobe of the cerebrum. Like the cerebrum, it has two hemispheres and a convoluted surface. The cerebellum also has an outer cortex made up of gray matter and an inner region of white matter.

The cerebellum provides the precise timing for skeletal muscle activity and controls our balance and equilibrium. Because of its activity, body movements are smooth and coordinated. Fibers reach the cerebellum from the equilibrium apparatus of the inner ear, the eye, the proprioceptors of the skeletal muscles and tendons, and many other areas. The cerebellum can be compared to an automatic pilot, continuously comparing the brain's "intentions" with actual body performance by monitoring body position and amount of tension in various body parts. When needed, it sends messages to initiate the appropriate corrective measures.

A CLOSER LOOK

PLEASURE ME, PLEASURE ME!

Deep inside the hypothalamus is a small bundle of neural tissue, called the *pleasure center*. We humans are programmed to eat, drink, and procreate, and it is the pleasure center that reinforces these behaviors. However, because of it, we are also perilously vulnerable. Few would deny that much of what we do and value is driven by the desire for pleasure, and herein lie the beginnings of addiction.

Our ability to feel good involves brain neurotransmitters such as norepinephrine and dopamine. For example, romantic love has been described as a "brain bath" of norepinephrine and dopamine, which stimulate the pleasure center and cause all of the decidedly pleasurable sensations associated with that emotion. Cole Porter knew what he was talking about when he wrote "I get a kick out of you," because these neurotransmitters are chemical cousins of the amphetamines. People who use "speed" and other amphetamines are artificially stimulating their brains to provide their pleasure flush. However, the pleasure is short-lived because when the brain is flooded with neurotransmitter-like chemicals from the outside, it begins to produce less and less of its natural chemicals (why bother?). The wisdom of the body "decides" against unnecessary effort.

Cocaine, another drug that targets the pleasure center, has been around for a long time in a granular form which is inhaled, or "snorted." The appearance of "crack," a cheaper, much more potent, smokable form of cocaine is more recent. Crack is treacherous. Intensely addictive, it produces not only a higher high than the inhaled form of cocaine, but also a deeper crash that leaves its users desperate for more.

Researchers are beginning to zero in on how cocaine produces its effects. It appears that this drug first stimulates the pleasure center and then "squeezes it dry." Cocaine produces the "rush" by blocking the reabsorption of dopamine and norepinephrine. Dopamine is the chemical most linked to the feeling of euphoria. Since the neurotransmitters remain in the synapse, they stimulate the receptor cells again and again, allowing the body to feel their effects over a prolonged period. This sensation is accompanied by increases in heart rate, blood pressure, and sexual appetite. But over time—and sometimes after just one use—the effect changes. As uptake of dopamine continues to be blocked, the dopamine accumulated in

Vials of crack cocaine and pipe.

the synapses is washed away, and the brain's supply becomes inadequate to maintain normal mood. The sending neurons cannot produce neural dopamine fast enough to make up for its loss, and the pleasure circuits go dry. Meanwhile, the receptor cells sprout new receptors in a desperate effort to pick up dopamine signals. The user becomes anxious and, in a very real sense, unable to experience pleasure without cocaine. A vicious cycle of addiction is established. Cocaine is needed to experience pleasure, but using it depletes the neurotransmitter supply even more. Heavy, prolonged cocaine use may deplete the supply so totally that pleasure is impossible to achieve and deep depression sets in. Cocaine addicts tend to lose weight, have trouble sleeping, develop cardiac and pulmonary abnormalities, and have frequent infections.

Crack use in the United States is a national problem. Unlike heroin use, which has been most prevalent among the poor and has had a sordid reputation, cocaine use has been seen as more acceptable, and it has permeated all economic strata. Even small-town middle America is "getting hooked." Cocaine-related cardiovascular deaths (due to acceleration of the heart and enhanced blood pressure) are increasing daily and the rising numbers of underweight, brain-damaged "crack babies" are a newer horror.

The pleasure center, with its cries of "Pleasure me! Pleasure me!," is a powerful Pied Piper. We want to feel young and powerful forever, and we want instant solutions to our problems. But drugs are only temporary solutions. The brain, with its complex biochemistry, always wins and circumvents attempts to keep it in a pleasant haze!

Skin of scalp
Periosteum
Bone of skull
Periosteal } Dura
Meningeal } mater
Arachnoid membrane
Pia mater
Arachnoid villus
Blood vessel
Falx cerebri (in longitudinal fissure only)

Superior sagittal sinus
Subdural space
Subarachnoid space

Figure 7.16   Meninges of the brain. Three-dimensional frontal section showing the meninges—the dura mater, arachnoid, and pia mater—that surround and protect the brain. The relationship of the dura mater to the falx cerebri and the superior sagittal (dural) sinus is also shown.

If the cerebellum is damaged (for example, by a blow to the head, a tumor, or a stroke), movements become clumsy and disorganized. Victims cannot keep their balance and may appear to be drunk because of the loss of muscle coordination. They are no longer able to touch their finger to their nose with eyes closed—a feat that normal individuals accomplish easily. ❏

Protection of the Central Nervous System

Nervous tissue is very soft and delicate, and the irreplaceable neurons are injured by even the slightest pressure. Nature has tried to protect the brain and spinal cord by enclosing them within bone (the skull and vertebral column), membranes (the meninges), and a watery cushion (cerebrospinal fluid). Protection from harmful substances in the blood is provided by the so-called blood-brain barrier. Since we have already considered the bony enclosures (Chapter 5), we will focus on the other protective devices here.

Meninges

The three connective tissue membranes covering and protecting the CNS structures are **meninges** (mě-nin'jēz) (Figure 7.16). The outermost layer, the leathery **dura mater** (du'rah ma'ter), meaning "tough or hard mother," is a double-layered membrane where it surrounds the brain. One of its layers is attached to the inner surface of the skull, forming the periosteum *(periosteal layer)*. The other, called the *meningeal layer,* forms the outermost covering of the brain and continues as the dura mater of the spinal cord. The dural layers are fused together except in three areas where the inner membrane extends inward to form a fold that attaches the brain to the cranial cavity. One of these folds, the *falx* (falks) *cerebri,* is shown in Figure 7.16.

The middle meningeal layer is the weblike **arachnoid** (ah-rak'noid) **mater.** *Arachnida* means "spider," and some think the arachnoid membrane looks like a cobweb. Its threadlike extensions span the subarachnoid space to attach it to the innermost membrane, the **pia** (pi'ah) **mater.** The delicate pia mater ("gentle mother")

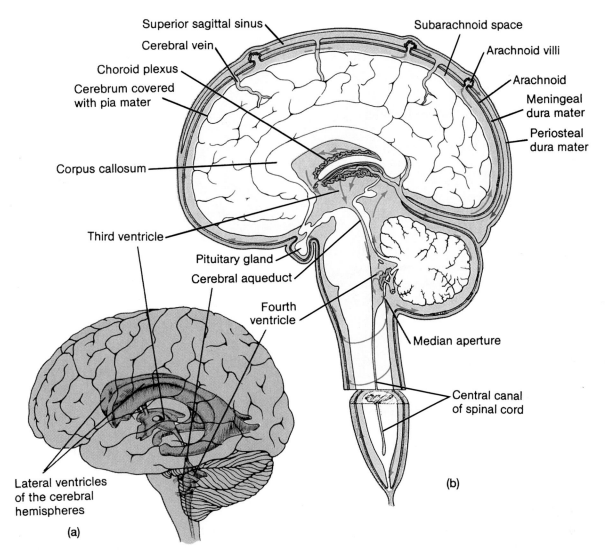

Superior sagittal sinus
Cerebral vein
Choroid plexus
Cerebrum covered with pia mater
Corpus callosum
Third ventricle
Pituitary gland
Cerebral aqueduct
Fourth ventricle
Lateral ventricles of the cerebral hemispheres
(a)

Subarachnoid space
Arachnoid villi
Arachnoid
Meningeal dura mater
Periosteal dura mater
Median aperture
Central canal of spinal cord
(b)

Figure 7.17   Ventricles and location of the cerebrospinal fluid.
(**a**) Three-dimensional view of the ventricles of the brain, left lateral view.
(**b**) Circulatory pathway of the cerebrospinal fluid (indicated by arrows)
within the central nervous system and the subarachnoid space.

clings tightly to the surface of the brain and spinal cord, following every fold.

The subarachnoid space is filled with cerebrospinal fluid. Specialized projections of the arachnoid membrane, **arachnoid villi** (vih′li), protrude through the dura mater. The cerebrospinal fluid is absorbed into the venous blood in the *dural sinuses* through the arachnoid villi.

Meningitis, an inflammation of the meninges, is a serious threat to the brain because bacterial or viral meningitis may spread into the nervous tissue of the CNS. This condition of brain inflammation is called *encephalitis* (en-sef-ah-li′tis). Meningitis is usually diagnosed by taking a sample of cerebrospinal fluid from the subarachnoid space. ❏

Cerebrospinal Fluid

Cerebrospinal (ser″e-bro-spi′nal) **fluid (CSF),** which is similar to plasma, is continually formed from blood by the choroid plexuses. Choroid plexuses are clusters of capillaries hanging from the "roof" in each of the brain's ventricles. The CSF in and around the brain and cord forms a watery cushion, which protects the fragile nervous tissue from blows and other trauma.

Inside the brain, CSF is continually moving (see Figure 7.17). It circulates from the two lateral ventricles (in the cerebral hemispheres) into the third ventricle (in the diencephalon), and then through the cerebral aqueduct of the midbrain into the fourth ventricle dorsal to the pons and medulla oblongata. Some of the fluid reaching the fourth ventricle continues down the *central canal* of the spinal cord, but most of it circulates into the *subarachnoid* space through three openings in the walls of the fourth ventricle. The fluid returns to the blood in the dural sinuses through the arachnoid villi.

Ordinarily, CSF forms and drains at a constant rate so that its normal pressure and volume (150 ml—about half a cup) are maintained. The major solutes in CSF are glucose, proteins, and sodium chloride, and any significant changes in its composition (or the appearance of blood cells in it) may help to diagnose meningitis or certain other brain pathologies (such as tumors and multiple sclerosis). The CSF sample for testing is obtained by a *lumbar (spinal) tap* (see Figure 7.18). Since the withdrawal of fluid for testing decreases CSF fluid pressure, the patient must remain in a horizontal position (lying down) for 6 to 12 hours after the procedure to prevent an agonizingly painful "spinal headache."

If something obstructs its drainage (for example, a tumor), CSF begins to accumulate and exert pressure on the brain. This condition is *hydrocephalus* (hi-dro-sef'ah-lus), literally, "water on the brain." Hydrocephalus in a newborn baby causes the head to enlarge as the brain increases in size. This is possible in an infant because the skull bones have not yet fused. However, in an adult this condition is likely to result in brain damage because the skull is hard, and the accumulating fluid crushes soft nervous tissue. Today hydrocephalus is treated surgically by inserting a shunt to direct the excess fluid into a vein in the neck. ❏

The Blood-Brain Barrier

No other body organ is so absolutely dependent on a constant internal environment as is the brain. Other body tissues can withstand the rather small fluctuations in the concentrations of hormones, ions, and nutrients that continually occur, particularly after eating or exercising. If the brain were exposed to such chemical changes, uncontrolled neural activity might result—remember that certain ions (sodium and potassium) are involved in initiating nerve impulses and certain amino acids serve as neurotransmitters. Consequently, neurons are kept separated from blood-borne substances by a so-called **blood-brain barrier,** composed of the

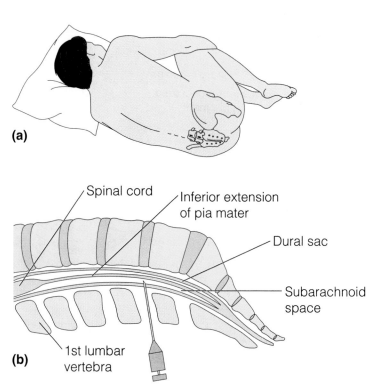

Figure 7.18   A lumbar puncture. (a) For the procedure, the patient is typically positioned on his or her side with the knees drawn toward the abdomen to increase the space between the vertebrae. **(b)** The spinal needle is inserted into the subarachnoid space between the L_3 and L_4 vertebrae (below the termination of the spinal cord) and a small amount of cerebrospinal fluid is withdrawn.

least permeable capillaries in the whole body. Of water-soluble substances, only water, glucose, and essential amino acids pass easily through the walls of these capillaries. Metabolic wastes, such as urea, and most drugs are prevented from entering the brain tissue. Nonessential amino acids and potassium ions are not only prevented from entering the brain, but also are actively pumped from the brain into the blood across the capillary wall. Although the bulbous "feet" of the astrocytes that cling to the capillaries may contribute to the barrier, the relative impermeability of the brain capillaries is most responsible for providing this protection.

The blood-brain barrier is virtually useless against fats, respiratory gases, and other fat-soluble molecules that diffuse easily through all plasma membranes. This explains why blood-borne alcohol, nicotine, and anesthetics can affect the brain.

Brain Dysfunctions

Brain dysfunctions are unbelievably varied. We have mentioned some of them already, and we will discuss developmental problems in the final section of this chapter. Here, we will focus on traumatic brain injuries and degenerative disorders. Techniques used to diagnose many brain disorders are described in the box on pages 220–221.

Traumatic Brain Injuries

Head injuries are a leading cause of accidental death in the United States. Consider, for example, what happens when you forget to fasten your seat belt and then crash into the rear end of another car. Your head is moving and then is suddenly stopped as it hits the windshield. Brain damage is caused not only by injury at the site of the blow, but also by the effect of the ricocheting brain hitting the opposite end of the skull.

A **concussion** occurs when brain injury is slight. The victim may be dizzy, "see stars," or lose consciousness briefly, but no permanent brain damage occurs. A brain **contusion** is the result of marked tissue destruction. If the cerebral cortex is injured, the individual may remain conscious, but severe brain stem contusions always result in a coma ranging in time from hours to a lifetime because of injury to the reticular activating system.

After head blows, death may result from intracranial hemorrhage (bleeding from ruptured vessels), or **cerebral edema** (swelling of the brain due to inflammatory response to injury). Individuals who are initially alert and lucid following head trauma and then begin to deteriorate neurologically later are most likely hemorrhaging or suffering the consequences of edema, both of which compress vital brain tissue.

Degenerative Brain Diseases

Two of the most important degenerative diseases of the central nervous system are cerebrovascular accidents and Alzheimer's disease.

Cerebrovascular Accident Commonly called *strokes,* **cerebrovascular** (ser″e-bro-vas′ku-lar) **accidents (CVAs),** are the third leading cause of death in the United States. CVAs occur when blood circulation to a brain area is blocked, as by a blood clot or a ruptured blood vessel, and vital brain tissue dies. After a CVA it is often possible to determine the area of brain damage by observing the symptoms of the patient. For example, if the patient has left-sided paralysis, the right motor cortex of the frontal lobe is most likely involved. *Aphasias* (ah-fa′ze-ahz) are a common result of damage to the left cerebral hemisphere, where the language areas are located. There are many types of aphasias, but the most common are *motor aphasia,* which involves damage to Broca's area and a loss of ability to speak, and *sensory aphasia,* in which a person loses the ability to understand written or spoken language. Aphasias are maddening to the victims because, as a rule, their intellect is unimpaired. Brain lesions can also cause marked changes in a person's disposition (for example, a change from a sunny to a foul personality). In such cases, a tumor as well as a CVA might be suspected.

Fewer than a third of those surviving a CVA are alive three years later. Even so, the picture is not hopeless. Some patients recover at least part of their lost faculties, because undamaged neurons spread into areas where neurons have died and take over some lost functions. Indeed, most of the recovery seen after brain injury is due to this phenomenon.

Not all strokes are "completed." Temporary brain ischemia, or restriction of blood flow, is called a *transient ischemic attack (TIA).* TIAs last from 5 to 50 minutes and are characterized by symptoms such as numbness, temporary paralysis, and impaired speech. Although these defects are not permanent, they do constitute "red flags" that warn of impending, more serious CVAs.

Alzheimer's Disease **Alzheimer's** (altz′hi″merz) **disease** is a progressive degenerative disease of the brain that ultimately results in dementia (mental deterioration). While CVA victims represent nearly half of nursing home patients, Alzheimer's sufferers represent most of the remaining half. Although this disorder is usually seen in elderly people (in fact, the condition was originally called *senile dementia*), it may begin in middle age. Victims of Alzheimer's disease have memory loss (particularly for recent events) and formerly good-natured individuals become irritable, moody, and confused. Ultimately, hallucinations occur.

In Alzheimer's disease, obvious structural changes occur in the brain, particularly in areas concerned with cognitive functions and memory. Abnormal protein deposits and twisted fibers appear within neurons, and there is localized brain atrophy. These degenerative changes develop over a period of several years, during which time the family members watch the person they love "disappear." It is a long and painful process. Alzheimer's disease appears to run in families. Its precise cause is still unknown, but it seems that the affected brain cells' inability to control calcium entry plays an important role in the havoc that occurs. At present, there is no therapy that reverses or stops the course of the disease. ❏

Spinal Cord

The cylindrical **spinal cord,** which is approximately 17 inches (42 cm) long, is a glistening white continuation of the brain stem. The spinal cord provides a two-way conduction pathway to and from the brain, and it is a major reflex center (the spinal reflexes are completed at this level). Enclosed within the vertebral column, the spinal cord extends from the foramen magnum of the skull to the first or second lumbar vertebra, where it ends just below the ribs (see Figure 7.19). Like the brain, the spinal cord is cushioned and protected by meninges. Meningeal coverings do not end at L_2, but instead extend well beyond the end of the spinal cord in the vertebral canal. Since there is no possibility of damaging the cord beyond L_3, the meningeal sac inferior to that point provides a nearly ideal spot for removing CSF for testing (see Figure 7.18).

In humans, 31 pairs of spinal nerves arise from the cord and exit from the vertebral column to serve the body area close by. The spinal cord is about the size of a thumb for most of its length, but it is obviously enlarged in the cervical and lumbar regions where the nerves serving the upper and lower limbs leave the cord. Because the spinal cord does not reach the end of the vertebral column, the spinal nerves leaving its inferior end must travel through the vertebral canal for some distance before exiting. This collection of spinal

Figure 7.19   Anatomy of the spinal cord, dorsal view.

nerves at the inferior end of the vertebral canal is called the **cauda equina** (kaw′da e-ki′nah) because it looks so much like a horse's tail (the literal translation of *cauda equina*).

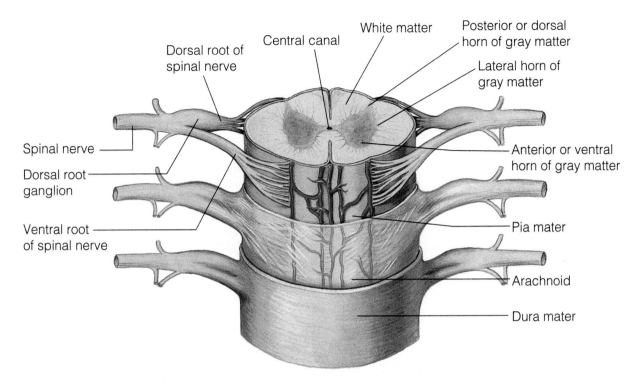

Figure 7.20 **Spinal cord with meninges, cross-sectional view.**

Gray Matter of the Spinal Cord and Spinal Roots

The gray matter of the spinal cord looks like a butterfly or the letter H in cross section (Figure 7.20). The two posterior projections are the **dorsal**, or **posterior, horns;** the two anterior projections are the **ventral**, or **anterior, horns.** The gray matter surrounds the **central canal** of the cord, which contains CSF.

Neurons with specific functions can be located in the gray matter. The dorsal horns contain association neurons, or interneurons. The cell bodies of the sensory neurons, whose fibers enter the cord by the **dorsal root,** are found in an enlarged area called the **dorsal root ganglion.** If the dorsal root or its ganglion is damaged, sensation from the body area served will be lost. The ventral horns of the gray matter contain cell bodies of motor neurons of the somatic (voluntary) nervous system, which send their axons out the **ventral root** of the cord. The dorsal and ventral roots fuse to form the spinal nerves.

Damage to the ventral root results in a *flaccid paralysis* of the muscles served. In flaccid paralysis, nerve impulses do not reach the muscles affected; thus, no voluntary movement of those muscles is possible. The muscles begin to atrophy because they are no longer stimulated. ❏

White Matter of the Spinal Cord

White matter of the spinal cord is composed of myelinated fiber tracts—some running to higher centers, some traveling from the brain to the cord, and some conducting impulses from one side of the spinal cord to the other.

Because of the irregular shape of gray matter, the white matter on each side of the cord is divided into three regions—the **posterior, lateral,** and **anterior columns.** Each of the columns contains a number of fiber tracts made up of axons with the same destination and function. Tracts conducting sensory impulses to the brain are *sensory,* or *afferent, tracts.* Those carrying impulses from the brain to skeletal muscles are *motor,* or *efferent, tracts.*

If the spinal cord is transected (cut crosswise) or crushed, *spastic paralysis* results. The affected muscles stay healthy because they are still stimulated by spinal reflex arcs, and movement

(Text continues on page 221)

A CLOSER LOOK

TRACKING DOWN CNS PROBLEMS

Anyone who has had a routine physical examination is familiar with the reflex tests done to assess neural function. The doctor taps your patellar or Achilles tendon with a reflex hammer and your leg muscles contract, resulting in the knee- or ankle-jerk response. These responses show that the spinal cord and brain centers are functioning normally. When reflex tests are abnormal or when brain cancer, intracranial hemorrhage, multiple sclerosis, or hydrocephalus are suspected, more sophisticated neurological tests may be ordered to try to localize and identify the problem.

An "oldie-but-goodie" procedure used to diagnose and localize many different types of brain lesions (such as epileptic lesions, tumors, and abscesses) is *electroencephalography* (e-lek″tro-en-sef-ah-law′grah-fe). Normal brain function involves the continuous transmission of electrical impulses by neurons. A recording of their activity, called an **electroencephalogram,** or **EEG,** can be made by placing electrodes at various points on the scalp and connecting these to a recording device. The patterns of electrical activity of the neurons are called *brain waves*. Because people differ genetically, and because everything we have ever experienced has left its imprint in our brain, each of us has a brain wave pattern that is as unique as our fingerprints. The four most commonly seen brain waves are illustrated and described in the figure.

As might be expected, brain-wave patterns typical of the alert wide-awake state differ from those that occur during relaxation or deep sleep. Interference with the function of the cerebral cortex is suggested by brain waves that are too fast or too slow, and unconsciousness occurs at both extremes. Sleep and coma result in brain-wave patterns that are

(a)

(b)

Electroencephalography and brain waves. (**a**) To obtain a recording of brain-wave activity (an EEG), electrodes are positioned on the patient's scalp and attached to a recording device called an *electroencephalograph*. (**b**) Typical EEGs. Alpha waves are typical of the awake, relaxed state; beta waves occur in the awake, alert state; theta waves are common in children but not in normal adults; and delta waves occur during deep sleep.

slower than normal, whereas fright, epileptic seizures, and some kinds of drug overdose cause abnormally fast brain waves. Since brain waves are seen even during coma, absence of brain waves (a flat EEG) is taken as evidence of clinical death.

Pneumoencephalography (nu″mo-en-sef″ah-lah′grah-fe) provides a fairly clear X ray picture of the brain ventricles and has long been the procedure of choice for diagnosing hydrocephalus. A small amount of cerebrospinal fluid is withdrawn by lumbar puncture. Then, air (or another gas) is injected into the subarachnoid space and allowed to float upward and into the ventricles of the brain, allowing them to be visualized. Although the procedure is remarkably simple, it can cause a blinding headache.

A **cerebral angiogram** (an′je-o-gram) is used to assess the condition of the cerebral arteries serving the brain (or the carotid arteries of the neck, which feed most of those vessels). A dye is injected into an artery, and time is allowed for the dye to become dispersed to the brain. Then, an X ray is taken of the arteries of interest. The dye allows arteries narrowed by arteriosclerosis to be localized. This procedure is commonly ordered for individuals who have suffered a stroke or who have a history of transient ischemic attacks.

The new imaging techniques described in Chapter 2 (pp. 30–31) have revolutionized the diagnosis of brain lesions. **CT scans** allow most tumors, intracranial lesions, and areas of dead brain tissue (infarcts) to be identified quickly. The CT scanner is also becoming an important tool to enhance the precision and safety of brain surgery. Multiple sclerosis plaques that are poorly revealed by CT scans are "flushed out" by **MRI scans,** which use powerful magnets to reveal the brain's problems. While CT scans and MRI excel at revealing detailed maps of brain anatomy, **PET scans** use high-energy gamma rays to monitor the brain's biochemical activity. PET scans are also being used to diagnose Alzheimer's disease.

(c)

PET (positron emission tomography) scan of the brain of a person with Alzheimer's disease. (c) This scan shows a predominance of blues in the cerebral cortex, which indicates a decline in glucose use and, hence, reduced cortical activity.

of those muscles does occur. However, movements are involuntary and not controllable; this can be as much of a problem as complete lack of mobility. In addition, since the spinal cord carries both sensory and motor impulses, a loss of feeling or sensory input occurs in the body areas below the point of cord destruction. Physicians often use a pin to see if a person can feel pain after spinal cord injury—to find out if regeneration is occurring. Pain is a hopeful sign in such cases. If the spinal cord injury occurs high in the cord, so that all four limbs are affected, the individual is a *quadriplegic* (kwod″-rĭ-ple′jik). If only the legs are paralyzed, the individual is a *paraplegic* (par″ă-ple′jik). ❏

PERIPHERAL NERVOUS SYSTEM

The **peripheral nervous system (PNS)** consists of nerves and scattered groups of neuronal cell bodies (ganglia) found outside the CNS. One type of ganglion has already been considered—the dorsal root ganglion of the spinal cord. Others will be covered in the discussion of the autonomic nervous system. Here, we will only concern ourselves with nerves.

Structure of a Nerve

A **nerve** is a bundle of neuron fibers found outside the CNS. Within a nerve, neuron fibers, or processes, are wrapped in protective connective tissue coverings. Each fiber is surrounded by a delicate connective tissue sheath, an **endoneurium** (en"do-nu′re-um). Groups of fibers are bound by a coarser connective tissue wrapping, the **perineurium** (per"ĭ-nu′re-um), to form fiber bundles, or **fascicles.** Finally, all the fascicles are bound together by a tough fibrous sheath, the **epineurium,** to form the cordlike nerve (see Figure 7.21).

Like neurons, nerves are classified according to the direction in which they transmit impulses. Nerves carrying both sensory and motor fibers are called **mixed nerves;** all spinal nerves are mixed nerves. Nerves that carry impulses toward the CNS only are called **afferent,** or **sensory, nerves,** whereas those that carry only motor fibers are **efferent,** or **motor, nerves.**

Cranial Nerves

The 12 pairs of **cranial nerves** primarily serve the head and neck. Only one pair (the vagus nerves) extends to the thoracic and abdominal cavities.

The cranial nerves are numbered in order, and in most cases their names reveal the most important structures they control. The cranial nerves are described by name, number, course, and major function in Table 7.1. The last column of the table describes how cranial nerves are tested, which is an important part of any neurologic examination. You do not need to memorize these tests, but this information may help you understand cranial

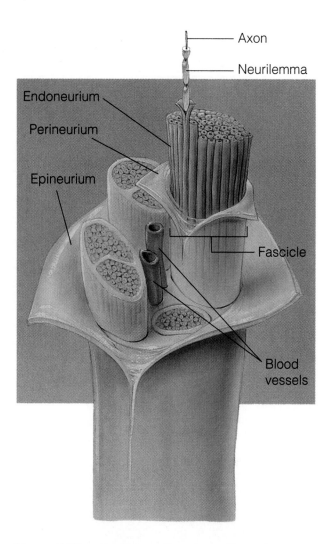

Figure 7.21 **Structure of a nerve.** Three-dimensional view of a portion of a nerve, showing its connective tissue wrapping.

nerve function. As you read through the table, also look at Figure 7.22, which shows the location of the cranial nerves on the brain's anterior surface.

Most cranial nerves are mixed nerves; however, three pairs, the optic, olfactory, and vestibulocochlear (ves-tib"u-lo-kok′le-ar) nerves, are purely sensory in function. (The older name for the vestibulocochlear nerve is *acoustic nerve,* a name which reveals its role in hearing but not in equilibrium.) I give my students the following little saying as a memory jog to help them remember the cranial nerves in order; perhaps it will help you, too. The first letter of each word in the saying (and both letters of "ah"), is the first letter of the cranial nerve to be remembered: "**O**h, **o**h, **o**h, **t**o **t**ouch **a**nd **f**eel **v**ery **g**ood **v**elvet, **ah.**"

Table 7.1   The Cranial Nerves

Name/Number	Origin/Course	Function	Test
I. Olfactory	Fibers arise from olfactory receptors in the nasal mucosa and synapse with the olfactory bulbs (which, in turn, send fibers to the olfactory cortex)	Purely sensory; carries impulses for the sense of smell	Subject is asked to sniff and identify aromatic substances, such as oil of cloves or vanilla
II. Optic	Fibers arise from the retina of the eye and form the optic nerve. The two optic nerves form the optic chiasma by partial crossover of fibers; the fibers continue to the optic cortex as the optic tracts	Purely sensory; carries impulses for vision	Vision and visual field are tested with an eye chart and by testing the point at which the subject first sees an object (finger) moving into the visual field; eye interior is viewed with an ophthalmoscope
III. Oculomotor	Fibers run from the midbrain to the eye	Supplies motor fibers to four of the six muscles (superior, inferior, and medial rectus, and inferior oblique) that direct the eyeball; to the eyelid; and to the internal eye muscles controlling lens shape and pupil size	Pupils are examined for size, shape, and size equality; pupillary reflex is tested with a penlight (pupils should constrict when illuminated); eye convergence is tested, as is the ability to follow moving objects
IV. Trochlear	Fibers run from the midbrain to the eye	Supplies motor fibers for one external eye muscle (superior oblique)	Tested in common with cranial nerve III for the ability to follow moving objects
V. Trigeminal	Fibers emerge from the pons and form three divisions that run to the face	Conducts sensory impulses from the skin of the face and mucosa of the nose and mouth; also contains motor fibers that activate the chewing muscles	Sensations of pain, touch, and temperature are tested with a safety pin and hot and cold objects; corneal reflex tested with a wisp of cotton; motor branch tested by asking the subject to open mouth against resistance and move jaw from side to side
VI. Abducens	Fibers leave the pons and run to the eye	Supplies motor fibers to the lateral rectus muscle, which rolls the eye laterally	Tested in common with cranial nerve III for the ability to move each eye laterally

➤

Table 7.1   The Cranial Nerves (continued)

Name/Number	Origin/Course	Function	Test
VII. Facial	Fibers leave the pons and run to the face	Activates the muscles of facial expression and the lacrimal and salivary glands; carries sensory impulses from the taste buds of anterior tongue	Anterior two-thirds of tongue is tested for ability to taste sweet, salty, sour, and bitter substances; subject is asked to close eyes, smile, whistle, etc.; tearing is tested with ammonia fumes
VIII. Vestibulocochlear	Fibers run from the equilibrium and hearing receptors of the inner ear to the brain stem	Purely sensory; vestibular branch transmits impulses for the sense of balance and cochlear branch transmits impulses for the sense of hearing	Hearing is checked by air and bone conduction, using a tuning fork
IX. Glossopharyngeal	Fibers emerge from the medulla and run to the throat	Supplies motor fibers to the pharynx (throat) that promote swallowing and saliva production; carries sensory impulses from taste buds of the posterior tongue and from pressure receptors of the carotid artery	Gag and swallowing reflexes are checked; subject is asked to speak and cough; posterior tongue may be tested for taste
X. Vagus	Fibers emerge from the medulla and descend into the thorax and abdominal cavity	Fibers carry sensory impulses from and motor impulses to the pharynx, larynx, and the abdominal and thoracic viscera; most motor fibers are parasympathetic fibers that promote digestive activity and help regulate heart activity	Tested in common with cranial nerve IX, since they both serve muscles of the throat
XI. Accessory	Fibers arise from the medulla and superior spinal cord and travel to muscles of the neck and back	Mostly motor fibers that activate the sternocleidomastoid and trapezius muscles	Sternocleidomastoid and trapezius muscles are checked for strength by asking the subject to rotate head and shrug shoulders against resistance
XII. Hypoglossal	Fibers run from the medulla to the tongue	Motor fibers control tongue movements; sensory fibers carry impulses from the tongue	Subject is asked to stick out tongue, and any position abnormalities are noted

I Olfactory

II Optic

III Oculomotor

IV Trochlear

VI Abducens

V Trigeminal

VII Facial

Vestibular branch

Cochlear branch

VIII Vestibulocochlear

X Vagus

IX Glossopharyngeal

XII Hypoglossal

XI Accessory

Figure 7.22 Distribution of cranial nerves. Sensory nerves are shown in blue, motor nerves in red. Although cranial nerves III, IV, and VI have sensory fibers, these are not shown because the sensory fibers account for only minor parts of these nerves.

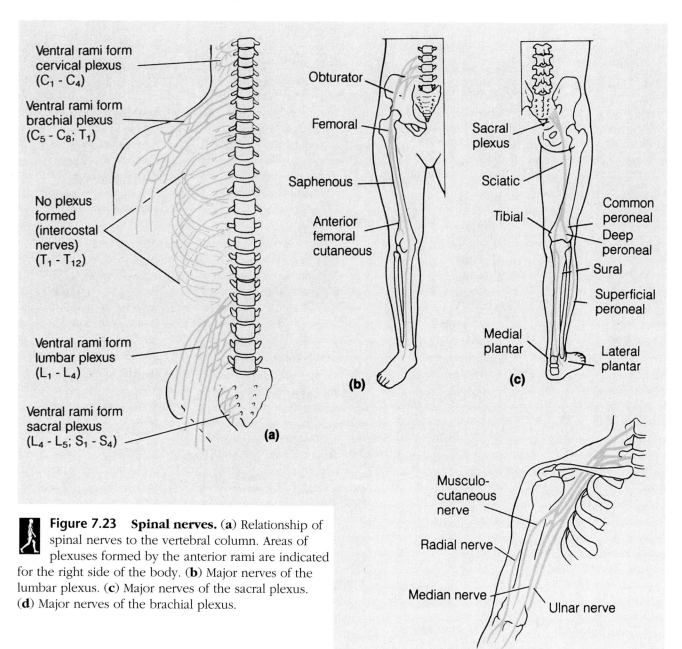

Ventral rami form cervical plexus (C_1 - C_4)

Ventral rami form brachial plexus (C_5 - C_8; T_1)

No plexus formed (intercostal nerves) (T_1 - T_{12})

Ventral rami form lumbar plexus (L_1 - L_4)

Ventral rami form sacral plexus (L_4 - L_5; S_1 - S_4)

(a)

Obturator

Femoral

Saphenous

Anterior femoral cutaneous

(b)

Sacral plexus

Sciatic

Tibial

Common peroneal

Deep peroneal

Sural

Superficial peroneal

Medial plantar

Lateral plantar

(c)

Musculo-cutaneous nerve

Radial nerve

Median nerve

Ulnar nerve

(d)

Figure 7.23   Spinal nerves. (a) Relationship of spinal nerves to the vertebral column. Areas of plexuses formed by the anterior rami are indicated for the right side of the body. (b) Major nerves of the lumbar plexus. (c) Major nerves of the sacral plexus. (d) Major nerves of the brachial plexus.

Spinal Nerves and Nerve Plexuses

The 31 pairs of human **spinal nerves** are formed by the combination of the ventral and dorsal roots of the spinal cord. Although each of the cranial nerves issuing from the brain is specifically named, the spinal nerves are named for the region of the cord from which they arise. Figure 7.23 shows how the nerves are named in this scheme.

Almost immediately after being formed, each spinal nerve divides into **dorsal** and **ventral rami** (ra′mi), making each spinal nerve only about 1/2-inch long. The rami, like the spinal nerves, contain both motor and sensory fibers. Thus, damage to a spinal nerve or either of its rami results both in loss of sensation and flaccid paralysis of the area of the body served. The smaller dorsal rami serve the skin and muscles of the posterior body trunk. The ventral rami of spinal nerves T_1 through T_{12} form the *intercostal nerves,* which supply the muscles between the ribs and the skin and muscles of the anterior and lateral trunk. The ventral rami of all other spinal nerves form complex networks of nerves called **plexuses,** which serve the motor and sensory needs of the limbs. The four nerve plexuses are shown in Figure 7.23 and described in Table 7.2.

Table 7.2   Spinal Nerve Plexuses

Plexus	Origin (from ventral rami)	Important nerves	Body areas served	Result of damage to plexus or its nerves
Cervical	C_1–C_4	Phrenic	Diaphragm and muscles of shoulder and neck	Respiratory paralysis (and death if not treated promptly)
Brachial	C_5–C_8 and T_1	Axillary	Deltoid muscle of shoulder	Paralysis and atrophy of deltoid muscle
		Radial	Triceps and extensor muscles of the forearm	Wristdrop—inability to extend hand at wrist
		Median	Flexor muscles of forearm and some muscles of hand	Decreased ability to flex and abduct hand and flex and abduct thumb and index finger—therefore, inability to pick up small objects
		Musculocutaneous	Flexor muscles of arm	Decreased ability to flex forearm on arm
		Ulnar	Wrist and many hand muscles	Clawhand—inability to spread fingers apart
Lumbar	L_1–L_4	Femoral	Lower abdomen, buttocks, anterior thighs, and skin of anteromedial leg and thigh	Inability to extend leg and flex hip
		Obturator	Adductor muscles of medial thigh and small hip muscles; skin of medial thigh and hip joint	Inability to adduct thigh
Sacral	L_4–L_5 and S_1–S_4	Sciatic (largest nerve in body; splits to common peroneal and tibial nerves)	Lower trunk and posterior surface of thigh (and leg)	Inability to extend hip and flex knee; sciatica
		Common peroneal (superficial, and deep branches)	Lateral aspect of leg and foot	Footdrop—inability to dorsiflex foot
		Tibial	Posterior aspect of leg and foot	Inability to plantar flex and invert foot; shuffling gait

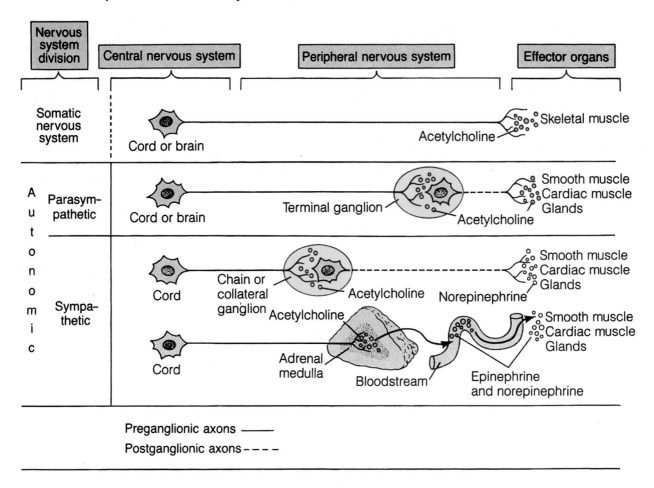

| Nervous system division | Central nervous system | Peripheral nervous system | Effector organs |

Figure 7.24 Comparison of the somatic and autonomic nervous systems.
Somatic division: Axons of somatic motor neurons extend from the CNS to skeletal muscle cells. Somatic motor fibers release acetylcholine; the effect is always stimulatory. Autonomic division: There is a chain of two motor neurons (presynaptic and postsynaptic) extending to the effectors, which are cardiac and smooth muscles and glands. Axons of most preganglionic neurons run from the CNS to synapse in a ganglion with a postganglionic neuron. (A few sympathetic preganglionic neurons synapse with cells of the adrenal medulla.) All preganglionic fibers and all parasympathetic postganglionic fibers release acetylcholine. Most sympathetic postganglionic fibers release norepinephrine. Autonomic effects can be stimulatory or inhibitory depending on which of its subdivisions is in control and on the target organs.

Autonomic Nervous System

The **autonomic nervous system (ANS)** is the motor subdivision of the PNS that controls body activities automatically. It is composed of a special group of neurons that regulate cardiac muscle (the heart), smooth muscles (found in the walls of the visceral organs and blood vessels), and glands. Although all body systems contribute to homeostasis, the relative stability of our internal environment depends largely on the workings of the ANS. At every moment, signals flood from the visceral organs into the CNS and the autonomic nerves make adjustments as necessary to best support body activities. For example, blood flow may be shunted to more "needy" areas, heart and breathing rate may be speeded or slowed, blood pressure may be adjusted, and stomach secretions may be increased or decreased. Most of this fine-tuning occurs without our awareness or attention—few of us realize when our pupils dilate or our arteries constrict—hence the ANS is also called the **involuntary nervous system.**

Comparison of the Somatic and Autonomic Nervous Systems

Our previous discussions of motor nerves have focused largely on the activity of the somatic nervous system, the motor subdivision that controls our skeletal muscles. So, before plunging into a description of autonomic nervous system anatomy, we will take the time to point out some important differences between the somatic and autonomic divisions.

Besides differences in their effector organs and in the neurotransmitters released, the patterns of their efferent pathways differ. In the somatic division, the cell bodies of the motor neurons are inside the CNS, and their axons (in spinal nerves) extend all the way to the skeletal muscles they serve. On the other hand, the autonomic nervous system has a chain of *two* motor neurons. The first motor neuron of each pair is in the brain or spinal cord. Its axon, the **preganglionic axon** (literally, the "axon before the ganglion"), leaves the CNS to synapse with the second motor neuron in a ganglion outside the CNS. The axon of this neuron, the **postganglionic axon,** then extends to the organ it serves. These differences are summarized in Figure 7.24.

The autonomic nervous system has two arms, the sympathetic and the parasympathetic (see Figure 7.25). Both serve the same organs but cause essentially opposite effects, counterbalancing each other's activities to keep body systems running smoothly. The sympathetic part mobilizes the body during extreme situations (such as fear, exercise, or rage), whereas the parasympathetic division allows us to "unwind" and conserve energy. These differences will be examined in more detail shortly, but first we will consider the structural characteristics of the two arms of the ANS.

Anatomy of the Sympathetic Division

The sympathetic division is also called the *thoracolumbar* (tho″rah-ko-lum′bar) *division* because its first neurons are in the gray matter of the spinal cord from T_1 through L_2 (Figure 7.25). The preganglionic axons leave the cord in the ventral root, enter the spinal nerve, and then pass through a *ramus communicans,* or small communicating branch, to enter a sympathetic chain ganglion (see Figure 7.26). The **sympathetic chain,** or **trunk,** lies alongside the vertebral column on each side. After it reaches the ganglion, the axon may synapse with the second neuron in the sympathetic chain at the same or a different level (the postganglionic axon then reenters the spinal nerve to travel to the skin), or the axon may pass through the ganglion without synapsing and form part of the *splanchnic* (splank′nik) *nerves.* The splanchnic nerves travel to the viscera to synapse with the second neuron, found in a **collateral ganglion** anterior to the vertebral column. The major collateral ganglia—the celiac and superior and inferior mesenteric ganglia—supply the abdominal and pelvic organs. The postganglionic axon then leaves the collateral ganglion and travels to serve a nearby visceral organ.

Anatomy of the Parasympathetic Division

The first neurons of the **parasympathetic division** are located in brain nuclei of several cranial nerves (the vagus being the most important of these) and in the S_2 through S_4 level of the spinal cord (see Figure 7.25). The neurons of the cranial region send their axons out in cranial nerves to serve the head and neck organs. There they synapse with the second motor neuron in a **terminal ganglion.** From the terminal ganglion, the postganglionic axon extends a short distance to the organ it serves. In the sacral region, the preganglionic axons leave the spinal cord and form the *pelvic nerves,* also called the *pelvic splanchnic nerves,* which travel to the pelvic cavity. In the pelvic cavity, the preganglionic axons synapse with the second motor neurons in terminal ganglia on, or close to, the organs they serve.

Now that the anatomical details have been described, we are ready to examine ANS functions in a little more detail.

Autonomic Functioning

Body organs served by the autonomic nervous system receive fibers from both divisions. Exceptions are most blood vessels and most structures of the

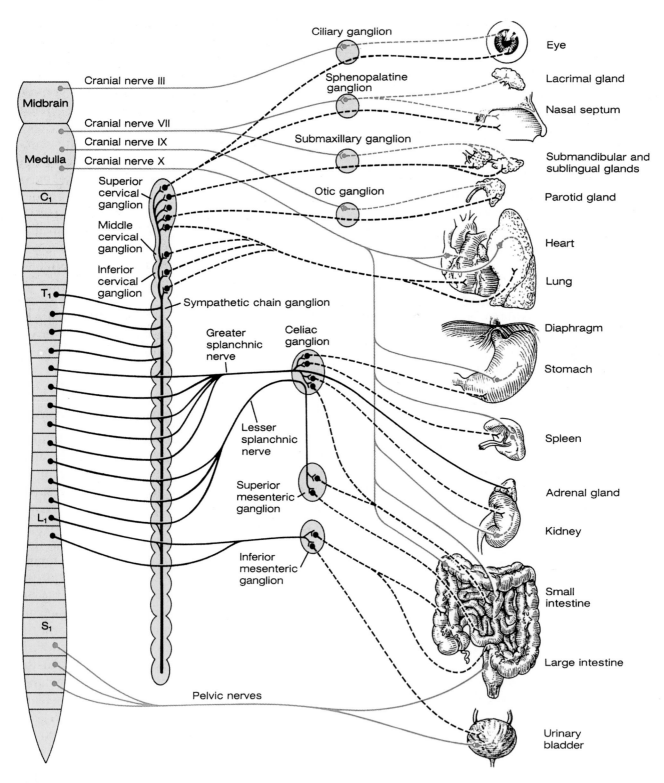

Figure 7.25 **Anatomy of the autonomic nervous system.**
Parasympathetic fibers are shown in red, sympathetic fibers in black.
Solid lines represent preganglionic fibers; dashed lines indicate post-
ganglionic fibers.

Figure 7.26 Sympathetic pathways. (**a**) Synapse in a sympathetic chain ganglion at the same level. (**b**) Synapse in a sympathetic chain ganglion at a different level. (**c**) Synapse in a collateral ganglion anterior to the vertebral column.

skin, some glands, and the adrenal medulla, all of which receive only sympathetic fibers. When both divisions serve the same organ they cause antagonistic effects, mainly because their postganglionic axons release different neurotransmitters. The parasympathetic fibers, called *cholinergic* (ko″lin-er′jik) *fibers,* release acetylcholine; the sympathetic postganglionic fibers, called *adrenergic* (ad″ren-er′jik) *fibers,* release norepinephrine (nor″ep-ĭ-nef′rin). The preganglionic axons of both divisions release acetylcholine. To emphasize the *relative* roles of the two arms of the ANS, we will focus briefly on situations in which each division is "in control."

The **sympathetic division** is often referred to as the "fight-or-flight" system. Its activity is evident when we are excited or find ourselves in emergency or threatening situations, such as being frightened by street toughs late at night. A pounding heart; rapid, deep breathing; cold, sweaty skin; a prickly scalp; and dilated eye pupils are sure signs of sympathetic nervous system activity. Under such conditions, the sympathetic nervous system increases heart rate, blood pressure, and blood glucose levels; dilates the bronchioles of the lungs; and brings about many other effects that help the individual cope with the stressor. Dilation of the blood vessels in skeletal muscles (so that one can run faster or fight better) and withdrawal of blood from the digestive organs (so that the bulk of the blood can be used to serve the heart, brain, and skeletal muscles) are other examples. The sympathetic nervous system is working at full speed, not only when you are emotionally upset, but also when you are physically stressed. For example, if you have just had surgery or run a marathon, your adrenal glands, activated by the sympathetic nervous system, would be pumping out epinephrine and norepinephrine (see Figure 7.24). The effects of sympathetic nervous system activation continue for several minutes until its hormones are destroyed by the liver. Thus, although sympathetic nerve impulses themselves may act only briefly, the hormonal effects they provoke linger. The widespread and prolonged effects of sympathetic activation helps explain why we need time to "come down" after an extremely stressful situation.

The sympathetic division generates a head of steam that enables the body to cope rapidly and vigorously with situations that threaten homeostasis. Its function is to provide the best conditions for responding to some threat, whether the best response is to run, to see better, or to think more clearly.

Some illnesses or diseases are believed to be at least aggravated, if not caused, by excessive sympathetic nervous system stimulation. Some individuals, called "Type A" people, always work at breakneck speed and push themselves continually. These are people who are likely to have heart disease, high blood pressure, and ulcers, all of which can result from prolonged sympathetic nervous system activity or the rebound from it. ❏

The **parasympathetic division** is most active when the body is at rest and not threatened in any way. This division, sometimes called the "resting and digesting" system, is chiefly concerned with promoting normal digestion and elimination, and with conserving body energy, particularly by decreasing demands on the cardiovascular system. (This explains why it is a good idea to relax after a heavy meal so that digestion is not inhibited or disturbed by sympathetic activity.) Its activity is best illustrated by a person who relaxes after a meal and reads the newspaper. Blood pressure and heart and respiratory rates are being regulated at low normal levels, the digestive tract is actively digesting food, and the skin is warm (indicating that there is no need to divert blood to skeletal muscles or vital organs). The eye pupils are constricted to protect the retinas from excessive damaging light, and the lenses of the eyes are "set" for close vision. We might also consider the parasympathetic division the "housekeeping" system of the body.

While it is easier to think of the operation of the sympathetic and parasympathetic divisions as working in an all-or-none fashion, this is rarely the case. A dynamic balance exists between the two divisions, and they are continuously making fine adjustments. Also, although we have described the parasympathetic division as the "at-rest" system, blood vessels are controlled only by the sympathetic fibers regardless of whether the body is "on alert" or relaxing. A summary of the major effects of each division appears in Table 7.3.

DEVELOPMENTAL ASPECTS OF THE NERVOUS SYSTEM

Because the nervous system is formed during the first month of embryonic development, any maternal infection early in pregnancy can have extremely harmful effects on the fetal nervous system. For example, maternal measles (rubella) often causes deafness and other types of CNS damage. Also, since nervous tissue has the highest metabolic rate in the body, lack of oxygen for even a few minutes leads to death of neurons. (Because smoking decreases the amount of oxygen that can be carried in the blood, a smoking mother may be sentencing her infant to possible brain damage.) Various drugs (alcohol, opiates, cocaine, and others) and radiation can also be very damaging if administered during early fetal development.

Table 7.3   Effects of the Sympathetic and Parasympathetic Divisions of the Autonomic Nervous System

Target organ/system	Parasympathetic effects	Sympathetic effects
Digestive system	Increases smooth muscle mobility (peristalsis) and amount of secretion by digestive system glands; relaxes sphincters	Decreases activity of digestive system and constricts digestive system sphincters (for example, anal sphincter)
Liver	No effect	Causes glucose to be released to blood
Lungs	Constricts bronchioles	Dilates bronchioles
Urinary bladder/urethra	Relaxes sphincters (allows voiding)	Constricts sphincters (prevents voiding)
Kidneys	No effect	Decreases urine output
Heart	Decreases rate; slows and steadies	Increases rate and force of heartbeat
Blood vessels	No effect on most blood vessels	Constricts blood vessels in viscera and skin (dilates those in skeletal muscle and heart); increases blood pressure
Glands—salivary, lacrimal	Stimulates; increases production of saliva and tears	Inhibits; result is dry mouth and dry eyes
Eye (iris)	Stimulates constrictor muscles; constricts pupils	Stimulates dilator muscles; dilates pupils
Eye (ciliary muscle)	Stimulates to increase bulging of lens for close vision	Inhibits; decreases bulging of lens; prepares for distant vision
Adrenal medulla	No effect	Stimulates medulla cells to secrete epinephrine and norepinephrine
Sweat glands of skin	No effect	Stimulates to produce perspiration
Arrector pili muscles attached to hair follicles	No effect	Stimulates; produces "goosebumps"
Penis	Causes erection due to vasodilation	Causes ejaculation (emission of semen)
Cellular metabolism	No effect	Increases metabolic rate; increases blood sugar levels; stimulates fat breakdown

In difficult deliveries, a temporary lack of oxygen may lead to *cerebral palsy* (pawl′ze), but this is only one of the suspected causes. Cerebral palsy is a neuromuscular disability in which the voluntary muscles are poorly controlled and spastic due to brain damage. About half of its victims have seizures, are mentally retarded, and/or have impaired hearing or vision. Cerebral palsy is the largest single cause of physical disabilities in children. A number of other congenital malformations, triggered by genetic or environmental factors, also plague the CNS. Most serious are hydrocephalus (see p. 216), *anencephaly*—a failure of the cerebrum to develop, resulting in a child that cannot hear, see, or process sensory inputs—and spina bifida. *Spina bifida* (spi′nah bi′f ĭ-dah) results when the vertebrae form incompletely (typically in the lumbosacral region). There are several varieties. In the least serious, a dimple, and perhaps a tuft of hair, appears over the site of malformation, but no neurological problems occur. In the most serious, meninges, nerve roots, and even parts of the spinal cord protrude from the spine, rendering the lower part of the spinal cord functionless. The child is unable to control the bowels or bladder, and the lower limbs are paralyzed. ❏

One of the last areas of the CNS to mature is the hypothalamus, which contains centers for regulating body temperature. For this reason, premature babies usually have problems in controlling their loss of body heat and must be carefully monitored. No more neurons are formed after birth (because neurons are amitotic), but growth and maturation of the nervous system continues all through childhood, largely as a result of myelination that goes on during this period. A good indication of the degree of myelination of particular neural pathways is the level of neuromuscular control in that body area. As described in Chapter 6, neuromuscular coordination progresses in a superior to inferior (cephalocaudal) direction and in a proximal to distal direction, and we know that myelination occurs in the same sequence.

The brain reaches its maximum weight in the young adult. Over the next 60 years or so, neurons are damaged and die, and since they cannot reproduce themselves, our store of neurons continually decreases. However, an unlimited number of neural pathways are always available and ready to be developed. We never run out of "recording tape" and can continue to learn throughout life.

As we grow older, the sympathetic nervous system gradually becomes less and less efficient, particularly in its ability to vasoconstrict blood vessels. When elderly people stand up quickly after sitting or lying down, they often become lightheaded or faint. This is because the sympathetic nervous system is not able to react quickly enough to counteract the pull of gravity by activating the vasoconstrictor fibers, and blood pools in the feet. This condition, *orthostatic hypotension,* is a type of low blood pressure resulting from changes in body position as described. Orthostatic hypotension can be prevented to some degree if changes in position are made slowly. This gives the sympathetic nervous system a little more time to adjust and react.

The usual cause of nervous system deterioration is circulatory system problems. For example, arteriosclerosis (ar-ter″e-o-skle-ro′sis) and high blood pressure result in a decreasing supply of oxygen to the brain neurons. A gradual lack of oxygen due to the aging process finally leads to *senility,* characterized by forgetfulness, irritability,

difficulty in concentrating and thinking clearly, and confusion. A rapid loss of blood and oxygen delivery to the brain results in a CVA, as described earlier. However, many people continue to enjoy intellectual lives and to work at mentally demanding tasks through their entire life. In fact, fewer than 5 percent of people over the age of 65 demonstrate true senility.

Sadly, many cases of "reversible senility," caused by certain drugs, low blood pressure, constipation, poor nutrition, depression, dehydration, and hormone imbalances, go undiagnosed. The best way to maintain one's mental abilities in old age may be to seek regular medical checkups throughout life.

Although eventual shrinking of the brain is normal, it seems that some individuals (professional boxers and chronic alcoholics, for example) accelerate the process long before aging plays its part. Whether a boxer wins the match or not, the likelihood of brain damage and atrophy increases with each fight as the brain bounces and rebounds within the skull with every blow. The expression "punch drunk" reflects the symptoms of slurred speech, tremors, abnormal gait, and dementia (mental illness) seen in many members of this group.

Everyone recognizes that alcohol has a profound effect on the mind as well as the body. However, these effects may not be temporary. CT scans of chronic alcoholics reveal reduced brain size at a fairly early age. Like boxers, chronic alcoholics tend to exhibit signs of senile (age-related) dementia unrelated to the aging process.

The human cerebral hemispheres—our "thinking caps"—are awesome in their complexity. No less amazing are the brain regions that oversee all our subconscious, autonomic body functions—the diencephalon and brain stem—particularly when you consider their relatively insignificant size. The spinal cord, which acts as a reflex center, and the peripheral nerves, which provide communication links between the CNS and body periphery, are equally important to body homeostasis.

A good deal of new terminology has been introduced in this chapter and, as you will see, much of it will come up again in later chapters as we study the other organ systems of the body and examine how the nervous system helps to regulate their activity. The terms *are* essential, so try to learn them as you go along. Use the Glossary at the back of the book as often as you find it helpful.

SYSTEMS IN SYNC

Homeostatic Relationships Between the Nervous System and Other Body Systems

Nervous System

Endocrine System
- Sympathetic division of the ANS activates the adrenal medulla; hypothalamus helps regulate the activity of the anterior pituitary gland and produces two hormones
- Hormones influence metabolism of neurons

Lymphatic System/Immunity
- Nerves innervate lymphoid organs; the brain plays a role in regulating immune function
- Lymphatic vessels carry away leaked tissue fluids from tissues surrounding nervous system structures; immune elements protect all body organs from pathogens (CNS has additional mechanisms)

Digestive System
- ANS (particularly the parasympathetic division) regulates digestive system activity
- Digestive system provides nutrients needed for health of neurons

Urinary System
- ANS regulates bladder emptying and renal blood pressure
- Kidneys help to dispose of metabolic wastes and maintain proper electrolyte composition and pH of blood for neural functioning

Muscular System
- Somatic division of nervous system activates skeletal muscles; maintains muscle health
- Skeletal muscles are the effectors of the somatic division

Respiratory System
- Nervous system regulates respiratory rhythm and depth
- Respiratory system provides life-sustaining oxygen; disposes of carbon dioxide

Cardiovascular System
- ANS helps regulate heart rate and blood pressure
- Cardiovascular system provides blood containing oxygen and nutrients to the nervous system; carries away wastes

Reproductive System
- ANS regulates sexual erection and ejaculation in males; erection of the clitoris in females
- Testosterone masculinizes the brain and underlies sex drive and aggressive behavior

Integumentary System
- Sympathetic division of the ANS regulates sweat glands and blood vessels of skin (therefore heat loss/retention)
- Skin serves as heat loss surface

Skeletal System
- Nerves innervate bones
- Bones serve as depot for calcium needed for neural function; protect CNS structures

235

Important Terms

afferent neurons	**myelin** (mi′ĕ-lin)
autonomic (aw″to-nom′ik) **nervous system**	**nerve**
axon (ak′son)	**neuroglia** (nu-rog′le-ah)
brain stem	**neurons**
central nervous system	**nodes of Ranvier** (rahn-vēr)
cerebellum (ser″e-bel′um)	**peripheral** (pĕ-rif′er-al) **nervous system**
cerebral (ser′e-bral) **hemispheres**	**plexus**
cerebrospinal (ser″e-bro-spi′nal) **fluid**	**polarization**
dendrites (den′drītz)	**receptor**
depolarization	**reflex**
efferent neurons	**repolarization**
ganglion	**somatic** (so-mat′ik) **nervous system**
gray matter	**synapse**
interneurons	**tracts**
meninges (mĕ-nin′jēz)	**white matter**

Summary

Organization of the Nervous System (pp. 197–198)

1. Structural: All nervous system structures are classified as part of the CNS (brain and spinal cord) or PNS (nerves and ganglia).

2. Functional: Motor nerves of the PNS are classified on the basis of whether they stimulate skeletal muscle (somatic division) or smooth/cardiac muscle and glands (autonomic division).

Nervous Tissue: Structure and Function (pp. 198–207)

1. Supportive connective tissue cells
   a. Neuroglia support and protect neurons in the CNS. Specific glial cells are phagocytes; others myelinate neuron processes in the CNS or line cavities.
   b. Schwann cells myelinate neuron processes in the PNS.

2. Neurons
   a. Anatomy: All neurons have a cell body containing the nucleus and processes (fibers) of two types: (1) axons (one per cell) typically generate and conduct impulses away from the cell body and release a neurotransmitter, and (2) dendrites (one to many per cell) typically conduct electrical currents toward the cell body. Most large fibers are myelinated; myelin increases the rate of nerve impulse transmission.
   b. Classification
      (1) On the basis of function (direction of impulse transmission) there are sensory (affer-

ent), motor (efferent), and association (interneuron) neurons. Dendritic endings of sensory neurons are bare (pain receptors) or are associated with sensory receptors.

(2) On the basis of structure, there are unipolar, bipolar, and multipolar neurons; the terminology reveals the number of processes extending from the cell body. Motor and association neurons are multipolar; most sensory neurons are unipolar. The exceptions are sensory neurons in certain special sense organs (ear, eye) which are bipolar.

  c. Physiology

(1) A nerve impulse is an electrochemical event (initiated by various stimuli) that causes a change in neuron plasma membrane permeability, allowing sodium ions (Na^+) to enter the cell (depolarization). Once begun, the action potential, or nerve impulse, continues over the entire surface of the cell. Electrical conditions of the resting state are restored by the diffusion of potassium ions (K^+) out of the cell (repolarization). Ion concentrations of the resting state are restored by the sodium-potassium pump.

(2) A neuron influences other neurons or effector cells by releasing neurotransmitters, chemicals that diffuse across the synaptic cleft and attach to membrane receptors. The result is opening of specific ion channels and activation or inhibition, depending on the neurotransmitter released and the target cell.

(3) A reflex is a rapid, predictable response to a stimulus. There are two types—autonomic and somatic. The minimum number of components of a reflex arc is four: receptor, effector, and sensory and motor neurons (most, however, have one or more interneurons). Normal reflexes indicate normal nervous system function.

Central Nervous System *(pp. 207–221)*

1. The brain is located within the cranial cavity of the skull and consists of the cerebral hemispheres, diencephalon, brain stem structures, and cerebellum.

  a. The two cerebral hemispheres form the largest part of the brain. Their surface, or cortex, is gray matter and their interior is white matter. The cortex is convoluted and has gyri, sulci, and fissures. The cerebral hemispheres are involved in logical reasoning, moral conduct, emotional responses, sensory interpretation, and the initiation of voluntary muscle activity. The functional areas of the cerebral lobes have been identified (see p. 000). The basal nuclei, regions of gray matter deep within the white matter of the cerebral hemispheres, modify voluntary motor activity. Parkinson's disease and Huntington's chorea are disorders of the basal nuclei.

  b. The diencephalon (interbrain) is superior to the brain stem and is enclosed by the cerebral hemispheres. The major structures include the following:

(1) The thalamus encloses the third ventricle and is the relay station for sensory impulses passing to the sensory cortex for interpretation.

(2) The hypothalamus makes up the "floor" of the third ventricle and is the most important regulatory center of the autonomic nervous system (regulates water balance, metabolism, thirst, temperature, and the like).

(3) The epithalamus includes the pineal body (an endocrine gland) and the choroid plexus of the third ventricle.

  c. The brain stem is the short region that is inferior to the hypothalamus and that merges with the spinal cord.

(1) The midbrain is most superior and is primarily fiber tracts.

(2) The pons is inferior to the midbrain and has fiber tracts and nuclei involved in respiration.

(3) The medulla oblongata is the most inferior part of the brain stem. In addition to fiber tracts, it contains many autonomic nuclei involved in the regulation of vital life activities (breathing, heart rate, blood pressure, etc.).

  d. The cerebellum is a large, cauliflowerlike part of the brain dorsal to the fourth ventricle. It coordinates muscle activity and body balance.

2. Protection of the CNS

  a. Bones of the skull and vertebral column are the most external protective structures.

  b. Meninges are three connective tissue membranes—dura mater (tough outermost), arachnoid mater (middle weblike), and pia mater (innermost delicate). The meninges extend beyond the end of the spinal cord.

  c. Cerebrospinal fluid (CSF) provides a watery cushion around the brain and cord. CSF is formed by the choroid plexuses of the brain. It is found in the subarachnoid space, ventricles, and central canal. CSF is continually formed and drained.

   d. The blood-brain barrier is composed of relatively impermeable capillaries.

3. Brain dysfunctions
   a. Head trauma may cause concussions (reversible damage) or contusions (nonreversible damage). When the brain stem is affected, unconsciousness (temporary or permanent) occurs. Trauma-induced brain injuries may be aggravated by intracranial hemorrhage or cerebral edema, both of which compress brain tissue.
   b. Cerebrovascular accidents (CVAs or strokes) result when blood circulation to brain neurons is blocked and brain tissue dies. The result may be visual impairment, paralysis, and aphasias.
   c. Alzheimer's disease is a degenerative brain disease in which abnormal protein deposits and other structural changes appear. It results in slow, progressive loss of memory and motor control plus increasing dementia.
   d. Techniques used to diagnose brain dysfunctions include the EEG, simple reflex tests, pneumoencephalography, angiography, and CT, PET, and MRI scans.

4. The spinal cord is a reflex center and conduction pathway. Found within the vertebral canal, the cord extends from the foramen magnum to L_1 or L_2. The cord has a central bat-shaped area of gray matter surrounded by columns of white matter, which carry motor and sensory tracts from and to the brain.

Peripheral Nervous System (pp. 222–232)

1. A nerve is a bundle of neuron processes wrapped in connective tissue coverings (endoneurium, perineurium, epineurium).

2. Cranial nerves: Twelve pairs of nerves that extend from the brain to serve the head and neck region. The exception is the vagus nerves, which extend into the thorax and abdomen.

3. Spinal nerves: Thirty-one pairs of nerves are formed by the union of the dorsal and ventral roots of the spinal cord on each side. The spinal nerve proper is very short and splits into dorsal and ventral rami. Dorsal rami serve the posterior body trunk; ventral rami (except T_1 through T_{12}) form plexuses (cervical, brachial, lumbar, sacral) that serve the limbs.

4. Autonomic nervous system: Part of the PNS, composed of neurons that regulate the activity of smooth and cardiac muscle and glands. This system differs from the somatic nervous system in that there is a chain of two motor neurons from the CNS to the effector. Two subdivisions serve the same organs with different effects.
   a. The sympathetic division is the "fight-or-flight" subdivision, which prepares the body to cope with some threat. Activation of the sympathetic nervous system results in an increased heart rate and blood pressure. The preganglionic neurons are in the gray matter of the cord. The postganglionic neurons are in sympathetic chains or in collateral ganglia. Postganglionic axons secrete norepinephrine.
   b. The parasympathetic division is the "housekeeping" system and is in control most of the time. This division maintains homeostasis by seeing that normal digestion and elimination occur, and that energy is conserved. The first motor neurons are in the brain or the sacral region of the cord. The second motor neurons are in the terminal ganglia close to the organ served. Postganglionic axons secrete acetylcholine.

Developmental Aspects of the Nervous System (pp. 232–234)

1. Maternal and environmental factors may impair embryonic brain development. Oxygen deprivation destroys brain cells. Severe congenital brain diseases include cerebral palsy, anencephaly, hydrocephalus, and spina bifida.

2. Premature babies have trouble regulating body temperature because the hypothalamus is one of the last brain areas to mature prenatally.

3. Development of motor control indicates the progressive myelination and maturation of a child's nervous system. Brain growth ends in young adulthood. Neurons die throughout life and are not replaced; thus, brain mass declines with age.

4. Healthy aged people maintain nearly optimal intellectual function. Disease—particularly cardiovascular disease—is the major cause of declining mental function with age.

Review Questions

1. What are the two great controlling systems of the body?

2. Explain both the structural and functional classifications of the nervous system. Include in your explanation the subdivisions of each.

3. List the structures of the CNS and PNS.

4. Two major cell groups make up the nervous system—neurons and connective tissue cells such as astrocytes and Schwann cells. Which are "nervous" cells? Why? What are the major functions of the other cell group?

5. Give the basis for the functional classification of neurons.

6. Briefly explain how nerve impulses are initiated and why one-way conduction at synapses always happens.

7. Name four types of cutaneous sensory receptors. Which of the cutaneous receptor types is most numerous? Why?

8. What is a reflex arc? Name its minimum components.

9. Make a rough drawing of the left cerebral hemisphere. On your drawing, locate at least five different functional areas and then indicate their specific functions.

10. Other than serving as a conduction pathway, what is a major function of the pons? Why is the medulla the most vital part of the brain?

11. What is the function of the thalamus? the hypothalamus? the cerebellum?

12. Discuss how the brain is protected by bone, membranes, fluid, and capillaries.

13. What is gray matter? white matter? How does the arrangement of gray and white matter differ in the cerebral hemispheres and the spinal cord?

14. What are two functions of the spinal cord?

15. How many pairs of cranial nerves are there? Which are purely sensory? Which three serve the external eye muscles? Which activates the chewing muscles? Which helps regulate heart rate and activity of the digestive tract?

16. Except for the vagus nerves, what general area of the body do the cranial nerves serve?

17. How many pairs of spinal nerves are there? How do they arise?

18. What region of the body is served by the dorsal rami of the spinal nerves? the ventral rami?

19. Name the four major nerve plexuses formed by the ventral rami and the body region served by each.

20. How does the autonomic nervous system differ from the somatic nervous system?

21. What is the difference in function of the sympathetic and parasympathetic divisions of the autonomic nervous system (a) generally speaking, and (b) as specifically relates to the operation of the cardiovascular and digestive systems?

22. Since the sympathetic and parasympathetic fibers serve the same organs, how can their opposing effects be explained?

23. Differentiate between a *concussion* and a *contusion*. Why do severe brain stem contusions result in unconsciousness?

24. Compare CVAs and TIAs in terms of causes, symptoms, and consequences.

25. Define *senility*. Name possible causes of permanent and reversible senility.

At the Clinic

1. Mrs. Jones has had a progressive decline in her mental capabilities in the past 5 or 6 years. At first, her family attributed her occasional memory lapses, confusion, and agitation to grief over her husband's death 6 years earlier. When examined, Mrs. Jones was aware of her cognitive problems and was shown to have an IQ score approximately 30 points lower than would be predicted by her work history. A CT scan showed diffuse cerebral atrophy. The physician prescribed a mild tranquilizer for Mrs. Jones and told her family that there was little else he could recommend. What is Mrs. Jones's problem?

2. Joseph, a man in his early 70s, was having problems chewing his food. He was asked to stick out his tongue. It deviated to the right, and its right side was quite wasted. What cranial nerve was impaired?

3. Jed, a couch potato, likes to eat a very large meal in the evening. After the meal, his wife asks him to

help clean the dishes, but Jed explains that he is "too tired" and promptly goes to sleep. What seems to be his problem?

4. A semiconscious young woman is brought to the hospital by friends after falling from a roof. She did not lose consciousness immediately, and she was initially lucid. After a while, though, she became confused and then unresponsive. What is a likely explanation of her condition?

5. Elderly Mrs. Barker has just suffered a stroke. She is able to understand verbal and written language, but when she tries to respond, her words are garbled. What cortical region was damaged by the stroke?

6. Mrs. Tonegawa, a new mother, brings her infant to the clinic because it has suffered repeated seizures. Upon questioning, she states that her labor was unusually long and difficult. What condition do you suspect? Will the infant's condition worsen?

7. Three-year-old Samantha is sobbing that her right arm is "gone" and the examination shows her to have little muscle strength in that limb. Questioning the parents reveals that her father had been swinging her by the arms. What part of the PNS has been damaged?

8. Mr. Dingledine is an 82-year-old bedridden gentleman who has discovered a new interest in learning about his body. While being tended by the visiting nurse, he remarks that the supporting cells in nervous tissue (such as Schwann cells and oligodendrocytes) act like the rubber coating around household wiring. What did he mean by this analogy?

8
Special Senses

YOUR GOALS

After completing this chapter, you will have working knowledge of the functions of special senses and will have mastered the objectives listed below.

FUNCTION PREVIEW

■ The special senses respond to different types of energetic stimuli involved in vision, hearing, balance, taste, and smell.

(continued)

People are responsive creatures. Hold freshly baked bread before us and our mouths water. A sudden clap of thunder makes us recoil. These "irritants" (the bread and the thunderclap) and many others are the stimuli that continually greet us and are interpreted by our nervous system.

We are usually told that we have five senses that keep us in touch with what is going on in the external world: touch, taste, smell, sight, and hearing. Actually touch is a mixture of the general senses that we considered in Chapter 7—the temperature, pressure, and pain receptors of the skin and the proprioceptors of muscles and joints. The other four "traditional" senses—*smell, taste, sight,* and *hearing*—are called **special senses.** Receptors for a fifth special sense, *equilibrium,* are housed in the ear, along with the organ of hearing. In contrast to the small and widely distributed general receptors, the **special sense receptors** are either large, complex sensory organs (eyes and ears) or localized clusters of receptors (taste buds and olfactory epithelium).

This chapter focuses on the functional anatomy of each of the special sense organs individually, but keep in mind that sensory inputs are overlapping. What we finally experience—our "feel" of the world—is a blending of stimulus effects.

THE EYE AND VISION

How we see has captured the curiosity of many researchers. Vision is the sense that has been studied most. Of all the sensory receptors in the body, 70 percent are in the eyes. The optic tracts that carry information from the eyes to the brain are massive bundles, containing over a million nerve fibers. Vision is the sense that requires the most "learning," and the eye appears to delight in being fooled. The old expression "You see what you expect to see" is often very true.

Anatomy of the Eye

External and Accessory Structures

The adult eye is a sphere that measures about 1 inch (2.5 cm) in diameter. Only the anterior one-sixth of the eye's surface can normally be seen. The rest of it is enclosed and protected by a cushion of fat and the walls of the bony orbit. The accessory structures of the eye include the extrinsic eye muscles, eyelids, conjunctiva, and lacrimal apparatus.

Anteriorly the eyes are protected by the **eyelids,** which meet at the medial and lateral corners of the eye, the *medial* and *lateral canthus* respectively (see Figure 8.1). Projecting from the border of each eyelid are the **eyelashes.** Modified sebaceous glands associated with the eyelid edges are the **meibomian** (mi-bo′me-an) **glands.** These glands produce an oily secretion that lubricates the eye. The **ciliary glands,** modified sweat glands, lie between the eyelashes (*cilium* = eyelash).

A delicate membrane, the **conjunctiva** (kon-junk″ti′vah), lines the eyelids and covers part of the outer surface of the eyeball. It ends at the edge of the cornea by fusing with the corneal epithelium. The conjunctiva secretes mucus, which helps to lubricate the eyeball and keep it moist.

Inflammation of the conjunctiva, called *conjunctivitis,* results in reddened, irritated eyes. *Pinkeye,* its infectious form caused by bacteria or viruses, is highly contagious. ❏

The **lacrimal apparatus** consists of the lacrimal gland and a number of ducts that drain the lacrimal secretions into the nasal cavity (Figure 8.1). The **lacrimal glands** are located above the lateral end of each eye. They continually release a dilute salt solution *(tears)* onto the anterior surface of the eyeball through several small ducts. The tears flush across the eyeball into the **lacrimal canals** medially, then into the **lacrimal sac,** and finally into the **nasolacrimal duct,** which empties

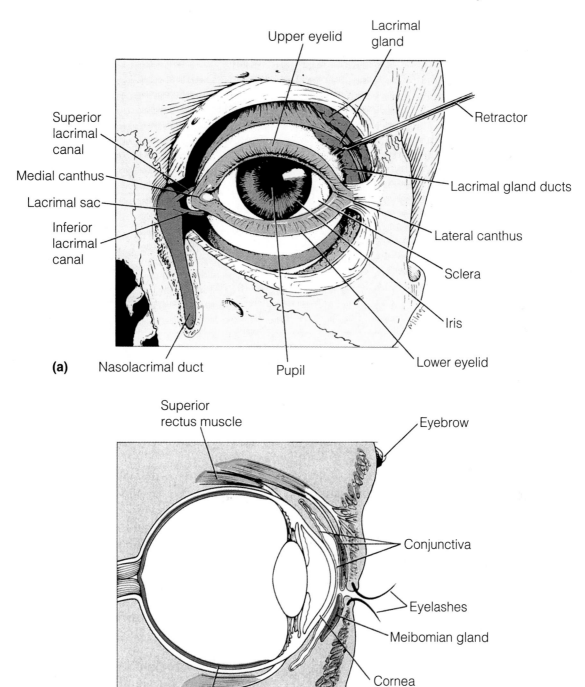

(a)

Upper eyelid

Lacrimal gland

Retractor

Superior lacrimal canal

Medial canthus

Lacrimal sac

Inferior lacrimal canal

Lacrimal gland ducts

Lateral canthus

Sclera

Iris

Lower eyelid

Nasolacrimal duct

Pupil

(b)

Superior rectus muscle

Eyebrow

Conjunctiva

Eyelashes

Meibomian gland

Cornea

Retina

Figure 8.1   External anatomy of the eye and accessory structures.
(**a**) Anterior view. Lacrimal gland is shown pulled superiorly by a retractor to expose its ducts. (**b**) Sagittal section.

into the nasal cavity. Lacrimal secretion also contains antibodies and *lysozyme* (li′so-zīm), an enzyme that destroys bacteria. Thus, it cleanses and protects the eye surface as it moistens and lubri-

cates it. When lacrimal secretion increases substantially, tears spill over the eyelids and empty into the nasal cavities so rapidly that you soon become congested. This happens when the eyes are irri-

Figure 8.2   Extrinsic muscles of the eye. (a) Lateral view of the right eye.
(b) Anterosuperior view of the right eye. The four rectus muscles originate
from the annular ring, a ringlike tendon at the back of the eye socket.
(c) Summary of cranial nerve supply and actions of the extrinsic eye muscles.

tated by foreign objects or chemicals, and when
we are emotionally upset. In the case of irritation,
the enhanced tearing acts to wash away or dilute
the irritating substance. The importance of "emotional tears" is poorly understood, but some suspect that crying is important in reducing stress.
Anyone who has had a good cry would probably
agree, but this has been difficult to prove scientifi-
cally.

Because the nasal cavity mucosa is continuous with that of the lacrimal duct system, a
cold or nasal inflammation often causes the lacrimal mucosa to become inflamed and swell. This
impairs the drainage of tears from the eye surface,
causing "watery" eyes. ❏

Six **extrinsic,** or **external, eye muscles** are
attached to the outer surface of each eye. These
muscles produce gross eye movements and make
it possible for the eyes to follow a moving object.
The names, locations and actions of the extrinsic
muscles are given in Figure 8.2.

Internal Structures: The Eyeball

The eye itself, commonly called the **eyeball,** is a
hollow sphere (see Figure 8.3). Its wall is composed of three *tunics,* or coats, and its interior is
filled with fluids called *humors* that help to maintain its shape. The lens, the main focusing apparatus of the eye, is supported upright within the eye

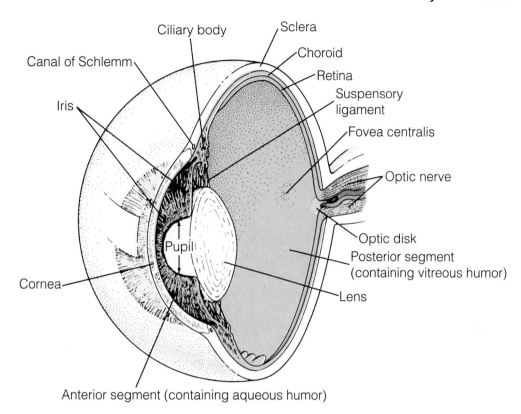

Figure 8.3   Internal anatomy of the eye (midsagittal section).

cavity, dividing it into two chambers. Now that we have covered the general anatomy of the eyeball, we are ready to get specific.

Tunics of the Eyeball   The outermost tunic, the protective **sclera** (skle′rah), is thick, white connective tissue. Also called the *fibrous tunic,* the sclera is seen anteriorly as the "white of the eye." Its central anterior portion is modified so that it is crystal clear. This transparent "window" is the **cornea** (kor′ne-ah) through which light enters the eye. The cornea is well supplied with nerve endings. Most are pain fibers, and when the cornea is touched, blinking and increased tearing occur. Even so, the cornea is the most exposed part of the eye, and it is very vulnerable to damage. Luckily, its ability to repair itself is extraordinary. Furthermore, the cornea is the only tissue in the body that can be transplanted from one person to another without the worry of rejection. Since it has no blood vessels, it is beyond the reach of the immune system.

The middle coat of the eyeball is the **choroid** (ko′roid), a blood-rich nutritive tunic that contains a dark pigment. The pigment prevents light from scattering inside the eye. Anteriorly, the choroid is modified to form two smooth muscle structures, the **ciliary** (sil′e-er-e) **body,** to which the **lens** is attached, and the **iris.** The pigmented iris has a rounded opening, the **pupil,** through which light passes. Circularly and radially arranged smooth muscle fibers form the iris, which acts like the diaphragm of a camera. That is, it regulates the amount of light entering the eye so that one can see as clearly as possible in the available light. In close vision and bright light, the circular muscles contract, and the pupil constricts. In distant vision and dim light, the radial fibers contract to enlarge (dilate) the pupil, which allows more light to enter the eye.

The innermost *sensory tunic* of the eye is the delicate white **retina** (ret′ĭ-nah), which extends anteriorly only to the ciliary body. The retina contains millions of receptor cells, the **rods** and **cones.** Rods and cones are called **photoreceptors,** because they respond to light. Electrical signals pass from the photoreceptors via a two-neuron chain—bipolar cells and then ganglion

Figure 8.4 **The three major types of neurons composing the retina.**
The axons of the ganglion cells form the optic nerve, which leaves the back
of the eyeball at the optic disk. Notice that light passes through the retina to
excite the rods and cones. The flow of electrical signals (black arrows) occurs in the opposite direction: from the photoreceptors to the bipolar cells
and finally to the ganglion cells.

cells—before leaving the retina as nerve impulses
that are transmitted to the optic cortex (see Figure
8.4). The result is vision.

The photoreceptor cells are distributed over
the entire retina, except where the **optic nerve**
(composed of ganglion cell axons) leaves the eyeball; this site is called the **optic disk,** or **blind
spot.** When light from an object is focused on the
optic disk, it disappears from our view and we cannot see it. To illustrate this, perform the blind spot
demonstration by holding Figure 8.5 about 18
inches from your eyes. Close your left eye and
stare at the **X** with your right eye. Move the figure
slowly toward your face, keeping your right eye
focused on the **X**. When the dot focuses on your
blind spot, which lacks photoreceptors, the dot
will disappear. Repeat the test for your left eye.
This time close your right eye and stare at the dot
with your left eye. Move the page as before until
the **X** disappears.

Figure 8.5 **Blind spot test figure.**

The rods and cones are not evenly distributed in the retina. The rods are most dense at the
periphery, or edge, of the retina and decrease in
number as the center of the retina is approached.

VISUAL PIGMENTS—THE ACTUAL PHOTORECEPTORS

The minute photoreceptor cells of the retina have names that reflect their general shapes. As shown in (a), rods are slender, elongated neurons, whereas the fatter cones taper to more pointed tips. In each type of photoreceptor, there is a region called an *outer segment*, attached to the cell body. The outer segment corresponds to a light-trapping dendrite, in which the disks containing the visual pigments are stacked like a row of pennies.

The behavior of the visual pigments is dramatic. When light strikes them, they lose their color, or are "bleached"; shortly afterward, they regenerate their pigment. Absorption of light and pigment bleaching cause electrical changes in the photoreceptor cells that ultimately cause nerve impulses to be transmitted to the brain for visual interpretation. Pigment regeneration ensures that one is not blinded and unable to see in bright sunlight.

A good deal is known about the structure and function of *rhodopsin,* the purple pigment found in rods (see **b**). It is formed from the union of a protein (opsin) and a modified vitamin A product (retinal). When combined in rhodopsin, retinal has a kinked shape that allows it to bind to opsin. But when light strikes rhodopsin, retinal straightens out and releases the protein. Once straightened out, the retinal continues its conversion until it is once again vitamin A. As these changes occur, the purple color of rhodopsin changes to the yellow of retinal and finally becomes colorless as the change to vitamin A occurs. Thus the term "bleaching of the pigment" accurately describes the color changes that occur when light hits the pigment. Regeneration of rhodopsin occurs as vitamin A is again converted to the kinked form of retinal and recombined with opsin. The cone pigments, while similar to rhodopsin, differ in the specific kinds of proteins they contain.

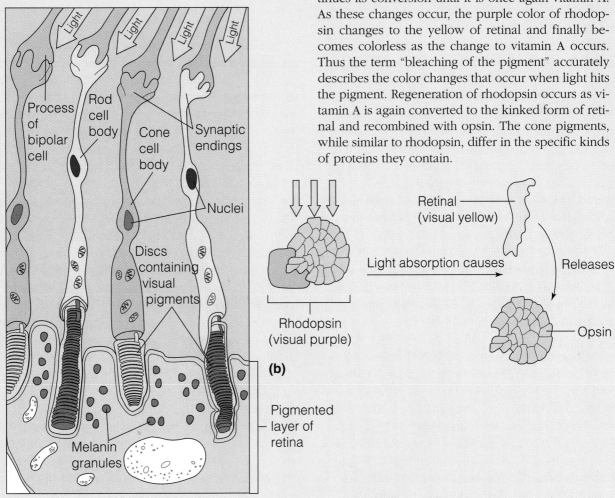

Process of bipolar cell

Rod cell body

Cone cell body

Synaptic endings

Nuclei

Discs containing visual pigments

Melanin granules

Pigmented layer of retina

(a)

Light   Light   Light   Light

Rhodopsin (visual purple)

Light absorption causes

Retinal (visual yellow)

Releases

Opsin

(b)

The rods allow us to see in gray tones in dim light, and they provide for our peripheral vision.

Anything that interferes with rod function hinders our ability to see at night, a condition called *night blindness*. Night blindness dangerously impairs the ability to drive safely at night. Its most common cause is prolonged vitamin A deficiency, which eventually results in deterioration of much of the neural retina. Vitamin A supplements will restore function if taken before the degenerative changes occur. ❏

The cones are discriminatory receptors that allow us to see the details of our world in color under bright light conditions. They are densest in the center of the retina and decrease in number toward the retinal edge. Lateral to each blind spot is the **fovea centralis** (fo′ve-ah sen-tra′lis), a tiny pit that contains only cones (see Figure 8.3). Consequently, this is the area of greatest *visual acuity,* or point of sharpest vision, and anything we wish to view critically is focused on the fovea centralis.

There are three varieties of cones; each is most sensitive to one of the *primary* colors of light. One type responds most vigorously to blue light, another to red light, and the third to green light. Impulses received at the same time from more than one type of cone by the visual cortex are interpreted as *intermediate* colors. For example, simultaneous impulses from blue and red color receptors are seen as purple or violet tones. When all three cone types are being stimulated, we see white. If someone shines red light into one of your eyes and green into the other, you will see yellow, indicating that the "mixing" and interpretation of colors occurs in the brain, not in the retina.

Lack of all three cone types results in total **color blindness,** whereas lack of one cone type leads to partial color blindness. Most common is the lack of red or green receptors, which leads to two varieties of red-green color blindness. Red and green are seen as the same color—either red or green, depending on the cone type *present*. Many color-blind people are unaware of their condition because they have learned to rely on other cues—such as differences in intensities of the same color—to distinguish something green from something red, on traffic signals, for example. Since the genes regulating color vision are on the X (female) sex chromosome, color blindness is a sex-linked condition. It occurs almost exclusively in males. ❏

Figure 8.6   Photograph of a cataract.

Lens   Light entering the eye is focused on the retina by the lens, a flexible biconvex crystal-like structure. The **lens** is held upright in the eye by suspensory ligaments attached to the ciliary body (see Figure 8.3).

In youth, the lens is perfectly transparent and has the consistency of hardened jelly, but as we age it becomes increasingly hard and opaque. **Cataracts,** which result from this process, cause vision to become hazy and eventually cause blindness in the affected eye (Figure 8.6). Current treatment of cataracts is surgical removal of the lens and replacement with a lens implant, or special cataract glasses. ❏

The lens divides the eye into two segments or chambers. The *anterior (aqueous) segment,* anterior to the lens, contains a clear watery fluid called **aqueous humor.** The *posterior (vitreous) segment,* posterior to the lens, is filled with a gel-like substance, called **vitreous** (vit′re-us) **humor,** or the **vitreous body** (see Figure 8.3). Vitreous humor helps prevent the eyeball from collapsing inward by reinforcing it internally. Aqueous humor is similar to blood plasma, and is continually secreted by a special area of the choroid. Like the vitreous humor, it helps maintain intraocular (in″trah-ok′u-lar) pressure, or the pressure inside the eye. It also provides nutrients for the lens and cornea, which lack a blood supply. Aqueous humor is reabsorbed into the venous blood through the **canal of Schlemm** (shlĕm), which is located at the junction of the sclera and cornea (see Figure 8.3).

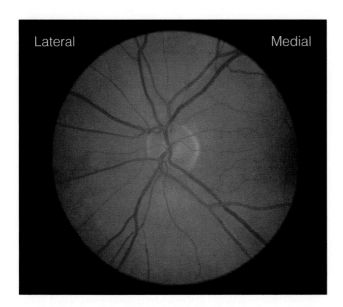

Figure 8.7 The posterior wall (fundus) of the retina as seen with an ophthalmoscope. Notice the optic disk from which the blood vessels radiate.

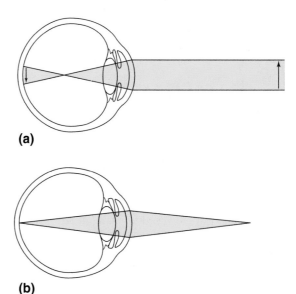

(a)

(b)

Figure 8.8 Relative convexity of the lens during focusing for close and distant vision. (**a**) Light rays from a distant object are nearly parallel as they reach the eye and can be focused without changes in lens convexity. (**b**) Diverging light rays from close objects require that the lens bulge more to focus the image sharply on the retina.

If drainage of aqueous humor is blocked, pressure within the eye increases dramatically and begins to compress the delicate retina and optic nerve. The condition, **glaucoma** (glaw-ko′mah), eventually results in pain and possible blindness. Glaucoma is a common cause of blindness in elderly persons. Since it progresses slowly and has almost no symptoms at first, it tends to occur without obvious signs. Later signs include seeing halos around lights, headaches, and blurred vision. A simple instrument called a *tonometer* (to-nom′e-ter) is used to measure the intraocular pressure. This examination should be performed yearly in people over 40. Glaucoma is treated with eyedrops (miotics), which increase the rate of aqueous humor drainage. ❑

The *ophthalmoscope* (of-thal′mo-skōp) is an instrument that illuminates the interior of the eyeball, allowing the retina, optic disk, and internal blood vessels to be viewed and examined. Certain pathological conditions such as diabetes, arteriosclerosis, and degeneration of the optic nerve and retina can be detected by such an examination. When the ophthalmoscope is correctly set, the **fundus** or posterior wall of the healthy eye should appear as shown in Figure 8.7.

Pathway of Light through the Eye and Light Refraction

When light passes from one substance to another substance that has a different density, its speed changes and its rays are bent, or **refracted.** Light rays are bent in the eye as they encounter the cornea, aqueous humor, lens, and vitreous humor.

The refractive, or bending, power of the cornea and humors is constant. However, that of the lens can be changed by changing its shape—that is, by making it more or less convex, so that light can be properly focused on the retina. The greater the lens convexity, or bulge, the more it bends the light. On the other hand, the flatter the lens, the less it bends the light. In general, light from a distant source (over 20 feet away) approaches the eye as parallel rays (see Figure 8.8a), and no change in lens shape is necessary for it to be focused properly on the retina. However, light from a close object tends to scatter and to diverge, or spread out, and the lens must bulge more to make close vision possible (Figure 8.8b). To achieve this, the ciliary body contracts, allowing

A CLOSER LOOK

IF I CAN'T SEE THINGS FAR AWAY, AM I NEARSIGHTED OR FARSIGHTED?

It seems that whenever people who wear glasses or contact lenses get together and discuss their vision, one of them says something like, "Nearby objects appear blurry to me, but I can't remember if that means I'm nearsighted or farsighted." Or, someone else may say, "My glasses allow me to see far-away objects more clearly, so does that mean I am farsighted?" Here we will explain the meaning of nearsightedness and farsightedness as we explore the basis of eye-focusing disorders.

The eye that focuses images correctly on the retina is said to have **emmetropia** (em″ĕ-tro′pe-ah), literally, "harmonious vision." Such an eye is shown in part (**a**) of the figure.

Nearsightedness is formally called **myopia** (mi″o′pe-ah; "short vision"). It occurs when the parallel light rays from distant objects fail to reach the retina and instead are focused in front of it (see part [**b**] in the figure). Therefore, *distant* objects appear blurry to myopic people. Nearby objects are in focus, however, because the lens "accommodates" (bulges) to focus the image properly on the retina. Myopia results from an eyeball that is too long, a lens that is too strong, or a cornea that is too curved. Correction requires *concave* corrective lenses that diverge the light rays before they enter the eye, so that they converge farther back. To answer the first question posed above, *near*sighted people see *near* objects clearly and need corrective lenses to focus distant objects.

Farsightedness is formally called **hyperopia** (hi″per-o′pe-ah; "far vision"). It occurs when the par-

allel light rays from distant objects are focused *behind* the retina—at least in the resting eye in which the lens is flat and the ciliary muscle is relaxed (see part [**c**] in the figure). Hyperopia usually results from an eyeball that is too short or a "lazy" lens. People with hyperopia can see distant objects clearly because their ciliary muscles contract continuously to increase the light-bending power of the lens, which moves the focal point forward onto the retina. However, the diverging rays from *nearby* objects are focused so far behind the retina that even at full "bulge," the lens cannot focus the image on the retina. Therefore, nearby objects appear blurry. Furthermore, hyperopic individuals are subject to eye-strain as their endlessly contracting ciliary muscles tire from overwork. Correction of hyperopia requires *convex* corrective lenses that converge the light rays before they enter the eye. To answer the second question posed at the beginning of this essay, *far*sighted people can see *far*away objects clearly and require corrective lenses to focus on nearby objects. Unequal curvatures in different parts of the cornea or lens cause **astigmatism** (ah-stig′mah-tizm). In this condition, blurry images occur because points of light are focused not as points on the retina but as lines (*astigma* = not a point). Special cylindrically-ground lenses or contacts are used to correct this problem. Eyes that are myopic or hyperopic *and* astigmatic require more complex correction.

---

the lens to become more convex. This ability of the eye to focus specifically for close objects (those less than 20 feet away) is called **accommodation.** The image formed on the retina as a result of the light-bending activity of the lens is a *real image*—that is, it is reversed from left to right, upside down (inverted), and smaller than the object (Figure 8.9). The normal eye is able to accommodate properly. However, vision problems occur when a lens is too strong or too weak (overconverging and under-converging, respectively) or from structural prob-

lems of the eyeball (as described in the "Closer Look" box on near- and farsightedness above).

Visual Fields and Visual Pathways to the Brain

Axons carrying impulses from the retina are bundled together at the posterior aspect of the eyeball and issue from the back of the eye as the optic nerve. At the **optic chiasma** (ki-as′mah; *chiasm* =

Figure 8.9   Real image (reversed left to right, and upside down) formed on the retina. Notice that the farther away the object, the smaller its image on the retina.

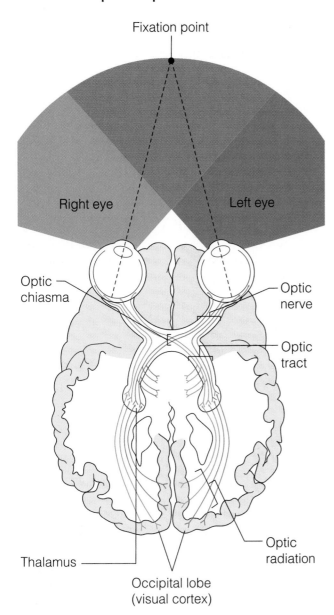

Fixation point

Right eye

Left eye

Optic chiasma

Optic nerve

Optic tract

Optic radiation

Thalamus

Occipital lobe (visual cortex)

Figure 8.10 **Visual fields of the eyes and visual pathway to the brain.** Notice that the visual fields overlap considerably (area of binocular vision). Notice also the retinal sites at which a real image would be focused when both eyes are fixed on a close, pointlike object.

they synapse with the cortical cells and visual interpretation, or seeing, occurs. The visual pathway from the eye to the brain is shown in Figure 8.10. As you can see, each side of the brain receives visual input from both eyes—from the lateral field of vision of the eye on its own side and the medial field of the other eye. Also notice that each eye "sees" a slightly different view, but their *visual fields* overlap quite a bit. As a result of these two facts, humans have **binocular vision.** Binocular vision, literally "two-eyed vision," provides for depth perception, also called "three-dimensional" vision, as our visual cortex fuses the two slightly different images delivered by the two eyes.

Hemianopia (hem″e-ah-no′pe-ah) is the loss of the same side of the visual field of both eyes, which results from damage to the visual cortex on one side only (as occurs in some CVAs). Thus, the person would not be able to see things past the middle of his or her visual field on either the right or left side, depending on the site of the CVA. Such individuals should be carefully attended and warned of objects in the nonfunctional (nonseeing) side of the visual field. Their food and personal objects should always be placed on their functional side, or they might miss them. ❑

Eye Reflexes

Both the internal and the external eye muscles are necessary for proper eye function. The internal muscles are controlled by the autonomic nervous system. These muscles include those of the ciliary body, which alters lens curvature, and the radial and circular muscles of the iris, which control pupil size. The external muscles are the rectus and oblique muscles attached to the eyeball exterior. The external (extrinsic) muscles control eye movements and make it possible to follow moving objects. They are also responsible for **convergence,** which is the reflexive movement of the eyes medially when we view close objects. When convergence occurs, both eyes are aimed toward the near object being viewed. The extrinsic muscles are controlled by somatic fibers of cranial nerves III, IV, and VI, as shown in Figure 8.2.

When the eyes are suddenly exposed to bright light, the pupil immediately constricts; this is the **photopupillary reflex.** This protective reflex prevents excessively bright light from damaging the

cross) the fibers from the medial side of each eye cross over to the opposite side. The fiber tracts that result are the **optic tracts.** Each optic tract contains fibers from the lateral side of the eye on the same side and the medial side of the opposite eye. The optic tract fibers synapse with neurons in the thalamus, whose axons form the **optic radiation,** which runs to the occipital lobe of the brain. There delicate photoreceptors. The pupils also constrict reflexively when we view close objects; this **accommodation pupillary reflex** provides for more acute vision.

Reading requires almost continuous work by both sets of muscles. The muscles of the ciliary body bring about the lens bulge, and the circular (or *constrictor*) muscles of the iris produce the accommodation pupillary reflex. In addition, the extrinsic muscles must converge the eyes as well as move them to follow the printed lines. This is why long periods of reading tire the eyes and often result in what is commonly called *eyestrain*. When you read for an extended time, it is helpful to look up from time to time and stare into the distance. This temporarily relaxes all the eye muscles.

THE EAR: HEARING AND BALANCE

At first glance, the machinery for hearing and balance appears very crude. Fluids must be stirred to stimulate the receptors of the ear: Sound vibrations move fluid to stimulate hearing receptors, whereas gross movements of the head disturb fluids surrounding the balance organs. Receptors that respond to such physical forces are called **mechanoreceptors** (mek″ah-no-re-sep′terz). Our hearing apparatus allows us to hear an extraordinary range of sound, and our highly sensitive equilibrium receptors keep our nervous system continually up to date on the position and movements of the head. Without this information, it would be difficult if not impossible to maintain our balance. Although these two sense organs are housed together in the ear, their receptors respond to different stimuli and are activated independently of one another.

Anatomy of the Ear

Anatomically, the ear is divided into three major areas: the outer, or external, ear; the middle ear; and the inner, or internal, ear (see Figure 8.11). The outer and middle ear structures are involved with hearing *only*. The inner ear functions in both equilibrium and hearing.

Outer (External) Ear

The **outer ear** is composed of the pinna and the external auditory canal. The **pinna** (pin′nah), or **auricle** (aw′ri-kul), is what most people call the "ear"—the shell-shaped structure surrounding the auditory canal opening. In many animals, it collects and directs sound waves into the auditory canal, but in humans this function is largely lost.

The **external auditory canal** is a short, narrow chamber (about 1 inch long by ¼ inch wide) carved into the temporal bone of the skull. In its skin-lined walls are the **ceruminous** (sĕ-roo′mĭ-nus) **glands,** which secrete a waxy yellow substance called **earwax,** or **cerumen.** Sound waves entering the external auditory canal eventually hit the **tympanic** (tim-pan′ik; *tympanum* = drum) **membrane,** or **eardrum,** and cause it to vibrate. The canal ends at the eardrum, which separates the outer from the middle ear.

Middle Ear

The **middle ear,** or **tympanic cavity,** is a small, air-filled cavity within the temporal bone. It is flanked laterally by the eardrum and medially by a bony wall with two openings, the **oval window** and the inferior, membrane-covered **round window.** The **auditory tube** runs obliquely downward to link the middle ear cavity with the throat, and the mucosae lining the two regions are continuous. Normally, the auditory tube is flattened and closed, but swallowing or yawning can open it briefly to equalize the pressure in the middle ear cavity with the external, or atmospheric, pressure. This is an important function because the eardrum does not vibrate freely unless the pressure on both of its surfaces is the same. When the pressures are unequal, the eardrum bulges inward or outward, causing hearing difficulty (voices may sound far away) and sometimes earaches. The ear-popping sensation of the pressures equalizing is familiar to anyone who has flown in an airplane.

Inflammation of the middle ear, *otitis media* (o-ti′tis me′de-ah), is a fairly common result of a sore throat, especially in children, whose auditory tubes run more horizontally. In otitis media, the eardrum bulges and often becomes inflamed. When large amounts of fluid or pus accumulate in the cavity, an emergency *myringotomy* (lancing of the eardrum) may be required to relieve the

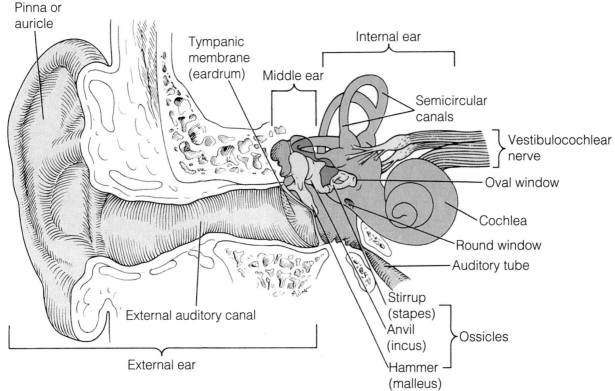

Figure 8.11   Anatomy of the ear.

pressure. During myringotomy, a tiny tube is implanted in the eardrum that allows pus formed in the middle ear to continue to drain into the external ear canal. The tube usually falls out by itself within the year. ❑

The more horizontal course of the auditory tube in infants also explains why it is never a good idea to "prop" a bottle or feed them when they are lying flat (a condition that favors the entry of the food into that tube).

The tympanic cavity is spanned by the three smallest bones in the body, the **ossicles** (os'sĭ-kulz), which transmit the vibratory motion of the eardrum to the fluids of the inner ear (see Figure 8.11). These bones, named for their shape, are the **hammer,** or **malleus** (mă'le-us), the **anvil,** or **incus** (in'kus), and the **stirrup,** or **stapes** (sta'pēz). When the eardrum moves, the hammer moves with it and transfers the vibration to the anvil. The anvil, in turn, passes it on to the stirrup, which presses on the oval window of the inner ear. The movement at the oval window sets the fluids of the inner ear into motion, eventually exciting the hearing receptors.

Inner (Internal) Ear

The **inner ear** is a maze of bony chambers called the **osseous,** or **bony, labyrinth** (lab'ĭ-rinth), located deep within the temporal bone, and just behind the eye socket. The three subdivisions of the bony labyrinth are the **cochlea** (kok'le-ah), the **vestibule** (ves'ti-būl), and the **semicircular canals**. The vestibule is situated between the semicircular canals and the cochlea. The views of the bony labyrinth typically seen in textbooks, including this one, are somewhat misleading because we are really talking about a cavity. The view seen in Figure 8.11 can be compared to a *cast* of the bony labyrinth; that is, a labyrinth that has been filled with plaster of paris and then had the bony walls removed after the plaster hardened. The shape of the plaster then reveals the shape of the *cavity* that worms through the temporal bone.

The bony labyrinth is filled with a plasmalike fluid called **perilymph** (per'ĭ-limf). Suspended in the perilymph is a **membranous labyrinth,** a system of membrane sacs that (more or less) follows the shape of the bony labyrinth. The membranous labyrinth itself contains a thicker fluid called **endolymph** (en'do-limf).

Mechanisms of Equilibrium

If a cat is released upside down from a height, it will land on its feet. If an infant is tilted backward, its eyes will roll downward so that its gaze remains fixed (the *doll's eyes reflex*). Both of these reactions, and countless others, are compensations for a disturbance in balance, reflexes that depend on the sensory receptors within the vestibule and semicircular canals.

The equilibrium sense is not easy to describe because it does not "see," "hear," or "feel." What it *does* is respond (frequently without our awareness) to the various movements of our head. The equilibrium receptors of the inner ear, sometimes called the **vestibular apparatus,** can be divided into two functional arms—one arm responsible for monitoring *static equilibrium*, and the other involved with *dynamic equilibrium.*

Static Equilibrium

Within the membrane sacs of the vestibule are receptors called **maculae** (mak'u-le; "spots") that are essential to our sense of **static equilibrium** (see Figure 8.12). The maculae report on the position of the head with respect to the pull of gravity when the body is not moving (*static* = at rest). Since they provide information on which way is up or down, they help us keep our head erect. They are extremely important to divers swimming in the dark depths (where other orienting cues are absent), enabling them to tell which way is up (to the surface). Each macula is a patch of receptor cells with their "hairs" embedded in a gel or jellylike material containing **otoliths** (o'to-lithz), tiny stones made of calcium salts. As the head moves, the otoliths roll in response to changes in the pull of gravity. This movement creates a pull on the gel, which in turn slides like a greased plate over the hair cells, bending their hairs. This event activates the hair cells, which send impulses along the **vestibular nerve** (a division of cranial nerve VIII) to the cerebellum of the brain, informing it of the position of the head in space.

Dynamic Equilibrium

The **dynamic equilibrium** receptors, found in the semicircular canals, respond to angular or rotatory

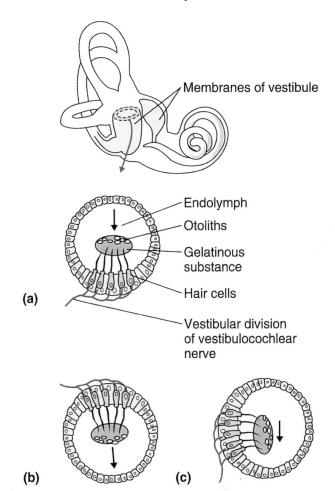

(a)
Membranes of vestibule
Endolymph
Otoliths
Gelatinous substance
Hair cells
Vestibular division of vestibulocochlear nerve

(b)   **(c)**

Figure 8.12   Receptors for static equilibrium within the inner ear. (**a**) Location of the maculae in the membrane sacs of the vestibule. (**b** and **c**) Stimulation of the maculae by movement of the otoliths in the gel. Arrows indicate the direction of gravitational pull.

movements of the head, rather than to straight-line movements. When you twirl on the dance floor or suffer through a rough boat ride, these receptors are working overtime. The semicircular canals (each about ½ inch around) are oriented in the three planes of space. Thus, regardless of which plane one moves in, there will be receptors to detect the movement.

Within each membranous semicircular canal is a receptor region called a **crista ampullaris** (kris'tah am″pu-lar'is) that consists of a tuft of hair cells covered with a gelatinous cap called the **cupula** (ku'pu-lah) (see Figure 8.13). When your head moves in an arclike or angular direction, the

(a)

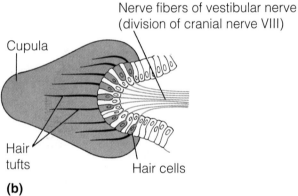

(b)

Figure 8.13   Receptors for dynamic equilibrium within the inner ear. (a) Location of the cristae in each semicircular canal. **(b)** Enlarged view of one crista ampullaris

endolymph in the canal lags behind and moves in the *opposite* direction, pushing the cupula—like a swinging door—in a direction opposite to the body's motion. This stimulates the hair cells and impulses are transmitted up the **vestibular nerve** to the cerebellum. When the angular motion stops suddenly, the endolymph again flows in the opposite direction and reverses the cupula's movement, which stimulates the hair cells again. It is this back-flow that causes the reversed motion sensation you feel when you stop suddenly after twirling. When you are moving at a constant rate, the receptors gradually stop sending impulses, and you no longer have the sensation of motion until your speed or direction of movement changes.

Although the receptors of the semicircular canals and vestibule are responsible for dynamic and static equilibrium, respectively, they usually act together. Besides these equilibrium senses, sight and the proprioceptors of the muscles and tendons are also important in providing information to the cerebellum that it uses to control balance.

Mechanism of Hearing

Within the membranes of the snail-like cochlea is the **organ of Corti** (kor′te), which contains the hearing receptors or **hair cells** (see Figure 8.14a). Sound waves that reach the cochlea through vibrations of the eardrum, ossicles, and the oval window set the cochlear fluids into motion (Figure 8.14b). As the sound waves are transmitted by the ossicles from the eardrum to the oval window, their force (amplitude) is increased by the lever activity of the ossicles. In this way, nearly the total force exerted on the much larger eardrum reaches the tiny oval window, which in turn sets the fluids of the inner ear into motion. The receptor cells, positioned on the basilar membrane in the organ of Corti, are stimulated when their "hairs" are bent or pulled by the movement of the gel-like **tectorial** (tek-to′re-al) **membrane** that lies over them. In general, high-pitch sounds disturb receptor cells close to the oval window whereas low-pitch sounds stimulate specific hair cells further along the cochlea. Once stimulated, the hair cells transmit impulses along the **cochlear nerve** (a division of the eighth cranial nerve—the vestibulocochlear nerve) to the auditory cortex in the temporal lobe where interpretation of the sound, or hearing, occurs. Since sound usually reaches the two ears at different times, we could say that we hear "in stereo." Functionally, this helps us to determine where sounds are coming from in our environment.

When the same sounds, or tones, keep reaching the ears, the auditory receptors tend to *adapt* or stop responding to those sounds, and we are no longer aware of them. This is why the drone of a continually running motor does not demand our attention after the first few seconds. However, hearing is the last sense to leave our awareness when we fall asleep or receive anesthesia (or die) and is the first to return as we awaken.

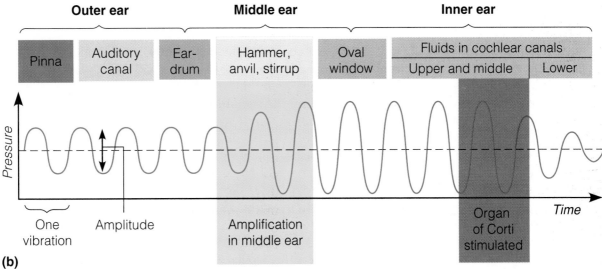

Figure 8.14 **Organ of Corti and route of sound waves through the ear.** (**a**) Cross section through the cochlea showing the position of the organ of Corti in the cochlear duct. (**b**) To exite the hair cells in the organ of Corti in the inner ear, sound wave vibrations must pass through air, membranes, and fluid.

Hearing and Equilibrium Deficits

Children with ear problems or hearing deficits often pull on their ears or fail to respond when spoken to. Under such conditions, tuning fork tests are done to try to diagnose the problem. **Deafness** is defined as *hearing loss of any degree*—from a slight loss to a total inability to hear sound. Generally speaking, there are two kinds of deafness, conduction and sensorineural. Temporary or permanent **conduction deafness** results when something interferes with the conduction of sound vibrations to the fluids of the inner ear. Something as simple as a buildup of earwax may be the cause. Other causes of conduction deafness include fusion of the ossicles (a problem called *otosclerosis* [o″to-sklĕ-ro′sis]), a ruptured eardrum, and otitis media.

Sensorineural deafness occurs when there is degeneration or damage to the receptor cells in the organ of Corti, to the cochlear nerve, or to neurons of the auditory cortex. This often results from extended listening to excessively loud sounds. Thus, whereas conduction deafness results from mechanical factors, sensorineural deafness is a problem of nervous system structures.

A person who has a hearing loss due to conduction deafness will still be able to hear by bone conduction, even though his or her ability to hear air-conducted sounds (the normal conduction route) is decreased or lost. On the other hand, individuals with sensorineural deafness cannot hear better by *either* conduction route. Hearing aids, which use skull bones to conduct sound vibrations to the inner ear, are generally very successful in helping those with conduction deafness to hear. They are less helpful for sensorineural deafness.

Equilibrium problems are usually obvious. Nausea, dizziness, and problems in maintaining balance are common symptoms, particularly when impulses from the vestibular apparatus "disagree" with what we see (visual input). There also may be strange (jerky or rolling) eye movements. Simple tests (for example, observing a person after he or she has been rotated, or watching a person stand on one leg) are used to screen for problems with the equilibrium apparatus. If these tests indicate that there are such problems, then more sophisticated tests are done to try to track down the exact cause.

A serious pathology of the inner ear is *Ménière's* (mān″e-airz′) *syndrome.* The exact cause of this condition is not fully known, but suspected causes are arteriosclerosis, degeneration of the eighth cranial nerve, and increased pressure of the inner ear fluids. In Ménière's syndrome, progressive deafness occurs. Affected individuals become nauseated and often have *vertigo* (a sensation of spinning), which is so severe that they cannot stand up without extreme discomfort. Anti-motion-sickness drugs are often prescribed to decrease the discomfort. ❏

CHEMICAL SENSES: TASTE AND SMELL

The receptors for taste and olfaction are classified as **chemoreceptors** (ke″mo-re-sep′terz), because they respond to chemicals in solution. Four types of taste receptors have been identified, but the olfactory receptors (for smell) are believed to be sensitive to a much wider range of chemicals. The receptors for smell and taste complement each other and respond to many of the same stimuli.

Taste Buds and the Sense of Taste

The word *taste* comes from the Latin word *taxare*, which means "to touch, estimate, or judge." When we taste things, we are, in fact, testing or judging our environment in an intimate way, and the sense of taste is considered by many to be the most pleasurable of our special senses. There is no question that what does not taste good to us will usually not be allowed to enter the body.

The **taste buds**, or specific receptors for the sense of taste, are widely scattered in the oral cavity. Of the 10,000 or so taste buds that we have, most are located on the tongue. A few are found on the soft palate and inner surface of the cheeks.

The dorsal tongue surface is covered with small peglike projections, or **papillae** (pah-pil′e), of three types: sharp *filiform* (fil′ĭ-form) papillae and the rounded *fungiform* (fun′ji-form) and *circumvallate* (ser″kum-val′at) papillae. The taste buds are found on the sides of the circumvallate papillae and on the more numerous fungiform papillae (see Figure 8.15). The specific cells that respond to chemicals dissolved in the saliva are epithelial cells called **gustatory cells**, which are surrounded by supporting cells in the taste bud. Their long microvilli protrude through the *taste pore*, and when they are stimulated, they depolarize and impulses are transmitted to the brain. Three cranial nerves—the seventh, ninth, and tenth—carry taste impulses from the various taste buds to the gustatory cortex. The **facial nerve** (VII) serves the anterior part of the tongue. The other two cranial nerves—the **glossopharyngeal** and **vagus** respectively—serve the other taste bud–containing areas.

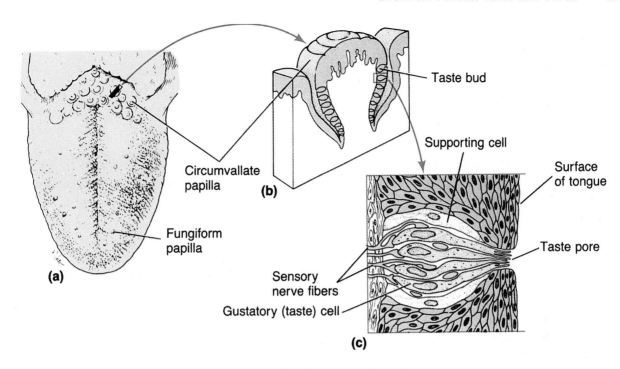

Figure 8.15   Location and structure of taste buds. (**a**) Taste buds on the tongue are associated with papillae, projections of the tongue mucosa. (**b**) A sectioned circumvallate papilla shows the position of the taste buds in its lateral walls. (**c**) An enlarged view of one taste bud.

There are four basic taste sensations, each corresponding to stimulation of one of the four major types of taste buds. The *sweet receptors* respond to substances such as sugars, saccharine, and some amino acids. Some believe that the common factor is the hydroxyl (OH^-) group. *Sour receptors* respond to hydrogen ions (H^+), or the acidity of the solution; *bitter receptors* to alkaloids; and *salty receptors* to metal ions in solution. A "taste map" showing the major location of each receptor type on the tongue is shown in Figure 8.16.

Taste likes and dislikes have homeostatic value. A liking for sugar and salt will satisfy the body's need for carbohydrates and minerals (as well as some amino acids). Many sour, naturally acidic foods (such as oranges, lemons, and tomatoes) are rich sources of vitamin C, an essential vitamin. Since many natural poisons and spoiled foods are bitter, our dislike for bitterness is protective. (The position of the bitter receptors on the most posterior part of the tongue seems strange, however, because usually by the time we actually taste a bitter substance, we have swallowed some of it.)

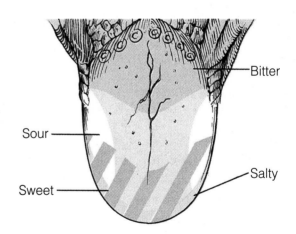

Figure 8.16   A "taste map" showing areas of the tongue most sensitive to the four basic taste sensations. (Note: Since the area most sensitive to sweet overlaps the sour and salty areas, the sweet area is indicated by pink stripes.)

Taste is affected by many factors, and what is commonly referred to as our sense of taste depends heavily on stimulation of our olfactory receptors by aromas. Think of how bland, almost tasteless, food is when your nasal passages are

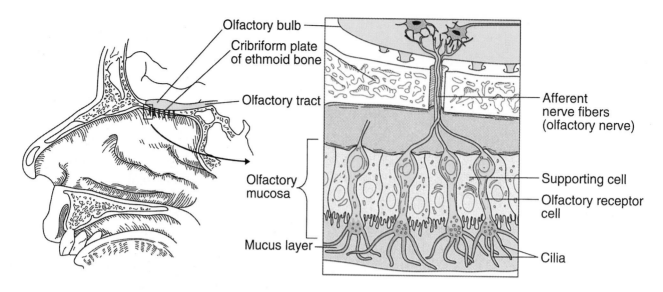

Figure 8.17   Location and cellular makeup of the olfactory epithelium.

congested by a cold. Without the sense of smell, our morning coffee would simply taste bitter. In addition to the sense of smell, the temperature and texture of food can enhance or spoil its taste for us. For example, some people will not eat foods that have a pasty texture (avocados) or that are gritty (pears), and almost everyone considers a cold greasy hamburger unfit to eat. "Hot" foods like chili peppers actually excite pain receptors in the mouth.

Olfactory Receptors and the Sense of Smell

Even though our sense of smell is far less acute than that of many other animals, the human nose is still much keener than any machine in picking up small differences in odors. Some people capitalize on this ability by becoming tea and coffee blenders, perfumers, or wine tasters.

The thousands of **olfactory receptors,** receptors for the sense of smell, occupy a postage stamp–sized area in the roof of each nasal cavity (Figure 8.17). Air entering the nasal cavities must make a hairpin turn to enter the respiratory passageway below, so sniffing, which causes more air to flow superiorly across the olfactory receptors, intensifies the sense of smell.

The **olfactory receptor cells** are neurons equipped with *olfactory hairs,* long cilia that pro-

trude outward from the nasal epithelium and are continually bathed by a layer of mucus secreted by underlying glands. When the receptors are stimulated by chemicals dissolved in the mucus, they transmit impulses along the **olfactory nerve** (first cranial nerve) to the olfactory cortex of the brain. There interpretation of the odor occurs, and a sort of "odor snapshot" is made. The olfactory pathways are closely tied into the limbic system (emotional-visceral part of the brain). Thus, olfactory impressions are long-lasting and very much a part of our memories and emotions. For example, the smell of chocolate chip cookies may remind you of your grandmother, and the smell of a special pipe tobacco may make you think of your father. There are hospital smells, school smells, mother smells, baby smells, travel smells. The list can be continued almost without end. Our reactions to odors are rarely neutral. We tend to either like or dislike certain odors, and we change, avoid, or add odors according to our preferences.

The olfactory receptors are exquisitely sensitive—just a few molecules can activate them. Like the auditory receptors, the olfactory neurons tend to adapt rather quickly when they are exposed to an unchanging stimulus, in this case, an odor. This is why a woman stops smelling her own perfume after a while but will quickly pick up the scent of another perfume on someone else.

While it is possible to have either taste or smell deficits, most people seeking medical help for loss of chemical senses have olfactory disorders, or *anosmias* (ah-noz′me-uz). Most anosmias result from head injuries, the aftereffects of nasal cavity inflammation (due to a cold, an allergy, or smoking), or aging. The culprit in fully one-third of all cases of chemical sense loss is zinc deficiency, and the cure is almost instant once the right form of zinc supplement is prescribed. Some brain disorders can destroy the sense of smell or mimic it. For example, *olfactory auras* (olfactory hallucinations) are experienced by some epileptics just before they go into seizures. ❏

DEVELOPMENTAL ASPECTS OF THE SPECIAL SENSES

The special sense organs, essentially part of the nervous system, are formed very early in embryonic development. For example, the eyes, which are literally outgrowths of the brain, are developing by the fourth week. All of the special senses are functional, to a greater or lesser degree, at birth.

Congenital eye problems are relatively uncommon, but some examples can be given. *Strabismus* (strah-biz′mus), commonly called "crossed eyes," results from unequal pulls by the external eye muscles that prevent the baby from coordinating movement of the two eyes. Exercises and/or surgery are always used to correct the condition because, if it is allowed to persist, the brain may stop recognizing signals from the deviating eye, causing that eye to become functionally blind.

Maternal infections, particularly *rubella* (measles), which occur during early pregnancy, may lead to congenital blindness or cataracts. If the mother has a type of sexually transmitted disease called *gonorrhea* (gon″o-re′ah), the baby's eyes will be infected by the bacteria during delivery. In the resulting *conjunctivitis*, specifically called *ophthalmia neonatorum* (of-thal′me-ah ne″o-na-to′-rum), the baby's eyelids become red and swollen, and pus is produced. All states now legally require that all newborn babies' eyes be routinely treated with silver nitrate or antibiotics shortly after birth. ❏

Generally speaking, vision is the only special sense that is not fully functional when the baby is born, and many years of "learning" are needed before the eyes are fully mature. The eyeballs continue to enlarge until the age of 8 or 9, but the lens grows throughout life. At birth, the eyeballs are foreshortened, and all babies are hyperopic. As the eyes grow, this condition usually corrects itself. The newborn infant sees only in gray tones, eye movements are uncoordinated, and often only one eye at a time is used. Because the lacrimal glands are not fully developed until about 2 weeks after birth, the baby is tearless for this period, even though he or she may cry lustily.

By 5 months, the infant is able to focus on objects within easy reach and to follow moving objects, but visual acuity is still poor (20/200). By the time the child is 5 years old, color vision is well developed, visual acuity has improved to about 20/30, and depth perception is present, providing a readiness to begin reading. By school age, the earlier hyperopia has usually been replaced by emmetropia. This condition continues until about age 40, when **presbyopia** (pres″be-o′pe-ah) begins to set in. Presbyopia (literally, "old vision") results from decreasing lens elasticity that accompanies aging. This condition makes it difficult to focus for close vision; it is basically farsightedness. The person who holds the newspaper at arm's length to read it provides the most familiar example of this developmental change in vision.

As aging occurs, the lacrimal glands become less active and the eyes tend to become dry and more vulnerable to bacterial infection and irritation. The lens loses its crystal clarity and becomes discolored. The dilator muscles of the iris become less efficient; thus, the pupils are always somewhat constricted. These last two conditions work together to decrease by half the amount of light reaching the retina, so that visual acuity is dramatically lower by one's 70s. In addition to these changes, elderly persons are susceptible to certain conditions that may result in blindness, such as glaucoma and cataracts. Other common aging-related problems, such as arteriosclerosis and diabetes, may lead to the death of the delicate photoreceptors because of an increasing lack of oxygen and nutrients.

Congenital abnormalities of the ears are fairly common. Examples include partly or completely missing pinnas and closed or absent external ear canals. Maternal infections can have a devastating effect on ear development, and maternal rubella during the early weeks of pregnancy results in sensorineural deafness. ❏

A newborn infant can hear after his or her first cry, but early responses to sound are mostly reflexive—for example, crying and clenching the eyelids in response to a loud noise. By the age of 3 or 4 months, the infant is able to localize sounds and will turn to the voices of family members. Critical listening occurs in the toddler as he or she begins to imitate sounds, and good language skills are very closely tied to an ability to hear well.

Except for ear inflammations (otitis) resulting from bacterial infections or allergies, few problems affect the ears during childhood and adult life. By the 60s, however, a gradual deterioration and atrophy of the organ of Corti begins and leads to a loss in the ability to hear high tones and speech sounds. This condition, **presbycusis** (pres″bĭ-ku′sis), is a type of sensorineural deafness. In some cases, the ear ossicles fuse (otosclerosis), which compounds the hearing problem by interfering with sound conduction to the inner ear. Because many elderly people refuse to accept their hearing loss and resist using hearing aids, they begin to rely more and more on their vision for clues as to what is going on around them, and may be accused of ignoring people. Although presbycusis was once considered a disability of old age, it is becoming much more common in younger people as our world grows noisier day by day. Noise pollution has become a major health problem, and the damage caused by excessively loud sounds is progressive and cumulative. Each insult causes a bit more damage. Rock music played and heard at deafening levels is definitely a contributing factor to the deterioration of the hearing receptors.

The chemical senses, taste and smell, are sharp at birth, and infants relish some food that adults consider bland or tasteless. Some researchers claim the sense of smell is just as important as the sense of touch in guiding a newborn baby to its mother's breast. However, very young children seem indifferent to odors and can play happily with their own feces. As they get older, their emotional responses to specific odors increase.

There appear to be few problems with the chemical senses throughout childhood and young adulthood. Beginning in the mid-40s, our ability to taste and smell diminishes, which reflects the gradual decrease in the number of these receptor cells. Almost half of people over the age of 80 cannot smell at all, and their sense of taste is poor. This may explain their inattention to formerly disagreeable odors, and why older adults often prefer highly seasoned (although not necessarily spicy) foods or lose their appetite entirely.

Important Terms

accommodation	**iris**
blind spot	**labyrinth** (lab′ĭ-rinth)
chemoreceptors (ke″mo-re-sep′terz)	**lens**
choroid (ko′roid)	**maculae** (mak′u-le)
ciliary (sil′e-er-e) **body**	**mechanoreceptors**
cochlea (kok′le-ah)	**olfactory receptors**
cornea (kor′ne-ah)	**organ of Corti** (kor′te)
crista ampullaris (kris′tah am″pu-lar′is)	**ossicles** (os′sĭ-kulz)
dynamic equilibrium	**photoreceptors**

refraction

retina (ret′ĭ-nah)

sclera (skle′rah)

semicircular canals

static equilibrium

taste buds

tympanic (tim-pan′ik) **cavity**

vestibule (ves′ti-būl)

Summary

The Eye and Vision (pp. 242-253)

1. External/accessory structures of the eye.
   a. Extrinsic eye muscles aim the eyes for following moving objects and for convergence.
   b. Lacrimal apparatus includes a series of ducts and the lacrimal glands that produce a saline solution, which washes and lubricates the eyeball.
   c. Eyelids protect the eyes. Associated with the eyelashes are the ciliary glands, modified sweat glands, and the meibomian glands, which produce an oily secretion that helps keep the eye lubricated.
   d. Conjunctiva is a mucous membrane that covers the anterior eyeball and lines the eyelids. It produces a lubricating mucus.

2. Three tunics form the eyeball.
   a. The sclera is the outer, tough, protective tunic. Its anterior portion is the cornea, which is transparent to allow light to enter the eye.
   b. The choroid is the middle coat that provides nutrition to the internal eye structures and prevents light's scattering in the eye. Anterior modifications include two smooth muscle structures, the ciliary body, and the iris (which controls the size of the pupil).
   c. The retina is the innermost (sensory) coat that contains the photoreceptors. Rods are dim light receptors. Cones are receptors that provide for color vision and high visual acuity. The fovea centralis, on which acute focusing occurs, contains only cones.

3. The blind spot (optic disk) is the point where the optic nerve leaves the back of the eyeball.

4. The lens is the major light-bending (refractory) structure of the eye. Its convexity is increased by the ciliary body for close focus. Anterior to the lens is aqueous humor; posterior to the lens is vitreous humor. Both humors reinforce the eye internally. The aqueous humor also provides nutrients to the avascular lens and cornea.

5. Errors of refraction include myopia, hyperopia, and astigmatism. All are correctable with specially ground lenses.

6. The pathway of light through the eye is cornea → aqueous humor → (through pupil) → aqueous humor → lens → vitreous humor → retina.

7. Overlap of the visual fields and inputs from both eyes to each optic cortex provide for depth perception.

8. The pathway of nerve impulses from the retina of the eye is optic nerve → optic chiasma → optic tract → thalamus → optic radiation → visual cortex in occipital lobe of brain.

9. Eye reflexes include the photopupillary, accommodation pupillary, and convergence.

The Ear: Hearing and Balance (pp. 253-258)

1. The ear is divided into three major areas:
   a. Outer ear structures are the pinna, external auditory canal, and tympanic membrane. Sound entering the external auditory canal sets the eardrum into vibration. These structures are involved with sound transmission only.
   b. Middle ear structures are the ossicles and auditory tube within the tympanic cavity. Ossicles transmit the vibratory motion from the eardrum to the oval window. The auditory tube allows pressure to be equalized on both sides of the eardrum. These structures are also involved with sound transmission only.
   c. Inner ear or bony labyrinth consists of bony chambers (cochlea, vestibule, and semicircular canals) in the temporal bone. The bony labyrinth contains perilymph and membranous sacs filled with endolymph. Within the membranous sacs of the vestibule and semicircular canals are equilibrium receptors. Hearing receptors are found within the membranes of the cochlea.

2. Receptors of the semicircular canals (cristae) are dynamic equilibrium receptors that respond to angular or rotational body movements. Receptors of the vestibule (maculae) are static equilibrium receptors that respond to the pull of gravity and report on head position. Visual and proprioceptor input are also necessary for normal balance.

3. Hair cells of the organ of Corti (the receptor for hearing within the cochlea) are stimulated by sound vibrations transmitted through air, membranes, and fluids.

4. Deafness is any degree of hearing loss. Conduction deafness results when the transmission of sound vibrations through the external and middle ears is hindered. Sensorineural deafness occurs when there is damage to the nervous system structures involved in hearing.

5. Symptoms of equilibrium apparatus problems include involuntary rolling of the eyes, nausea, vertigo, and an inability to stand erect.

Chemical Senses: Taste and Smell (pp. 258–261)

1. Chemical substances must be dissolved in water to excite the receptors for smell and taste.

2. Gustatory (taste) cells are located in the taste buds, primarily on the tongue. The four major taste sensations are: sweet, salt, sour, bitter.

3. Taste and appreciation of foods is influenced by the sense of smell and the temperature and texture of foods.

4. The olfactory (smell) receptors are located in the superior aspect of each nasal cavity. Sniffing helps to bring more air (containing odors) over the olfactory mucosa.

5. Olfactory pathways are closely linked to the limbic system; odors recall memories and arouse emotional responses.

Developmental Aspects of the Special Senses (pp. 261–262)

1. Special sense organs are formed early in embryonic development. Maternal infections during the first five or six weeks of pregnancy may cause visual abnormalities as well as sensorineural deafness in the developing child. An important congenital eye problem is strabismus. The most important congenital ear problem is lack of the external auditory canal.

2. Vision requires the most learning. The infant has poor visual acuity (is farsighted) and lacks color vision and depth perception at birth. The eye continues to grow and mature until the eighth or ninth year of life.

3. Problems of aging associated with vision include presbyopia, glaucoma, cataracts, and arteriosclerosis of the eye's blood vessels.

4. The newborn infant can hear sounds, but initial responses are reflexive. By the toddler stage, the child is listening critically and beginning to imitate sounds as language development begins.

5. Sensorineural deafness (presbycusis) is a normal consequence of aging.

6. Taste and smell are most acute at birth and decrease in sensitivity after the age of 40 as the number of olfactory and gustatory receptors decreases.

Review Questions

1. Name three accessory eye structures that help to lubricate the eyeball, and name the secretion of each.

2. Why do you often have to blow your nose after crying?

3. Diagram and label the internal structures of the eye and give the major function of each structure.

4. Name the external eye muscles that allow you to direct your eyes.

5. Locate and describe the functions of the two humors of the eye.

6. What is the blind spot, and why is it called this?

7. How do the functions of the rods and cones differ?

8. What is the fovea centralis, and why is it important?

9. Trace the pathway of light from the time it hits the cornea until it excites the rods and cones.

10. Trace the pathway of nerve impulses from the photoreceptors in the retina to the visual cortex of the brain.

11. How is the right optic *tract* anatomically different from the right optic *nerve*?

12. Define *refraction* and name the refractory structures or substances of the eye.

13. Define *hyperopia, myopia,* and *emmetropia.*

14. Why do most people develop presbyopia as they age? Which of the conditions in Question 13 does it most resemble?

15. Since there are only three types of cones, how can you explain the fact that we see many more colors?

16. Why are ophthalmoscopic examinations important?

17. Many students struggling through mountains of reading assignments are told that they need glasses for eyestrain. Why is it more of a strain on the extrinsic and intrinsic muscles to look at close objects than at far objects?

18. When a light is shone into one eye, the pupils constrict. Why is this an important protective reflex?

19. Name the structures of the outer, middle, and inner ears and give the general function of each structure and each group of structures.

20. Sound waves hitting the eardrum set it into motion. Trace the pathway of vibrations from the eardrum to the organ of Corti, where the hair cells are stimulated.

21. Explain the difference between *sensorineural* deafness and *conductive* deafness, and then give two causes of each type of deafness.

22. Distinguish between *static* and *dynamic* types of equilibrium.

23. Normal balance depends on information transmitted from a number of sensory receptor types. Name at least three of these receptors.

24. What name is given to the taste receptors, and where are they found?

25. Name the four primary taste sensations.

26. Where are the olfactory receptors located, and why is that site poorly suited for their job?

27. What common name is used to describe both the taste and the olfactory receptors, and why?

28. Describe the effects or results of aging on the special sense organs.

29. Which special sense requires the most learning?

At the Clinic

1. An engineering student has been working in a disco to earn money to pay for his education. After about 8 months, he notices that he is having problems hearing high-pitched tones. What is the cause-and-effect relationship here?

2. Nine children attending the same day-care center developed red, inflamed eyes and eyelids. What is the most likely cause and name of this condition?

3. Dr. Nakvarati used an instrument to press on Mr. Cruz's eye during his annual physical examination on his sixtieth birthday. The eye deformed very little, indicating the intraocular pressure was too high. What was Mr. Cruz's probable condition?

4. Lionel suffered a ruptured artery in his middle cranial fossa, and a pool of blood compressed his left optic tract, destroying its axons. What part of the visual field was blinded?

5. Sylvia Marcus, aged 70, recently underwent surgery for otosclerosis. The operation was a failure and did not improve her condition. (a) What was the purpose of the surgery, and exactly what was it trying to accomplish? (b) Why do you think it is a difficult procedure with a relatively high failure rate?

6. Janie is referred to the eye clinic by her teacher, who suspects a need for glasses. Examination demonstrates that Janie is myopic. Will she need concave or convex lenses? Explain.

9
The Endocrine System

YOUR GOALS

After completing this chapter, you will have a working knowledge of the functions of the endocrine system and will have mastered the objectives listed below.

FUNCTION PREVIEW

- The endocrine system maintains homeostasis by releasing chemicals called *hormones,* and controls prolonged or continuous processes such as growth and development, reproduction, and metabolism.

OBJECTIVE CHECKLIST

THE ENDOCRINE SYSTEM AND HORMONE FUNCTION—AN OVERVIEW (pp. 268–271)

- ❏ Define *hormone* and *target organ.*
- ❏ Describe how hormones bring about their effects in the body.
- ❏ Explain how various endocrine glands are stimulated to release their hormonal products.
- ❏ Define *negative feedback* and describe its role in regulating blood levels of the various hormones.
- ❏ Describe the difference between endocrine and exocrine glands.

THE MAJOR ENDOCRINE ORGANS (pp. 271–286)

- ❏ On an appropriate diagram, identify the major endocrine glands and tissues.
- ❏ List hormones produced by the endocrine glands and discuss their general functions.
- ❏ Discuss ways in which hormones promote body homeostasis by giving examples of hormonal actions.
- ❏ Describe the functional relationship between the hypothalamus and the pituitary gland.
- ❏ Describe major pathologic consequences of hypersecretion and hyposecretion of the hormones considered in this chapter.

(continued)

When insulin molecules, carried passively along in the blood, bind tightly to protein receptors of nearby cells, the response is dramatic: Blood-borne glucose molecules begin to disappear into the cells, and cellular activity accelerates. Such is the power of the second great controlling system of the body, the **endocrine system.** Along with the nervous system, it acts to coordinate and direct the activity of the body's cells. However, the speed of control in these two great regulating systems is very different. The nervous system is "built for speed." It uses nerve impulses to prod the muscles and glands into immediate action so that rapid adjustments can be made in response to changes occurring both inside and outside the body. On the other hand, the more slowly acting endocrine system uses chemical messengers called **hormones,** which are released into the blood to be transported leisurely throughout the body.

Although hormones have widespread and varied effects, the major processes controlled by hormones are reproduction; growth and development; mobilization of body defenses against stressors; maintenance of electrolyte, water, and nutrient balance of the blood; and regulation of cellular metabolism and energy balance. As you can see, the endocrine system regulates processes that go on for relatively long periods and, in some cases, continuously.

THE ENDOCRINE SYSTEM AND HORMONE FUNCTION— AN OVERVIEW

Compared to other organs of the body, the organs of the endocrine system are small and unimpressive. Indeed, to collect 1 kg (about 2.2 pounds) of hormone-producing tissue, you would need to collect *all* the endocrine tissue from eight or nine adults! The endocrine organs also lack the structural or anatomical continuity typical of most organ systems. Instead, bits and pieces of endocrine tissue are tucked away in widely separated regions of the body (see Figure 9.3, p. 271). However, functionally the endocrine organs are very impressive, and when their role in maintaining body homeostasis is considered, they are true giants.

The Chemistry of Hormones

The key to the incredible power of the endocrine glands is the hormones they produce and secrete. Hormones may be defined as chemical substances, secreted by cells into the extracellular fluids, that regulate the metabolic activity of other cells in the body. Although many different hormones are produced, nearly all of them can be classified chemically as either *amino acid–based* molecules (including proteins, peptides, and amines) or *steroids.* Steroid hormones (made from cholesterol) include the sex hormones made by the gonads (ovaries and testes) and the hormones produced by the adrenal cortex. All others are nonsteroidal amino acid derivatives. If we also consider the local hormones called **prostaglandins** (pros″tah-glan′dinz), described later in the chapter (Table 9.2, p. 287), we must add a third chemical class, because the prostaglandins are made from highly active lipids found in the cells' plasma membranes.

Mechanisms of Hormone Action

Although the blood-borne hormones circulate to all the organs of the body, a given hormone affects only certain tissue cells or organs, referred to as its

(a) Steroid hormone action   **(b) Nonsteroid hormone action**

Figure 9.1 **Mechanisms of hormone action.** (**a**) Direct gene activation: the steroid hormone mechanism. (**b**) A second messenger system: the nonsteroid (amino acid-based) hormone mechanism.

target cells or **target organs.** In order for a target cell to respond to a hormone, specific protein receptors must be present on its plasma membrane or in its interior, to which *that* hormone can attach. Only when this binding occurs can the hormone influence the workings of a cell.

The term *hormone* comes from a Greek word meaning "to arouse." In fact, the body's hormones do just that. They "arouse" or bring about their effects on the body's cells primarily by *altering* cellular activity—that is, by increasing or decreasing the rate of a normal, or usual, metabolic process rather than by stimulating a new one. The precise changes that follow hormone binding depend on the specific hormone and the target cell type, but typically one or more of the following occurs:

1. Changes in plasma membrane permeability or electrical state.
2. Synthesis of proteins or certain regulatory molecules (such as enzymes) in the cell.
3. Activation or inactivation of enzymes.
4. Stimulation of mitosis.

Despite the huge variety of hormones, there are really only two mechanisms by which hormones trigger changes in cells. Steroidal hormones use the mechanism shown in Figure 9.1a. As lipid-soluble molecules, the steroid hormones can (1) diffuse through the plasma membranes of their target cells. Once inside, the steroid hormone (2) enters the nucleus, and (3) binds to a specific receptor protein there. The hormone-receptor complex then (4) binds to specific sites on the cell's DNA, (5) activating certain genes to transcribe messenger RNA (mRNA). The mRNA then is (6) translated in the cytoplasm, resulting in the synthesis of new proteins.

Nonsteroidal hormones are unable to enter the target cells and instead bind to receptors situated on the target cell's plasma membrane. In these cases, (Figure 9.1b), (1) the hormone binds to the membrane receptor, (2) setting off a series of reactions that activates an enzyme. The enzyme, in turn, (3) catalyzes a reaction that produces a second messenger molecule (in this case, *cyclic AMP*) that (4) oversees additional intracellular changes that promote the typical response of the target cell to the hormone. As you might guess, there are a variety of possible second messengers (including *G proteins* and *calcium ions*) and many possible target cell responses to the same hormone, depending on the tissue type stimulated.

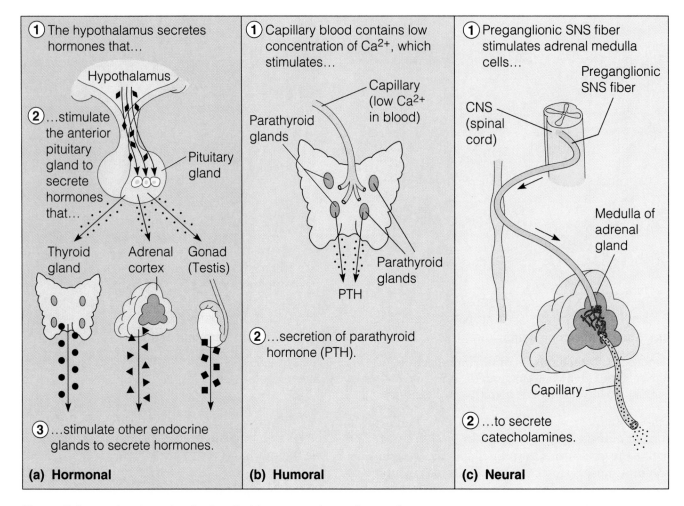

① The hypothalamus secretes hormones that...

Hypothalamus

② ...stimulate the anterior pituitary gland to secrete hormones that...

Pituitary gland

Thyroid gland   Adrenal cortex   Gonad (Testis)

③ ...stimulate other endocrine glands to secrete hormones.

(a) Hormonal

① Capillary blood contains low concentration of Ca²⁺, which stimulates...

Capillary (low Ca²⁺ in blood)

Parathyroid glands

Parathyroid glands

PTH

② ...secretion of parathyroid hormone (PTH).

(b) Humoral

① Preganglionic SNS fiber stimulates adrenal medulla cells...

Preganglionic SNS fiber

CNS (spinal cord)

Medulla of adrenal gland

Capillary

② ...to secrete catecholamines.

(c) Neural

Figure 9.2 **Endocrine gland stimuli.** (**a**) Hormonal stimulus. In this example, hormones released by the hypothalamus stimulate the anterior pituitary to release hormones that stimulate other endocrine organs to secrete hormones. (**b**) Humoral stimulus. Low blood calcium levels trigger parathyroid hormone (PTH) release from the parathyroid glands. PTH causes blood calcium levels to rise by stimulating release of Ca^{2+} from bone. Consequently, the stimulus for PTH secretion ends. (**c**) Neural stimulus. The stimulation of adrenal medullary cells by sympathetic nervous system (SNS) fibers triggers the release of catecholamines (epinephrine and norepinephrine) to the blood.

Control of Hormone Release

Now that we've discussed *how* hormones work, the next question is, "What prompts the endocrine glands to release or not release their hormones?" Let's take a look.

Negative feedback mechanisms are the chief means of regulating blood levels of nearly all hormones (see Chapter 1, p. 10). In such systems, hormone secretion is triggered by some internal or external stimulus; then rising hormone levels inhibit further hormone release (even while promoting responses in their target organs). As a result, blood levels of many hormones vary only within a very narrow range.

The stimuli that activate the endocrine organs fall into three major categories—hormonal, humoral, and neural (see Figure 9.2). The most common stimulus is a *hormonal stimulus,* in which endocrine organs are prodded into action by other hormones. For example, hypothalamic hormones stimulate the anterior pituitary gland to secrete its hormones, and many anterior pituitary hormones stimulate other endocrine organs to release their hormones into the blood (Figure 9.2a). As the hormones produced by the final target glands increase in the blood, they "feed back" to inhibit the release of anterior pituitary hormones and thus their own release. Hormone release promoted by this mechanism tends to be rhythmic, with hormone blood levels rising and falling again and again.

Changing blood levels of certain ions and nutrients may also stimulate hormone release. Such stimuli are referred to as *humoral* (hyoo-mor′al) *stimuli* to distinguish them from hormonal stimuli, which are also blood-borne chemicals. The term "humoral" refers to the ancient use of *humor* to indicate the various body fluids (blood, bile, and others). For example, the release of parathyroid hormone (PTH) by cells of the parathyroid glands is prompted by decreasing blood calcium levels. Because PTH acts by several routes to reverse that decline, blood Ca^{2+} levels soon rise, ending the stimulus for PTH release (Figure 9.2b). Other hormones released in response to humoral stimuli include calcitonin, released by the thyroid gland, and insulin, produced by the pancreas.

In isolated cases, nerve fibers stimulate hormone release and the target cells are said to respond to *neural stimuli*. The classic example is sympathetic nervous system stimulation of the adrenal medulla to release norepinephrine and epinephrine during periods of stress (Figure 9.2c). Although these three mechanisms typify most systems that control hormone release, they by no means explain all of them, and some endocrine organs respond to many different stimuli.

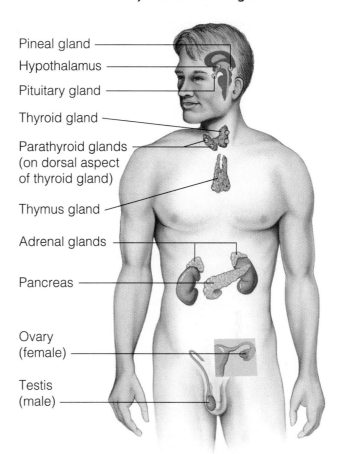

Pineal gland
Hypothalamus
Pituitary gland
Thyroid gland
Parathyroid glands (on dorsal aspect of thyroid gland)
Thymus gland
Adrenal glands
Pancreas
Ovary (female)
Testis (male)

Figure 9.3   Location of the major endocrine organs of the body.

THE MAJOR ENDOCRINE ORGANS

The major endocrine organs of the body include the pituitary, thyroid, parathyroid, adrenal, pineal, and thymus glands, the pancreas, and the gonads (ovaries and testes) (Figure 9.3). The hypothalamus, which is part of the nervous system, is also recognized as a major endocrine organ because it produces several hormones. Although the function of some hormone-producing glands (the anterior pituitary, thyroid, adrenals, and parathyroids) is purely endocrine, the function of others (pancreas and gonads) is mixed—both endocrine and exocrine. Both types of glands are formed from epithelial tissue, but the endocrine glands are **ductless glands** that produce hormones which they release into the blood or lymph. (As you might expect, the endocrine glands have a very rich blood supply.) Conversely, the exocrine glands release their products at the body's surface or into body cavities through ducts. The formation of and differences and similarities between these two types of glands have already been discussed in Chapter 3. Here we will direct our attention to the endocrine glands only. Besides the more detailed descriptions of the endocrine organs provided next, a summary of their hormones' main actions and regulatory factors appears in Table 9.1 (p. 272–273).

Table 9.1 Major Endocrine Glands and Some of Their Hormones

Gland	Hormone	Chemical class	Representative actions	Regulated by
Pineal gland	Melatonin	Amine	Involved in rhythmic activities (daily and seasonal)	Light/dark cycles
Hypothalamus	Hormones released by the posterior pituitary inhibiting and releasing hormones that regulate the anterior pituitary (see below)			
Pituitary gland				
• Posterior lobe (releases hormones made by the hypothalamus)	Oxytocin	Peptide	Stimulates contraction of uterus and the milk "letdown" reflex	Nervous system (hypothalamus)
	Antidiuretic hormone (ADH)	Peptide	Promotes retention of water by kidneys	Water/salt balance
• Anterior lobe	Growth hormone (GH)	Protein	Stimulates growth (especially of bones and muscles) and metabolism	Hypothalamic hormones
	Prolactin (PRL)	Protein	Stimulates milk production	Hypothalamic hormones
	Follicle-stimulating hormone (FSH)	Protein	Stimulates production of ova and sperm	Hypothalamic hormones
	Luteinizing hormone (LH)	Protein	Stimulates ovaries and testes	Hypothalamic hormones
	Thyroid-stimulating hormone (TSH)	Protein	Stimulates thyroid gland	Thyroxine in blood; hypothalamic hormones
	Adrenocorticotropic hormone (ACTH)	Protein	Stimulates adrenal cortex to secrete glucocorticoids	Glucocorticoids; hypothalamic hormones
Thyroid gland	Thyroxine (T_4) and tri-iodothyronine (T_3)	Amine	Stimulates metabolism	TSH
	Calcitonin	Peptide	Reduces blood calcium level	Calcium in blood
Parathyroid glands	Parathyroid hormone (PTH)	Peptide	Raises blood calcium level	Calcium in blood
Thymus	Thymosin	Peptide	"Programs" T cells	Not known
Adrenal glands				
• Adrenal medulla	Epinephrine and norepinephrine	Amines	Increase blood glucose; increase metabolic activities; constrict certain blood vessels	Nervous system
• Adrenal cortex	Glucocorticoids	Steroids	Increase blood glucose	ACTH
	Mineralocorticoids	Steroids	Promote reabsorption of Na^+ and excretion of K^+ in kidneys	K^+ (potassium) in blood
			Reduces blood glucose	

Table 9.1   (continued)

Gland	Hormone	Chemical class	Representative actions	Regulated by
Pancreas	Insulin	Protein	Raises blood glucose	Glucose in blood
	Glucagon	Protein		Glucose in blood
Gonads				
• Testes	Androgens	Steroids	Support sperm formation; development and maintenance of male secondary sex characteristics	FSH and LH
• Ovaries	Estrogens	Steroids	Stimulate uterine lining growth; development and maintenance of female secondary sex characteristics	FSH and LH
	Progesterone	Steroid	Promotes growth of uterine lining	FSH and LH

Pituitary Gland

The **pituitary** (pĭ-tu′ĭ-tār″e) **gland** is approximately the size of a grape. It hangs by a stalk from the inferior surface of the hypothalamus of the brain where it is snugly surrounded by the "turk's saddle" of the sphenoid bone. It has two functional lobes—the anterior pituitary (glandular tissue) and the posterior pituitary (nervous tissue).

Hormones of the Anterior Pituitary

As shown in Figure 9.4 on page 274, there are several anterior pituitary hormones that affect many body organs. Two of the six anterior pituitary lobe hormones—growth hormone and prolactin—exert their major effects on nonendocrine targets. The remaining four—thyrotropic hormone, adrenocorticotropic hormone, and the two gonadotropic hormones—are all **tropic** (tro′pik) **hormones.** Tropic hormones stimulate their target organs, which are also endocrine glands, to secrete their hormones, which in turn exert their effects on other body organs and tissues. All anterior pituitary hormones (1) are proteins (or peptides), (2) act

through second-messenger systems, and (3) are regulated by hormonal stimuli and, in most cases, negative feedback.

Growth hormone (GH) is a general metabolic hormone. However, its major effects are directed to the growth of skeletal muscles and long bones of the body, and thus it plays an important role in determining final body size. GH is a protein-sparing and anabolic hormone that causes amino acids to be built into proteins and stimulates most target cells to grow in size and divide. At the same time, it causes fats to be broken down and used for energy, and spares glucose, helping to maintain blood sugar homeostasis.

Both deficits and excesses of GH may result in structural abnormalities. Hyposecretion of GH during childhood leads to *pituitary dwarfism*. Body proportions are fairly normal, but the person as a whole is a living miniature (with a maximum adult height of 4 feet). Hypersecretion during childhood results in *gigantism*. The individual becomes extremely tall; 8–9 feet is common. Again, body proportions are fairly normal. If hypersecretion occurs after long bone growth has ended, *acromegaly* (ak″ro-meg′ah-le) results. The facial

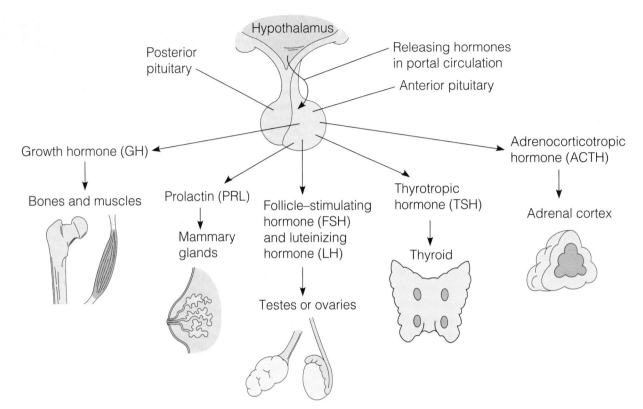

Figure 9.4   Hormones of the anterior pituitary and their major target organs. Secretion of anterior pituitary hormones is stimulated by releasing hormones secreted by hypothalamic neurons. The releasing hormones are secreted into a capillary network that connects via portal veins to a second capillary bed in the anterior lobe of the pituitary gland.

bones, particularly the lower jaw and the bony ridges underlying the eyebrows, enlarge tremendously, as do the feet and hands (see Figure 9.5). Thickening of soft tissues leads to coarse or malformed facial features. Most cases of hypersecretion by endocrine organs (the pituitary and the other endocrine organs) result from tumors of the affected gland. The tumor cells act in much the same way as the normal glandular cells do; that is, they produce the hormones normally made by that gland. ❑

Prolactin (PRL) is a protein hormone structurally similar to growth hormone. Its only known target in humans is the breast (*pro* = for; *lact* = milk). After childbirth, it stimulates and maintains milk production by the mother's breasts. Its function in males is not known.

Adrenocorticotropic (ad-re″no-kor″tĭ-ko-trop′ik) **hormone (ACTH)** regulates the endocrine activity of the cortex portion of the adrenal gland. **Thyroid-stimulating hormone (TSH),** also called **thyrotropic hormone (TH),** influences the growth and activity of the thyroid gland.

The **gonadotropic** (gon″ă-do-trop′ik) **hormones** regulate the hormonal activity of the gonads (ovaries and testes). In females, **follicle-stimulating hormone (FSH)** stimulates follicle development in the ovaries. As the follicles mature, they produce estrogen, and eggs are readied for ovulation. In males, FSH stimulates sperm development by the testes. **Luteinizing** (lu′te-nīz-ing) **hormone (LH)** triggers ovulation of an egg from the female ovary and causes the ruptured follicle to be converted to a **corpus luteum** (lu′te-um). It then stimulates the corpus luteum to produce progesterone and some estrogen. In men, LH is also referred to as **interstitial cell–stimulating hormone (ICSH)** because it stimulates testosterone production by the interstitial cells of the testes.

Hyposecretion of FSH or LH leads to sterility in both males and females. In general, hypersecretion does not appear to cause any problems. However, some drugs used to promote fertility stimulate the release of the gonadotropic hormones, and multiple births (indicating multiple

Figure 9.5   An individual with acromegaly. This condition results from hypersecretion of growth hormone in the adult. Note the enlarged jaw, nose, and hands. Left to right, the same person is shown at ages 16, 33, and 52.

ovulations at the same time rather than the usual single ovulation each month) are fairly common after their use. ❏

Pituitary-Hypothalamus Relationship

Despite its insignificant size, the anterior pituitary gland controls the activity of so many other endocrine glands that it has often been called the "master endocrine gland." Its removal or destruction has a dramatic effect on the body. The adrenal and thyroid glands and gonads atrophy, and results of hyposecretion by those glands quickly become obvious. However, the anterior pituitary is not as all-powerful in its control as it might appear because the release of each of its hormones is controlled by **releasing** and **inhibiting hormones** produced by the hypothalamus. The hypothalamus liberates these regulatory hormones into the blood of the *portal circulation,* which connects the blood supply of the hypothalamus with that of the anterior pituitary.

The hypothalamus also makes two additional hormones, oxytocin and antidiuretic hormone, which are transported along the axons of the hypothalamic **neurosecretory cells** to the posterior pituitary for storage (see Figure 9.6). They are later released into the blood in response to nerve impulses from the hypothalamus.

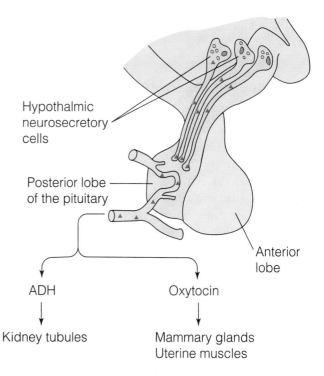

Figure 9.6   Hormones released by the posterior lobe of the pituitary, and their target organs. Neurosecretory cells in the hypothalamus synthesize oxytocin and antidiuretic hormone (ADH) and transport them down their axons to the posterior pituitary. There, the hormones are stored until their release is triggered by nerve impulses from the hypothalamus.

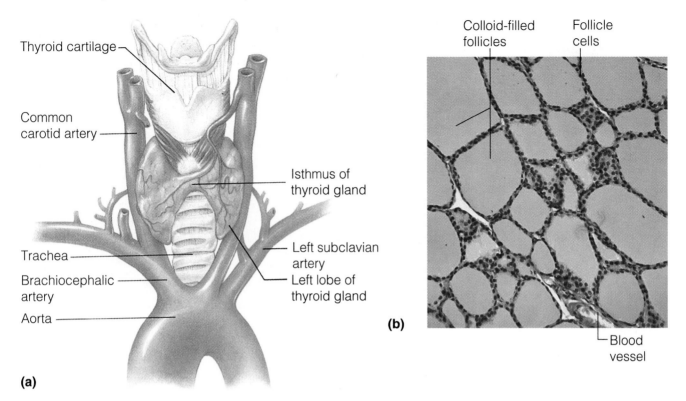

Figure 9.7   Gross anatomy of the thyroid gland. (**a**) Location of the thyroid gland, anterior view. (**b**) Photomicrograph of the thyroid gland (150×).

Hormones of the Posterior Pituitary

The posterior pituitary is not an endocrine gland in the strict sense because it *does not make* the peptide hormones it releases. Instead, as mentioned above, it simply acts as a storage area for hormones made by hypothalamic neurons.

Oxytocin (ok″se-to′sin) is released in significant amounts only during childbirth and in nursing women. It stimulates powerful contractions of the uterine muscle during labor, sexual relations, and when a woman breast feeds her baby. It also causes milk ejection (the *letdown reflex*) in a nursing woman. Both natural and synthetic oxytocic drugs (Pitocin and others) are used to induce labor or to hasten labor that is progressing normally but at a slow pace. Less frequently, oxytocics are used to stop postpartum bleeding (by causing constriction of the ruptured blood vessels at the placental site) and to stimulate the milk ejection reflex.

The second hormone released by the posterior pituitary is **antidiuretic** (an″ti-di″u-ret′ik) **hormone (ADH).** *Diuresis* is urine production. Thus, an antidiuretic is a chemical that inhibits or prevents urine production. ADH causes the kidneys to reabsorb more water from the forming urine; as a

result, urine volume decreases and blood volume increases. In large amounts, ADH also increases blood pressure by causing constriction of the arterioles (small arteries). For this reason, it is sometimes referred to as **vasopressin** (vas″o-pres′in).

Drinking alcoholic beverages inhibits ADH secretion and results in output of large amounts of urine. The dry mouth and intense thirst experienced "the morning after" reflects this dehydrating effect of alcohol. Certain drugs, classed together as *diuretics,* antagonize the effects of ADH, causing water to be flushed from the body. These drugs are used to manage the edema (water retention in tissues) typical of congestive heart failure.

Hyposecretion of ADH leads to a condition of excessive urine output called *diabetes insipidus* (di″ah-be′tez in-sip′ĭ-dus). People with this problem are continually thirsty and drink huge amounts of water. ❏

Thyroid Gland

The **thyroid gland** is a hormone-producing gland that is familiar to most people primarily because many obese individuals blame their overweight

Figure 9.8   Goiter. An enlarged thyroid (goiter) of a Malaysian woman.

condition on their "glands" (meaning the thyroid). Actually, the effect of thyroid hormones on body weight is not as great as many believe it to be.

The thyroid gland is located at the base of the throat, just inferior to the Adam's apple, where it is easily palpated during a physical examination. It is a fairly large gland consisting of two lobes joined by a central mass, or *isthmus* (see Figure 9.7a). The thyroid gland makes two hormones, one called *thyroid hormone,* the other called *calcitonin.* Internally, the thyroid gland is composed of hollow structures called *follicles* that store a sticky colloidal material (Figure 9.7b). Thyroid hormone is derived from this colloid.

Thyroid hormone, often referred to as the body's major metabolic hormone, is actually two active iodine-containing hormones, **thyroxine** (thi-rok′sin), or **T₄,** and **triiodothyronine** (tri″i″o-do-thi′ro-nēn), or **T₃.** Thyroxine is the major hormone secreted by the thyroid follicles. Most T_3 is formed at the target tissues by conversion of T_4 to T_3. These two hormones are very much alike. Each is constructed from two tyrosine amino acids linked together, but thyroxine has four bound iodine atoms, whereas triiodothyronine has three (thus, T_4 and T_3).

Thyroid hormone controls the rate at which glucose is "burned," or oxidized, and converted to body heat and chemical energy. Since all body cells depend on a continuous supply of chemical energy to power their activities, every cell in the body is a target. Thyroid hormone is also important for normal tissue growth and development, especially in the reproductive and nervous systems.

Without iodine, functional hormones cannot be made. The source of iodine is our diet, and the foods richest in iodine are seafoods. Years ago many people who lived in the Midwest, in areas with iodine-deficient soil that were far from the seashore (and a supply of fresh seafood), developed **goiters** (goy′terz). That region of the country came to be known as the "goiter belt." A goiter is an enlargement of the thyroid gland (Figure 9.8) which results when the diet is deficient in iodine. TSH keeps "calling" for thyroxine, and the thyroid gland continues to enlarge so that it can put out more thyroxine. Since there is no iodine, however, only the peptide part of the molecule is produced, which is nonfunctional and so fails to provide negative feedback to inhibit TSH release. Simple goiter is uncommon in the United States today because most of our salt is iodized, but it is still a problem in some other areas of the world.

Hyposecretion of thyroxine may also reflect problems other than iodine deficiency, such as lack of TH stimulation. If it occurs in early childhood, the result is *cretinism* (kre′tin-izm). Cretinism results in dwarfism in which adult body proportions remain childlike. The head and trunk are about 1½ times the length of the legs rather than being approximately the same length, as in normal adults. Untreated cretins are mentally retarded. Their hair is scanty, and their skin is very dry. If the hyposecretion problem is discovered early, hormone replacement will prevent mental retardation and other signs and symptoms of the deficiency. Hypothyroidism occurring in adults results in *myxedema* (mik″se-de′mah), in which there is both physical and mental sluggishness (however, mental retardation does not occur). Other signs are a puffiness of the face, fatigue, poor muscle tone, low body temperature (the person is always cold), obesity, and dry skin. Oral thyroxine is prescribed to treat this condition.

Hyperthyroidism generally results from a tumor of the thyroid gland. Extreme overproduction of thyroxine results in a high basal metabolic rate, intolerance of heat, rapid heart beat, weight loss, nervous and agitated behavior, and a general inability to relax. *Graves' disease* is one form of hyperthyroidism. In addition to the symptoms of hyperthyroidism given earlier, the thyroid gland enlarges and the eyes may bulge, or protrude anteriorly (a condition called *exophthalmos* (ek″sof-thal′mos). Hyperthyroidism may be treated surgically by removal of part of the thyroid (and/or a tumor if present) or chemically by administration of thyroid-blocking drugs or radioactive iodine, which destroys some of the thyroid cells. ❑

The second important hormone product of the thyroid gland, **calcitonin,** or **thyrocalcitonin,** decreases blood calcium levels by causing calcium to be deposited in the bones. It acts antagonistically to parathyroid hormone, the hormone produced by the parathyroid glands. Whereas thyroxine is made and stored in follicles before it is released to the blood, calcitonin is made by the so-called *C cells* found in the connective tissue *between* the follicles. It is released directly to the blood in response to increasing levels of blood calcium. Few effects of hypo- or hypersecretion of calcitonin are known, but it is believed that calcitonin production ceases in elderly adults. This may help to explain (at least in part) the progressive decalcification of bones that accompanies aging.

Parathyroid Glands

The **parathyroid glands** are tiny masses of glandular tissue found on the posterior surface of the thyroid gland (see Figure 9.3). Typically, there are two glands on each thyroid lobe; that is, a total of four glands. They secrete **parathyroid hormone (PTH),** or **parathormone** (par″ah-thor′mōn), which is the most important regulator of calcium ion (Ca^{2+}) homeostasis of the blood. When blood calcium levels drop below a certain level, the parathyroids release PTH, which stimulates bone destruction cells (osteoclasts) to break down bone matrix and release calcium into the blood. Thus, PTH is a *hypercalcemic* hormone (that is, it acts to increase blood levels of calcium), whereas calcitonin is a *hypocalcemic* hormone. The negative feedback interaction between these two hormones as they control blood calcium level is illustrated in Figure 9.9. Although the skeleton is the major PTH target, PTH also stimulates the kidneys and intestine to absorb more calcium (from forming urine and foodstuffs, respectively).

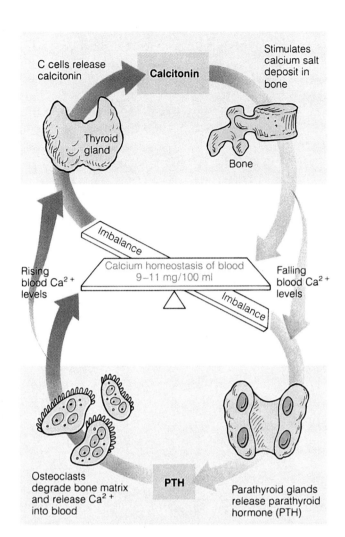

Figure 9.9   Hormonal controls of ionic calcium levels in the blood. PTH and calcitonin operate in negative feedback control systems that influence each other.

If blood calcium levels fall too low, neurons become extremely irritable and overactive. They deliver impulses to the muscles at such a rapid rate that the muscles go into uncontrollable spasms *(tetany),* which may be fatal. Before surgeons knew the importance of these tiny glands on the backside of the thyroid, they would remove a hyperthyroid patient's gland entirely. Many times this resulted in death. Once it was revealed that the parathyroids are functionally very different from the thyroid gland, surgeons began to leave at least some parathyroid-containing tissue (if at all possible) to take care of blood calcium homeostasis.

Severe hyperparathyroidism causes massive bone destruction—an X-ray examination of the bones shows large punched-out holes in the bony matrix. The bones become very fragile, and spontaneous fractures begin to occur. ❑

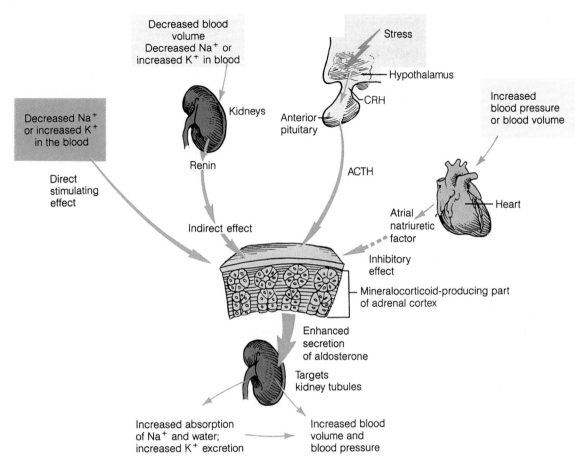

Figure 9.10   Major mechanisms controlling aldosterone release from the adrenal cortex. Solid arrows indicate factors that stimulate aldosterone release; dashed arrow indicates an inhibitory factor.

Adrenal Glands

As illustrated in Figure 9.3, the two bean-shaped **adrenal glands** curve over the top of the kidneys. Although the adrenal gland looks like a single organ, it is structurally and functionally two endocrine organs in one. Much like the pituitary gland, it has glandular (cortex) and neural tissue (medulla) parts. The central medulla region is enclosed by the adrenal cortex, which contains three separate layers of cells.

Hormones of the Adrenal Cortex

The **adrenal cortex** produces three major groups of steroid hormones collectively called **corticosteroids** (kor″ti-ko-ster′oidz)—mineralocorticoids, glucocorticoids, and sex hormones.

The **mineralocorticoids,** mainly **aldosterone** (al″dos-ter′ōn), are produced by the outermost adrenal cortex cell layer. As their name suggests, the mineralocorticoids are important in regulating the mineral (or salt) content of the blood, particularly the concentrations of sodium and potassium ions. Their target is the kidney tubules that selectively reabsorb the minerals or allow them to be flushed out of the body in urine. When blood levels of aldosterone rise, the kidney tubule cells reclaim increasing amounts of sodium ions and allow potassium ions to go out in urine. When sodium is reabsorbed, water follows. Thus, the mineralocorticoids help regulate both water and electrolyte balance in body fluids. As shown in Figure 9.10, the release of aldosterone is stimulated by humoral factors such as fewer sodium ions or more potassium ions in the blood (and by ACTH to a lesser degree). **Renin,** an enzyme produced by the kidneys when blood pressure drops, also causes the release of aldosterone. A hormone released by the heart, *atrial natriuretic* (na″tre-u-ret′ik) *factor,* prevents aldosterone release, its goal being to *reduce* blood volume and blood pressure.

The middle cortical layer produces the **gluco-corticoids,** which include **cortisone** and **cortisol.** Glucocorticoids promote normal cell metabolism and help the body to resist *long-term stressors,* primarily by increasing blood glucose levels. When blood levels of glucocorticoids are high, fats and even proteins are broken down by body cells and converted to glucose, which is released to the blood. For this reason, glucocorticoids are said to be *hyperglycemic hormones.* Glucocorticoids also seem to control the more unpleasant effects of inflammation by decreasing edema, and they reduce pain by inhibiting some pain-causing molecules called *prostaglandins* (see Table 9.2). Because of their anti-inflammatory properties, glucocorticoids are often prescribed as drugs for patients with rheumatoid arthritis, to suppress inflammation. Glucocorticoids are released from the adrenal cortex in response to increased blood levels of ACTH.

Regardless of one's gender, both male and female **sex hormones** are produced by the adrenal cortex throughout life in relatively small amounts. Although the bulk of the sex hormones produced by the innermost cortex layer are **androgens** (male sex hormones), some **estrogens** (female sex hormones) are also formed.

A generalized hyposecretion of all the adrenal cortex hormones leads to *Addison's disease.* A major sign of Addison's disease is a peculiar bronze tone of the skin. Because aldosterone levels are low, sodium and water are lost from the body, which leads to problems with electrolyte and water balance. This, in turn, causes the muscles to become weak, and shock is a possibility. Other signs and symptoms of Addison's disease include those resulting from deficient levels of glucocorticoids, such as hypoglycemia, a lessened ability to cope with stress (burnout), and suppression of the immune system and increased susceptibility to infection. A complete lack of glucocorticoids is incompatible with life.

Hypersecretion problems are generally the result of tumors, and the resulting condition depends on the cortex area involved. Hyperactivity of the outermost cortical area results in *hyperaldosteronism* (hi″per-al″dos-ter′on-izm). Excessive water and sodium are retained, leading to high blood pressure and edema, and potassium is lost to such an extent that the activity of the heart and nervous system may be disrupted. When the tumor is in the middle cortical area, **Cushing's syndrome** oc-curs. Excessive output of glucocorticoids results in a "moon face," and the appearance of a "buffalo hump" of fat on the upper back (see Figure 9.11). Other common and undesirable effects include high blood pressure, hyperglycemia and possible diabetes, weakening of the bones (as protein is withdrawn to be converted to glucose), and severe depression of the immune system. Hypersecretion of the sex hormones leads to *masculinization,* regardless of sex. In adult males these effects may be masked, but in females the results are often dramatic. A beard develops and a masculine pattern of body hair distribution occurs, among other things. ❏

Hormones of the Adrenal Medulla

The **adrenal medulla,** like the posterior pituitary, develops from a knot of nervous tissue. When the medulla is stimulated by sympathetic nervous system neurons, its cells release two similar hormones, **epinephrine** (ep″ĭ-nef′rin), also called *adrenaline,* and **norepinephrine** *(noradrenaline),* into the bloodstream. Collectively, these hormones are referred to as **catecholamines** (kat″ĕ-kol-ah′menz). Since some sympathetic neurons also release norepinephrine as a neurotransmitter, the adrenal medulla is often thought of as a "misplaced sympathetic nervous system ganglion."

When you are (or feel) threatened physically or emotionally, your sympathetic nervous system brings about the "fight-or-flight" response to help you cope with the stressful situation. One of the organs it stimulates is the adrenal medulla, which literally pumps its hormones into the bloodstream to enhance and prolong the effects of the neurotransmitters of the sympathetic nervous system. Basically, the catecholamines increase heart rate, blood pressure, and blood glucose levels, and dilate the small passageways of the lungs. These events result in more oxygen and glucose in the blood and a faster circulation of blood to the body organs (most importantly, to the brain, muscles, and heart). Thus, the body is better able to deal with a short-term stressor whether the job at hand is to fight, begin the inflammatory process, or make you more alert so you think more clearly.

The catecholamines of the adrenal medulla prepare the body to cope with a brief or short-term stressful situation and cause the so-called *alarm stage* of the stress response. Glucocorticoids, by

(a)

(b)

Figure 9.11   Appearance of a woman (a) before and (b) during Cushing's disease.

contrast, are produced by the adrenal cortex and are more important in helping the body to cope with prolonged or continuing stressors, such as dealing with the death of a family member or having a major operation. Glucocorticoids operate primarily during the *resistance stage* of the stress response. If they are successful in protecting the body, the problem will eventually be resolved without lasting damage to the body. When the stress continues on and on, the adrenal cortex may simply "burn out," which is usually fatal. The relationship of catecholamines and glucocorticoids in the stress response is shown in Figure 9.12 on p. 282.

Damage or destruction of the adrenal medulla has no major effects as long as the sympathetic nervous system neurons continue to function normally. However, hypersecretion of catecholamines leads to symptoms typical of excessive sympathetic nervous system activity—a rapidly beating heart, high blood pressure, and a tendency to perspire and be very irritable. Surgical removal of the adrenal gland corrects this condition. ❏

Pancreatic Islets

The **pancreas,** located close to the stomach in the abdominal cavity (see Figure 9.3), is a mixed gland. Probably the best-hidden endocrine glands in the body are the **pancreatic islets,** formerly called the *islets of Langerhans* (lahng′er-hanz). These little masses of hormone-producing tissue are scattered among the enzyme-producing tissue of the pancreas. The exocrine (enzyme-producing) part of the pancreas, which acts as part of the digestive system, will be discussed later. Only the pancreatic islets will be considered here.

Although there are more than a million islets, separated by exocrine cells, each of these tiny clumps of cells goes busily about its business manufacturing its hormones and working like an organ within an organ. Two important hormones produced by the islet cells are **insulin** and **glucagon** (gloo′kah-gon). Both help regulate the amount of sugar (glucose) in the blood, but in opposite ways,

Figure 9.12 **Roles of the hypothalamus, adrenal medulla, and adrenal cortex in the stress response.** (Note that ACTH is only a weak stimulator of mineralocorticoid release under normal conditions.)

(see Figure 9.13). The islets also produce small amounts of other hormones, but those will not be discussed here.

High blood glucose levels stimulate the release of insulin from the **beta** (ba′tah) **cells** of the islets. Insulin acts on just about all body cells and increases their ability to transport glucose across their plasma membranes. Once inside the cells, glucose is oxidized for energy or converted to glycogen or fat for storage. These activities are also speeded up by insulin. Since insulin sweeps the glucose out of the blood, its effect is said to be *hypoglycemic*. As blood glucose levels fall, the stimulus for insulin release ends—another classic case of negative feedback control. Many hormones

have hyperglycemic effects (glucagon, glucocorticoids, and epinephrine, to name a few), but insulin is the only hormone that decreases blood glucose levels. Insulin is absolutely necessary for the use of glucose by the body cells. Without it, essentially no glucose can get into the cells to be used.

Without insulin, blood levels of glucose (which normally range from 80 to 120 mg/100 ml of blood) rise to dramatically high levels (for example, 600 mg/100 ml of blood). In such instances, glucose begins to spill into the urine because the kidney tubule cells cannot reabsorb it fast enough. As glucose flushes from the body, water follows, leading to dehydration. The clinical name for this condition is **diabetes mellitus** (me-

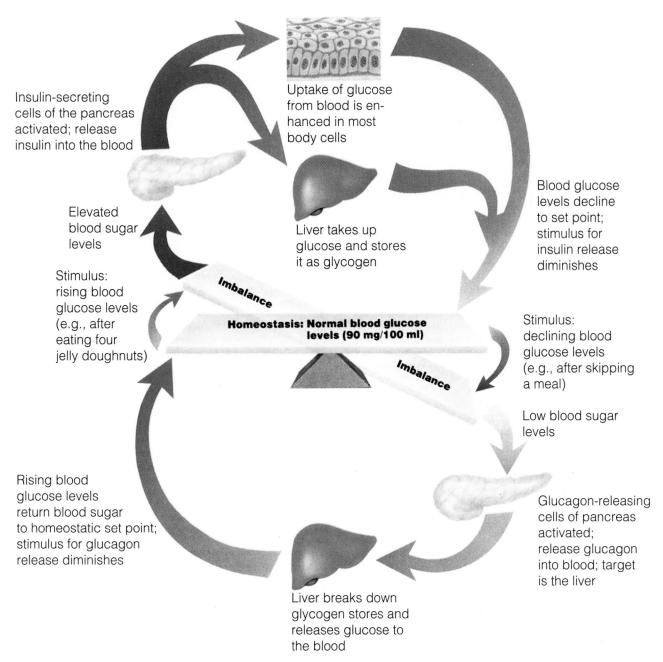

Figure 9.13   Regulation of blood glucose levels by a negative feedback mechanism involving pancreatic hormones.

li'tus), which literally means that something sweet (*mel* = honey) is passing through or siphoning (*diabetes* = Greek "siphon") from the body. Since cells cannot use glucose, fats and even proteins are broken down and used to meet the energy requirements of the body. As a result, weight loss begins. Loss of body proteins leads to a decreased ability to fight infections, so diabetics must be careful with their hygiene and in caring for even small cuts and bruises. When large amounts of fats (instead of sugars) are used for energy, the blood becomes very acidic (*acidosis* [as″ĭ-do′sis]) as ketones (intermediate products of fat breakdown) appear in the urine. Unless corrected, coma and death result. The three cardinal signs of diabetes mellitus are: (1) *polyuria* (pol″e-u′re-ah)—excessive urination to flush out the glucose and ketones; (2) *polydipsia* (pol″e-dip′se-ah)—excessive thirst resulting

A CLOSER LOOK

SUNLIGHT AND THE CLOCK WITHIN

It has been known for a long time that many body rhythms move in step with one another. Body temperature, pulse, and the sleep-wake cycles seem to follow the same "beat" over approximately 24-hour cycles, while other processes follow "a different drummer." What can throw these rhythms out of whack? Illness, drugs, jet travel, and changing to the night shift are all candidates. So is sunlight.

Light exerts its internal biochemical effects through the eye. Light hitting the retina generates nerve impulses along the visual pathway to the visual cortex (where "seeing" occurs) and also to the suprachiasmatic nucleus (SCN), the so-called biological clock of the hypothalamus. Among other things, the SCN regulates the rhythmic melatonin output of the pineal gland and the output of various anterior pituitary hormones. Generally speaking, melatonin release is inhibited by light and triggered by darkness. Historically, the pineal gland has been called the "third eye."

Light produces melatonin-related effects on mating, eating, and sleeping patterns of other animals but, until very recently, humans were believed to have evolved free of such effects. Thanks to the work of many scientists, however, we now know that a number of human processes are influenced by light and melatonin is a major player in many of these responses.

- *Sleep.* If there is one thing that the various researchers studying melatonin agree on, it is that melatonin is the sleep "trigger." They suggest that the calcification of the pineal gland that occurs with age can probably explain the problems that some elderly have getting a good night's sleep. Some pharmacies and health food stores are stocking melatonin as a sleep aid. Happily, unlike most sedative drugs, melatonin does not alter the natural sleep pattern.

- *Mood.* Many people have seasonal mood rhythms, particularly those of us who live far from the equator, where the length of a day changes dramatically during the course of a year. We seem to feel better during the summer and become cranky and depressed in the short, gray days of winter. Is this just our imagination?

Apparently not. Researchers have found a relatively rare emotional disorder called *seasonal affective disorder,* or *SAD,* in which these mood swings are grossly exaggerated. As the days grow shorter each fall, people with SAD become irritable, anxious, sleepy, and withdrawn. They crave carbohydrates and gain weight. Phototherapy, the use of very bright lights for two hours daily, reversed these symptoms in nearly 90 percent of patients studied in 2 to 4 days (considerably faster than any antidepressant drug could do it). When patients stopped receiving phototherapy or were given melatonin, their symptoms returned as quickly as they had lifted, indicating that melatonin may be the key to seasonal mood changes.

- *Night work schedules and jet lag.* People who work at night (the "graveyard shift") have reversed melatonin-secretion patterns, with no hormone released during the night (when they are exposed to light) and high levels secreted during daytime sleeping hours. Waking such people and exposing them to bright light causes their melatonin levels to drop. The same sort of melatonin inversion (but much more precipitous) occurs in those who fly across the United States from coast to coast.

A clinician measures the light intensity before administering phototherapy for treatment of SAD.

■ *Immunity*. Ultraviolet (UV) light activates white blood cells called suppressor T cells, which partially block the immune response. UV therapy has been found to stop rejection of tissue transplants from unrelated donors in animals. This technique offers the hope that diabetics who have become immune to their own pancreas islet tissue may be treated with pancreas transplants in the future. Light, of course, dampens melatonin secretion. So the other side of the same coin is that darkness, via its stimulation of melatonin release, enhances and revitalizes the immune system. Indeed, animal experiments have shown that melatonin supplements can reverse some of the age-related shrinking of the thymus (an important programming organ for some immune cells).

■ *Aging*. One of the theories of aging and chronic disease is that free radicals, naturally generated in the body as cells destroy unwanted materials (such as bacteria), are less effectively disarmed as we age, and accelerate the degeneration of the body. The timing of this decline in free radical scavenging correlates closely with the decrease in melatonin release as the pineal gland calcifies.

It now appears that melatonin might well be the most effective antioxidant or destroyer of free radicals. Experiments in which animals were exposed to a toxic herbicide revealed that those injected with melatonin before and after showed virtually none of the lung and liver damage seen in the unprotected animals.

People have worshipped sunlight since the earliest times. Scientists are just now beginning to understand the reasons for this and, as they do, they are increasingly distressed about windowless offices, artificial lighting of work areas, and the growing numbers of institutionalized elderly who rarely feel the sun's warm rays. Artificial lights do not provide the full spectrum of sunlight. Bulbs used in homes provide primarily the red wavelengths, and fluorescent bulbs of institutions provide yellow-green; neither provides the invisible UV or infrared wavelengths that are also components of sunlight. Animals exposed for prolonged periods to artificial lighting exhibit reproductive abnormalities and an enhanced susceptibility to cancer. Could it be that some of us are unknowingly expressing the same effects?

---

from water loss; and (3) *polyphagia* (pol″e-fa′je-ah)—hunger due to inability to use sugars and the loss of fat and proteins from the body. Mild cases of diabetes mellitus (most cases of type II or adult-onset diabetes) are treated with special diets or oral medications that prod the sluggish islets into action. To regulate blood glucose levels in the more severe type I (juvenile or brittle) diabetic, insulin is infused continuously by an insulin pump worn externally, or a regimen of carefully planned insulin injections are administered throughout the day. ❑

As explained earlier, glucagon acts as an antagonist of insulin. Its release by the **alpha** (al′fah) **cells** of the islets is stimulated by low blood levels of glucose. Its action is basically hyperglycemic. Its primary target organ is the liver, which it stimulates to break down stored glycogen to glucose and to release the glucose into the blood. No important disorders resulting from hypo- or hypersecretion of glucagon are known.

Pineal Gland

The **pineal** (pin′e-al) **gland** is a small, cone-shaped gland found in the roof of the third ventricle of the brain (see Figure 9.3). The endocrine function of this tiny gland is still somewhat of a mystery. Although many chemical substances have been identified in the pineal gland, only the hormone **melatonin** (mel″ah-to′nin) appears to be secreted in substantial amounts. The levels of melatonin rise and fall during the course of the day and night, and many believe that melatonin plays an important role in establishing the body's day-night cycle (see the "Closer Look" box). In some animals, melatonin also helps regulate mating behavior and rhythms. In humans, it is believed to coordiate the hormones of fertility and to inhibit the reproductive system (especially the ovaries of females) so that sexual maturation is prevented from occurring during childhood, before adult body size has been reached.

Thymus

The **thymus** is located in the upper thorax, posterior to the sternum. Large in infants and children, it decreases in size throughout adulthood. By old age, it is composed mostly of fibrous connective tissue and fat. The thymus produces a hormone called **thymosin** (thi′mo-sin), and during childhood the thymus acts as an "incubator" for the maturation of a special group of white blood cells (*T lymphocytes*) that are important in the immune response. The role of the thymus (and its hormones) in immunity is described in Chapter 12.

Gonads

The female and male gonads (see Figure 9.3) produce sex hormones that are identical to those produced by adrenal cortex cells. The major differences are the source and relative amounts produced.

Hormones of the Ovaries

The female **gonads** (go′nadz), or **ovaries,** are paired, almond-sized organs located in the pelvic cavity. Besides producing female sex cells (ova, or eggs), ovaries produce two groups of steroid hormones, estrogens and progesterone. The ovaries do not really begin to function until puberty, when the anterior pituitary gonadotropic hormones stimulate them into activity. This results in the rhythmic ovarian cycles in which ova develop and blood levels of ovarian hormones rise and fall.

Estrogens, primarily **estrone** (es′trōn) and **estradiol** (es″trah-di′ol), produced by the **Graafian follicles** of the ovaries, stimulate the development of the secondary sex characteristics in females (primarily growth and maturation of the reproductive organs and the appearance of hair in the pubic and axillary regions). In addition, the estrogens work with progesterone to prepare the uterus to receive a fertilized egg. This results in cyclic changes in the uterine lining, which is called the *menstrual cycle.* Estrogens also help maintain pregnancy and prepare the breasts to produce milk (lactation). However, the placenta and not the ovaries is the source of the estrogens at this time.

Progesterone (pro-jes′tĕ-rōn), as already noted, acts with estrogen to bring about the menstrual cycle. During pregnancy, it quiets the muscles of the uterus so that an implanted embryo will not be aborted and helps prepare breast tissue for lactation. Progesterone is produced by another glandular structure of the ovaries, the **corpus luteum.** The corpus luteum produces both estrogen and progesterone, but progesterone is secreted in larger amounts.

Ovaries are stimulated to release their estrogens and progesterone in a cyclic way by the anterior pituitary gonadotropic hormones. More detail on this feedback cycle and on the structure and function of the ovaries is given in Chapter 16, but it should be obvious that hyposecretion of the ovarian hormones severely hampers the ability of a woman to conceive and bear children.

Hormones of the Testes

The paired oval **testes** of the male are suspended in a sac, the *scrotum,* outside the pelvic cavity. In addition to male sex cells, or *sperm,* the testes also produce male sex hormones, or **androgens,** of which **testosterone** (tes-tos′tĕ-rōn) is the most important. Testosterone, made by the **interstitial cells** of the testes, causes the development of the adult male sex characteristics. It promotes the growth and maturation of the reproductive system organs to prepare the young man for reproduction. It also causes development of the male's secondary sex characteristics (growth of the beard, development of heavy bones and muscles, and lowering of the voice) as well as stimulating the male sex drive.

In adulthood, testosterone is necessary for continuous production of sperm. In cases of hyposecretion, the man becomes sterile; such cases are usually treated by testosterone injections. Both the endocrine and exocrine functions of the testes begin at puberty under the influence of the anterior pituitary gonadotropic hormones. Testosterone production is specifically stimulated by LH. Chapter 16, which deals with the reproductive system, contains more information on the structure and exocrine function of the testes.

OTHER HORMONE-PRODUCING TISSUES AND ORGANS

Besides the major endocrine organs, pockets of hormone-producing cells are found in the walls of

Table 9.2   Hormones Produced by Organs Other Than the Major Endocrine Organs

Hormone	Chemical composition	Source	Stimulus for secretion	Target organ/Effects
Prostaglandins (PGs); several groups indicated by letters A–I (PGA–PGI)	Derived from fatty acid molecules	Plasma membranes of virtually all body cells	Various (local irritation, hormones, etc.)	Have many targets, but act locally at site of release/Examples of effects include: increase blood pressure by acting as vasoconstrictors; cause constriction of respiratory passageways; stimulate muscle of the uterus promoting labor; enhance blood clotting; increase inflammation and pain; increase output of digestive secretions by stomach; cause fever
Gastrin	Peptide	Stomach	Food	Stomach/Stimulates glands to release hydrochloric acid (HCl)
Intestinal gastrin	Peptide	Duodenum of small intestine	Food, especially fats	Stomach/Inhibits HCl secretion and gastrointestinal tract mobility
Secretin	Peptide	Duodenum	Food	Pancreas/Stimulates release of bicarbonate-rich juice Liver/Increases release of bile Stomach/Inhibits secretory activity
Cholecystokinin	Peptide	Duodenum	Food	Pancreas/Stimulates release of enzyme-rich juice Gallbladder/Stimulates expulsion of stored bile Duodenal papilla/Causes sphincter to relax, allowing bile and pancreatic juice to enter duodenum
Erythropoietin	Glycoprotein	Kidney	Hypoxia	Bone marrow/Stimulates production of red blood cells
Active vitamin D_3	Steroid	Kidney (activates vitamin D made by epidermal cells of skin)	PTH	Intestine/Stimulates active transport of dietary calcium across intestinal cell membranes
Atrial natriuretic factor	Peptide	Heart	Stretching of heart	Kidney/Inhibits sodium ion reabsorption and renin release Adrenal cortex/Inhibits secretion of aldosterone

the small intestine, stomach, kidneys, and heart—organs whose chief functions have little to do with hormone production. The placenta, a temporary organ formed during pregnancy, produces hormones generally thought of as ovarian hormones (estrogen and progesterone). Additionally, certain tumor cells, such as those of some lung and pancreatic cancers, make hormones identical to those made in normal endocrine glands, but in an excessive and uncontrolled fashion.

Because most of these hormones are described in later chapters, their chief characteristics are only summarized in Table 9.2. Only the placenta will be considered further here.

Placenta

The **placenta** (plah-sen'tah) is a remarkable organ formed temporarily in the uterus of pregnant women. In addition to its roles as the respiratory, excretory, and nutrition-delivery systems for the fetus, it also produces hormones that help to maintain the pregnancy and pave the way for delivery of the baby.

During very early pregnancy, a hormone called **human chorionic** (ko"re-on'ik) **gonadotropin (hCG)** is produced by the fetal part of the placenta. hCG is similar to LH (luteinizing hormone). It stimulates the corpus luteum of the ovary to *continue* producing estrogen and progesterone so that the lining of the uterus is not sloughed off in menses. In the third month, the placenta assumes the job of producing *estrogen* and *progesterone,* and the ovaries become inactive for the rest of the pregnancy. The high estrogen and progesterone blood levels maintain the lining of the uterus (thus, the pregnancy), and prepare the breasts for producing milk. *Human placental lactogen (hPL)* works cooperatively with estrogen and progesterone in preparing the breasts for lactation, and *relaxin,* another placental hormone, causes the mother's pelvic ligaments and the pubic symphysis to relax and become more flexible, which eases birth passage. The placenta also produces a placental growth hormone and a thyroid-stimulating hormone.

DEVELOPMENTAL ASPECTS OF THE ENDOCRINE SYSTEM

The embryonic development of the endocrine glands varies. The pituitary gland is derived from epithelium of the oral cavity and a neural tissue projection of the hypothalamus. The pineal gland is entirely composed of neural tissue. Most of the strictly epithelial glands appear to develop as little saclike outpocketings of the epithelial lining of the digestive tract. These would include the thyroid,

thymus, and pancreas. The formation of the gonads and the adrenal and parathyroid glands is much more complex, and will not be considered here.

Barring outright malfunctions of the endocrine glands (hypoactivity and hyperactivity), which have already been discussed, most endocrine organs seem to operate smoothly until old age. In late middle age, the efficiency of the ovaries begins to decline. This causes the onset of **menopause** (commonly called "change of life"). During this period, a woman's reproductive organs begin to atrophy, and the ability to bear children ends. Problems associated with estrogen deficiency begin to occur, such as arteriosclerosis, osteoporosis, decreased skin elasticity, and changes in the operation of the sympathetic nervous system that result in "hot flashes." In addition, fatigue, nervousness, and mood changes such as depression are common. No such dramatic changes seem to happen in men. In fact, many men remain fertile throughout their lifespan, indicating that testosterone is still being produced in adequate amounts.

While striking changes occur in aging women, due to decreasing levels of female hormones, many other much less noticeable hormone-related changes occur in both sexes. The efficiency of the endocrine system as a whole gradually declines in old age. There is no question that growth hormone output by the anterior pituitary declines, which partially explains muscle atrophy in old age. Since the anterior pituitary affects so many other endocrine glands through its tropic hormones, it is assumed that its target organs also become less productive; for example, older people are often mildly hypothyroid. It is common knowledge that elderly persons are less able to resist stress and infection. This may result from overproduction or underproduction of the defensive hormones, since both "derail" the stress defense equilibrium and alter general body metabolism. However, it now appears likely that at least part of the decline in immune function typical of the elderly results from the waning melatonin production of the pineal gland, which calcifies with age. There is little doubt that melatonin plays a positive role in promoting the immune response. All older people have some decline in insulin production, and adult-onset diabetes is most common in the elderly.

SYSTEMS IN SYNC

Homeostatic Relationships Between the *Endocrine System* and Other Body Systems

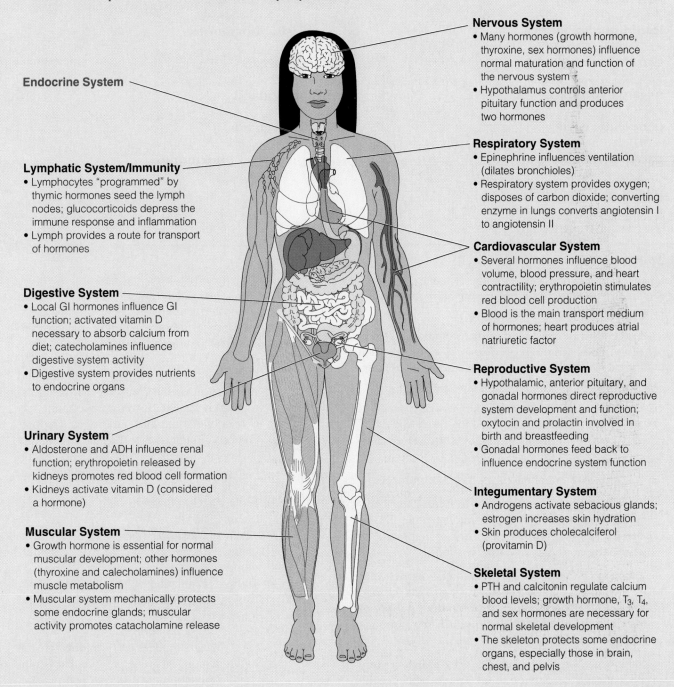

Nervous System
- Many hormones (growth hormone, thyroxine, sex hormones) influence normal maturation and function of the nervous system
- Hypothalamus controls anterior pituitary function and produces two hormones

Respiratory System
- Epinephrine influences ventilation (dilates bronchioles)
- Respiratory system provides oxygen; disposes of carbon dioxide; converting enzyme in lungs converts angiotensin I to angiotensin II

Cardiovascular System
- Several hormones influence blood volume, blood pressure, and heart contractility; erythropoietin stimulates red blood cell production
- Blood is the main transport medium of hormones; heart produces atrial natriuretic factor

Reproductive System
- Hypothalamic, anterior pituitary, and gonadal hormones direct reproductive system development and function; oxytocin and prolactin involved in birth and breastfeeding
- Gonadal hormones feed back to influence endocrine system function

Integumentary System
- Androgens activate sebacious glands; estrogen increases skin hydration
- Skin produces cholecalciferol (provitamin D)

Skeletal System
- PTH and calcitonin regulate calcium blood levels; growth hormone, T_3, T_4, and sex hormones are necessary for normal skeletal development
- The skeleton protects some endocrine organs, especially those in brain, chest, and pelvis

Endocrine System

Lymphatic System/Immunity
- Lymphocytes "programmed" by thymic hormones seed the lymph nodes; glucocorticoids depress the immune response and inflammation
- Lymph provides a route for transport of hormones

Digestive System
- Local GI hormones influence GI function; activated vitamin D necessary to absorb calcium from diet; catecholamines influence digestive system activity
- Digestive system provides nutrients to endocrine organs

Urinary System
- Aldosterone and ADH influence renal function; erythropoietin released by kidneys promotes red blood cell formation
- Kidneys activate vitamin D (considered a hormone)

Muscular System
- Growth hormone is essential for normal muscular development; other hormones (thyroxine and calecholamines) influence muscle metabolism
- Muscular system mechanically protects some endocrine glands; muscular activity promotes catacholamine release

Important Terms

adrenal cortex

adrenal medulla

inhibiting hormones

gonads (go′nadz)

negative feedback

pancreatic islets

parathyroid glands

pineal (pin′e-al) gland

pituitary (pĭ-tu′ĭ-tār″e) gland

placenta (plah-sen′tah)

releasing hormones

target organ

thymus gland

thyroid gland

tropic (tro′pik) hormones

Summary

The Endocrine System and Hormone Function—An Overview (pp. 268-271)

1. The endocrine system is a major controlling system of the body. Through hormones, it stimulates such long-term processes as growth and development, metabolism, reproduction, and body defense.

2. Endocrine organs are small and widely separated in the body. Some are mixed glands (both endocrine and exocrine in function). Others are purely hormone producing.

3. All hormones are amino acid–based or steroids.

4. Endocrine organs are activated to release their hormones into the blood by hormonal, humoral, or neural stimuli. Negative feedback is important in regulating hormone levels in the blood.

5. Blood-borne hormones alter the metabolic activities of their target organs. The ability of a target organ to respond to a hormone depends on the presence of receptors in or on its cells to which the hormone binds or attaches.

6. Amino acid–based hormones act through second messengers. Steroid hormones directly influence the DNA.

The Major Endocrine Organs (pp. 271-286)

1. Pituitary gland
   a. The pituitary gland hangs from the base of the brain by a stalk and is enclosed by bone. It consists of a glandular (anterior) portion and a neural (posterior) portion.

   b. Except for growth hormone and prolactin, hormones of the anterior pituitary are all tropic hormones.
   (1) Growth hormone (GH): An anabolic and protein-conserving hormone that promotes total body growth. Its most important effect is on skeletal muscles and bones. Hyposecretion during childhood results in pituitary dwarfism; hypersecretion produces gigantism (in childhood) and acromegaly (in adulthood).
   (2) Prolactin (PRL): Stimulates production of breast milk.
   (3) Adrenocorticotropic hormone (ACTH): Stimulates the adrenal cortex to release its hormones.
   (4) Thyroid-stimulating hormone (TSH): Stimulates the thyroid gland to release thyroid hormone.
   (5) Gonadotropic hormones
      (a) Follicle-stimulating hormone (FSH): Beginning at puberty, stimulates follicle development and estrogen production by the female ovaries; promotes sperm production in the male.
      (b) Luteinizing hormone (LH): Beginning at puberty, stimulates ovulation, converts the ruptured ovarian follicle to a corpus luteum, and causes the corpus luteum to produce progesterone; stim-

ulates the male's testes to produce testosterone.

c. Releasing and inhibiting hormones made by the hypothalamus regulate release of hormones made by the anterior pituitary. Hypothalamus also makes two hormones that are transported to the posterior pituitary for storage and later release.

d. The posterior pituitary stores and releases hypothalamic hormones on command.
   (1) Oxytocin: Stimulates powerful uterine contractions and causes milk ejection in the nursing woman.
   (2) Antidiuretic hormone (ADH): Causes kidney tubule cells to reabsorb and conserve body water and increases blood pressure by constricting arterioles. Hyposecretion leads to diabetes insipidus.

2. Thyroid gland
   a. The thyroid gland is located in the anterior throat.
   b. Thyroid hormone (thyroxine [T_4] and triiodothyronine [T_3]) is released from the thyroid follicles when blood levels of TSH rise. Thyroid hormone is the body's metabolic hormone. It increases the rate at which cells oxidize glucose and is necessary for normal growth and development. Lack of iodine leads to goiter. Hyposecretion of thyroxine results in cretinism in children and myxedema in adults. Hypersecretion results from Graves' disease or other forms of hyperthyroidism.
   c. Calcitonin is released by C cells surrounding the thyroid follicles in response to high blood levels of calcium. It causes calcium to be deposited in bones.

3. Parathyroid glands
   a. The parathyroid glands are four small glands located on the posterior aspect of the thyroid gland.
   b. Low blood levels of calcium stimulate the parathyroid glands to release parathyroid hormone (PTH). It causes bone calcium to be liberated into the blood. Hyposecretion of PTH results in tetany; hypersecretion leads to extreme bone wasting and fractures.

4. Adrenal glands
   a. The adrenal glands are paired glands perched on the kidneys. Each gland has two functional endocrine portions, cortex and medulla.
   b. Three groups of hormones are produced by the adrenal cortex.
      (1) Mineralocorticoids, primarily aldosterone, regulate sodium ion (Na^+) and potassium ion (K^+) reabsorption by the kidneys. Their release is stimulated primarily by low Na^+ and/or high K^+ levels in blood.
      (2) Glucocorticoids enable the body to resist long-term stress by increasing blood glucose levels and decreasing the inflammatory response.
      (3) Sex hormones (mainly male sex hormones) are produced in small amounts throughout life.
   c. Generalized hypoactivity of the adrenal cortex results in Addison's disease. Hypersecretion can result in hyperaldosteronism, Cushing's disease, and/or masculinization.
   d. The adrenal medulla produces catecholamines (epinephrine and norepinephrine) in response to sympathetic nervous system stimulation. Its catecholamines enhance and prolong the effects of the fight-or-flight (sympathetic nervous system) response to short-term stress. Hypersecretion leads to symptoms typical of sympathetic nervous system overactivity.

5. Pancreatic islets
   a. Located in the abdomen close to the stomach, the pancreas is both an exocrine and endocrine gland. The endocrine portion (islets) releases insulin and glucagon to blood.
   b. Insulin is released when blood levels of glucose are high. It increases the rate of glucose uptake and metabolism by body cells. Hyposecretion of insulin results in diabetes mellitus, which severely disturbs body metabolism. Cardinal signs are polyuria, polydipsia, and polyphagia.
   c. Glucagon is released when blood levels of glucose are low. It stimulates the liver to release glucose to blood, thus increasing blood glucose levels.

6. The pineal gland, located in the third ventricle of the brain, releases melatonin, which affects biological rhythms and reproductive behavior.

7. The thymus gland, located in the upper thorax, functions during youth, but atrophies in old age. Its hormone, thymosin, is believed to cause the maturation of T lymphocytes, important in body defense.

8. Gonads
   a. The ovaries of the female, located in the pelvic cavity, release two hormones.
      (1) Estrogens: Release of estrogens by ovarian follicles begins at puberty under the influence of FSH. Estrogens stimulate maturation of the female reproductive organs and development of secondary sex characteristics of the female. With progesterone, they cause the menstrual cycle.

(2) Progesterone: Progesterone is released from the corpus luteum of the ovary in response to high blood levels of LH. It works with estrogens in establishing the menstrual cycle.

b. The testes of the male begin to produce testosterone at puberty in response to LH stimulation. Testosterone promotes maturation of the male reproductive organs, male secondary sex characteristics, and production of sperm by the testes.

c. Hyposecretion of gonadal hormones results in sterility in both females and males.

Other Hormone-Producing Tissues and Organs
(pp. 286–288)

1. The placenta is a temporary organ formed in the uterus of pregnant women. Its primary endocrine role is to produce estrogen and progesterone, which maintain pregnancy and ready breasts for lactation.

2. Several organs that are generally nonendocrine in overall function, such as the stomach, small intestine, kidneys, and heart, have cells that secrete hormones.

3. Certain cancer cells secrete hormones.

Developmental Aspects of the Endocrine System
(p. 288)

1. Excluding pathologic excesses and lack of hormones, efficiency of the endocrine system remains high until old age.

2. Decreasing function of female ovaries at menopause leads to osteoporosis, increased chance of heart disease, and possible mood changes.

3. Efficiency of all endocrine glands gradually decreases with aging, which leads to a generalized increase in incidence of diabetes mellitus, immune system depression, and lower metabolic rate.

Review Questions

1. Explain how the nervous and endocrine systems differ in: (a) the rate of their control, (b) the way in which they communicate with body cells, and (c) the types of body processes they control.

2. Which endocrine organs are actually mixed (endocrine and exocrine) glands? Which are purely endocrine?

3. Define *hormone* and describe the chemical nature of hormones.

4. Name three ways in which endocrine glands are stimulated to release their hormones, and give one example for each way.

5. Define *negative feedback* and explain how it regulates blood levels of the various hormones.

6. Define *target organ* and explain why all organs are not target organs for all hormones.

7. Describe the body location for each of the following endocrine organs: anterior pituitary, pineal gland, thymus, pancreas, ovaries, testes. Then, for each organ, name its hormones and their effect(s) on body processes. Finally, for each hormone, list the important results of its hypersecretion or hyposecretion.

8. Name two endocrine-producing glands (or regions) that are important in the stress response and explain *why* they are important.

9. The anterior pituitary is often referred to as the master endocrine gland, but it too has a "master." What controls the release of hormones by the anterior pituitary?

10. What are tropic hormones?

11. The posterior pituitary is not really an endocrine gland. Why not? What is it?

12. What is the most common cause of hypersecretion by endocrine organs?

13. Name three hormone antagonists of insulin and one of PTH.

14. Two hormones are closely involved in the regulation of the fluid and electrolyte balance of the body. Name them and explain their effects on their common target organ.

15. What causes a simple goiter?

16. In general, the endocrine system becomes less efficient as we age. List some examples of problems that elderly individuals have as a result of decreasing hormone production.

At the Clinic

1. A woman with excessive body hair and a deep voice shows the outward symptoms of which hormonal dysfunction?

2. The parents of 14-year-old Megan are concerned about her height because she is only 4 feet tall and they are both close to 6 feet tall. After tests by their doctor, certain hormones are prescribed for the girl. What are the probable diagnosis, hormones prescribed, and the reason why the girl might expect to reach normal height?

3. Paula, a 28-year-old, has been in the first stage of labor for 15 hours. Her uterine contractions are weak, and her labor is not progressing normally. Since Paula and her doctor desire a vaginal delivery, the physician orders that Pitocin (a synthetic oxytocin) be infused. What will the effect of this hormone be?

4. Mr. Holdt brings his wife to the clinic, concerned about her nervousness, heart palpitations, and excessive sweating. Tests show hyperglycemia and hypertension. What hormones are probably being hypersecreted? What is the cause? What physical factors allow you to rule out thyroid problems?

10

Blood

YOUR GOALS

After completing this chapter, you will have a working knowledge of the functions of blood and will have mastered the objectives listed below.

FUNCTION PREVIEW

◼ Blood serves as a vehicle for transporting nutrients, respiratory gases, and other substances throughout the body, and distributes body heat.

OBJECTIVE CHECKLIST

COMPOSITION AND FUNCTIONS OF BLOOD (pp. 296–303)

❏ Describe the composition and volume of whole blood.

❏ Describe the composition of plasma and discuss its importance in the body.

❏ List the cell types comprising the formed elements and describe the major functions of each type.

❏ Define *anemia, polycythemia, leukopenia,* and *leukocytosis,* and list possible causes for each condition.

❏ Explain the role of the hemocytoblast.

HEMOSTASIS (pp. 303–306)

❏ Describe the blood-clotting process.

❏ Name some factors that may inhibit or enhance the blood-clotting process.

BLOOD GROUPS AND TRANSFUSIONS (pp. 306–308)

❏ Describe the ABO and Rh blood groups.

❏ Explain the basis for a transfusion reaction.

DEVELOPMENTAL ASPECTS OF BLOOD (pp. 308–309)

❏ Explain the basis of physiological jaundice seen in some newborn babies.

❏ Indicate blood disorders that increase in frequency in the aged.

Blood is the "river of life" that surges within us. It transports everything that must be carried from one place to another within the body—nutrients, wastes (headed for elimination from the body), and body heat—through blood vessels. For centuries, long before modern medicine, people recognized that blood was vital (some believed "magical"), and its loss was always considered to be a possible cause of death. In this chapter, we consider the composition and function of this life-sustaining fluid. The means by which it is propelled throughout the body is discussed in Chapter 11.

COMPOSITION AND FUNCTIONS OF BLOOD

Components, Physical Characteristics, and Volume

Among all of the body's tissues, blood is unique: It is the only *fluid* tissue. Blood is a sticky, opaque fluid with both solid and liquid components. Essentially, blood is a complex connective tissue in which living blood cells, the **formed elements,** are suspended in a nonliving fluid matrix called **plasma** (plaz'muh).

If a sample of blood is spun in a centrifuge, the heavier formed elements are packed down by centrifugal force and the plasma rises to the top (see Figure 10.1). Most of the reddish mass at the bottom of the tube consists of *erythrocytes* (eh-rith'ro-sīts), the red blood cells that function in oxygen transport. Although it is barely visible in Figure 10.1, there is a thin, whitish layer called the *buffy coat* at the junction between the formed elements and the plasma. This layer contains *leukocytes* (lu'ko-sīts), the white blood cells which act in various ways to protect the body, and *platelets,* cell fragments that function in the blood-clotting process. Erythrocytes normally account for about 45 percent of the total volume of a blood sample, a percentage known as the **hematocrit.** White blood cells and platelets contribute less than 1 percent, and plasma makes up most of the remaining 55 percent of whole blood.

Depending on the amount of oxygen it is carrying, the color of blood varies from scarlet (oxy-gen-rich) to a dull red (oxygen-poor). Blood is heavier than water and about five times thicker or more viscous, largely because of its formed elements. Blood is slightly alkaline, with a pH between 7.35 and 7.45.

Blood accounts for approximately 8 percent of body weight, and its volume in healthy adults is 5 to 6 liters, or approximately 6 quarts.

Plasma

Plasma, which is approximately 90 percent water, is the liquid part of the blood. Over 100 different substances are dissolved in this straw-colored fluid. As children, we discover its salty taste (due to the metal ions blood contains) the first time we stick a cut finger into our mouth. Examples of other dissolved substances include nutrients, respiratory gases, hormones, plasma proteins, various wastes, and products of cell metabolism.

Plasma proteins are the most abundant solutes in plasma. Except for antibodies and protein-based hormones, most plasma proteins are made by the liver. The plasma proteins serve a variety of functions (for instance, albumin contributes to the osmotic pressure of blood, which acts to keep water in the bloodstream, clotting proteins help to stem blood loss when a blood vessel is injured, and antibodies help protect the body from pathogens), but they are *not* taken up by cells to be used as food fuels or metabolic nutrients, as are other solutes such as glucose, fatty acids, and oxygen.

The composition of plasma varies continuously as cells remove or add substances to the blood. Assuming a healthy diet, however, the composition of plasma is kept relatively constant by various homeostatic mechanisms of the body. For example, when blood protein drops to undesirable levels, the liver is stimulated to make more blood proteins; when the blood starts to become too acid *(acidosis)* or too basic *(alkalosis),* both the respiratory system and kidneys are called into action to restore it to its normal, slightly alkaline pH range of 7.35 to 7.45. Various body organs make literally dozens of adjustments day in and day out to maintain the many plasma solutes at life-sustaining levels. Besides transporting various substances around the body, plasma helps to distribute body heat evenly throughout the body.

Figure 10.1   The composition of blood.

Formed Elements

If you observe a smear of human blood under a light microscope, you will see smooth, disk-shaped red blood cells, a variety of gaudily stained white blood cells and, most likely, some scattered platelets that look like debris (see Figure 10.2). However, erythrocytes vastly outnumber the other types of formed elements. Table 10.2, p. 301 provides a summary of the important characteristics of the various formed elements that make up about 45 percent of whole blood.

Erythrocytes

Erythrocytes, or **red blood cells (RBCs),** function primarily to ferry oxygen in blood to all cells of the body. They are superb examples of the "fit" between cell structure and function. RBCs differ from other blood cells because they are *anucleate*

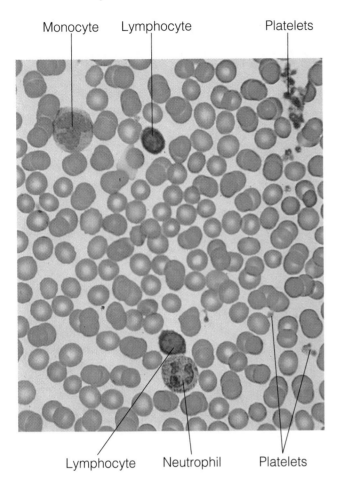

Monocyte   Lymphocyte   Platelets

Lymphocyte   Neutrophil   Platelets

Figure 10.2   Photomicrograph of a blood smear.
Most of the cells in this view are erythrocytes (red blood cells). Three kinds of leukocytes (white blood cells) are also present: monocytes, lymphocytes, and neutrophils. Also note the platelets. (750×)

(a-nu′kle-at); that is, they lack a nucleus. They also contain very few organelles. In fact, mature RBCs circulating in the blood are literally sacs of hemoglobin molecules. **Hemoglobin** (he″mo-glo′bin) **(Hb),** an iron-containing protein, transports the bulk of the oxygen that is carried in the blood. (It also binds with a small amount of carbon dioxide.) Moreover, because erythrocytes lack mitochondria and make ATP by anaerobic mechanisms, they do not use up any of the oxygen they are transporting, making them very efficient oxygen transporters indeed.

Erythrocytes are small cells shaped like biconcave disks—flattened disks with depressed centers (see Figure 10.2). They look like miniature dough-

nuts when viewed with a microscope. Their small size and peculiar shape provide a large surface area relative to their volume, making them ideally suited for gas exchange.

RBCs outnumber white blood cells by about 1000 to 1, and are the major factor contributing to blood viscosity. Although the numbers of RBCs in the circulation do vary, there are normally about 5 million cells per cubic millimeter of blood. (A cubic millimeter [mm³] is a very tiny drop of blood, almost not enough to be seen.) When the number of RBC/mm³ increases, blood viscosity increases. Similarly, as the number of RBCs decreases, blood thins and flows more rapidly. However, let's not get carried away talking about RBC *numbers*. Although their numbers are important, it is the amount of hemoglobin in the bloodstream at any time that best determines how well the erythrocytes are performing their role of oxygen transport.

The more hemoglobin molecules the RBCs contain, the more oxygen they will be able to carry. Perhaps the most accurate way of measuring the oxygen-carrying capacity of the blood is to determine how much hemoglobin it contains. Since a single red blood cell contains about 250 million hemoglobin molecules, each capable of binding 4 molecules of oxygen, each of these tiny cells can carry about 1 billion molecules of oxygen! This information is astounding but not very practical. Much more important clinically is the fact that normal blood contains 12–18 g hemoglobin per 100 ml blood. The hemoglobin content is slightly higher in men (13–18 g) than in women (12–16 g).

A decrease in the oxygen-carrying ability of the blood, whatever the reason, is **anemia** (ah-ne′me-ah). Anemia may be the result of: (a) a lower than normal *number* of RBCs, or (b) abnormal or deficient *hemoglobin content* in the RBCs. Several types of anemia are classified and described briefly in Table 10.1, but one of these, *sickle-cell anemia,* deserves a little more attention because people with this genetic disorder are frequently seen in hospital emergency rooms.

In sickle-cell anemia, the abnormal hemoglobin formed becomes spiky and sharp when the RBCs unload oxygen molecules or when the oxygen content of the blood is lower than normal, as during vigorous exercise, anxiety, or other stressful situations. The deformed (crescent-shaped) erythrocytes rupture easily and dam up in small blood

Table 10.1 Types of Anemia

Direct cause	Resulting from	Leading to
Decrease in RBC number	Sudden hemorrhage	Hemorrhagic anemia
	Lysis of RBCs—due to bacterial infections	Hemolytic (he″mo-lit′ik) anemia
	Lack of vitamin B_{12} (Usually due to lack of intrinsic factor required for absorption of vitamin. Intrinsic factor is formed by stomach mucosa cells.)	Pernicious (per-nish′us) anemia
	Depression/destruction of bone marrow by cancer, radiation, or certain medications	Aplastic anemia
Inadequate hemoglobin content in RBCs	Lack of iron in diet or slow/prolonged bleeding (such as results from heavy menstrual flow or bleeding ulcer), which depletes iron reserves needed to make hemoglobin; RBCs are small and pale because they lack hemoglobin	Iron-deficiency anemia
Abnormal hemoglobin in RBCs	Genetic defect leads to abnormal hemoglobin, which becomes sharp and sickle-shaped under conditions of increased oxygen use by body; occurs mainly in people of African descent	Sickle-cell anemia

vessels. These events interfere with oxygen delivery and cause extreme pain. It is amazing that this havoc results from a change in just *one* of the 287 amino acids in the globin molecule!

Sickle-cell anemia occurs chiefly in black people who live in the malaria belt of Africa and among their descendants. Apparently, the same gene that causes sickling makes red blood cells infected by the malaria-causing parasite stick to the capillary walls and then lose potassium, an essential nutrient for survival of the parasite. Hence, the malaria-causing parasite is prevented from multiplying within the red blood cells, and individuals with the sickle-cell gene have a better chance of surviving where malaria is prevalent. Only those carrying two copies of the defective gene have sickle-cell anemia. Those carrying only one sickling gene have *sickle-cell trait;* they generally do not display the symptoms, but can pass on the gene to their offspring.

An excessive or abnormal increase in the number of erythrocytes is **polycythemia** (pol″e-si-the′me-ah). Polycythemia may result from bone marrow cancer *(polycythemia vera)*. It may also represent a normal physiological (homeostatic) response to living at high altitudes where the air is thinner and less oxygen is available *(secondary polycythemia)*. The major problem that results from excessive numbers of RBCs is increased viscosity of blood, which causes it to flow sluggishly in the body and impairs circulation. ❑

Leukocytes

Although **leukocytes,** or **white blood cells (WBCs),** are far less numerous than red blood cells, they are crucial to body defense against disease. On average, there are 4000 to 11,000 WBCs per cubic millimeter, and they account for less than 1 percent of total blood volume. White blood cells are the only "complete cells" in blood; that is, they contain nuclei and the usual organelles.

Leukocytes form a protective, movable army that helps defend the body against damage by bacteria, viruses, parasites, and tumor cells. As such, they have some very special characteristics. Red blood cells are confined to the bloodstream, and carry out their functions in the blood. White blood cells, by contrast, are able to slip into and out of the blood vessels—a process called *diapedesis* (di″ah-peh-de′sis; "leaping across")—and the circulatory system is simply their means of transportation to areas of the body where their services are needed for inflammatory or immune responses (as described in Chapter 12).

In addition, WBCs can locate areas of tissue damage and infection in the body by responding to certain chemicals that diffuse from the damaged cells. This capability is called *positive chemotaxis* (ke″mo-tax′is). Once they have "caught the scent," the WBCs move through the tissue spaces by *ameboid* (ah-me′boid) *motion* (they form flowing cytoplasmic extensions that help move them along). By following the diffusion gradient, they pinpoint areas of tissue damage and rally round in large numbers to destroy foreign substances or dead cells.

Whenever WBCs mobilize for action, the body speeds up their production, and as many as twice the normal number of WBCs may appear in the blood within a few hours. A total WBC count above 11,000 cells/mm^3 is referred to as **leukocytosis** (lu″ko-si-to′sis). Leukocytosis generally indicates that a bacterial or viral infection is stewing in the body. The opposite condition, **leukopenia** (lu′ko-pe′ne-ah), is an abnormally low WBC count. It is commonly caused by certain drugs, such as corticosteroids and anticancer agents.

Leukocytosis is a normal and desirable response to infectious threats to the body. By contrast, the excessive production of abnormal WBCs that occurs in infectious mononucleosis and leukemia is distinctly pathological. In **leukemia** (lu-ke′me-ah), literally "white blood," the bone marrow becomes cancerous, and huge numbers of WBCs are turned out rapidly. Although this might not appear to present a problem, the "newborn" WBCs are immature and incapable of carrying out their normal protective functions. Consequently, the body becomes the easy prey of disease-causing bacteria and viruses. ❑

WBCs are classified into two major groups, depending on whether or not they contain visible granules in their cytoplasm.

Granulocytes (gran′u-lo-sītz″) are granule-containing WBCs. They have lobed nuclei, which typically consist of several rounded nuclear areas connected by thin strands of nuclear material. The granules in their cytoplasm stain specifically with Wright's stain. The granulocytes include the **neutrophils** (nu′tro-filz), **eosinophils** (e″o-sin′o-filz), and **basophils** (ba′so-filz).

The second group, **agranulocytes,** lack visible cytoplasmic granules. Their nuclei are closer to the norm—that is, they are spherical, oval, or kidney-shaped. The agranulocytes include **lymphocytes** (lim′fo-sīts) and **monocytes** (mon′o-sīts).

Specific characteristics of the leukocytes are given in Table 10.2. Microscopic views of neutrophils, monocytes, and lymphocytes can be seen in Figure 10.2.

Platelets

Platelets are not cells in the strict sense. They are fragments of extraordinarily large multinucleate cells called **megakaryocytes** (meg″ah-kar′e-o-sītz), which rupture, releasing 50 or more anucleate "pieces" that quickly seal themselves off from the surrounding fluids. The platelets appear as darkly staining, irregularly shaped bodies scattered among the other blood cells. The normal platelet count in blood is about 300,000/mm^3. As indicated in Table 10.2, platelets are needed for the clotting process that occurs in plasma when blood vessels are ruptured or broken. (This process is explained shortly.)

Hematopoiesis (Blood Cell Formation)

Blood cell formation, or **hematopoiesis,** (hem″ah-to-poi-e′sis), occurs in red bone marrow, or *myeloid* tissue. In adults, this tissue is found chiefly in the flat bones of the skull and pelvis, the ribs, sternum, and proximal epiphyses of the humerus and femur. Each type of blood cell is produced in different numbers in response to changing body needs and different stimuli. After they mature, they are discharged into the blood vessels surrounding the area.

All the formed elements arise from a common type of *stem cell,* the **hemocytoblast** (he″mo-si′to-blast), which resides in the red bone marrow. Their development differs, however, and once a cell is

Table 10.2   Characteristics of Formed Elements of the Blood

Cell type	Occurrence in blood (per mm³)	Cell anatomy*	Function
Erythrocytes (red blood cells or RBCs)	4–6 million	Salmon-colored biconcave disks; anucleate; literally, sacs of hemoglobin; most organelles have been ejected	Transport oxygen bound to hemoglobin molecules; also transport small amount of carbon dioxide
Leukocytes (white blood cells or WBCs) *Granulocytes*	4000–11,000		
• Neutrophils	3000–7000 (40–70% of WBCs)	Cytoplasm stains pale pink and contains fine granules, which are difficult to see; deep purple nucleus consists of three to seven lobes connected by thin strands of nucleoplasm	Active phagocytes; number increases rapidly during short-term or acute infections
• Eosinophils	100–400 (1–4% of WBCs)	Red coarse cytoplasmic granules, figure-8 or bilobed nucleus stains blue-red	Exact function not known; increase during allergy attacks; might phagocytize antigen-antibody complexes and inactivate some inflammatory chemicals
• Basophils	20–50 (0–1% of WBCs)	Cytoplasm has a few large blue-purple granules; U- or S-shaped nucleus with constrictions, stains dark blue	Granules contain histamine (vasodilator chemical) which is discharged at sites of inflammation
Agranulocytes			
• Lymphocytes	1500–3000 (20–45% of WBCs)	Cytoplasm pale blue and appears as thin rim around nucleus; spherical (or slightly indented) dark purple nucleus	Part of immune system; one group (B lymphocytes) produces antibodies; other group (T lymphocytes) involved in graft rejection, fighting tumors and viruses; and activates B lymphocytes
• Monocytes	100–700 (4–8% of WBCs)	Abundant gray-blue cytoplasm; dark blue-purple nucleus often kidney-shaped	Active phagocytes that become macrophages in the tissues; long-term "clean-up team"; increase in number during chronic infections such as tuberculosis
Platelets	250,000–500,000	Essentially irregularly shaped cell fragments; stain deep purple	Needed for normal blood clotting; initiate clotting cascade by clinging to broken area; help to control blood loss from broken blood vessels

*Appearance when stained with Wright's stain.

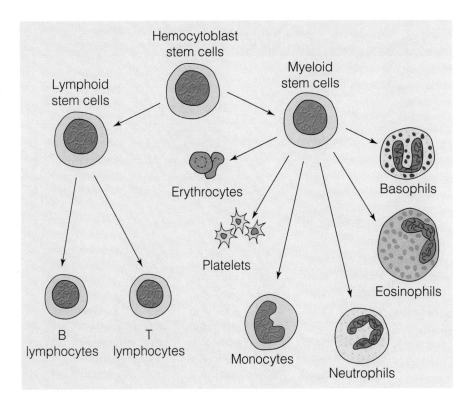

Figure 10.3   The development of blood cells. All blood cells differentiate from a common source, hemocytoblast stem cells in red bone marrow. The population of stem cells renews itself by mitosis. Some of these cells become lymphoid stem cells, which then develop into B cells and T cells, two classes of lymphocytes that function in the immune response. All other blood cells differentiate from myeloid stem cells, also derived from the stem cells.

committed to a specific blood pathway it cannot change. As indicated in the flowchart in Figure 10.3, the hemocytoblast forms two types of descendants—the *lymphoid stem cell,* which produces lymphocytes, and the *myeloid stem cell,* which can produce all other classes of formed elements.

Because they are anucleate, RBCs are unable to divide and have a limited life span of 100 to 120 days. Then they begin to fragment, or fall apart, and their remains are eliminated by the phagocytes in the spleen, liver, and other body tissues. Lost cells are replaced more or less continuously by the division of hemocytoblasts in the red bone marrow. The developing RBCs divide many times and then begin synthesizing huge amounts of hemoglobin. Suddenly, when enough hemoglobin has been accumulated, the nucleus and most organelles are ejected and the cell collapses inward. The result is the young RBC, called a *reticulocyte*

(rě-tik′u-lo-sīt) because it still contains some rough endoplasmic reticulum (ER). The reticulocytes enter the bloodstream to begin their task of transporting oxygen. Within two days of release, they have ejected the remaining ER and have become fully functional erythrocytes. The entire developmental process from hemocytoblast to mature RBC takes 3 to 5 days.

The rate of erythrocyte production is controlled by a hormone called **erythropoietin** (ě-rith″ro-poi-e′tin). Normally a small amount of erythropoietin circulates in the blood at all times, and red blood cells are formed at a fairly constant rate. Although the liver produces some, the kidneys play the major role in producing this hormone. When blood levels of oxygen begin to decline for any reason, the kidneys step up their release of erythropoietin (Figure 10.4). Erythropoietin targets the bone marrow, prodding it into "high gear" to turn out more RBCs. As you might expect, an over-

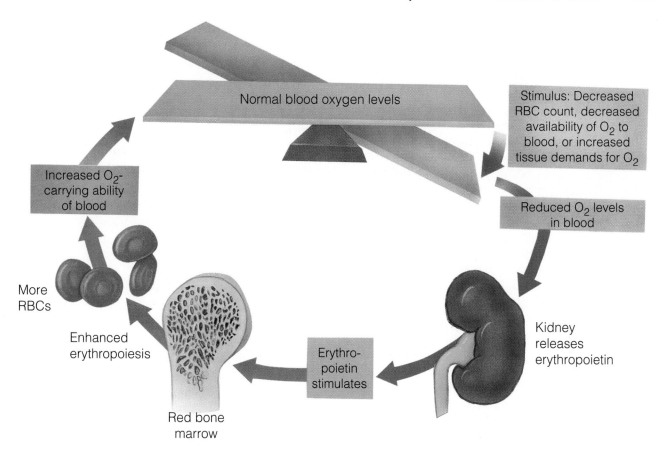

Figure 10.4   Mechanism for regulating the rate of RBC production.
Increased erythropoietin release, which stimulates RBC production in bone
marrow, occurs when oxygen levels in the blood become inadequate to support normal cellular activity, whatever the cause.

abundance of erythrocytes, or an excessive
amount of oxygen in the bloodstream, depresses
erythropoietin release and red blood cell production. An important point to remember is that it is
not the relative number of RBCs in the blood that
controls RBC production. Control is based on their
ability to transport enough oxygen to meet the
body's demands.

Like erythrocyte production, the formation of
leukocytes and platelets is stimulated by hormones. These *colony stimulating factors* (*CSFs*) not
only prompt red bone marrow to turn out leukocytes, but also enhance the ability of mature leukocytes to protect the body. Apparently, the CSFs are
released in response to specific chemical signals in
the environment such as inflammatory chemicals.
Additionally, exposure to certain bacteria or their
toxins stimulate macrophages and lymphocytes
(and probably other cell types) to release CSFs that

marshal an army of WBCs to ward off the attack.
The production of platelets is accelerated by the
hormone *thrombopoietin,* but little is known about
the regulation of platelet formation.

When bone marrow problems or disease conditions such as aplastic anemia or leukemia are
suspected, a special needle is used to withdraw a
small sample of red marrow from one of the flat
bones (ilium or sternum) close to the body surface.
This procedure provides cells for a microscopic examination called a *bone marrow biopsy.*

HEMOSTASIS

Normally, blood flows smoothly past the intact lining (endothelium) of blood vessel walls. But if a
blood vessel wall breaks, a series of reactions is set

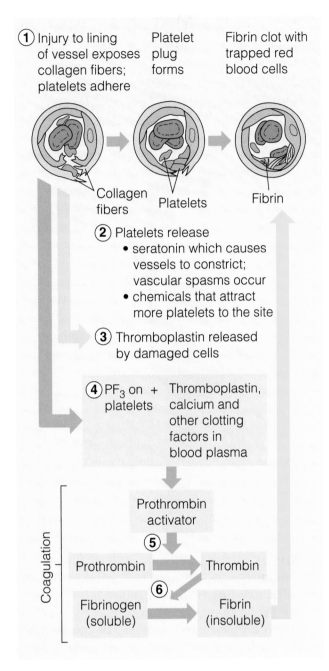

① Injury to lining of vessel exposes collagen fibers; platelets adhere

Platelet plug forms

Fibrin clot with trapped red blood cells

Collagen fibers

Platelets

Fibrin

② Platelets release
- seratonin which causes vessels to constrict; vascular spasms occur
- chemicals that attract more platelets to the site

③ Thromboplastin released by damaged cells

④ PF₃ on platelets + Thromboplastin, calcium and other clotting factors in blood plasma

Coagulation

Prothrombin activator

⑤

Prothrombin → Thrombin

⑥

Fibrinogen (soluble) → Fibrin (insoluble)

Figure 10.5 Hemostasis. The multistep process, detailed in the text, begins when a blood vessel is damaged and connective tissue in the vessel wall is exposed to blood. (1) Platelets become sticky and adhere to collagen fibers at the site of injury, forming the platelet plug. (2) Anchored platelets release serotonin, which promotes blood vessel spasms. The platelet plug and vessel spasms temporarily prevent blood loss. (3) Thromboplastin is released by damaged tissue cells. (4) PF₃ on platelets mixes with thromboplastin, plasma clotting factors, and Ca²⁺ to form an activator. (5) Prothombin activator converts prothrombin (in plasma) to thrombin. Thrombin catalyzes the final step (6) of the clotting process, conversion of fibrinogen to fibrin. The fibrin threads trap blood cells, forming a clot that seals the injured vessel until the wound heals by regeneration.

in motion to accomplish **hemostasis** (*hem* = blood; *stasis* = standing still), or stoppage of blood flow. This response, which is fast and localized, involves many substances normally present in plasma, as well as some that are released by platelets and injured tissue cells.

Hemostasis involves three phases that occur in rapid sequence: **vascular spasms, platelet plug formation**, and **coagulation, or blood clotting.** Blood loss at the site is permanently prevented when fibrous tissue grows into the clot and seals the hole in the blood vessel.

Basically, hemostasis occurs as follows (see Figure 10.5):

1. Platelets are repelled by an intact endothelium, but when it is broken so that the underlying collagen fibers are exposed, the platelets become "sticky" and cling to the damaged site. Anchored platelets release chemicals that attract more platelets to the site, and as more and more platelets pile up, a small mass called a *platelet plug* or *white thrombus* is formed.

2. Once anchored, the platelets release **serotonin** (ser″o-to′nin), which causes that blood vessel to go into spasms. The spasms narrow the blood vessel at that point, decreasing blood loss until clotting can occur. (Other factors causing vessel spasms include direct injury to the smooth muscle cells and stimulation of local pain receptors.)

3. At the same time, the injured tissues are releasing **thromboplastin** (throm″bo-plas′tin), a factor that plays an important role in the clotting process.

4. PF₃, a phospholipid which coats the surfaces of the platelets, interacts with thromboplastin, other blood protein clotting factors, and calcium ions (Ca²⁺) to form an activator that triggers the *clotting cascade.*

5. This **prothrombin activator** converts **prothrombin** (pro-throm′bin), present in the plasma, to **thrombin,** an enzyme.

6. Thrombin then joins soluble **fibrinogen** (fibrin′o-jen) proteins into long hairlike molecules of insoluble **fibrin,** which forms a meshwork that traps the RBCs and forms the basis of the clot (Figure 10.6). Within the hour, the clot begins to retract, squeezing **serum** (plasma minus the clotting proteins) from the mass and pulling the ruptured edges of the blood vessel closer together.

Figure 10.6   Scanning electron micrograph (artificially colored) of red blood cells trapped in a mesh of fibrin threads.

Normally, blood clots within 3 to 6 minutes. As a rule, once the clotting cascade has started, the triggering factors are rapidly inactivated to prevent widespread clotting ("solid blood"). Eventually, the endothelium regenerates and the clot is broken down. Once these events of the clotting cascade were understood, it became clear that placing a sterile gauze over a cut or applying pressure to a wound would speed up the clotting process. The gauze provides a rough surface to which the platelets can adhere and the pressure fractures cells, increasing the release of thromboplastin locally.

Disorders of Hemostasis

The two major disorders of hemostasis—undesirable clot formation and bleeding disorders—are at opposite poles.

Undesirable Clotting

Despite the body's safeguards against abnormal clotting, undesirable clots sometimes form in intact blood vessels, particularly in the legs. A clot that develops and persists in an unbroken blood vessel is called a **thrombus** (throm′bus). If large enough, it may prevent blood flow to the cells beyond the blockage. For example, if a thrombus forms in the blood vessels serving the heart *(coronary thrombosis)*, the consequences may be the death of the heart muscle and a fatal heart attack. If a thrombus breaks away from the vessel wall and floats freely in the bloodstream, it becomes an **embolus** (em′bo-lus) (plural, *emboli*). An embolus is usually no problem unless or until it lodges in a blood vessel too narrow for it to pass through. For example, a *cerebral embolus* may cause a stroke.

Undesirable clotting may be caused by anything that roughens the endothelium of a blood vessel and encourages clinging of platelets, such as severe burns, physical blows, or an accumulation of fatty material. Slowly flowing blood, or blood pooling, is another risk factor, especially in immobilized patients. In this case, clotting factors are not washed away as usual and accumulate so that clot formation becomes possible. Anticoagulants such as heparin and dicumarol are usually prescribed for thrombus-prone patients.

Bleeding Disorders

The most common causes of abnormal bleeding are platelet deficiency (thrombocytopenia) and deficits of some of the clotting factors, such as might result from impaired liver function or certain genetic conditions.

Thrombocytopenia results from an insufficient number of circulating platelets. Even normal movement causes spontaneous bleeding from small blood vessels. This is evidenced by many small purplish blotches, called *petechiae* (pe-te′ke-e), on the skin. It can arise from any condition that suppresses myeloid tissue, such as bone marrow cancer, radiation, or certain drugs.

When the liver is unable to synthesize its usual supply of clotting factors, abnormal and often severe bleeding episodes occur. If vitamin K (needed by the liver cells to produce the clotting factors) is deficient, the problem is easily corrected with vitamin K supplements. However, when liver function is severely impaired (as in hepatitis and cirrhosis), only whole blood transfusions are helpful.

The term **hemophilia** (he″mo-fil′e-ah) applies to several different *hereditary* bleeding disorders that result from a lack of any of the factors needed for clotting. Commonly called "bleeder's disease," the hemophilias have similar signs and symptoms that begin early in life. Even minor tissue trauma results in prolonged bleeding and can be life-threatening. Repeated bleeding into joints causes them to become disabled and painful. When a bleeding episode occurs, hemophiliacs are given a transfusion of fresh plasma or injections of the purified clotting factor they lack. Because hemophiliacs are absolutely dependent on one or the other of these therapies, some have become the victims of blood-transmitted viral diseases such as hepatitis and AIDS. (AIDS is a condition of depressed immunity and is described in Chapter 12.) These problems have been largely resolved because of the availability of genetically engineered clotting factors. ❑

BLOOD GROUPS AND TRANSFUSIONS

As we have seen, blood is vital for transporting substances through the body. When blood loss occurs, the blood vessels constrict and the bone marrow steps up blood cell formation in an attempt to keep the circulation going. However, the body can compensate for a loss of blood volume only up to a certain limit. Losses of 15 to 30 percent lead to pallor and weakness. Losses of over 30 percent cause severe shock, which can be fatal. Whole blood transfusions are routinely given to replace substantial blood loss and to treat severe anemia or thrombocytopenia. The usual blood bank procedure involves collecting blood from a donor and mixing it with an anticoagulant to prevent clotting. The treated blood can be stored (refrigerated) for several weeks until needed.

Human Blood Groups

Although whole blood transfusion can save lives, people have different blood groups, and transfusing incompatible or mismatched blood can be fatal. How so? The plasma membranes of RBCs, like those of all body cells, bear genetically determined proteins (antigens), which identify each person as unique. An *antigen* (an′tĭ-jen) is a substance that the body recognizes as foreign; it stimulates the immune system to release antibodies or use other means to mount a defense against it. Most antigens

are foreign proteins, such as those that are part of viruses or bacteria that have managed to invade the body. Although each of us tolerates our own cellular (self) antigens, one person's RBC proteins will be recognized as foreign if transfused into another person with different RBC antigens. The "recognizers" are antibodies present in the plasma that attach to RBCs bearing surface antigens different from those on the patient's (blood recipient's) RBCs. Binding of the antibodies causes the RBCs to clump, a phenomenon called **agglutination*** (ah-gloo″tĭ-na′shun) which leads to the clogging of small blood vessels throughout the body. During the next few hours, the foreign RBCs are lysed (ruptured) and their hemoglobin is released into the bloodstream. Although the transfused blood is unable to deliver the increased oxygen-carrying capacity hoped for and some tissue areas may be deprived of blood, the most devastating consequence of severe transfusion reactions is that the freed hemoglobin molecules may block the kidney tubules and cause kidney failure. Transfusion reactions can also cause fever, chills, nausea, and vomiting, but in the absence of kidney shutdown these reactions are rarely fatal. Treatment is aimed at preventing kidney damage by infusing alkaline fluids to dilute and dissolve the hemoglobin and diuretics to flush it out of the body in urine.

There are over 30 common RBC antigens in humans, allowing each person's blood cells to be classified into several different blood groups. However, it is the antigens of the ABO and Rh blood groups that cause the most vigorous transfusion reactions, and it is these two blood groups that we will describe here.

As shown in Table 10.3, the **ABO blood groups** are based on which of two antigens, type A or type B, a person inherits. Absence of both antigens results in type O blood, presence of both antigens leads to type AB, and the possession of either A or B antigen yields type A or B blood, respectively. In the ABO blood group, antibodies are formed during infancy against the ABO antigens *not* present on your own RBCs. As shown in the

---

*The RBC antigens that promote this clumping are sometimes called *agglutinogens* (ag″loo-tin′o-jenz), and the antibodies that bind them together *agglutinins* (ag″loo′tĭ-ninz).

Table 10.3	ABO Blood Groups						

Blood group	Frequency (% U.S. population)			RBC antigens (agglutinogens)	Illustration	Plasma antibodies (agglutinins)	Blood that can be received
	White	Black	Asian				
AB	4	4	5	A B		None	A, B, AB, O Universal recipient
B	11	20	27	B		Anti-A	B, O
A	40	27	28	A		Anti-B	A, O
O	45	49	40	None		Anti-A Anti-B	O Universal donor

table, a baby with neither the A nor the B antigen (group O) forms both anti-A and anti-B antibodies, while those with type A antigens (group A) form anti-B antibodies, and so on.

The **Rh blood groups** are so named because one of the eight Rh antigens (agglutinogen D) was originally identified in *Rh*esus monkeys. Later the same antigen was discovered in human beings. Most Americans are Rh^+ (Rh positive), meaning that their RBCs carry the Rh antigen. Unlike the antibodies of the ABO system, anti-Rh antibodies are *not* automatically formed and present in the blood of Rh^- (Rh negative) individuals. However, if an Rh^- person receives mismatched blood (that is, Rh^+), shortly after the transfusion his or her immune system becomes sensitized and begins producing antibodies (anti-Rh^+ antibodies) against the foreign blood type. Hemolysis (rupture of RBCs) does not occur with the first transfusion because it takes time for the body to react and start making antibodies. But the second time and every time

thereafter, a typical transfusion reaction occurs in which the patient's antibodies attack and rupture the donor's RBCs.

An important Rh-related problem occurs in pregnant Rh^- women who are carrying Rh^+ babies. The *first* such pregnancy usually results in the delivery of a healthy baby. But because the mother is sensitized by Rh^+ antigens that have passed through the placenta into her bloodstream, she will form anti-Rh^+ antibodies unless treated with RhoGAM shortly after giving birth. RhoGAM is an immune serum that prevents this sensitization and prevents her immune response. If she is not treated and becomes pregnant again with an Rh^+ baby, her antibodies will cross through the placenta and destroy the baby's RBCs, producing a condition known as *hemolytic disease of the newborn*. The baby is anemic and becomes hypoxic. Brain damage and even death may result unless fetal transfusions are done before birth to provide more RBCs for oxygen transport.

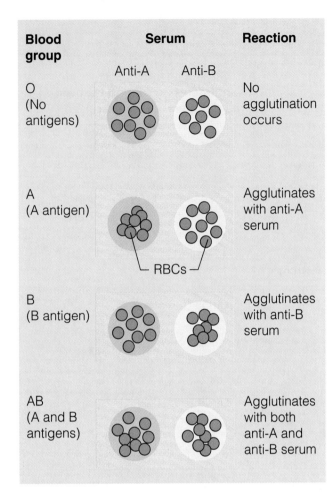

Figure 10.7   Blood typing of ABO blood groups.
When serum containing anti-A or anti-B antibodies is added to a blood sample diluted with saline, agglutination will occur between the antibody and the corresponding antigen (if present).

Blood Typing

The importance of determining the blood group of both the donor and the recipient *before* blood is transfused is glaringly obvious. The general procedure for determining ABO blood type is briefly outlined in Figure 10.7. Essentially, it involves testing the blood by mixing it with two different types of immune serum—anti-A and anti-B. Agglutination occurs when RBCs of a group A person are mixed with the anti-A serum, but not when they are mixed with the anti-B serum. Likewise, RBCs of type B blood are clumped by anti-B serum, but not by anti-A serum. A photograph of just-typed AB and O blood appears in Figure 10.8. Because it is critical that blood groups be compatible, cross

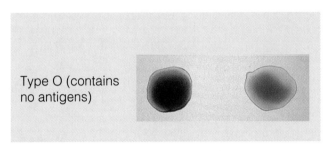

Figure 10.8   Photograph of typed AB and O blood.

matching is also done. *Cross matching* involves testing for agglutination of donor RBCs by the recipient's serum, and of the recipient's RBCs by the donor serum. Typing for the Rh factors is done in the same manner as ABO blood typing.

DEVELOPMENTAL ASPECTS OF BLOOD

In the young embryo, development of the entire circulatory system occurs early. Before birth, there are many sites of blood cell formation—the fetal liver and spleen, among others—but by the seventh month of development, the red marrow has become the chief site of hematopoiesis and remains so throughout life. Generally, embryonic blood cells are circulating in the newly formed blood vessels by day 28 of development. Fetal hemoglobin (HbF) differs from the hemoglobin formed after birth. It has a greater ability to pick up oxygen, a characteristic that is highly desirable since fetal blood is less oxygen rich than that of the mother. After birth, fetal blood cells are gradually replaced by RBCs that contain the more typical hemoglobin A. In situations in which the fetal RBCs are destroyed at such a rapid rate that the immature infant liver cannot rid the body of hemoglobin breakdown products in the bile fast enough, the infant becomes *jaundiced* (jawn′dist). This type of jaundice generally causes no major problems and is referred to as *physiologic jaundice,* to distinguish it from more serious disease conditions that result in jaundiced or yellowed tissues.

Various congenital diseases result from genetic factors (such as hemophilia and sickle-cell anemia) and from interactions with maternal blood factors (such as hemolytic disease of the newborn). All these conditions have already been discussed, and will not be reconsidered here.

As indicated earlier, dietary factors can lead to abnormalities in blood cell formation as well as hemoglobin production. Iron-deficiency anemia is especially common in women because of their monthly blood loss during menses. The young and the old are particularly at risk for leukemia. With increasing age, chronic types of leukemias, anemias, and diseases involving undesirable clot formation are more prevalent. However, these are usually secondary to disorders of the heart, blood vessels, or immune system. For example, the increased incidence of leukemias in old age is believed to result from the waning efficiency of the immune system, whereas abnormal thrombus and embolus formation reflects the insidious progress of atherosclerosis, which roughens blood vessel walls. Anemias that appear in the elderly commonly result from nutritional deficiencies, drug therapy, or some disorder such as cancer or leukemia. The elderly are particularly at risk for pernicious anemia because the stomach mucosa (which produces intrinsic factor) frequently atrophies with age. ❑

A CLOSER LOOK

CONCOCTING BLOOD: ARTIFICIAL BLOOD SUBSTITUTES

The term "blood substitute" is somewhat misleading. Blood has many components that play such a wide variety of roles—from fighting infection to transporting oxygen—that no single artificial substitute yet engineered can fulfill all those functions. However, substitute liquids are available that can transport oxygen from the lungs through the body and can "stretch" a limited blood supply, while sidestepping transfusion reactions. An important benefit of these blood substitutes is that the recipient is not at risk for transmission of blood-borne disease factors. Three of these oxygen-transporting products—Fluosol, chemically altered hemoglobin, and artificial red blood cells—are described briefly.

FLUOSOL

The main ingredient of *Fluosol,* a milky artificial blood substitute, is a chemical related to the nonstick coating used in cookware. Developed in Japan, the product was first tested in the United States in 1982. Many of the early recipients were people who needed surgery but refused blood transfusions on religious grounds.

Fluosol serves as a dissolving medium for oxygen. In order to "load" sufficient amounts of oxygen into it, patients must breathe pure oxygen by mask or must be in a hyperbaric (high-pressure) chamber. The developers claim that oxygen transported by Fluosol is used more easily by the body tissues because the slippery particles are much smaller than erythrocytes and glide through the capillaries at a faster rate. Although initially promising for treatment of heart attack, carbon monoxide poisoning, and sickle-cell anemia, its use has been clouded by recent research which indicates that Fluosol may depress the patient's immune system. Nonetheless, use of Fluosol to keep donor organs oxygenated while they are being readied for transplant, and as an agent to protect heart muscle from ischemia during a heart attack, still appears promising.

(continued)

Normal red blood cell and smaller hemoglobin-carrying artificial red cells (neohemocytes) in front of a capillary.

CHEMICALLY ALTERED HEMOGLOBIN

In their search to find a substitute to boost oxygen delivery, scientists have altered the chemistry of hemoglobin by creating a chemical bridge between two of its four peptide chains and by linking several hemoglobin molecules together. Animal tests have shown that the altered hemoglobin gives up more oxygen to the tissues than normal hemoglobin, even at low temperatures (10°C). Since body temperature is routinely lowered in patients undergoing heart surgery, using the modified hemoglobin during such procedures might boost oxygen delivery to the patient's tissues. An additional plus is that because the cross-linked form of hemoglobin is prevented from fragmenting in the blood (as natural hemoglobin does), the necessity for a RBC plasma membrane "container" is eliminated.

Although test results are promising, there are still important problems to be solved. For example, bacterial endotoxins, potent poisons produced by some bacteria, tend to cling to the modified hemoglobin and there is some evidence that the free hemoglobin provokes a generalized constriction of blood vessels, making oxygen delivery more difficult. Additionally, since the chemically modified hemoglobin is derived from erythrocytes, the risk of transmitting viral and bacterial diseases still exists.

NEOHEMOCYTES

Researchers at the University of California at San Francisco have created artificial RBCs, which they call *neohemocytes,* by packaging natural hemoglobin molecules in fat bubbles made from phospholipids and cholesterol. The resulting "red cells" are about one-twelfth the size of human erythrocytes. Although the neohemocytes are destroyed and cleared more rapidly from the bloodstream than are real RBCs, they have a shelf life of 6 months (versus less than 2 months for whole blood). This would make neohemocyte infusion a viable choice for trauma patients in immediate need of blood, but clinical trials on humans are still in the distant future.

Although the U.S. Food and Drug Administration has encouraged the development of artificial blood for over 20 years, no marketable product, including those described above, has been approved for anything but experimental use in humans. Blood is still a priceless commodity, and its beautiful complexity has yet to be replaced by modern medical technology.

Important Terms

agglutination (ag″loo-tĭ-na′shun)

anemia (ah-ne′me-ah)

basophils (ba′so-filz)

coagulation

eosinophils (e″o-sin′o-filz)

erythrocytes (eh-rith′ro-sīts)

fibrin (fi′brin)

formed elements

hemoglobin (he″mo-glo′bin)

leukocytes (lu′ko-sīts)

lymphocytes (lim′fo-sīts)

monocytes (mon′o-sīts)

neutrophils (nu′tro-filz)

plasma (plaz′muh)

platelets

prothrombin (pro-throm′bin)

thromboplastin (throm″bo-plas′tin)

Summary

Composition and Functions of Blood (pp. 296–303)

1. Blood is composed of a nonliving fluid matrix (plasma) and formed elements. It is scarlet to dull red, depending on the amount of oxygen carried. Normal adult blood volume is 5 to 6 liters.

2. Dissolved in plasma (primarily water) are nutrients, gases, hormones, wastes, proteins, salts, and so on. Plasma composition changes as body cells remove or add substances to it, but homeostatic mechanisms act to keep it relatively constant. Plasma makes up 55 percent of whole blood.

3. Formed elements, the living blood cells that make up about 45 percent of whole blood, include:
   a. Erythrocytes, or RBCs—disk-shaped anucleate cells that transport oxygen bound to their hemoglobin molecules. Their life span is 100 to 120 days.
   b. Leukocytes, or WBCs—ameboid cells involved in protection of the body.
   c. Platelets—cell fragments that act in blood clotting.

4. A decrease in oxygen-carrying ability of blood is anemia. Possible causes are decrease in number of functional RBCs or decrease in amount of hemoglobin they contain. Polycythemia is an excessive number of RBCs that may result from bone marrow cancer or a move to a location where less oxygen is available in the air (at high altitude, for example).

5. Leukocytes are nucleated cells, classed into two groups:
   a. Granulocytes include neutrophils, eosinophils, and basophils.
   b. Agranulocytes include monocytes and lymphocytes.

6. When bacteria, viruses, or other foreign substances invade the body, WBCs increase in number (leukocytosis) and fight them in various ways.

7. An abnormal decrease in number of WBCs is leukopenia; an abnormal increase is seen in infectious mononucleosis and leukemia (cancer of leukocytes).

8. All formed elements arise in red bone marrow from a common stem cell, the hemocytoblast. However, their developmental pathways differ. The stimulus for hematopoiesis is hormonal (erythropoietin in the case of RBCs).

Hemostasis (pp. 303–306)

1. Stoppage of blood loss from an injured blood vessel, or hemostasis, involves three steps: vascular spasms, platelet plug formation, blood-clot formation.

2. Hemostasis is started by a tear or interruption in the blood vessel lining. Platelets adhere to the damaged site and release serotonin, which causes vasoconstriction. Platelet PF_3 and tissue cell thromboplastin initiate the clotting cascade, leading to formation of fibrin threads. Fibrin traps RBCs as they flow past, forming the clot.

3. Normally, clots are digested when a vessel has been permanently repaired. An attached clot that forms or persists in an unbroken blood vessel is a *thrombus;* a clot traveling in the bloodstream is an *embolus.*

4. Abnormal bleeding may reflect a deficit of platelets (thrombocytopenia), genetic factors (hemophilia), or inability of the liver to make clotting factors.

Blood Groups and Transfusions (pp. 306–308)

1. Blood groups are classified on the basis of proteins (antigens) on RBC membranes. Complementary antibodies may (or may not) be present in blood. Antibodies act to agglutinate (clump) and lyse foreign RBCs.

2. The blood group most commonly typed for is ABO. Type O is most common; least common is AB. ABO antigens are accompanied by preformed antibodies in plasma, which act against RBCs with "foreign" antigens.

3. Rh factor is found in most Americans. Rh^- people do not have preformed antibodies to Rh^+ RBCs, but form them once exposed to Rh^+ blood.

Developmental Aspects of Blood (p. 308–309)

1. Congenital blood defects include various types of hemolytic anemias and bleeder's disease. Incompatibility between maternal and fetal blood can result in fetal cyanosis, resulting from destruction of fetal blood cells.

2. Fetal hemoglobin (HbF) binds more readily with oxygen than does HbA.

3. Physiologic jaundice in a newborn reflects immaturity of the infant's liver.

4. Excessive leukocytosis may be indicative of malignancy of blood-forming organs or leukemia. Leukemias are most common in the very young and very old.

5. The elderly are at risk for anemia and clotting disorders.

Review Questions

1. What is the blood volume of an average-sized adult?

2. What determines whether blood is bright red or dull red in color?

3. Name as many different categories of substances carried in plasma as you can.

4. Define *formed elements* and list their three major categories. Which category is most numerous? Which comprises the buffy coat?

5. What is the average lifespan of a RBC? How does the fact that it has no nucleus affect its lifespan?

6. Define anemia and give three possible causes.

7. Name the granular and agranular WBCs. Give the major function of each type in the body.

8. If you had a severe infection, would you expect your total WBC count to be closest to 5000, 10,000, or 15,000/mm^3? Why? What is this condition called?

9. Name the stem cell that produces virtually all formed elements. Name the formed elements that arise from the myeloid stem cell. Name those arising from the lymphoid stem cell.

10. Describe the process of hemostasis. Indicate what starts the process.

11. How can liver dysfunction cause bleeding disorders?

12. What is the basis of blood groups? What are agglutinins?

13. Name the four ABO blood groups. Which is most common? Which is least common?

14. What is a transfusion reaction? Why does it happen?

15. Explain why an Rh$^-$ person does not have a transfusion reaction on the first exposure to Rh$^+$ blood. Why is there a transfusion reaction the second time he or she receives the Rh$^+$ blood?

16. If you had a high hematocrit, would you expect your hemoglobin determination to be high or low? Why?

17. What blood-related problems are most common in the aged?

At the Clinic

1. A patient on renal dialysis has a low RBC count. What hormone, secreted by the kidney, can be assumed to be deficient?

2. A bone marrow biopsy of Mr. Bongalonga, a man on a long-term drug therapy, shows an abnormally high percentage of nonhematopoietic connective tissue. What condition does this indicate? If the symptoms are critical, what short-term and long-term treatments are indicated? Will infusion of whole blood or packed red cells be more likely?

3. A woman comes to the clinic complaining of fatigue, shortness of breath, and chills. Blood tests show anemia, and a bleeding ulcer is diagnosed. What type of anemia is this?

4. A patient is diagnosed with bone marrow cancer and has a hematocrit of 70 percent. What is this condition called?

5. A middle-aged college professor from Boston is in the Swiss Alps studying astronomy. He arrived two days ago and plans to stay the entire year. However, he notices that he is short of breath when he walks up steps and that he tires easily with any physical activity. His symptoms gradually disappear; after two months, he feels fine. Upon returning to the United States, he has a complete physical exam and is told that his erythrocyte count is higher than normal. (a) Attempt to explain this finding. (b) Will his RBC count remain at this higher-than-normal level? Why or why not?

6. Why is someone more likely to bleed to death when an artery is cleanly severed than when it is crushed and torn?

11

The Circulatory System

YOUR GOALS

After completing this chapter, you will have a working knowledge of the functions of the circulatory system and will have mastered the objectives listed below.

FUNCTION PREVIEW

- ■ The heart pumps blood.
- ■ Blood vessels provide the conduits within which blood circulates to all body tissues.
- ■ The lymphatic system returns leaked plasma to the blood vessels after cleansing it of bacteria and other foreign matter.

OBJECTIVE CHECKLIST

CARDIOVASCULAR SYSTEM: THE HEART (pp. 315–325)

- ❑ Describe the location of the heart in the body and identify its major anatomical areas on an appropriate model or diagram.
- ❑ Trace the pathway of blood through the heart.
- ❑ Compare the pulmonary and systemic circuits.
- ❑ Name the functional blood supply of the heart.
- ❑ Define *systole, diastole, stroke volume,* and *cardiac cycle.*
- ❑ Explain the operation of the heart valves.
- ❑ Define *heart sounds* and *murmur.*
- ❑ Name the elements of the intrinsic conduction system of the heart and describe the pathway of impulses through this system.
- ❑ Explain what information can be gained from an electrocardiogram.
- ❑ Describe the effect of the following on heart rate: stimulation by the vagus nerve, exercise, epinephrine, and various ions.

CARDIOVASCULAR SYSTEM: BLOOD VESSELS (pp. 325–343)

- ❑ Compare and contrast the structure and function of arteries, veins, and capillaries.
- ❑ Identify the body's major arteries and veins and name the body region supplied by each.

(continued)

❏   Discuss the unique features of special circulations of the body: arterial circulation of the brain, hepatic portal circulation, and fetal circulation.

❏   Define *blood pressure* and *pulse* and name several pulse points.

❏   List factors affecting and/or determining blood pressure.

❏   Define *hypertension* and *atherosclerosis* and describe possible health consequences of these conditions.

❏   Describe the exchanges that occur across capillary walls.

LYMPHATIC SYSTEM (pp. 345–347)

❏   Name the two major types of structures composing the lymphatic system and explain how the lymphatic system is functionally related to the cardiovascular and immune systems.

❏   Describe the composition of lymph and explain how it is formed and transported through the lymphatic vessels.

❏   Describe the function(s) of lymph nodes, tonsils, the thymus, Peyer's patches, and the spleen.

DEVELOPMENTAL ASPECTS OF THE CIRCULATORY SYSTEM (pp. 347–349)

❏   Describe briefly the development of the organs of the cardiovascular system and the origin of the lymphatic vessels.

❏   Name the fetal vascular modifications or "fetal shunts" and describe their function before birth.

❏   Explain how regular exercise and a diet low in fats and cholesterol may help maintain cardiovascular health.

When most people hear the term **circulatory system,** they immediately think of the heart. We have all felt our own heart "pound" from time to time, and we tend to get a bit nervous when this happens. The crucial importance of the heart has been recognized for a long time, and both serious and comical songs referring to it have been written. However, the circulatory system is much more than just the heart, and from a scientific and medical standpoint, it is important to understand *why* the circulatory system is so vital to life.

The almost continuous traffic into and out of a busy factory at rush hour occurs at a snail's pace compared to the endless activity going on within our bodies. Night and day, minute after minute, our trillions of cells take up nutrients and excrete wastes. Although the pace of these exchanges slows during sleep, they must go on continuously, because when they stop we die. Cells can make such exchanges only with the tissue fluid in their immediate environment. Thus, some means of changing and "refreshing" these fluids is necessary to renew the nutrients and prevent pollution caused by the buildup of wastes. Like the bustling factory, the body must have a transportation system to carry its various "cargos" back and forth. Instead of roads, railway tracks, and airways, the body's delivery routes are its hollow blood vessels.

Most simply stated, the major function of the circulatory system is transportation. Using blood as the transport vehicle, the system carries oxygen, nutrients, cell wastes, hormones, and many other

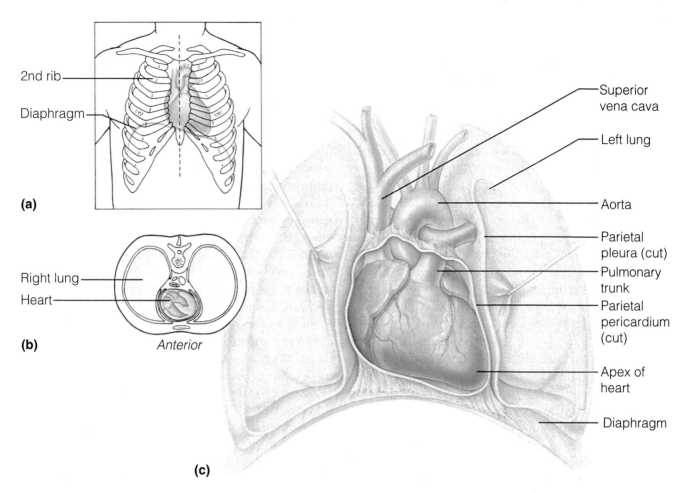

2nd rib

Diaphragm

(a)

Right lung

Heart

(b)   *Anterior*

Superior vena cava

Left lung

Aorta

Parietal pleura (cut)

Pulmonary trunk

Parietal pericardium (cut)

Apex of heart

Diaphragm

(c)

Figure 11.1   Location of the heart within the thorax. (**a**) Relationship of the heart to the sternum and ribs. (**b**) Cross-sectional view showing relative position of the heart in the thorax. (**c**) Relationship of the heart and great vessels to the lungs.

substances vital for body homeostasis to and from the cells. The force to move the blood around the body is provided by the beating heart.

The circulatory system has two major subdivisions—the **cardiovascular** (kar″de-o-vas′ku-lar) **system** and the **lymphatic** (lim-fat′ik) **system.** The cardiovascular system can be compared to a muscular pump equipped with one-way valves and a system of large and small plumbing tubes within which the blood travels. Blood (the substance transported) is discussed in Chapter 10. Here we will consider the heart (the pump) and the blood vessels (the network of tubes). The lymphatic system, a pumpless system of vessels (and lymphoid organs) that aids the cardiovascular system in its function, is the final topic of this chapter.

CARDIOVASCULAR SYSTEM: THE HEART

Anatomy

Location and Size

The relative size and weight of the heart give few hints of its incredible strength. Approximately the size of a person's fist, the hollow, cone-shaped heart weighs less than a pound. The heart is located within the bony thorax and is flanked on each side by the lungs (see Figure 11.1). Its more pointed **apex** is directed toward the left hip and rests on the diaphragm, approximately at the level

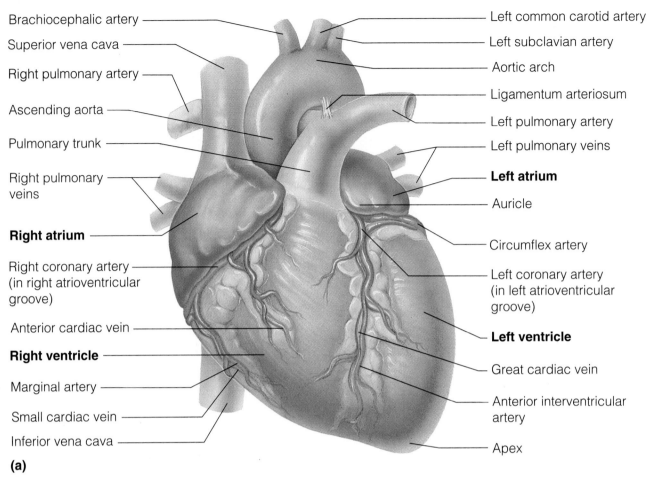

Brachiocephalic artery

Superior vena cava

Right pulmonary artery

Ascending aorta

Pulmonary trunk

Right pulmonary veins

Right atrium

Right coronary artery (in right atrioventricular groove)

Anterior cardiac vein

Right ventricle

Marginal artery

Small cardiac vein

Inferior vena cava

(a)

Left common carotid artery

Left subclavian artery

Aortic arch

Ligamentum arteriosum

Left pulmonary artery

Left pulmonary veins

Left atrium

Auricle

Circumflex artery

Left coronary artery (in left atrioventricular groove)

Left ventricle

Great cardiac vein

Anterior interventricular artery

Apex

Figure 11.2   Gross anatomy of the heart. (a) Anterior view. **(b)** Frontal section showing interior chambers and valves.

of the fifth intercostal space. (This is exactly where one would place a stethoscope to count the heart rate for an apical pulse.) Its broader posterosuperior aspect, or **base,** from which the great vessels of the body emerge, points toward the right shoulder and lies beneath the second rib.

Coverings and Wall

The heart is enclosed by a double sac of serous membrane, the **pericardium** (per″i-kar′de-um). The thin **visceral pericardium,** or **epicardium,** tightly hugs the external surface of the heart and is actually part of the heart wall. It is continuous at the heart apex with the loosely applied **parietal pericardium,** which is reinforced on its superficial face by dense connective tissue. This fibrous layer helps protect the heart and anchors it to surround-

ing structures, such as the diaphragm and sternum. A slippery lubricating fluid (serous fluid) is produced by the serous pericardial membranes. This fluid allows the heart to beat easily in a relatively frictionless environment as the pericardial layers slide smoothly across each other.

Inflammation of the pericardium, *pericarditis* (per″ĭ-kar-di′tis), often results in a decrease in the amount of serous fluid. This causes the pericardial layers to bind and stick to each other, forming painful *adhesions* that interfere with heart movements. ❑

The heart walls are composed of three layers: the outer *epicardium* (described above), the *myocardium,* and the innermost *endocardium.* The **myocardium** (mi″o-kar′de-um) consists of thick bundles of cardiac muscle twisted and whorled into ringlike arrangements. It is the layer that actu-

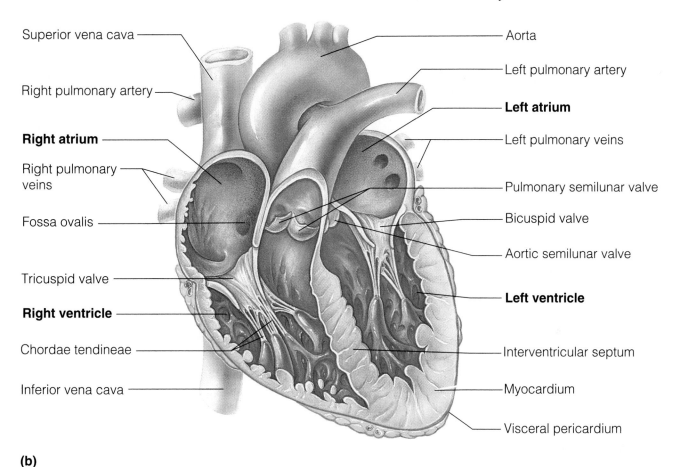

Superior vena cava

Right pulmonary artery

Right atrium

Right pulmonary veins

Fossa ovalis

Tricuspid valve

Right ventricle

Chordae tendineae

Inferior vena cava

Aorta

Left pulmonary artery

Left atrium

Left pulmonary veins

Pulmonary semilunar valve

Bicuspid valve

Aortic semilunar valve

Left ventricle

Interventricular septum

Myocardium

Visceral pericardium

(b)

ally contracts. The myocardium is reinforced internally by a dense, fibrous connective tissue network called the "skeleton of the heart." The **endocardium** (en″do-kar′de-um) is a thin, glistening sheet of endothelium that lines the heart chambers. It is continuous with the linings of the blood vessels leaving and entering the heart. Figure 11.2 shows two views of the heart—an external anterior view and a frontal section. As the anatomical areas of the heart are described in the next section, keep referring to Figure 11.2 to locate each of the heart structures or regions.

Chambers and Associated Great Vessels

The heart has four hollow chambers or cavities—two **atria** (a′tre-ah) and two **ventricles** (ven′trĭ-kulz). Each of these chambers is lined with endocardium, which helps blood flow smoothly through the heart. The superior atria are primarily *receiving chambers*. As a rule, they are not important in the pumping activity of the heart. Blood flows into the atria under low pressure from the

veins of the body and then continues on to fill the ventricles below. The inferior, thick-walled ventricles are the *discharging chambers,* or actual pumps of the heart. When they contract, blood is propelled out of the heart and into the circulation. As illustrated in Figure 11.2, the heart is somewhat twisted; the right ventricle forms most of its anterior surface, whereas the left ventricle forms its apex. The septum that divides the heart longitudinally is referred to as the **interventricular** or **interatrial septum,** depending on which chamber it divides and separates.

Although it is a single organ, the heart functions as a double pump. The right side works as the pulmonary circuit pump. It receives oxygen-poor blood from the veins of the body through the large **superior** and **inferior venae cavae** (ka′ve) and pumps it out through the **pulmonary trunk.** The pulmonary trunk splits into the right and left **pulmonary arteries,** which carry blood to the lungs, where oxygen is picked up and carbon dioxide is unloaded. Oxygen-rich blood drains from the lungs and is returned to the left side of

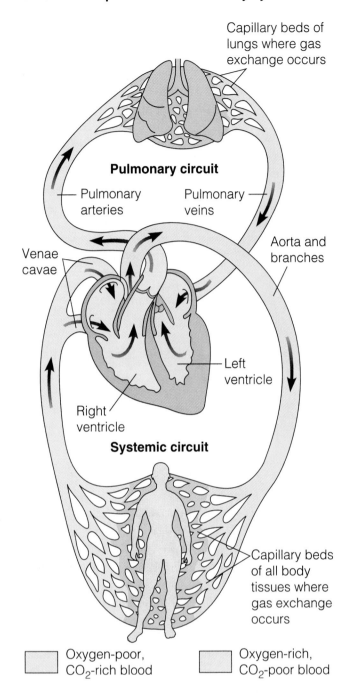

Capillary beds of lungs where gas exchange occurs

Pulmonary circuit

Pulmonary arteries

Pulmonary veins

Aorta and branches

Venae cavae

Left ventricle

Right ventricle

Systemic circuit

Capillary beds of all body tissues where gas exchange occurs

Oxygen-poor, CO_2-rich blood

Oxygen-rich, CO_2-poor blood

Figure 11.3   The systemic and pulmonary circulations. The left side of the heart is the systemic pump; the right side is the pulmonary circuit pump. (Although there are two pulmonary arteries, one each to the right and left lung, for simplicity only one is shown.)

the heart through the four **pulmonary veins.** The circulation just described, from the right side of the heart to the lungs and back to the left side of the heart, is called the **pulmonary circulation** (Figure 11.3). Its only function is to carry blood to the lungs for gas exchange and then return it to the heart.

Blood returned to the left side of the heart is pumped out of the heart into the **aorta** (a-or'tah), from which the systemic arteries branch to supply essentially all body tissues. Oxygen-poor blood circulates from the tissues through systemic veins, which return the blood to the right atrium through the superior and inferior venae cavae, as mentioned earlier. This second circuit, from the left side of the heart through the body tissues and back to the right side of the heart, is called the **systemic circulation** (see Figure 11.3). It supplies oxygen- and nutrient-rich blood to all body organs. Because the left ventricle is the systemic pump that must pump blood over a much longer pathway through the body, its walls are substantially thicker than those of the right ventricle and it is a much more powerful pump.

Valves

The heart is equipped with four valves that allow blood to flow in only one direction through the heart chambers—from the atria through the ventricles and out the great arteries leaving the heart. The **atrioventricular** (a"tre-o-ven-trik'u-lar) or **AV valves** are located between the atrial and ventricular chambers on each side. The AV valves prevent backflow into the atria when the ventricles contract. The left AV valve—the **bicuspid** or **mitral** (mi'tral) **valve**—consists of two cusps, or flaps, of endocardium. The right AV valve, the **tricuspid valve,** has three cusps (see Figure 11.2b). Tiny white cords, the **chordae tendineae** (kor'de tendin'e)—literally, "heart strings"—anchor the cusps to the walls of the ventricles. When the heart is relaxed and blood is passively filling its chambers, the AV-valve flaps hang limply into the ventricles. As the ventricles contract, they press on the blood in their chambers, and the intraventricular pressure (pressure inside the ventricles) begins to rise. This causes the AV-valve flaps to be forced upward, closing the valves. At this point the chordae tendineae are working to anchor the flaps in a closed position. If the flaps were unanchored, they would

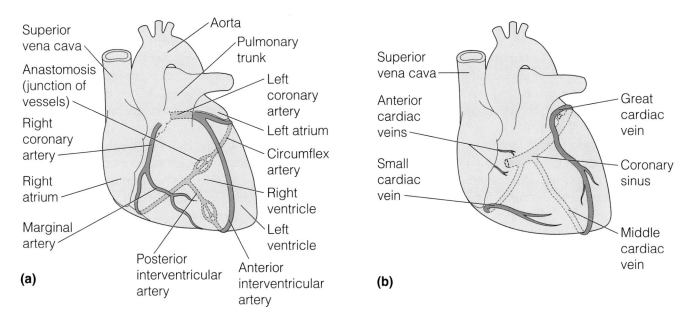

Figure 11.4   Coronary blood vessels. (a) The main coronary arteries.
(**b**) The major cardiac veins.

blow upward into the atria like an umbrella being turned inside out by a gusty wind. In this manner, the AV valves prevent backflow into the atria when the ventricles are contracting.

The second set of valves, the **semilunar** (sem″e-lu′nar) **valves,** guards the bases of the two large arteries leaving the ventricular chambers. Thus, they are known as the **pulmonary** and **aortic semilunar valves** (see Figure 11.2b). Each semilunar valve has three cusps that fit tightly together when the valves are closed. When the ventricles are contracting and forcing blood out of the heart, the cusps are forced open and flattened against the walls of the arteries by the tremendous force of rushing blood. Then, when the ventricles relax, the blood begins to flow backward toward the heart and the cusps fill with blood, closing the valves. This prevents arterial blood from reentering the heart.

Each set of valves operates at a different time. The AV valves are open during heart relaxation and closed when the ventricles are contracting. The semilunar valves are closed during heart relaxation and are forced open when the ventricles contract. As they open and close in response to pressure changes in the heart, the valves force blood to continually move forward in its journey through the heart.

Heart valves are basically simple devices, and the heart—like any mechanical pump—can function with "leaky" valves as long as the damage is not too great. However, severe valve deformities can seriously hamper cardiac function. For example, an *incompetent valve* forces the heart to pump and repump the same blood because the valve does not close properly and blood backflows. In *valvular stenosis,* the valve flaps become stiff, often because of repeated bacterial infection of the endocardium (endocarditis). This compels the heart to contract more forcibly than normal. In each case, the heart's workload increases, and ultimately the heart weakens and may fail. Under such conditions, the faulty valve is replaced with a synthetic valve or a valve taken from a pig heart. ❑

Cardiac Circulation

Although the heart chambers are bathed with blood almost continuously, the blood contained in the heart does not nourish the myocardium. The blood supply that oxygenates and nourishes the heart is provided by the right and left coronary arteries. The **coronary arteries** branch from the base of the aorta and encircle the heart in the *atrioventricular groove* at the junction of the atria and ventricles (see Figure 11.4). The coronary arteries and their major branches (the *anterior interventricular* and *circumflex arteries* on the left, and the *posterior interventricular* and *marginal* arteries on the right) are compressed when the ventricles are contracting and fill when the heart is relaxed. The myocardium is drained by several **cardiac veins,** which empty into the **coronary sinus,** which in turn empties into the right atrium.

When the heart beats at a very rapid rate, the myocardium may receive an inadequate blood supply because the relaxation periods (when the blood is able to flow to the heart tissue) are shortened. Situations in which the myocardium is deprived of oxygen often result in crushing chest pain called **angina pectoris** (an-ji′nah pek′tor-is). This pain is a warning that should *never* be ignored because, if angina is prolonged, the ischemic heart cells may die, forming an *infarct.* The resulting **myocardial infarction** (in-fark′shun) is commonly called a "heart attack" or "coronary." ❑

Physiology

As the heart beats or contracts, the blood makes continuous round trips—in and out of the heart, through the rest of the body, and then back to the heart—only to be sent out again. The amount of work that a heart does is almost too incredible to believe. In one day it pushes the body's supply of 6 quarts or so of blood through the blood vessels over 1000 times, meaning that it actually pumps about 6000 quarts of blood in a single day!

Conduction System of the Heart

Unlike skeletal muscle cells that must be stimulated by nerve impulses before they will contract, cardiac muscle cells can and do contract spontaneously and independently, even if all nervous connections are severed. Moreover, these spontaneous contractions occur in a regular and continuous way. Although cardiac muscle *can* beat independently, the muscle cells in different areas of the heart have different rhythms. For example, the atrial cells beat about 60 times per minute, whereas the ventricular cells contract much more slowly (20–40/min). Therefore, without some type of unifying control system, the heart would be an uncoordinated and inefficient pump.

Two types of controlling systems act to regulate heart activity. One of these involves the nerves of the autonomic nervous system that act like "brakes" and "accelerators" to decrease or increase the heart rate depending on which division is activated (see p. 324). The second system is the **intrinsic conduction system,** or **nodal system,** that is built into the heart tissue (Figure 11.5). The intrinsic conduction system is composed of a special tissue found nowhere else in the body; it is much like a cross between muscle and nervous tissue. This system causes heart muscle depolarization in only one direction—from the atria to the ventricles. In addition, it enforces a contraction rate of approximately 75 beats per minute on the heart; thus, the heart beats as a coordinated unit.

One of the most important parts of the intrinsic conduction system is a crescent-shaped node of tissue called the **sinoatrial** (si″no-a′tre-al) **(SA) node,** located in the right atrium. Other components include the **atrioventricular (AV) node** at the junction of the atria and ventricles, the **atrioventricular (AV) bundle (bundle of His)** and the right and left **bundle branches** located in the interventricular septum, and finally the **Purkinje** (pur-kin′je) **fibers,** which spread within the muscle of the ventricle walls.

The SA node is a tiny cell mass with a mammoth job. Because it has the highest rate of depolarization in the whole system, it starts each heartbeat and sets the pace for the whole heart. Consequently, the SA node is often called the **pacemaker.** From the SA node, the impulse spreads through the atria to the AV node, and then the atria contract. At the AV node, the impulse is delayed briefly to give the atria time to finish contracting. It then passes rapidly through the AV bundle, the bundle branches, and the Purkinje fibers, resulting in a "wringing" contraction of the ventricles that begins at the heart apex and moves toward the atria. This contraction effectively ejects blood superiorly into the large arteries leaving the heart. The "Closer Look" box on p. 323 describes *electrocardiography,* the clinical procedure for mapping the electrical activity of the heart.

Because the atria and ventricles are separated from one another by "insulating" connective tissue, which is part of the fibrous skeleton of the heart, depolarization waves can reach the ventricles only by traveling through the AV node. Thus, any damage to the AV node can partially or totally release the ventricles from the control of the SA node. When this occurs, the ventricles (thus the heart) begin to beat at their own rate, which is much slower, some or all of the time. This condition is called *heart block.*

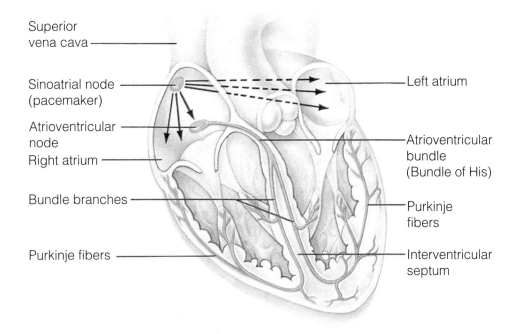

Superior
vena cava

Sinoatrial node
(pacemaker)

Atrioventricular
node

Right atrium

Bundle branches

Purkinje fibers

Left atrium

Atrioventricular
bundle
(Bundle of His)

Purkinje
fibers

Interventricular
septum

Figure 11.5 **The intrinsic conduction system of the heart.** The depolarization wave is initiated by the sinoatrial (SA) node and then passes successively through the atrial myocardium to the atrioventricular (AV) node, the AV bundle, the right and left bundle branches, and the Purkinje fibers in the ventricular walls.

There are other conditions that can interfere with the regular conduction of impulses across the heart—for example, damage to the SA node results in a slower heart rate. When this is a problem, artificial pacemakers are usually installed surgically. *Ischemia* (is-ke′me-ah), or lack of an adequate blood supply to the heart muscle, may lead to **fibrillation**—a rapid uncoordinated shuddering of the heart muscle (it looks like a bag of worms). Fibrillation makes the heart totally useless as a pump and is a major cause of death from heart attacks in adults. ❑

Tachycardia (tak″e-kar′de-ah) is a rapid heart rate (over 100 beats per minute). *Bradycardia* (brad″e-kar′de-ah) is a heart rate that is substantially slower than normal (less than 60 beats per minute). Neither condition is pathological, but prolonged tachycardia may progress to fibrillation.

Cardiac Cycle and Heart Sounds

In a healthy heart, the atria contract simultaneously. Then, as they start to relax, contraction of the ventricles begins. **Systole** (sis′to-le) and **diastole** (di-as′to-le) mean heart *contraction* and *relaxation,* respectively. Since most of the pumping work is done by the ventricles, these terms always refer to the contraction and relaxation of the *ventricles* unless otherwise stated.

The term **cardiac cycle** refers to the events of one complete heartbeat, during which both atria and ventricles contract and then relax. Since the average heart beats approximately 75 times per minute, the length of the cardiac cycle is normally about 0.8 seconds. We will consider the cardiac cycle in terms of events occurring during three periods—mid-to-late diastole, ventricular systole, and early diastole (see Figure 11.6).

1. *Mid-to-late diastole.* Our discussion begins with the heart in complete relaxation. At this point, the pressure in the heart is low, and blood is flowing passively into and through the atria into the ventricles from the pulmonary and systemic circulations. The semilunar valves are closed and the AV valves are open. Then the atria contract and force the blood remaining in their chambers into the ventricles.

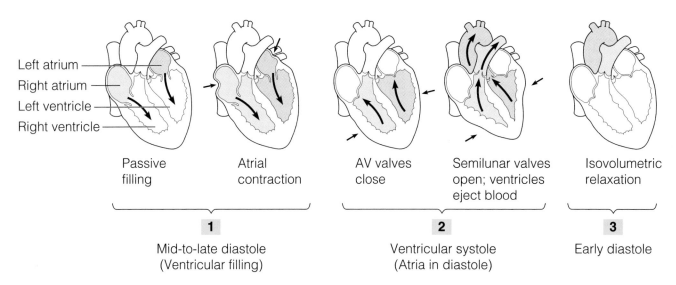

Left atrium
Right atrium
Left ventricle
Right ventricle

| Passive filling | Atrial contraction | AV valves close | Semilunar valves open; ventricles eject blood | Isovolumetric relaxation |

1

Mid-to-late diastole
(Ventricular filling)

2

Ventricular systole
(Atria in diastole)

3

Early diastole

Figure 11.6   Summary of events occurring during the cardiac cycle.
(Small black arrows indicate the regions of the heart that are contracting.)

2. *Ventricular systole.* Shortly after, ventricular contraction (systole) begins and the pressure within the ventricles increases rapidly, closing the AV valves. When the intraventricular pressure (pressure in the ventricles) is higher than the pressure in the large arteries leaving the heart, the semilunar valves are forced open and blood rushes through them out of the ventricles. During ventricular systole, the atria are relaxed, and their chambers are again filling with blood.

3. *Early diastole.* At the end of systole, the ventricles relax, the semilunar valves snap shut (preventing backflow) and, for a moment, the ventricles are completely closed chambers. During early diastole, the intraventricular pressure drops. When it drops below the pressure in the atria (which has been increasing as blood has been filling their chambers), the AV valves are forced open and the ventricles again begin to refill rapidly with blood, completing the cycle.

When using a stethoscope, you can hear two distinct sounds during each cardiac cycle. These **heart sounds** are often described by the two syllables "lub" and "dup," and the sequence is lub-dup, pause, lub-dup, pause, and so on. The first heart sound (lub) is caused by the closing of the AV valves. The second heart sound (dup) occurs when the semilunar valves close at the end of systole. The first heart sound is longer and louder than the second heart sound, which tends to be short and sharp.

Abnormal or unusual heart sounds are called *murmurs*. Blood flows silently so long as the flow is smooth and uninterrupted. If it strikes obstructions, its flow becomes turbulent and generates sounds, such as heart murmurs, that can be heard with a stethoscope. Heart murmurs are fairly common in young children (and some elderly people) with perfectly healthy hearts, probably because their heart walls are relatively thin and vibrate with rushing blood. However, most often, murmurs indicate valve problems. For example, if a valve does not close tightly (is *incompetent*), a swishing sound will be heard *after* that valve has (supposedly) closed, as the blood flows back through the partially open valve. Distinct sounds also can be heard when blood flows turbulently through *stenosed* (narrowed) valves. ❑

Cardiac Output

Cardiac output (CO) is the amount of blood pumped out by *each* side of the heart (actually each ventricle) in 1 minute. It is the product of the heart rate (HR) and the stroke volume (SV). **Stroke volume** is the volume of blood pumped out by a ventricle with each heartbeat. In general, stroke volume increases as the force of ventricular contraction increases. If we use the normal resting values for heart rate (75 beats per minute) and stroke

A CLOSER LOOK

ELECTROCARDIOGRAPHY: (DON'T) BE STILL MY HEART

When impulses pass through the heart, electrical currents are generated that spread throughout the body. These impulses can be detected on the body surface and recorded with an *electrocardiograph*. The recording that is made, the *electrocardiogram (ECG)*, traces the flow of current through the heart. The illustration shows a normal ECG tracing.

The typical ECG has three recognizable waves. The first wave, the *P wave*, is small and signals the depolarization of the atria immediately before they contract. The large *QRS complex*, which results from the depolarization of the ventricles, has a complicated shape. It precedes the contraction of the ventricles. The *T wave* results from currents flowing during the repolarization of the ventricles. (Atrial repolarization is generally hidden by the large QRS complex, which is being recorded at the same time.)

Abnormalities in the shape of the waves and changes in their timing send signals that something may be wrong with the intrinsic conduction system or may indicate a *myocardial infarct* (present or past). A myocardial infarct is an area of heart tissue in which the cardiac cells have died; it is generally a result of *ischemia*. During *fibrillation*, the normal pattern of the ECG is totally lost and the heart ceases to act as a functioning pump.

An electrocardiogram tracing showing the three normally recognizable deflection waves—P, QRS, and T.

volume (70 ml per beat), the average adult cardiac output can be easily figured:

$$CO = HR \text{ (75 beats/min)} \times SV \text{ (70 ml/beat)}$$
$$CO = 5250 \text{ ml/min}$$

Since the normal adult blood volume is about 5000 ml, the entire blood supply passes through the body once each minute. Cardiac output varies with the demands of the body. It rises when the stroke volume is increased or the heart beats faster or both; it drops when either or both of these factors decrease. Since this is so, let's take a look at how stroke volume and heart rate are regulated.

Regulation of Stroke Volume A healthy heart pumps out about 60 percent of the blood that enters it. As noted above, this is approximately 70 ml (about 2 ounces) with each heartbeat. According to *Starling's law of the heart,* the critical factor controlling stroke volume is how much the cardiac muscle cells are stretched just before they contract.

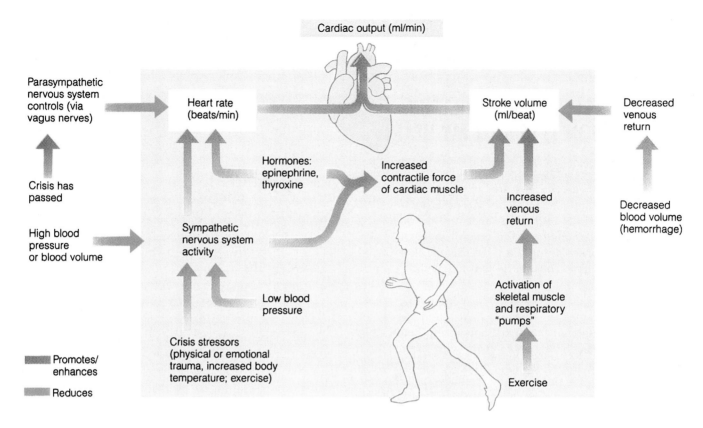

Figure 11.7   Influence of selected factors on cardiac output. Note that the direction of the flowchart is from the bottom up.

The more they are stretched, the stronger the contraction will be. The important factor stretching the heart muscle is *venous return,* the amount of blood entering the heart and distending its ventricles. If one side of the heart suddenly begins to pump more blood than the other, the increased venous return to the opposite ventricle will force it to pump out an equal amount, thus preventing backup of blood in the circulation.

Anything that increases the volume or speed of venous return also increases stroke volume and force of contraction (see Figure 11.7). For example, a slow heartbeat allows more time for the ventricles to fill. Exercise speeds venous return because it results in increased heart rate and force. The enhanced squeezing action of active skeletal muscles on the veins returning blood to the heart, the so-called muscular pump, also plays a major role in increasing the venous return. On the other hand, low venous return, such as might result from severe blood loss or an extremely rapid heart rate, decreases stroke volume, causing the heart to beat less forcefully.

Regulation of Heart Rate   In healthy people, stroke volume tends to be relatively constant. However, when blood volume drops suddenly or when the heart has been seriously weakened, stroke volume declines and cardiac output is maintained by a faster heartbeat. Although heart contraction does not depend on the nervous system, its rate *can* be changed temporarily by the autonomic nerves. Indeed, the most important external influence on heart rate is the activity of the autonomic nervous system. Heart rate is also modified by various chemicals, hormones, and ions. Some of these factors are summarized in Figure 11.7.

During times of physical or emotional stress, the nerves of the *sympathetic division* of the autonomic nervous system stimulate the SA and AV nodes and the cardiac muscle itself. As a result, the heart beats more rapidly. This is a familiar phenomenon to anyone who has ever been frightened or has had to run to catch a bus. As fast as the heart pumps under ordinary conditions, it really speeds up when special demands are placed on it. Since a faster blood flow increases the rate at which fresh blood reaches body cells, more oxygen and glucose are made available to them during periods of stress. When demand declines, the heart adjusts. *Parasympathetic nerves,* primarily the vagus nerves, slow and steady the heart, giving it more time to rest during noncrisis times. In patients with *congestive heart failure,* a condition in which the heart is nearly "worn out" due to age or hypertensive heart disease, the heart pumps weakly. For those patients, the drug digitalis is routinely prescribed. It acts in much the same way as the vagus nerves; that is, it slows and steadies the heart, resulting in a stronger heartbeat.

Various hormones and ions can have a dramatic effect on heart activity. *Epinephrine,* which mimics the effect of the sympathetic nerves, and *thyroxine* both cause increased heart rate. Electrolyte imbalances pose a real threat to the heart. For example, reduced levels of ionic calcium in the blood depress the heart, whereas hypercalcemia causes such prolonged contractions that the heart may stop entirely. Excesses or a lack of needed ions such as sodium and potassium also modify heart activity. A deficit of potassium ions in the blood, for example, causes the heart to beat feebly and abnormal heart rhythms appear.

A number of physical factors, including age, gender, exercise, and body temperature, influence heart rate. Resting heart rate is fastest in the fetus (140–160 beats per minute) and then gradually decreases throughout life. The average heart rate is faster in females (72–80 beats per minute) than in males (64–72 beats per minute). Heat increases heart rate by boosting the metabolic rate of heart cells. This explains the rapid, pounding heartbeat you feel when you have a high fever, and accounts in part for the effect of exercise on heart rate (remember, working muscles generate heat). Cold has the opposite effect; it directly decreases heart rate. As noted above, exercise acts through nervous system controls (sympathetic division) to increase heart rate (and also, through the action of the muscular pump, to increase stroke volume).

The pumping action of the healthy heart maintains a balance between cardiac output and venous return. But when the pumping efficiency of the heart is depressed so that circulation is inadequate to meet tissue needs, **congestive heart failure (CHF)** occurs. Congestive heart failure is usually a progressive condition that reflects weakening of the heart by coronary atherosclerosis (clogging of the coronary vessels with fatty buildup), persistent high blood pressure, or multiple myocardial infarcts (leading to repair with noncontracting scar tissue).

Because the heart is a double pump, each side can fail independently of the other. If the left heart fails, *pulmonary congestion* occurs. The right side of the heart continues to propel blood to the lungs, but the left side is unable to eject the returning blood into the systemic circulation. As blood vessels within the lungs become swollen with blood, the pressure within them increases, and fluid leaks from the circulation into the lung tissue, causing pulmonary edema. If untreated, the person suffocates.

If the right side of the heart fails, *peripheral congestion* occurs as blood backs up in the systemic circulation. Edema is most noticeable in the distal parts of the body: The feet, ankles, and fingers become swollen and puffy. Failure of one side of the heart puts a greater strain on the opposite side, and eventually the whole heart fails. ❑

CARDIOVASCULAR SYSTEM: BLOOD VESSELS

Blood circulates inside the blood vessels, which form a closed transport system. The idea that blood circulates or "makes rounds" through the body is only about 300 years old. The ancient Greeks believed that blood moved through the body like an ocean tide, first moving out from the heart and then ebbing back to it in the same vessels to get rid of its impurities in the lungs. It was not until the seventeenth century that William Harvey, an English physician, proved that blood did, in fact, move in circles.

Like a system of roads, the circulatory system has its freeways, secondary roads, and alleys. As the heart beats, blood is propelled into the large **arteries** leaving the heart. It then moves into successively smaller and smaller arteries and then into the **arterioles** (ar-ter′e-ōlz), which feed the **capillary** (kap′ĭ-lar″e) **beds** in the tissues. Capillary beds are drained by **venules** (ven′ulz), which in turn empty into **veins** that finally empty into the great veins entering the heart. Thus arteries, which carry blood away from the heart, and veins, which

(a)

Figure 11.8   Structure of blood vessels. (a) The walls of arteries and veins are composed of three tunics: the tunica intima (endothelium underlain by a scanty layer of loose connective tissue), tunica media (smooth muscle and elastic fibers), and tunica externa (largely collagen fibers). Capillaries—between arteries and veins in the circulatory pathway—are composed only of the tunica intima. Notice that the tunica media is thick in arteries and relatively thin in veins. **(b)** Scanning electron micrograph of an artery and vein in cross section (120×).

(b)

drain the tissues and return the blood to the heart, are simply conducting vessels—the freeways and secondary roads. Only the tiny hairlike capillaries, which extend and branch through the tissues and connect the smallest arteries (arterioles) to the smallest veins (venules), directly serve the needs of the body cells. The capillaries are the alleys that intimately intertwine among the body cells. It is only through their walls that exchanges between the tissue cells and the blood can occur.

Microscopic Anatomy

Except for the microscopic capillaries, the walls of blood vessels have three coats, or tunics (see Figure 11.8). The **tunica intima** (tu′nĭ-kah in′tĭ-mah), which lines the lumen or interior of the vessels, is a thin layer of endothelium (squamous epithelial cells) resting on a scanty layer of loose connective tissue. Its cells fit closely together and form a slick surface that decreases friction as blood flows through the vessel lumen.

The **tunica media** (me′de-ah) is the bulky middle coat. It is mostly smooth muscle and elastic tissue. The smooth muscle, which is controlled by the sympathetic nervous system, is active in changing the diameter of the vessels. As the vessels constrict or dilate, the blood pressure increases or decreases, respectively.

The **tunica externa** (eks′tern-ah) is the outermost tunic; it is composed largely of fibrous connective tissue. Its function is basically to support and protect the vessels.

The walls of arteries are usually much thicker than the walls of veins. Their tunica media, in particular, tends to be much heavier. This structural difference is related to a difference in function of these two types of vessels. Arteries, which are closer to the pumping action of the heart, must be able to expand as blood is forced into them and then recoil passively as the blood flows off into the circulation during diastole. Their walls must be strong and stretchy enough to take these continuous changes in pressure.

On the other hand, veins are far from the heart in the circulatory pathway, and the pressure in them tends to be low all the time. Thus veins have thinner walls. However, since the blood pressure in veins is usually too low to force the blood back to the heart and blood returning to the heart often flows against gravity, veins are modified to ensure that the amount of blood returning to the heart *(venous return)* equals the amount being pumped out of the heart *(cardiac output)* at any time. The lumens of veins tend to be much larger than those of corresponding arteries, and the larger veins have valves that prevent backflow of blood (see Figure 11.8). Skeletal muscle activity also enhances venous return. As the muscles surrounding the veins contract and relax, the blood is "milked" through the veins toward the heart (Figure 11.9). Finally, when we inhale, the drop in pressure that occurs in the thorax causes the large veins near the heart

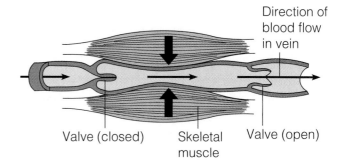

Figure 11.9   Operation of the muscular pump. When skeletal muscles contract and press against the flexible veins, the valves proximal to the area of contraction are forced open and blood is squeezed toward the heart. The valves distal to the point of contraction are closed by the backflowing blood.

to expand and fill. Thus, the "respiratory pump" also helps return blood to the heart (Figure 11.7).

To demonstrate the efficiency of the venous valves in preventing backflow of blood, perform the following simple experiment. Allow one hand to hang by your side for a minute or two, until the blood vessels on its dorsal aspect become distended or swollen with blood. Place two fingertips side by side against one of the distended veins. Then pressing firmly, move your proximal finger along the vein toward your heart. Now release that finger. As you can see, the vein remains flattened and collapsed in spite of gravity. Now remove your distal finger and watch the vein fill rapidly with blood.

The transparent walls of the tiny capillaries are only one cell layer thick—just the tunica intima. Because of this exceptional thinness, exchanges are easily made between the blood and the tissue cells.

Gross Anatomy

Major Arteries of the Systemic Circulation

The **aorta** is the largest artery of the body and it is a truly splendid vessel. In adults, the aorta is about the size of a garden hose (with an internal diameter the size of your thumb) where it issues from the left ventricle of the heart. It decreases only slightly in size as it runs to its terminus. Different parts of the aorta are named for their location or shape. The aorta curves upward from the left ventricle of the heart as the **ascending aorta,** arches to the left as the **aortic arch,** and then plunges downward through the thorax following the spine **(thoracic aorta)** to finally pass through the diaphragm into the abdominopelvic cavity where it becomes the **abdominal aorta** (see Figure 11.10).

The major branches of the aorta and the organs they serve are listed next in sequence from the heart. Figure 11.10 shows the course of the aorta and its major branches. As you locate the arteries on the figure, be aware of ways to make your learning easier. In many cases the name of the artery tells you the body region or organs served (renal artery, brachial artery, and coronary artery) or the bone followed (femoral artery and ulnar artery).

Arterial Branches of the Ascending Aorta

- The only branches of the ascending aorta are the **right [R.]** and **left [L.] coronary arteries,** which serve the heart.

Arterial Branches of the Aortic Arch

- The **brachiocephalic** (bra"ke-o-se-fal'ik) **artery** (the first branch off the aortic arch) splits into the **R. common carotid** (kah-ro'tid) **artery** and **R. subclavian** (sub-kla've-an) **artery.** (See same named vessels on left side of body for organs served.)

- The **left common carotid artery** is the second branch off the aortic arch. It divides, forming the **L. internal carotid** that serves the brain, and the **L. external carotid** that serves the skin and muscles of the head and neck.

- The third branch of the aortic arch, the **left subclavian artery,** gives off an important branch—the **vertebral artery,** which serves part of the brain. In the axilla, the subclavian artery becomes the **axillary artery** and then continues into the arm as the **brachial artery,** which supplies the arm. At the elbow, the brachial artery splits to form the **radial** and **ulnar arteries,** which serve the forearm.

Arterial Branches of the Thoracic Aorta

- The **intercostal arteries** (ten pairs) supply the muscles of the thorax wall. Other branches of the thoracic aorta supply the lungs **(bronchial arteries),** the esophagus **(esophageal arteries),** and the diaphragm **(phrenic arteries).** These arteries are not illustrated in Figure 11.10.

Arterial Branches of the Abdominal Aorta

- The **celiac trunk** is the first branch of the abdominal aorta. It is a single vessel that has three branches: (1) the **left gastric artery** supplies the stomach, (2) the **splenic artery** supplies the spleen, and (3) the **common hepatic artery** supplies the liver.

- The unpaired **superior mesenteric** (mes"enter'ik) **artery** supplies most of the small intestine and the first half of the large intestine or colon.

- The **renal** (R. and L.) **arteries** serve the kidneys.

- The **gonadal** (R. and L.) **arteries** supply the gonads. They are called the **ovarian arteries** in females (serving the ovaries) and the **testicular arteries** in males (serving the testes).

- The **lumbar arteries** (not illustrated in Figure 11.10) are several pairs of arteries serving the heavy muscles of the abdomen and trunk walls.

- The **inferior mesenteric artery** is a small, unpaired artery supplying the second half of the large intestine.

- The **common iliac** (R. and L.) **arteries** are the final branches of the abdominal aorta. Each divides into an **internal iliac artery,** which supplies the pelvic organs (bladder, rectum, and so on), and an **external iliac artery,** which enters the thigh, where it becomes the **femoral artery.** The femoral artery and its branch, the **deep femoral artery,** serve the thigh. At the knee, the femoral artery becomes the **popliteal artery,** which then splits into the **anterior** and **posterior tibial arteries,** which supply the leg and foot. The anterior tibial artery terminates in the **dorsalis pedis artery,** which supplies the dorsum of the foot. (The dorsalis pedis is often palpated in patients with circulatory problems of the legs to determine if the distal part of the leg has adequate circulation.)

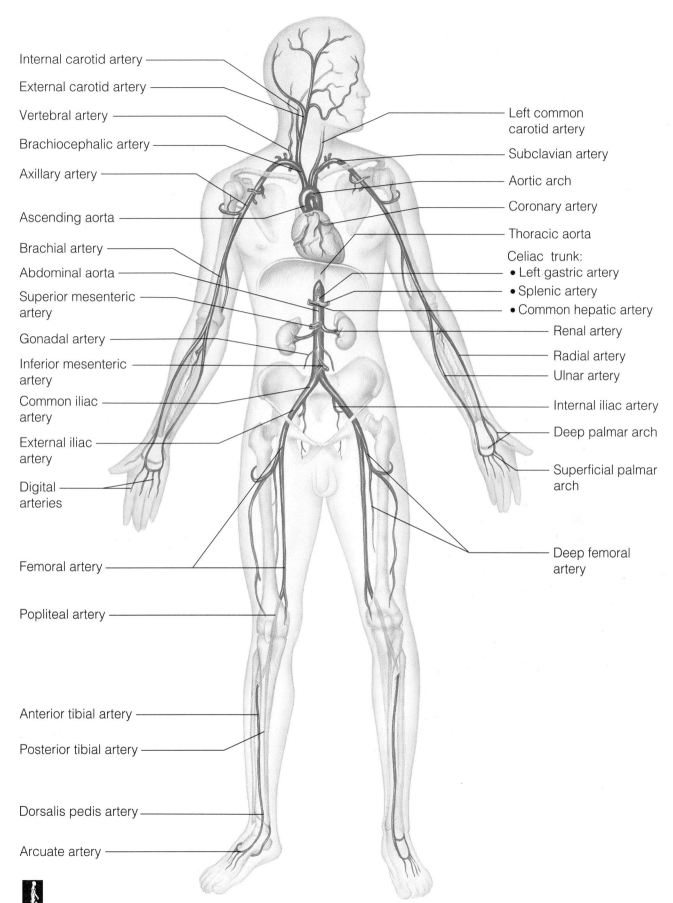

Internal carotid artery

External carotid artery

Vertebral artery

Brachiocephalic artery

Axillary artery

Ascending aorta

Brachial artery

Abdominal aorta

Superior mesenteric artery

Gonadal artery

Inferior mesenteric artery

Common iliac artery

External iliac artery

Digital arteries

Femoral artery

Popliteal artery

Anterior tibial artery

Posterior tibial artery

Dorsalis pedis artery

Arcuate artery

Left common carotid artery

Subclavian artery

Aortic arch

Coronary artery

Thoracic aorta

Celiac trunk:
• Left gastric artery
• Splenic artery
• Common hepatic artery

Renal artery

Radial artery

Ulnar artery

Internal iliac artery

Deep palmar arch

Superficial palmar arch

Deep femoral artery

Figure 11.10   Major arteries of the systemic circulation, anterior view.

Major Veins of the Systemic Circulation

Although arteries are generally located in deep, well-protected body areas, many veins are more superficial and some are easily seen and palpated on the body surface. Most deep veins follow the course of the major arteries and, with a few exceptions, the naming of these veins is identical to that of their companion arteries. Major systemic arteries branch off the aorta, whereas the veins converge on the venae cavae, which enter the right atrium of the heart. Veins draining the head and arms empty into the **superior vena cava** and those draining the lower body empty into the **inferior vena cava.** These veins are described next and shown in Figure 11.11. As before, locate the veins on the figure as you read their descriptions.

Veins Draining into the Superior Vena Cava   Veins draining into the superior vena cava are named in a distal/proximal direction; that is, in the same direction the blood flows into the superior vena cava.

- The **radial** and **ulnar veins** are deep veins draining the forearm. They unite to form the deep **brachial vein,** which drains the arm and empties into the **axillary vein** in the axillary region.

- The **cephalic** (se-fal′ik) **vein** provides for the superficial drainage of the lateral aspect of the arm and empties into the axillary vein.

- The **basilic** (bah-sil′ik) **vein** is a superficial vein that drains the medial aspect of the arm and empties into the brachial vein proximally. The basilic and cephalic veins are joined at the anterior aspect of the elbow by the **median cubital vein.** (The median cubital vein is often chosen as the site for blood removal for the purpose of blood testing.)

- The **subclavian vein** receives venous blood from the arm through the axillary vein and from the skin and muscles of the head through the **external jugular vein.**

- The **vertebral vein** drains the posterior part of the head.

- The **internal jugular vein** drains the dural sinuses of the brain.

- The **brachiocephalic** (R. and L.) **veins** are large veins that receive venous drainage from the subclavian, vertebral, and internal jugular veins on their respective sides. The brachiocephalic veins join to form the **superior vena cava,** which enters the heart.

- The **azygos** (az′ĭ-gos) **vein** is a single vein that drains the thorax and enters the superior vena cava just before it joins the heart. (This vein is not illustrated in Figure 11.11.)

Veins Draining into the Inferior Vena Cava   The inferior vena cava, which is much longer than the superior vena cava, returns blood to the heart from all body regions below the diaphragm. As before, we will trace the venous drainage in a distal/proximal direction.

- The **anterior** and **posterior tibial veins** and the **peroneal vein** drain the leg (calf and foot). The posterior tibial vein becomes the **popliteal vein** at the knee and then the **femoral vein** in the thigh. The femoral vein becomes the **external iliac vein** as it enters the pelvis.

- The **great saphenous** (sah-fe′nus) **veins** are the longest veins in the body. They receive the superficial drainage of the leg. They begin at the **dorsal venous arch** in the foot and travel up the medial aspect of the leg to empty into the femoral vein in the thigh.

- Each **common iliac** (R. and L.) **vein** is formed by the union of the **external iliac vein** and the **internal iliac vein** (which drains the pelvis) on its own side. The common iliac veins join to form the inferior vena cava, which then ascends superiorly in the abdominal cavity.

- The **R. gonadal vein** drains the right ovary in females and the right testicle in males. (The **L. gonadal vein** empties into the left renal vein superiorly.) (The gonadal veins are not illustrated in Figure 11.11.)

- The **renal** (R. and L.) **veins** drain the kidneys.

- The **hepatic portal vein** is a single vein that drains the digestive tract organs and carries this blood through the liver before it enters the systemic circulation. (The hepatic portal system is discussed in the next section.)

- The **hepatic** (R. and L.) **veins** drain the liver.

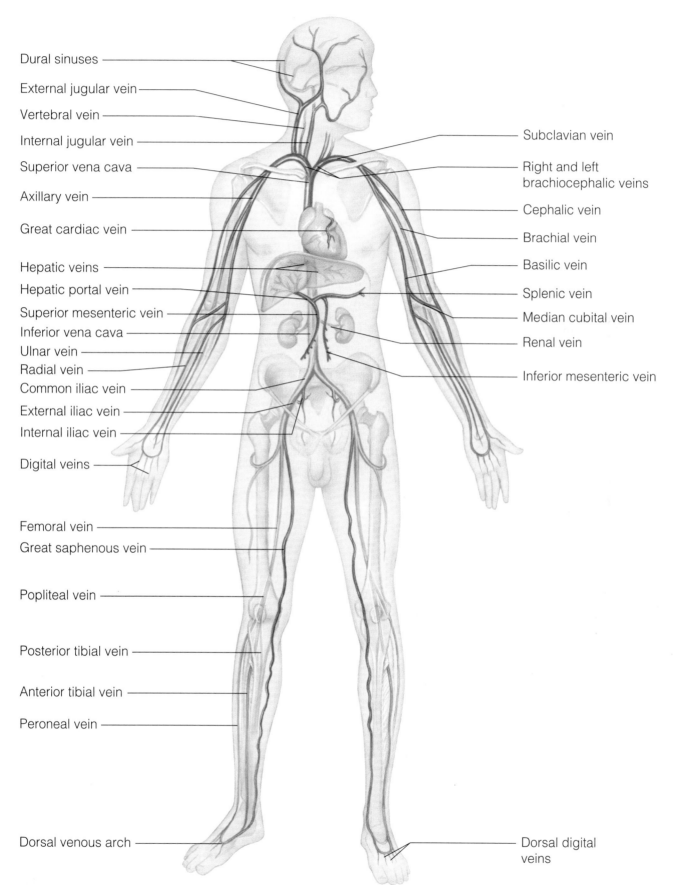

Dural sinuses

External jugular vein

Vertebral vein

Internal jugular vein

Superior vena cava

Axillary vein

Great cardiac vein

Hepatic veins

Hepatic portal vein

Superior mesenteric vein

Inferior vena cava

Ulnar vein

Radial vein

Common iliac vein

External iliac vein

Internal iliac vein

Digital veins

Femoral vein

Great saphenous vein

Popliteal vein

Posterior tibial vein

Anterior tibial vein

Peroneal vein

Dorsal venous arch

Subclavian vein

Right and left
brachiocephalic veins

Cephalic vein

Brachial vein

Basilic vein

Splenic vein

Median cubital vein

Renal vein

Inferior mesenteric vein

Dorsal digital
veins

Figure 11.11   Major veins of the systemic circulation, anterior view.
The vessels of the pulmonary circulation are not illustrated, accounting for
the incomplete appearance of the circulation from the heart.

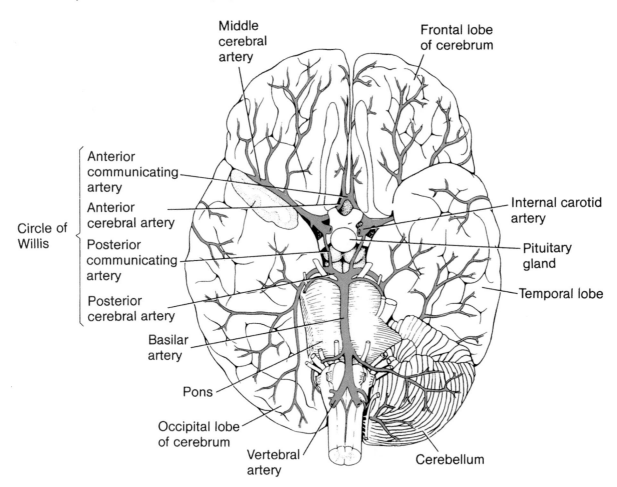

Figure 11.12   Arterial supply of the brain.

Special Circulations

Arterial Supply of the Brain and the Circle of Willis

A continuous blood supply to the brain is crucial, since a lack of blood for even a few minutes causes the delicate brain cells to die. The brain is supplied by two pairs of arteries, the internal carotid arteries and the vertebral arteries (see Figure 11.12).

The **internal carotid arteries,** branches of the common carotid arteries, run through the neck and enter the skull through the temporal bone. Once inside the cranium, each divides into the **anterior** and **middle cerebral arteries,** which supply most of the cerebrum.

The paired **vertebral arteries** pass upward from the subclavian arteries at the base of the neck. Within the skull, the vertebral arteries join to form the single **basilar artery,** which serves the brain stem and cerebellum as it travels upward. At the base of the cerebrum, the basilar artery divides

to form the **posterior cerebral arteries,** which supply the posterior part of the cerebrum.

The anterior and posterior blood supplies of the brain are united by small *communicating arterial branches.* The result is a complete circle of connecting blood vessels called the **circle of Willis,** which surrounds the base of the brain. The circle of Willis protects the brain, because it provides more than one route for blood to reach brain tissue in case of a clot or impaired blood flow anywhere in the system.

Hepatic Portal Circulation

The veins of the hepatic portal circulation drain the digestive organs, spleen, and pancreas, and deliver this blood to the liver through the **hepatic portal vein** (Figure 11.13). After you have just eaten, the hepatic portal blood contains large amounts of nutrients. Since the liver is a key body organ involved in maintaining the proper glucose, fat, and protein concentra-

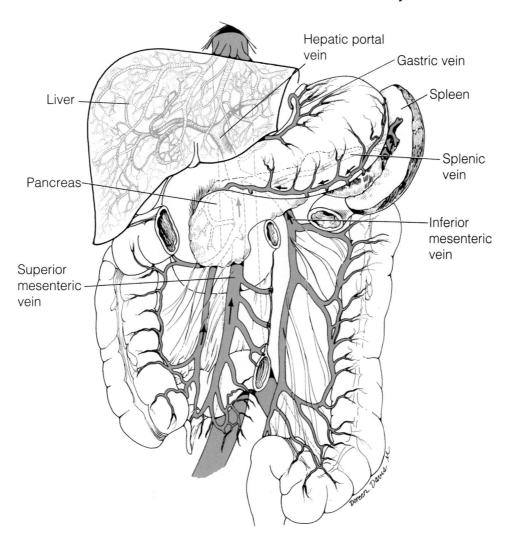

Figure 11.13   The hepatic portal circulation.

tions in the blood, this system "takes a detour" to ensure that the liver processes these substances before they enter the systemic circulation. As blood flows slowly through the liver, some of the nutrients are removed to be stored or processed in various ways for later release to the blood. The liver is drained by the hepatic veins that enter the inferior vena cava. Like the portal circulation that links the hypothalamus of the brain and the anterior pituitary gland (see Chapter 9), the hepatic portal circulation is a unique and unusual circulation. Normally, arteries feed capillary beds, which in turn drain into veins. Here we see *veins* feeding the liver circulation.

The **inferior mesenteric vein,** draining the terminal part of the large intestine, drains into the **splenic vein,** which itself drains the spleen, pan-

creas, and the left side of the stomach. The splenic vein and **superior mesenteric vein** (which drains the small intestine and the first part of the colon) join to form the hepatic portal vein. The **gastric vein,** which drains the right side of the stomach, drains directly into the hepatic portal vein.

Fetal Circulation   Since the lungs and digestive system are not yet functioning in a fetus, all nutrient, excretory, and gas exchanges occur through the placenta. Nutrients and oxygen move from the mother's blood into the fetal blood, and fetal wastes move in the opposite direction. As shown in Figure 11.14, the umbilical cord contains three blood vessels: one large **umbilical vein** and two smaller **umbilical arteries.** The umbilical vein carries blood rich in nutrients and oxygen to the

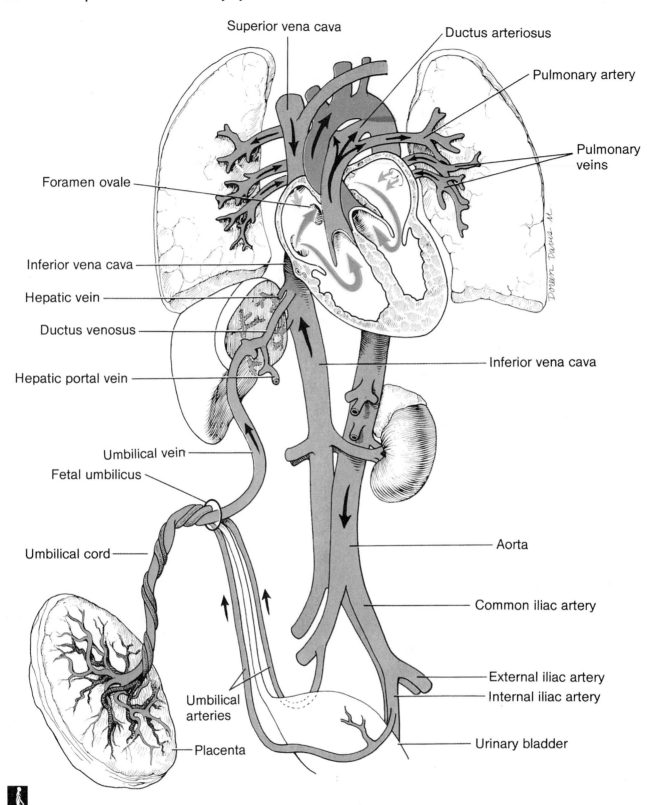

Figure 11.14 **The fetal circulation.**

fetus. The umbilical arteries carry carbon dioxide and debris-laden blood from the fetus to the placenta. As blood flows superiorly toward the heart of the fetus, most of it bypasses the immature liver through the **ductus venosus** (duk′tus ve-no′sus) and enters the inferior vena cava, which carries the blood to the right atrium of the heart.

Since fetal lungs are nonfunctional and collapsed, two shunts see to it that they are almost entirely bypassed. Some of the blood entering the right atrium is shunted directly into the left atrium through a flaplike opening in the interatrial septum, the **foramen ovale** (fo-ra′men o-val′e). Blood that does manage to enter the right ventricle is pumped out the pulmonary trunk, where it meets a second shunt, the **ductus arteriosus** (ar-ter″e-o′sus), a short vessel that connects the aorta and the pulmonary trunk. Because the collapsed lungs are a high-pressure area, blood tends to enter the systemic circulation through the ductus arteriosus. The aorta carries blood to the tissues of the fetal body and ultimately back to the placenta through the umbilical arteries.

At birth, or shortly after, the foramen ovale closes and the ductus arteriosus collapses and is converted to fibrous **ligamentum arteriosum** (lig″ah-men′tum ar-ter″e-o′sum) (see Figure 11.2a). As blood stops flowing through the umbilical vessels, they become obliterated, and the circulatory pattern becomes that of an adult.

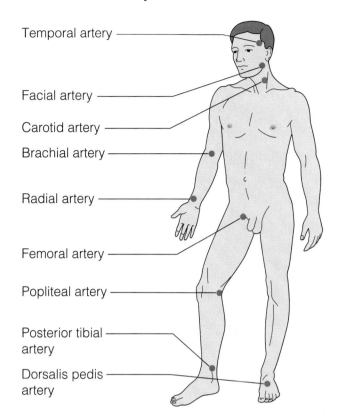

Figure 11.15   Body sites where the pulse is most easily palpated. (The specific arteries indicated are discussed on p. 328)

Physiology

A fairly good indication of the efficiency of a person's circulatory system can be obtained by taking arterial pulse and blood pressure measurements. These measurements, along with those of respiratory rate and body temperature, are referred to collectively as **vital signs** in clinical settings.

Arterial Pulse

The alternating expansion and recoil of an artery that occurs with each beat of the left ventricle creates a pressure wave—a **pulse**—that travels through the entire arterial system. Normally the pulse rate (pressure surges per minute) equals the heart rate (beats per minute). The pulse averages 70–76 beats per minute in a normal resting person. It is influenced by activity, postural changes, and emotions.

You can feel a pulse in any artery lying close to the body surface by compressing the artery against firm tissue; this provides an easy way of counting heart rate. Because it is so accessible, the point where the radial artery surfaces at the wrist (the radial pulse) is routinely used to take a pulse measurement, but there are several other clinically important arterial pulse points (see Figure 11.15). Because these same points are compressed to stop blood flow into distal tissues during hemorrhage, they are also called **pressure points.** For example, if you seriously cut your hand, you can stop the bleeding somewhat by compressing the radial artery or the brachial artery.

Try to palpate each of the pulse points shown in Figure 11.15 by placing the tips of your first two or three fingers of one hand over the artery at the site indicated. It helps to compress the artery firmly as you begin, and then immediately ease up on your pressure slightly. In each case, notice the regularity of the pulse and its relative strength.

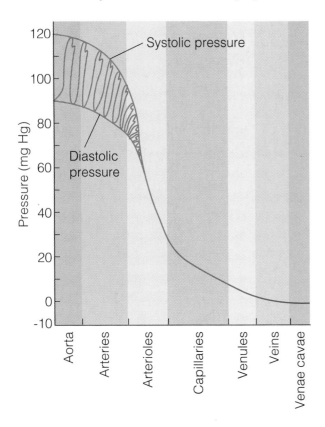

Figure 11.16   Blood pressure in various areas of the cardiovascular system.

Blood Pressure

Any system equipped with a pump that forces fluid through a network of closed tubes operates under pressure. The closer you get to the pump, higher the pressure. **Blood pressure** is the pressure the blood exerts against the inner walls of the blood vessels, and it is the force that keeps blood circulating continuously even between heartbeats. Unless stated otherwise, the term *blood pressure* is understood to mean the pressure within the large systemic arteries near the heart.

Blood Pressure Gradient   When the ventricles contract, they force blood into large, thick-walled elastic arteries that expand as the blood is pushed into them. The high pressure in these arteries forces the blood to continually move into areas where the pressure is lower. The pressure is highest in the large arteries and continues to drop throughout the pathway, reaching zero or negative pressure at the venae cavae (see Figure 11.16). Thus, the blood flows into the smaller arteries, then arterioles, capillaries, venules, veins, and fi-

nally back to the large venae cavae entering the right heart. Blood thus flows continually along a pressure gradient (from high to low pressure) as it makes its circuits day in and day out. Notice that if venous return depended entirely on a high blood pressure throughout the system, blood would probably never be able to complete its circuit back to the heart. This is why the valves in the larger veins, the milking activity of the skeletal muscles, and pressure changes in the thorax are so important.

The pressure differences between arteries and veins become very clear when these vessels are cut. If a vein is cut, the blood flows evenly from the wound; a lacerated artery produces rapid spurts of blood.

Continual blood flow absolutely depends on the stretchiness of the larger arteries and their ability to recoil and keep the pressure on the blood as it flows off into the circulation. To illustrate this, think of a garden hose with relatively hard walls. When the water is turned on, the water spurts out under high pressure because the hose walls don't expand. However, when the water faucet is suddenly turned off, the flow of water stops just as abruptly. This is because the walls of the hose cannot recoil to keep pressure on the water; therefore, the pressure drops and the flow of water stops. The importance of the elasticity of the arteries is best appreciated when it is lost, as happens in *arteriosclerosis*. Arteriosclerosis, also called "hardening of the arteries," is discussed in the "Closer Look" box on pp. 340–341.

Measuring Blood Pressure   Because the heart alternately contracts and relaxes, the off-and-on flow of blood into the arteries causes the blood pressure to rise and fall during each beat. Thus, two arterial blood pressure measurements are usually made: **systolic** (sis-to′lik) **pressure,** the pressure in the arteries at the peak of ventricular contraction, and **diastolic** (di″us-to′lik) **pressure,** the pressure when the ventricles are relaxing. Blood pressures are reported in millimeters of mercury (mm Hg), with the systolic pressure written first—120/80 (120 over 80) translates to a systolic pressure of 120 mm Hg and a diastolic pressure of 80 mm Hg. Most often, systemic arterial blood pressure is measured indirectly by the *auscultatory* (os-kul′tuh-tor-e) *method.* This procedure, as used to measure blood pressure in the brachial artery of the arm, is illustrated and described in Figure 11.17.

Figure 11.17 **Measurement of blood pressure.** (**a**) The course of the brachial artery of the arm. Assume a blood pressure of 120/70 in a young healthy person. (**b**) The blood pressure cuff is wrapped snugly around the arm just above the elbow and inflated until the cuff pressure exceeds the systolic blood pressure. At this point, blood flow into the arm is stopped and a brachial pulse cannot be felt or heard. (**c**) The pressure in the cuff is gradually reduced while the examiner listens (auscultates) carefully for sounds in the brachial artery with a stethoscope. The pressure read as the first soft tapping sounds are heard (the first point at which a small amount of blood is spurting through the constricted artery) is recorded as the systolic pressure. (**d**) As the pressure is reduced still further, the sounds become louder and more distinct, but when the artery is no longer constricted and blood flows freely, the sounds can no longer be heard. The pressure at which the sounds disappear is recorded as the diastolic pressure.

Effects of Various Factors on Blood Pressure Arterial blood pressure is directly related to cardiac output (the amount of blood pumped out of the left ventricle per minute) and peripheral resistance. Since regulation of cardiac output has already been considered (pp. 322–325), we will concentrate on peripheral resistance here.

Peripheral resistance is the amount of friction encountered by the blood as it flows through the blood vessels. It is increased by many factors, but probably the most important is the constriction, or narrowing, of blood vessels, especially arterioles, as a result of sympathetic nervous system activity or atherosclerosis. Increased blood volume or blood viscosity (thickness) also raises the peripheral resistance. Any factor that increases either the cardiac

output or peripheral resistance causes an almost immediate reflex rise in blood pressure. Many factors can alter blood pressure—age, weight, time of day, exercise, body position, emotional state, and various drugs, to name a few. The influence of a few of these factors is discussed next.

1. **Neural factors: the autonomic nervous system.** The parasympathetic division of the autonomic nervous system has little or no effect on blood pressure, but the sympathetic division is important and is responsive to many different factors. The major action of the sympathetic nerves on the vascular system is to cause *vasoconstriction* (vas″o-kon-strik′shun), or narrowing of the blood vessels, which increases the

blood pressure. The sympathetic center in the medulla of the brain is activated to cause vasoconstriction in many different circumstances (Figure 11.18). For example, when we stand up suddenly after lying down, the effect of gravity causes blood to pool in the vessels of the legs and feet and the blood pressure to drop. This activates *pressoreceptors* in the large arteries of the neck and chest. They send off warning signals that result in reflexive vasoconstriction that increases blood pressure back to homeostatic levels.

When blood volume suddenly decreases, as in hemorrhage, blood pressure drops and the heart begins to beat more rapidly (as it tries to compensate). However, because venous return is reduced by blood loss, the heart also beats weakly and inefficiently. In such cases, the sympathetic nervous system causes vasoconstriction to increase the blood pressure so that (hopefully) the venous return is increased and the circulation can be continued.

The final example concerns sympathetic nervous system activity when we exercise vigorously or are frightened and have to make a hasty escape. Under these conditions, there is a generalized vasoconstriction *except* in the skeletal muscles. The vessels of the skeletal muscles dilate to increase the blood flow to the working muscles. (It should be noted that the sympathetic nerves *never* cause vasoconstriction of blood vessels of the heart or brain.)

2. **Renal factors: the kidneys.** The kidneys play a major role in regulating arterial blood pressure by altering blood volume. As blood pressure (and/or blood volume) increases beyond normal, the kidneys allow more water to leave the body in the urine. Since the source of this water is the bloodstream, blood volume decreases, which in turn decreases blood pressure. However, when arterial blood pressure falls, the kidneys retain body water, increasing blood volume, and blood pressure rises (see Figure 11.18).

In addition, when arterial blood pressure is low, certain kidney cells release the enzyme **renin** into the blood. Renin triggers a series of chemical reactions that result in the formation of *angiotensin II,* a potent vasoconstrictor chemical. Angiotensin also stimulates the adrenal cortex to release aldosterone, a hormone that enhances sodium ion reabsorption by the kidneys (see p. 459). As sodium moves into the blood, water follows. Thus, blood volume and blood pressure both rise.

3. **Temperature.** In general, cold has a vasoconstricting effect. This is why your exposed skin feels cold to the touch on a winter day and why cold compresses are recommended to prevent swelling of a bruised area. On the other hand, since heat has a vasodilating effect, warm compresses are used to speed the circulation into an inflamed area.

4. **Chemicals.** The effects of chemical substances, many of which are drugs, on blood pressure are widespread and well known in many cases. Just a few examples will be given here. *Epinephrine* increases both heart rate and blood pressure. *Nicotine* increases blood pressure by causing vasoconstriction. Both *alcohol* and *histamine* cause vasodilation and decrease the blood pressure. The reason a person who has "one too many" becomes flushed is that the skin vessels have been dilated by alcohol.

5. **Diet.** Although medical opinions tend to change and are at odds from time to time, it is generally believed that a diet low in salt, saturated fats, and cholesterol helps to prevent *hypertension,* or high blood pressure.

Variations in Blood Pressure   In normal adults at rest, systolic blood pressure varies between 110 and 140 mm Hg, and diastolic pressure between 75 and 80 mm Hg—but blood pressure varies considerably from one person to another. What is normal for you may not be normal for your grandfather or your neighbor. Blood pressure varies with age, weight, race, mood, physical activity, and posture. Nearly all these variations can be explained in terms of the factors affecting blood pressure that have already been discussed.

Hypotension, or low blood pressure, is generally considered to be a systolic blood pressure below 100 mm Hg. In many cases, it simply reflects individual differences and is no cause for concern. In fact, low blood pressure is an expected result of physical conditioning and is often associated with long life and an old age free of illness.

Figure 11.18 Summary of factors causing an increase in arterial blood pressure. (SV = stroke volume; HR = heart rate)

Elderly people may experience temporary low blood pressure and dizziness when they rise suddenly from a reclining or sitting position—a condition called *orthostatic hypotension*. Because an aging sympathetic nervous system reacts more slowly to postural changes, blood pools briefly in the lower limbs, reducing blood pressure and, consequently, blood delivery to the brain. Making postural changes more slowly to give the nervous system time to make the necessary adjustments usually prevents this problem. ❏

Chronic hypotension (not explained by physical conditioning) may hint at poor nutrition and in-adequate levels of blood proteins. Because blood viscosity is low, blood pressure is also lower than normal. Acute hypotension is one of the most important warnings of *circulatory shock,* a condition in which the blood vessels are inadequately filled and blood cannot circulate normally. The most common cause is blood loss.

A brief elevation in blood pressure is a normal response to fever, physical exertion, and emotional upset, such as anger and fear. Persistent **hypertension,** or **high blood pressure,** is pathological, and is defined as a condition of sustained elevated arterial pressure of 140/90 or higher.

A CLOSER LOOK

ATHEROSCLEROSIS? GET OUT THE CARDIOVASCULAR DRĀNO!

When pipes get clogged, it is usually because something gets stuck in them—a greasy mass of bone bits or a hair ball in the sink drain. But in arteries narrowed by **atherosclerosis,** the damming-up process occurs from the inside out: The walls of the vessels thicken and protrude into the vessel lumen. Once this happens, it does not take much to close the vessel completely. A roaming blood clot or arterial spasms can do it.

Although all blood vessels are susceptible to this serious degenerative condition of blood vessel walls, for some unknown reason the aorta and the coronary arteries serving the heart are most often affected. The disease progresses through many stages before the arterial walls actually become hard and approach the stage of the rigid tube system described in the text, but some of the earlier stages are just as lethal as those that come later.

What triggers this scourge of blood vessels? According to the *response to injury* hypothesis, the initial event is damage to the tunica intima caused by blood-borne chemicals such as carbon monoxide (present in cigarette smoke or auto exhaust); by viruses; or by physical factors such as a blow or persistent hypertension. Once a break has occurred, blood platelets cling to the injured site and initiate clotting to prevent blood loss. Injured endothelial cells release chemotactic agents and chemicals that increase the permeability of the endothelium to fats and cholesterol, allowing them to take up residence just deep to the tunica intima. Monocytes attracted to the area become macrophages that gorge themselves on the fat, as do migrating local smooth muscle cells from the media of the blood vessel wall. The result is the so-called *fatty streak stage* characterized by greasy gray to yellow lesions called *atherosclerotic plaques.* When these small, fatty mounds of muscle begin to protrude into the vessel lumen, the condition is called *atherosclerosis* (see photo b; compare it to photo a).

Arteriosclerosis is the end stage of the disease. As enlarging plaques hinder diffusion of nutrients

(a)

(b)

Comparison of a normal artery and atherosclerotic artery. (a) Cross section of a normal artery (170×). **(b)** Cross section of an artery that is partially occluded by atherosclerotic plaque (30×).

from the blood to the deeper tissues of the artery wall, smooth muscle cells in the tunica media die, and are gradually replaced by nonelastic scar tissue. Then, calcium salts are deposited in the lesions. Collectively, these events cause the arterial wall to become frayed and often ulcerated. The increased rigidity of the vessels leads to hypertension. Together, these events increase the risk of myocardial infarcts, strokes, and kidney failure.

So what can help when the damage is done and the heart is at risk because of atherosclerotic coronary vessels? In the past, the only choice has been coronary artery bypass surgery, in which vessels removed from the legs or thoracic cavity are implanted in the heart to restore circulation. More recently, devices threaded through blood vessels to obstructed sites have become part of the ammunition of cardiovascular medicine. *Balloon angioplasty* uses a catheter with a balloon packed into its tip (see the two photos in c). When the catheter reaches the blockage, the balloon is inflated, and the fatty mass is compressed against the vessel wall. However, this procedure is useful only when a few, very localized obstructions are present. A newer catheter device uses a laser beam to vaporize the arterial clogs. Although these intravascular devices are faster, cheaper, and much less risky than bypass surgery, they carry with them the same major shortcoming: They do nothing to stop the underlying disease, and in time new blockages occur.

It was hoped that cholesterol-lowering drugs such as lovastatin would act as "cardiovascular Drāno" and simply wash the fatty plaques off the walls. However, these drugs are only used for those whose diet changes have failed to reduce blood cholesterol levels, and their side effects—nausea, bloating, and constipation—create such misery that many people simply stop taking them. When a blood clot is trapped by the diseased vessel walls, the answer may be a *clot-dissolving agent*. Examples include the powerful enzyme *streptokinase* and *tissue plasminogen activator* (*t-PA*), a naturally occurring substance now being produced by genetic engineering techniques. Injecting the enzyme or t-PA directly into the heart restores blood flow quickly and puts an early end to many heart attacks in progress.

There is little doubt that lifestyle factors—emotional stress, smoking, obesity, high-fat and high-cholesterol diets, and lack of exercise—contribute to both atherosclerosis and hypertension. If these are the risk factors (and indeed they are), then why not just have patients at risk change their lifestyle? This is not as easy as it seems. Although taking antioxidants (E and C vitamins and beta carotene) and exercising more may "undo" some of the damage, old habits die hard, and Americans like their burgers and butter. Can atherosclerosis be reversed to give the heart a longer and healthier life? One thing is certain: If so, many more people with diseased arteries may be more willing to trade lifelong habits for a healthy old age!

(c)

(**c**) Angiographs of an occluded artery (left) and the same artery after being cleared by balloon angioplasty (right). These views were prepared by a computer imaging technique, digital subtraction angiography (DSA).

Chronic hypertension is a common and dangerous disease that warns of increased peripheral resistance. Although it progresses without symptoms for the first 10 to 20 years, it slowly and surely strains the heart and damages the arteries. For this reason, hypertension is often called the "silent killer." Because the heart is forced to pump against increased resistance, it must work harder and, in time, the myocardium enlarges. When finally strained beyond its capacity to respond, the heart weakens and its walls become flabby. Hypertension also ravages blood vessels, causing small tears in the endothelium that accelerate the progress of atherosclerosis (see the "Closer Look" box on pp. 340–341).

Although hypertension and atherosclerosis are often linked, it is difficult to blame hypertension on any distinct anatomical pathology. In fact, about 90 percent of hypertensive people have *primary,* or *essential, hypertension,* which cannot be accounted for by any specific organic cause. However, factors such as diet, obesity, heredity, race, and stress appear to be involved. For instance, more women than men and more blacks than whites are hypertensive. Hypertension runs in families; the child of a hypertensive parent is twice as likely to develop high blood pressure as is a child of parents with normal blood pressure. Dietary factors presumed to contribute to hypertension include high cholesterol, saturated fat, and sodium intakes. High blood pressure is common in obese people because the total length of their blood vessels is relatively greater than that in thinner individuals. For each pound of fat, miles of additional blood vessels are required, making the heart work harder to pump blood over longer distances. ❏

Capillary Exchange

Capillaries form an intricate network among the body's cells such that no substance has to diffuse very far to enter or leave a cell. The substances exchanged first diffuse through an intervening space (Figure 11.19a) filled with *interstitial fluid (tissue fluid).*

Substances tend to move to and from body cells according to their concentration gradients. Thus, oxygen and nutrients leave the blood and move into the tissue cells, and carbon dioxide and other wastes exit the tissue cells and enter the blood. Basically, substances entering or leaving the bloodstream may take one of four routes across the plasma membranes of the single layer of endothelial cells forming the capillary wall (Figure 11.19b).

1. As with all cells, substances can diffuse directly through (cross) their plasma membranes if they are lipid-soluble (like the respiratory gases).

2. Certain lipid-insoluble substances may enter or leave the blood and/or endothelial cells within vesicles, that is, by endocytosis or exocytosis.

Diffusion of substances via the other two routes depends on the specific structural (and permeability) characteristics of the capillary.

3. Limited passage of fluid and small solutes is allowed by *intercellular clefts* (gaps or areas of plasma membrane not joined by tight junctions). It is safe to say that, with the exception of brain capillaries, which are entirely secured together by tight junctions (the basis of the blood-brain barrier described in Chapter 7), most of our capillaries have intercellular clefts.

4. Very free passage of small solutes and fluids is allowed by *fenestrated capillaries.* These unique capillaries are found where absorption is a priority (intestinal capillaries or those serving endocrine glands) or where filtration occurs (the kidney). A fenestra is an oval pore (*fenestra* = window) or opening and is usually covered by a delicate membrane (see Figure 11.19b). Even so, a fenestra is much more permeable than other regions of the plasma membrane.

Only substances unable to pass by one of these routes are prevented from leaving (or entering) the capillaries. This includes protein molecules (in plasma or interstitial fluid) and blood cells.

Besides the exchanges made via vesicles, and by passive diffusion through endothelial cell plasma membranes, clefts, or fenestrations, there are active forces operating at capillary beds. Because of their intercellular clefts, and fenestrations in some cases, capillaries are leaky. Hence, blood pressure tends to force fluid (and solutes) outward, and osmotic pressure tends to draw fluid back into the bloodstream because blood has a higher solute concentration (due to its plasma proteins) than does interstitial fluid. (To review the principles of

(a)

(b)

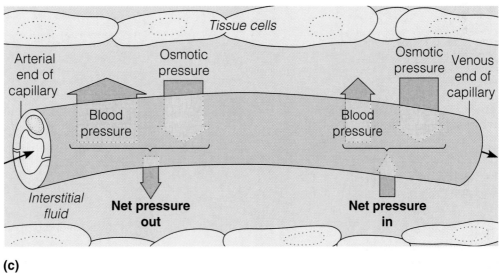

(c)

Figure 11.19   Capillary transport mechanisms. (a) Substances exchanged between the blood and the tissue cells must diffuse through intervening interstitial fluid. **(b)** The four possible pathways or routes of transport across the wall of an endothelial cell. (The endothelial cell is illustrated as if cut in cross section.) **(c)** Bulk fluid flows across capillary walls depend largely on the difference between the hydrostatic (blood) pressure and the osmotic pressure at different regions of the capillary bed. Blood pressure, which predominates at the arterial end of the bed, provides an outward force. Osmotic pressure, which predominates at the venous end of the bed, draws the fluid back. Hence, fluid is forced out of a capillary through clefts at the arterial end and returns to the blood at the venous end.

osmosis, see p. 65.) Whether fluid moves out of or into the capillary depends on the difference between the two pressures. As a rule, blood pressure is higher at the arterial end of the capillary bed and osmotic pressure is higher at the venous end. Consequently, fluid moves out of the capillaries at the beginning of the bed and is reclaimed at the opposite (venule) end (Figure 11.19c).

SYSTEMS IN SYNC

Homeostatic Relationships Between the Cardiovascular System and Other Body Systems

Endocrine System
- The cardiovascular system delivers oxygen and nutrients; carries away wastes; blood serves as a transport vehicle for hormones
- Several hormones influence blood pressure (epinephrine, ANF, thyroxine, ADH); estrogen maintains vascular health in women

Lymphatic System/Immunity
- The cardiovascular system delivers oxygen and nutrients to lymphatic organs, which house immune cells; transports lymphocytes and antibodies; carries away wastes
- The lymphatic system picks up leaked fluid and plasma proteins and returns them to the cardiovascular system; its immune cells protect cardiovascular organs from specific pathogens

Digestive System
- The cardiovascular system delivers oxygen and nutrients; carries away wastes
- The digestive system provides nutrients to the blood including iron and B vitamins essential for RBC (and hemoglobin) formation

Urinary System
- The cardiovascular system delivers oxygen and nutrients; carries away wastes; blood pressure maintains kidney function
- The urinary system helps regulate blood volume and pressure by altering urine volume and releasing renin

Muscular System
- The cardiovascular system delivers oxygen and nutrients; carries away wastes
- Aerobic exercise enhances cardiovascular efficiency and helps prevent arteriosclerosis; the muscle "pump" aids venus return

Nervous System
- The cardiovascular system delivers oxygen and nutrients; carries away wastes
- ANS regulates cardiac rate and force; sympathetic division maintains blood pressure and controls blood distribution according to need

Respiratory System
- The cardiovascular system delivers oxygen and nutrients; carries away wastes
- The respiratory system carries out gas exchange: loads oxygen and unloads carbon dioxide from the blood; respiratory "pump" aids venous return

Cardiovascular System

Reproductive System
- The cardiovascular system delivers oxygen and nutrients; carries away wastes
- Estrogen maintains vascular health in women

Integumentary System
- The cardiovascular system delivers oxygen and nutrients; carries away wastes
- The skin's blood vessels provide an important blood reservoir and a site for heat loss from body

Skeletal System
- The cardiovascular system delivers oxygen and nutrients and carries away wastes
- Bones are the site of hematopoiesis; protect cardiovascular organs by enclosure; provide a calcium depot

344

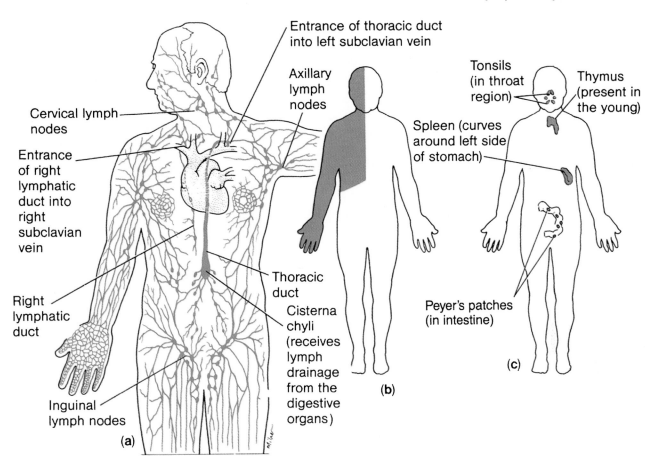

Figure 11.20   Lymphatic system. (**a**) Distribution of lymphatic vessels and lymph nodes. (**b**) Shading shows the body area drained by the right lymphatic duct; the rest of the body is drained by the thoracic duct. (**c**) Body location of the tonsils, thymus, spleen, and Peyer's patches.

LYMPHATIC SYSTEM

When we mentally tick off the names of the body's organ systems, some—like the nervous, cardiovascular, and digestive systems—trip instantly into our thoughts. We are not so likely to remember the **lymphatic system,** a meandering network of lymphatic vessels, lymph nodes, and various other lymphoid organs and tissues scattered throughout the body (Figure 11.20). Yet without this quietly working system, our cardiovascular system would cease working and our immune system would be hopelessly impaired. Let us examine how the organs of the lymphatic system accomplish their vitally important functions.

Lymphatic Vessels

As just described, fluid is forced out of the blood at the arterial ends of the capillary beds and reabsorbed into the bloodstream at the venous end. However, *not all* the lost fluid is returned to the blood at the venous end. The excess fluid that lags behind in the tissue spaces must eventually return to the blood if the vascular system is to have sufficient blood volume to operate properly. If it does not, fluid accumulates in the tissues, producing *edema,* which, if excessive, impairs the ability of tissue cells to make exchanges with the blood. The function of the **lymphatic vessels** is to pick up this excess tissue fluid, now called **lymph,** and return it to the bloodstream.

The lymphatic vessels, also called *lymphatics,* form a one-way system, and lymph flows only toward the heart. The microscopic, blind-ended *lymph capillaries* spiderweb through all the tissues of the body and absorb the leaked fluid (primarily water and a small amount of dissolved proteins). Lymph is transported through successively larger lymphatic vessels until it is finally returned to the venous system through one of the two large ducts in the thoracic region. The **right lymphatic duct** drains the lymph from the right arm and the right side of the head and thorax. The large **thoracic duct** receives lymph from the rest of the body, as shown in Figure 11.20. Both ducts empty the lymph into the subclavian vein on their own side of the body.

Like the veins of the cardiovascular system, the lymphatic vessels are thin-walled, and the larger ones have valves. Since the lymphatic system is a low-pressure but pumpless system, it was assumed that transport of lymph depends entirely on the milking action of the skeletal muscles and the pressure changes in the thorax during breathing (that is, the muscular and respiratory "pumps"). However, recent research has revealed that the smooth muscle in the lymphatic vessel walls contracts rhythmically and actually helps "pump" the lymph along.

Lymph Nodes

More closely related to the immune system, the **lymph nodes** help protect the body by removing foreign material such as bacteria and tumor cells from the lymphatic stream and by producing lymphocytes that function in the immune response.

As lymph is transported toward the heart, it is filtered through the thousands of lymph nodes that cluster along the lymphatic vessels. Particularly large clusters are found in the inguinal, axillary, and cervical regions of the body (see Figure 11.20). Within the lymph nodes are *macrophages* (mak'ro-fāj-ez), which engulf and destroy bacteria, viruses, and other foreign substances in the lymph before it is returned to the blood. Collections of *lymphocytes* (a type of white blood cell) are also strategically located in the lymph nodes and respond to foreign substances in the lymphatic stream. Although we are not usually aware of the protective nature of

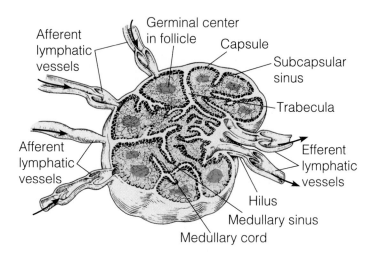

Figure 11.21   Structure of a lymph node. Longitudinal view of the internal structure of a lymph node and associated lymphatics. Notice that several afferent lymphatics enter the node, whereas fewer efferent lymphatics exit at its hilus.

the lymph nodes, most of us have had swollen glands during an active infection. This swelling is a result of the trapping function of the nodes.

Lymph nodes vary in shape and size, but most are kidney-shaped, less than 1 inch long, and "buried" in the connective tissue that surrounds them. Each node is surrounded by a fibrous capsule from which strands called *trabeculae* (trah-bek'yu-le) extend inward to divide the node into a number of compartments (see Figure 11.21). The internal framework is a network of soft reticular connective tissue that supports a continually changing population of lymphocytes. (As described in Chapter 10, lymphocytes arise from the red bone marrow, but then migrate to the lymphatic organs, where they proliferate further.)

The outer part of the node, its *cortex,* contains collections of lymphocytes called *follicles.* Many of the follicles have dark-staining centers called *germinal centers.* These centers enlarge when the lymphocytes (the B cells) are generating daughter cells called *plasma cells,* which release antibodies. The rest of the cortical cells are lymphocytes "in transit," the so-called T cells that circulate continuously between the blood, lymph nodes, and lymphatic stream, performing their surveillance role. The phagocytic macrophages are located in the central *medulla* area of the lymph node. (The precise roles of the cells of the lymph nodes in immunity are discussed in Chapter 12.)

Lymph enters the convex side of a lymph node through *afferent lymphatic vessels*. It then flows through a number of *sinuses* that cut through the lymph node, and finally exits from the node at its *hilus* (hi'lus), the indented region, via *efferent lymphatic vessels*. Because there are fewer efferent vessels draining the node than afferent vessels feeding it, the flow of lymph through the node is very slow. This allows time for the lymphocytes and macrophages to perform their protective functions. Lymph nodes help rid the body of infectious agents and cancer cells, but sometimes they are overwhelmed by the very agents they are trying to destroy. For example, when large numbers of bacteria or viruses are trapped in the nodes, the nodes become inflamed and tender to the touch. The lymph nodes can also become secondary cancer sites, particularly in cancers that use lymphatic vessels to spread throughout the body. The fact that cancer-infiltrated lymph nodes are swollen but not painful helps to distinguish cancerous lymph nodes from those infected by microorganisms. ❏

Other Lymphoid Organs

Lymph nodes are just one example of the many types of **lymphoid organs** in the body. Others are the spleen, thymus gland, tonsils, and Peyer's patches of the intestine (see Figure 11.20c), as well as bits of lymphatic tissue scattered in the epithelial and connective tissues. The common feature of all these organs is a predominance of reticular connective tissue and lymphocytes. Although all lymphoid organs have roles in protecting the body, only the lymph nodes filter lymph.

The **spleen** is a blood-rich organ that filters blood. It is located in the left side of the abdominal cavity and extends to curl around the anterior aspect of the stomach. Instead of filtering lymph, the spleen filters and cleanses blood of bacteria, viruses, and other debris. Its most important function is to destroy worn-out red blood cells and return some of their breakdown products to the liver. For example, iron is used again for making hemoglobin, and the rest of the hemoglobin molecule is secreted in bile. Other functions of the spleen include storing platelets and acting as a blood reservoir (as does the liver). During hemorrhage, both the spleen and liver contract and empty their contained blood into the circulation to help bring the blood volume back to normal levels. In the fetus, the spleen is an important hematopoietic (blood-cell forming) site, but only lymphocytes are produced by the adult spleen.

The **thymus,** which functions at peak levels only during youth, is a lymphatic mass found low in the throat overlying the heart. As described in Chapter 9, the thymus produces hormones, *thymosin* and others, that function in the programming of certain lymphocytes so they can carry out their protective roles in the body.

The **tonsils** are small masses of lymphatic tissue that ring the pharynx (the throat), where they are found in the mucosa. Their job is to trap and remove any bacteria or other foreign pathogens entering the throat. They carry out this function so efficiently that sometimes they become congested with bacteria and become red, swollen, and sore.

Peyer's patches, which resemble tonsils, are found in the wall of the small intestine. The macrophages of Peyer's patches are in an ideal position to capture and destroy bacteria (always present in tremendous numbers in the intestine), thereby preventing them from penetrating the intestinal wall. Peyer's patches and the tonsils are part of the collection of small lymphoid tissues referred to as **mucosa-associated lymphatic tissue (MALT).** Collectively, MALT acts as a sentinel to protect the upper respiratory and digestive tracts from the never-ending attacks of foreign matter entering those cavities.

DEVELOPMENTAL ASPECTS OF THE CIRCULATORY SYSTEM

The heart begins as a simple tube in the embryo. It is beating and busily pumping blood by the fourth week of pregnancy. During the next three weeks, the heart continues to change and mature, finally becoming a four-chambered structure capable of acting as a double pump—all without missing a beat! During fetal life, the collapsed lungs and nonfunctional liver are mostly bypassed by the blood, through special vascular shunts. After the seventh

SYSTEMS IN SYNC

Homeostatic Relationships Between the *Lymphatic System/Immunity* and Other Body Systems

Endocrine System
- Lymphatic vessels pick up leaked fluids and proteins; lymph distributes hormones; immune cells protect endocrine organs from pathogens
- The thymus produces hormones that promote development of lymphatic organs and "program" T lymphocytes

Lymphatic System/Immunity

Respiratory System
- Lymphatic vessels pick up leaked fluid and proteins from respiratory organs; immune cells protect respiratory organs from specific pathogens; lymphocytes populate the tonsils; plasma cells in the respiratory mucosa secrete IgA to prevent pathogen invasion of deeper tissues
- The lungs provide oxygen needed by lymphoid/immune cells and eliminates carbon dioxide; the pharynx houses some lymphoid organs (tonsils); the respiratory "pump" aids lymph flow

Digestive System
- Lymphatic vessels pick up leaked fluids and proteins from digestive organs; lymph transports some products of fat digestion to the blood; lymphoid nodules in the wall of the intestine prevent invasion of pathogens
- The digestive system digests and absorbs nutrients needed by cells of lymphatic organs; gastric acidity inhibits pathogens' entry into blood

Urinary System
- Lymphatic vessels pick up leaked fluid and proteins from urinary organs; immune cells protect urinary organs from specific pathogens
- Urinary system eliminates wastes and maintains homeostatic water/acid-base/electrolyte balance of the blood for lymphoid/immune cell functioning; urine flushes some pathogens out of the body

Muscular System
- Lymphatic vessels pick up leaked fluid and proteins from skeletal muscle; immune cells protect muscles from specific pathogens
- The skeletal muscle "pump" aids the flow of lymph; muscles protect superficial lymph nodes

Nervous System
- The lymphatic vessels pick up leaked plasma fluid and proteins in the peripheral nervous system structures; immune cells protect peripheral nervous system structures from specific pathogens
- The nervous system innervates larger lymphatic vessels; the brain helps regulate immune response

Cardiovascular System
- Lymphatic vessels pick up leaked plasma and proteins; spleen destroys aged RBCs, stores iron, and removes debris from blood; immune cells protect cardiovascular organs from specific pathogens
- Blood is the source of lymph; lymphatics develop from veins; blood provides the route for circulation of immune elements

Reproductive System
- Lymphatic vessels pick up leaked fluid and proteins from reproductive organs; immune cells protect the organs from specific pathogens
- Acidity of vaginal secretions prevents bacterial multiplication

Integumentary System
- Lymphatic vessels pick up leaked plasma fluid and proteins from the dermis; lymphocytes in lymph enhance the skin's protective role by defending against specific pathogens via the immune response
- The skin's keratinized epithelium provides a mechanical barrier to pathogens; acid pH of skin secretions inhibits growth of bacteria on skin

Skeletal System
- Lymphatic vessels pick up leaked plasma fluid and proteins from the periostea; immune cells protect bones from specific pathogens
- The bones house hematopoietic tissue (red marrow) which produces the lymphocytes (and macrophages) that populate the lymphoid organs and provide body immunity

week of development, few changes other than growth occur in the fetal circulation until birth. Shortly after birth, the bypass structures become blocked, and the special umbilical vessels stop functioning.

The lymphatic vessels, which bud from the veins of the blood vascular system, are obvious by the fifth week of development. Except for the spleen, the lymphoid organs are poorly developed before birth. However, shortly after birth, they become heavily populated with lymphocytes as the immune system begins to function.

Congenital heart defects account for about half of infant deaths resulting from all congenital defects. Environmental interferences, such as maternal infection and ingested drugs during the first three months of pregnancy (when the embryonic heart is forming), seem to be the major causes of such problems. Congenital heart defects may include a ductus arteriosus that does not close, septal openings, and other structural abnormalities of the heart. Such problems can usually be corrected surgically. ❑

In the absence of congenital heart problems, the heart usually functions smoothly throughout a long lifetime for most people. Homeostatic mechanisms are so effective that we rarely are aware of when the heart is working harder. The heart will hypertrophy and its cardiac output will increase substantially if we exercise regularly and aerobically (that is, vigorously enough to force it to beat at a higher than normal rate for extended periods of time). Not only does the heart become a more powerful pump, it also becomes more efficient: Pulse rate and blood pressure decrease. An added benefit of aerobic exercise is that it clears fatty deposits from the blood vessel walls, helping to slow the progress of atherosclerosis. However, let's raise a caution flag here: The once-a-month or once-a-year tennis player or snow shoveler has not built up this type of heart endurance and strength. When such an individual pushes his or her heart too much, it may not be able to cope with the sudden demand. This is why many weekend athletes are myocardial infarct victims.

As we get older, more and more signs of circulatory system disturbances start to appear. In some, weakness of the valves of the veins, leading to *varicose veins,* is a major problem. Varicose veins are common in people who stand for long periods

of time (for example, dentists and hairdressers) and in obese (or pregnant) individuals. The common factor is the pooling of blood in the feet and legs and inefficient venous return resulting from inactivity or pressure on the veins. In any case, the overworked valves give way and the veins become twisted and dilated. A serious complication of varicose veins is *thrombophlebitis* (throm″bo-fle-bi′tis), an inflammation of a vein that results when a clot forms in a vessel with poor circulation. Since all venous blood must pass through the pulmonary circulation before traveling through the body tissues again, a common consequence of thrombophlebitis is clot detachment and *pulmonary embolism,* which is a life-threatening condition.

Not everyone has varicose veins, but we all have progressive atherosclerosis. Some say the process begins at birth, and there's an old saying that goes, "You are only as old as your arteries," referring to this degenerative process. The gradual loss in elasticity in the blood vessels leads to hypertension and hypertensive heart disease. The insidious filling of the blood vessels with fatty, calcified deposits leads most commonly to **coronary artery disease.** Also, as described in Chapter 10, the roughening of the vessel walls encourages thrombus formation. At least 30 percent of the population in the United States has hypertension by the age of 50, and cardiovascular disease causes more than one-half of the deaths in those over age 65. Although the aging process itself contributes to changes in the walls of the blood vessels that can lead to strokes or myocardial infarcts, most researchers feel that diet, not aging, is the single most important contributing factor to cardiovascular diseases. There is some agreement that the risk is lowered if people eat less animal fat, cholesterol, and salt. Other recommendations include avoiding stress, eliminating cigarette smoking, and taking part in a regular, moderate exercise program.

Lymphatic system problems are relatively uncommon, but when they do occur, they are painfully obvious. For example, when the lymphatic vessels are blocked, as in *elephantiasis* (a tropical disease in which the lymphatics become clogged with parasitic worms), or when lymphatics are removed (as in radical breast surgery), severe edema is the result. However, lymphatic vessels removed surgically do grow back in time.

Important Terms

arteriole (ar-ter′e-ōl)

artery

atrioventricular (a″tre-o-ven-trik′u-lar) **valves**

atrium (a′tre-um)

capillary (kap′ĭ-lar″e)

cardiac cycle

cardiac output

diastolic (di″us-to′lik) **pressure**

myocardium (mi″o-kar′de-um)

pericardium (per″i-kar′de-um)

peripheral resistance

pulmonary circulation

pulse

Purkinje (pur-kin′je) **fibers**

SA node

semilunar (sem″e-lu′nar) **valves**

systemic circulation

systolic (sis-to′lik) **pressure**

vein

ventricles (ven′trĭ-kulz)

Summary

Cardiovascular System: The Heart (pp. 315–325)

1. The heart, located in the thorax, is flanked laterally by the lungs and enclosed in a pericardium.

2. The bulk of the heart (myocardium) is composed of cardiac muscle. The heart has four hollow chambers, the two atria (receiving chambers), and two ventricles (discharging chambers), each lined with endocardium. The heart is divided longitudinally by a septum.

3. The heart functions as a double pump. The right heart is the pulmonary pump (right heart to lungs to left heart). The left heart is the systemic pump (left heart to body tissues to right heart).

4. Four valves prevent backflow of blood in the heart. The AV valves (mitral and tricuspid) prevent backflow into the atria when the ventricles are contracting. The semilunar valves prevent backflow into the ventricles when the heart is relaxing. The valves open and close in response to pressure changes in the heart.

5. The myocardium is nourished by the coronary circulation, which consists of the right and left coronary arteries and their branches, and drained by the cardiac veins and the coronary sinus.

6. The time and events occurring from one heartbeat to the next is the cardiac cycle.

7. As the heart beats, sounds resulting from the closing of the valves ("lub-dup") can be heard. Faulty valves reduce the efficiency of the heart as a pump and result in abnormal heart sounds (murmurs).

8. Cardiac muscle is able to initiate its own contraction in a regular way, but its rate is influenced by both intrinsic and extrinsic factors. The intrinsic conduction system increases the rate of heart contraction and ensures that the heart beats as a unit. The SA node is the heart's pacemaker.

9. Cardiac output, the amount of blood pumped out by each ventricle in one minute, is the product of heart rate (HR) × stroke volume (SV). SV is the amount of blood ejected by a ventricle with each beat.

10. SV rises or falls with the volume of venous return. HR is influenced by the nerves of the autonomic nervous system, drugs (and other chemicals), and ion levels in the blood.

Cardiovascular System: Blood Vessels (pp. 325–343)

1. Arteries, which transport blood from the heart, and veins, which conduct blood back to the heart, are conducting vessels. Only capillaries play a role in actual exchanges with tissue cells.

2. Except for capillaries, blood vessels are composed of three tunics: The tunica intima forms a friction-reducing lining for the vessel. The tunica media is the bulky middle layer of muscle and elastic tissue. The tunica externa is the protective, outermost connective tissue layer. Capillary walls are formed of the intima only.

3. Artery walls are thick and strong to withstand pressure fluctuations. They expand and recoil as the heart beats. Vein walls are thinner, their lumens are larger, and they are equipped with valves. These modifications reflect the low-pressure nature of veins.

4. All the major arteries of the systemic circulation are all branches of the aorta, which leaves the left ventricle. They branch into smaller arteries and then into the arterioles, which feed the capillary beds of the body tissues. For the names and locations of the systemic arteries, see pp. 327–329.

5. The major veins of the systemic circulation ultimately converge on one of the venae cavae. All veins above the diaphragm drain into the superior vena cava, and those below the diaphragm drain into the inferior vena cava. Both venae cavae enter the right atrium of the heart. See pp. 330–331 for the names and locations of the systemic veins.

6. The arterial circulation of the brain is formed by branches of paired vertebral and internal carotid arteries. The circle of Willis provides alternate routes for blood flow in case of a blockage in the brain's arterial supply.

7. The hepatic portal circulation is formed by veins draining the digestive organs, which empty into the hepatic portal vein. The hepatic portal vein carries the nutrient-rich blood to the liver, where it is processed before the blood is allowed to enter the systemic circulation.

8. The fetal circulation is a temporary circulation seen only in the fetus. It consists primarily of three special vessels: The single umbilical vein that carries nutrient- and oxygen-laden blood to the fetus from the placenta, and the two umbilical arteries that carry carbon dioxide and waste-laden blood from the fetus to the placenta. Shunts bypassing the lungs and liver are also present.

9. Blood pressure is the pressure that blood exerts on the walls of the blood vessels. It is the force that causes blood to continue to flow in the blood vessels. It is high in the arteries, lower in the capillaries, and lowest in the veins. Blood is forced along a descending pressure gradient. Both systolic and diastolic pressures are recorded.

10. The pulse is the alternate expansion and recoil of a blood vessel wall (the pressure wave) that occurs as the heart beats. It may be felt easily over any superficial artery; such sites are called pressure points.

11. Arterial blood pressure is directly influenced by heart activity (increased heart rate leads to increased blood pressure) and by resistance to blood flow. The most important factors increasing the peripheral resistance are a decrease in the diameter or stretchiness of the arteries and arterioles, and an increase in blood viscosity.

12. Many factors influence blood pressure. Some of these factors are the activity of the sympathetic nerves and kidneys, drugs, and diet.

13. Hypertension, which reflects an increase in peripheral resistance, strains the heart and damages blood vessels. In most cases, the precise cause is unknown.

14. Substances move to and from the blood and tissue cells through capillary walls. Some substances are trasnported in vesicles but most move by diffusion—directly through the endothelial cell plasma membranes, through intercellular clefts, or through fenestrations. Fluid is forced from the bloodstream by blood pressure and drawn back into the blood by osmotic pressure.

Lymphatic System (pp. 345–347)

1. The lymphatic system consists of the lymphatic vessels, lymph nodes, and certain other lymphoid organs in the body.

2. Blind-ended lymphatic capillaries pick up excess tissue fluid leaked from the blood capillaries. The fluid (lymph) flows into the larger lymphatics and finally into the blood vascular system through the right lymphatic duct and the left thoracic duct.

3. Lymph nodes are clustered along lymphatic vessels, and the lymphatic stream flows through them. Lymph nodes form agranular WBCs (lymphocytes), and phagocytic cells within them remove bacteria, viruses, and the like from the lymph stream before it is returned to the blood.

4. Other lymphoid organs include the tonsils (in the throat), which remove bacteria trying to enter the digestive or respiratory tracts; the thymus, a programming region for some lymphocytes of the body;

Peyer's patches, which prevent bacteria in the intestine from penetrating deeper into the body; and the spleen, a RBC graveyard and blood reservoir.

Developmental Aspects of the Circulatory System (pp. 347–349)

1. The heart begins as a tubelike structure that is beating and pumping blood by the fourth week of embryonic development.

2. Lymphatic vessels form by budding off veins.

3. Congenital heart defects account for half of all infant deaths resulting from congenital problems.

4. Varicose veins, a structural defect due to incompetent valves, is a common vascular problem, especially in the obese and people who stand for long hours. It is a predisposing factor for thrombophlebitis.

5. Arteriosclerosis is an expected consequence of aging. Gradual loss of elasticity in the arteries leads to hypertension and hypertensive heart disease, and clogging of the vessels with fatty substances leads to coronary artery disease and stroke. Cardiovascular disease is an important cause of death in individuals over age 65.

6. Modifications in diet (decreased fats, cholesterol, and salt), stopping smoking, and regular aerobic exercise may help to reverse the atherosclerotic process and prolong life.

Review Questions

1. Describe the location and position of the heart in the thorax.

2. Draw a diagram of the heart showing the three layers composing its wall, and its four chambers. Label each. Show where the AV and semilunar valves would be. Show and label all blood vessels entering and leaving the heart chambers.

3. Trace one drop of blood from the time it enters the right atrium of the heart until it enters the left atrium. What is this circuit called?

4. Explain the difference in function of the systemic and pulmonary circulations.

5. Why are the heart valves important? Can the heart function with leaky valves?

6. Why might a thrombus in a coronary artery cause sudden death?

7. What is the function of the fluid that fills the pericardial sac?

8. Define *systole, diastole, stroke volume,* and *cardiac cycle.*

9. To which heart chamber(s) do the terms *systole* and *diastole* most often apply?

10. How does the heart's ability to contract differ from that of other muscles of the body?

11. What is the function of the intrinsic conduction system of the heart? Name its elements, *in order,* beginning with the pacemaker.

12. What monosyllables are used to describe heart sounds? What causes heart sounds?

13. Name three different factors that increase heart rate.

14. Name and describe from inside out the three tunics making up the walls of arteries and veins and give the most important function of each layer.

15. Describe the structure of capillary walls. How is their structure related to their function in the body?

16. Why are artery walls so much thicker than those of corresponding veins?

17. Since veins are low-pressure vessels, blood pressure is not very important in helping to return the blood to the heart. Name three factors that are important in promoting venous return.

18. Arteries are often described as vessels that carry oxygen-rich blood, and veins are said to carry oxygen-poor (carbon dioxide-rich) blood. Name two sets of exceptions to this rule that were discussed in this chapter.

19. Trace a drop of blood from the left ventricle of the heart to the wrist of the right hand and back to the heart. Now trace it to the dorsum of the right foot and back to the right heart.

20. What two paired vessels provide arterial blood to the brain? What name is given to the communication between them?

21. What is the function of the hepatic portal circulation? In what way is a portal circulation a "strange" circulation?

22. The liver and lungs are almost entirely bypassed by blood in a fetus. Why is this? Name the vessel that bypasses the liver. Name two lung bypasses. Three vessels travel in the umbilical cord; which of these carries oxygen- and nutrient-rich blood?

23. Define *pulse*.

24. Which artery is palpated at the following pressure points: Wrist? Front of the ear? Side of the neck? The groin? Back of the knee?

25. Define *blood* pressure, *systolic* pressure, and *diastolic* pressure.

26. What vital role does blood pressure play?

27. Two elements determine blood pressure—the cardiac output of the heart and the peripheral resistance or friction in the blood vessels. Name two factors that increase cardiac output. Name two factors that increase peripheral resistance.

28. What is the effect of hemorrhage on blood pressure? Why? In which position—sitting, lying down, or standing—is the blood pressure normally highest? lowest?

29. What is different about the capillary exchanges seen in a capillary with fenestrations and intercellular clefts and the exchanges seen in a capillary lacking those modifications?

30. What is the most important function of the lymphatic vessels? of the lymph nodes?

31. Where are the lymph nodes most dense?

32. What is the special role of the tonsils? The spleen?

33. What are varicose veins? What factors seem to promote their formation?

At the Clinic

1. Define *hypertension* and *arteriosclerosis*. How are they often related? Why is hypertension called the "silent killer"? Name three changes in your lifestyle that might help prevent cardiovascular disease in your old age.

2. A middle-aged woman is admitted to the coronary care unit with a diagnosis of left ventricular failure resulting from a myocardial infarction. Her chart indicates that she was awakened in the middle of the night by severe chest pain. Her skin is pale and cold, and moist sounds of pulmonary edema are heard over the lower regions of both lungs. Explain how failure of the left ventricle might cause these signs and symptoms.

3. Linda, a 14-year-old girl undergoing a physical examination before being admitted to summer camp, was found to have a loud heart murmur at the second intercostal space on the left side of the sternum. The murmur takes the form of a swishing sound with no high-pitched whistle. What, exactly, is producing the murmur?

4. Mrs. Johnson is brought to the emergency room after being involved in an auto accident. She is hemorrhaging and has a rapid, thready pulse, but her blood pressure is still within normal limits. Describe the compensatory mechanisms that are maintaining her blood pressure in the face of blood loss.

5. A 59-year-old woman has undergone a left radical mastectomy (removal of the left breast and left axillary lymph nodes and vessels). Her left arm is severely swollen and painful, and she is unable to raise it more than shoulder height. (a) Explain her signs and symptoms. (b) Can she expect relief from these symptoms in time? How so?

6. During a lethal heart attack, a blood clot lodges in the first part of the circumflex branch of the left coronary artery, blocking blood flow through this vessel. What parts of the heart will become ischemic and die?

12
Body Defenses

YOUR GOALS

After completing this chapter, you will have a working knowledge of the functions of body defenses and will have mastered the objectives listed below.

FUNCTION PREVIEW

■ The nonspecific defenses hinder pathogen entry, prevent the spread of disease-causing microorganisms, and strengthen the immune response. ■

■ The immune system protects against disease by destroying "foreign" cells, and by inactivating toxins and other foreign chemicals with its antibodies.

Every second of every day, an army of hostile bacteria, viruses, and fungi swarms on our skin and invades our inner passageways—yet we stay amazingly healthy most of the time. The body seems to have developed a single-minded approach toward such foes—if you're not with us, then you're against us!

The body's defenders against these tiny but mighty enemies are two systems, simply called the *nonspecific* and the *specific defense systems*. The **nonspecific defense system** responds immediately to protect the body from all foreign substances, whatever they are. The nonspecific defenses are provided by intact skin and mucous membranes, the inflammatory response, and a number of proteins produced by body cells. This system reduces the workload of the second protective arm, the specific defense system, by preventing entry and spread of microorganisms throughout the body.

The **specific defense system,** more commonly called the **immune system,** mounts the attack against *particular* foreign substances. Although certain body organs (lymphatic organs and blood vessels) are intimately involved with the immune response, the immune system is a *functional system* rather than an organ system in an anatomical sense. Its "structures" are a variety of molecules and trillions of immune cells, which inhabit lymphatic tissues and circulate in body fluids. The most important of the immune cells are *lymphocytes* and *macrophages.*

When our immune system is operating effectively, it protects us from most bacteria, viruses, transplanted organs or grafts, and even our own cells that have turned against us. The immune system does this both directly, by cell attack, and indirectly, by releasing mobilizing chemicals and protective antibody molecules. The resulting highly specific resistance to disease is called **immunity** (*immun* = free).

Unlike the nonspecific defenses which are always prepared to defend the body, the immune system must first "meet" or be primed by an initial exposure to a foreign substance (antigen) before it can protect the body against it. Nonetheless, what it lacks in speed it makes up for in the precision of its counterattacks. Although we will consider them separately, keep in mind that specific and nonspecific defenses always work hand-in-hand to protect the body.

NONSPECIFIC BODY DEFENSES

Some nonspecific resistance to disease is inherited. For instance, there are certain diseases that humans never get, such as some forms of tuberculosis that affect birds. Most often, however, the term *nonspecific body defense* refers to the mechanical barriers that cover body surfaces and to cells and chemicals that act on the initial battlefronts to protect the body from invading **pathogens** (harmful or disease-causing microorganisms). Table 12.1 provides a summary of the most important nonspecific defenses.

Surface Membrane Barriers

The body's *first line of defense* against the invasion of disease-causing microorganisms is the *skin* and *mucous membranes*. As long as the skin is unbroken, its keratinized epidermis is a strong physical barrier to most microorganisms that swarm on the skin. Intact mucous membranes provide similar mechanical barriers within the body. Recall that mucous membranes line all body cavities open to the exterior: the digestive, respiratory, urinary, and reproductive tracts. Besides serving as physical barriers, these membranes produce a variety of protective chemicals:

1. The acid pH of skin secretions inhibits bacterial growth, and sebum contains chemicals that are toxic to bacteria. Vaginal secretions of adult females are also very acidic.

2. The stomach mucosa secretes hydrochloric acid and protein-digesting enzymes. Both kill pathogens.

3. Saliva and lacrimal fluid contain *lysozyme,* an enzyme that destroys bacteria.

4. Sticky mucus traps many microorganisms that enter digestive and respiratory passageways.

Some mucosae also have structural modifications that counteract potential invaders. Mucus-coated hairs inside the nasal cavity trap inhaled particles, and the respiratory tract mucosa is ciliated. The cilia sweep dust- and bacteria-laden mucus superiorly toward the mouth, preventing it

Table 12.1   Summary of Nonspecific Body Defenses

Category and associated elements	Protective mechanism
Surface Membrane Barriers	
Intact skin (epidermis)	Forms mechanical barrier that prevents entry of pathogens and other harmful substances into body.
• Acid mantle	Skin secretions make epidermal surface acidic, which inhibits bacterial growth; sebum also contains bacteria-killing chemicals.
• Keratin	Provides resistance against acids, alkalis, and bacterial enzymes.
Intact mucous membranes	Form mechanical barrier that prevents entry of pathogens.
• Mucus	Traps microorganisms in respiratory and digestive tracts.
• Nasal hairs	Filter and trap microorganisms in nasal passages.
• Cilia	Propel debris-laden mucus away from lower respiratory passages.
• Gastric juice	Contains concentrated hydrochloric acid and protein-digesting enzymes that destroy pathogens in stomach.
• Acid mantle of vagina	Inhibits growth of bacteria and fungi in female reproductive tract.
• Lacrimal secretion (tears); saliva	Continuously lubricate and cleanse eyes (tears) and oral cavity (saliva); contain lysozyme, an enzyme that destroys microorganisms.
Nonspecific Cellular and Chemical Defenses	
Phagocytes	Engulf and destroy pathogens that breach surface membrane barriers; macrophages also contribute to immune response.
Natural killer cells	Promote cell lysis by direct cell attack against virus-infected or cancerous body cells; do not depend on specific antigen recognition.
Inflammatory response	Prevents spread of injurious agents to adjacent tissues, disposes of pathogens and dead tissue cells, and promotes tissue repair; chemical mediators released attract phagocytes (and immunocompetent cells) to the area.
Antimicrobial chemicals	
• Interferons	Proteins released by virus-infected cells that protect uninfected tissue cells from viral takeover; mobilize immune system.
• Complement	Lyses microorganisms, enhances phagocytosis by opsonization, and intensifies inflammatory response.
• Urine	Normally acid pH inhibits bacterial growth; cleanses the lower urinary tract as it flushes from the body.
Fever	Systemic response triggered by pyrogens; high body temperature inhibits multiplication of bacteria and enhances body repair processes.

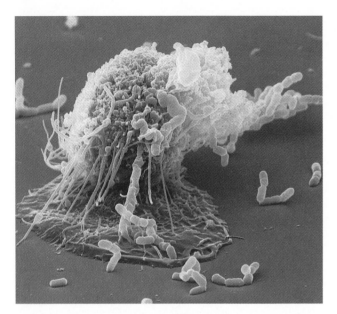

Figure 12.1   Phagocytosis by a macrophage. This scanning electron micrograph (4300×) in computer-generated color, shows a macrophage pulling sausage-shaped *E. coli* bacteria toward it with its long cytoplasmic extensions. Several bacteria on the macrophage's surface are being engulfed.

from entering the lungs, where the warm, moist environment would provide an ideal site for bacterial growth.

Although the surface barriers are very effective, they are broken from time to time by small nicks and cuts resulting, for example, from brushing your teeth or shaving. When this happens and microorganisms do invade deeper tissues, other non-specific mechanisms come into play to defend the body.

Cells and Chemicals

The body uses an enormous number of cells and chemicals to protect itself. These defenses rely on the destructive powers of phagocytes and natural killer cells, the inflammatory response, and a variety of chemical substances that kill pathogens and help repair tissue. Fever is also considered to be a nonspecific protective response.

Phagocytes

Pathogens that make it through the mechanical barriers are confronted by **phagocytes** (fa′go-sītz″;

phago = eat) in nearly every body organ. A phagocyte, such as a *macrophage* or *neutrophil,* engulfs a foreign particle much the way an amoeba ingests a food particle (Figure 12.1). Flowing cytoplasmic extensions bind to the particle and then pull it inside, enclosed in a vacuole. The vacuole is then fused with a *lysosome* and its contents are broken down or digested.

Natural Killer Cells

Natural killer (NK) cells, which "police" the body in blood and lymph, are a unique group of defensive cells that can lyse and kill cancer cells and virus-infected body cells well before the immune system is enlisted in the fight. Unlike lymphocytes of the immune system, which can recognize and react only against *specific* virus-infected or tumor cells, natural killer cells can act spontaneously against *any* such target. NK cells are not phagocytic. They attack the target cell's membrane and release several chemicals. Shortly thereafter, the target cell's membrane and nucleus disintegrate rapidly.

Inflammatory Response

The **inflammatory response,** the body's *second line of defense,* is a nonspecific response that is triggered whenever body tissues are injured (Figure 12.2). For example, it occurs in response to physical trauma, intense heat, and irritating chemicals as well as to infection by viruses and bacteria. The four *cardinal signs* and major symptoms of an acute inflammation are *redness, heat* (*inflamm* = set on fire), *swelling,* and *pain.* It is easy to understand why these symptoms occur, once the events of the inflammatory response are understood.

The inflammatory process begins with a chemical "alarm." When cells are injured, they release inflammatory chemicals, including *histamine* and *kinins,* that (1) cause blood vessels in the involved area to dilate and capillaries to become leaky, (2) activate pain receptors, and (3) attract phagocytes and white blood cells to the area. Dilation of the blood vessels increases the blood flow to the area, accounting for the redness and heat observed. Increased permeability of the capillaries allows plasma to leak from the bloodstream into the tissue spaces, causing local edema (swelling) that also activates pain receptors in the area. If the swollen,

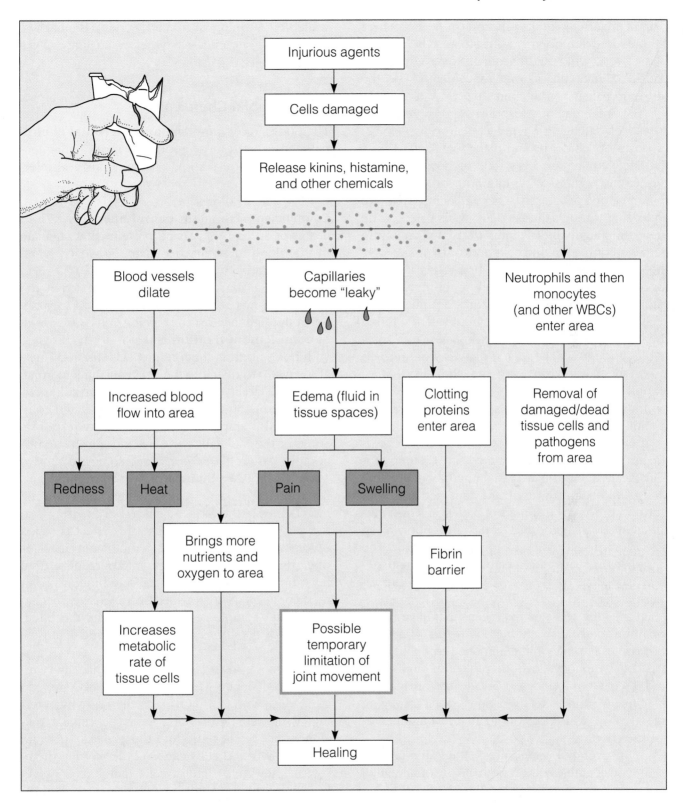

Figure 12.2   Flowchart of inflammatory events. The four cardinal signs of acute inflammation are shown in red boxes.

painful area is a joint, its function (movement) may be impaired temporarily. This forces the injured part to rest, which aids healing. Some authorities consider limitation of joint movement to be the fifth cardinal sign of inflammation.

The inflammatory response (1) prevents the spread of damaging agents to nearby tissues, (2) disposes of cell debris and pathogens, and (3) sets the stage for repair. Let's look at how it accomplishes these tasks. Within an hour or so after the inflammatory process has begun, neutrophils are squeezing through the capillary walls to enter the area and begin the cleanup detail by engulfing damaged or dead tissue cells and/or pathogens. As the counterattack continues, monocytes begin to leave the bloodstream and follow the neutrophils into the inflamed area. Monocytes are fairly poor phagocytes, but within 8 to 12 hours after entering the tissues they become macrophages with insatiable appetites. The macrophages continue to wage the battle, replacing the short-lived neutrophils on the battlefield. Macrophages are the central actors in the final disposal of cell debris as the inflammation subsides.

Besides phagocytosis, other protective events are also occurring at the inflamed site. Clotting proteins, leaked into the area from the blood, are activated and begin to wall off the damaged area with fibrin to prevent the spread of pathogens or harmful agents to neighboring tissues. The fibrin mesh also forms a scaffolding for permanent repair. The local heat increases the metabolic rate of the tissue cells, speeding up their defensive actions and repair processes.

If the area contains pathogens that have previously invaded the body, the *third line of defense* also comes into play—the immune response mediated by lymphocytes. Both protective antibodies and T cells (lymphocytes) invade the area to act specifically and directly against the damaging agents. (We will speak more of the immune response shortly.)

In severely infected areas, the battle takes a considerable toll on both sides, and creamy, yellow pus may be formed in the wound. **Pus** is a mixture of dead or dying neutrophils, broken-down tissue cells, and living and dead pathogens. If the inflammatory mechanism fails to fully clear the area of debris, the sac of pus may become walled off, forming an *abscess*. Surgical drainage of abscesses is often necessary before healing can occur. ❏

Antimicrobial Chemicals

The body's most important **antimicrobial chemicals,** apart from those produced in the inflammatory reaction, are complement proteins and interferon.

Complement   The term **complement** refers to a group of at least 20 plasma proteins that circulate in the blood in an inactive state. However, when complement becomes attached, or *fixed,* to foreign cells such as bacteria, fungi, or mismatched red blood cells, it is activated and becomes a major factor in the fight against the foreign cells. One result of **complement fixation** is that lesions, complete with holes, form in the foreign cell's surface. These allow water to rush into the cell, causing it to burst. Besides its ability to rupture, or lyse, invading microorganisms or other foreign cells, activated complement also amplifies the inflammatory response. Some of the molecules released during the activation process are *vasodilators*, and some are *chemotaxis chemicals* that attract neutrophils and macrophages into the region. Others cause the cell membranes of the foreign cells to become sticky so they are easier to phagocytize; this effect is called *opsonization*. Although the complement attack is often directed against specific microorganisms that have been "identified" by antibody binding, complement itself is a nonspecific defensive mechanism that "complements" or enhances the effectiveness of *both* nonspecific and specific defenses.

Interferon   Viruses lack the cellular machinery required to generate ATP or make proteins. They do their "dirty work" or damage in the body by entering tissue cells and taking over the cellular machinery needed to reproduce themselves. Although the virus-infected cells can do little to save themselves, they help defend cells that have not yet been infected by secreting small proteins called **interferons** (in-ter-fer′onz). The interferon molecules diffuse to nearby cells and bind to their membrane receptors. Somehow this binding hinders the ability of viruses to multiply within these cells.

Fever

Fever, or abnormally high body temperature, is a systemic response to invading microorganisms. As described further in Chapter 14, body temperature is regulated by a part of the hypothalamus, commonly referred to as the body's "thermostat." Normally the thermostat is set at approximately 36.2°C (98.2°F), but it can be reset upward in response to **pyrogens** (*pyro* = fire), chemicals secreted by white blood cells and macrophages exposed to bacteria and other foreign substances in the body.

Although high fevers are dangerous because excess heat begins to "scramble" enzymes and other body proteins, mild or moderate fever seems to benefit the body. Bacteria require large amounts of iron and zinc to multiply, but during a fever the liver and spleen gather up these nutrients, making them less available. Fever also increases the metabolic rate of tissue cells in general, speeding up repair processes.

SPECIFIC BODY DEFENSES: THE IMMUNE SYSTEM

Most of us would find it wonderfully convenient if we could walk into a single clothing store and buy a complete wardrobe—hat to shoes—that fit us "to a T" regardless of any special figure problems. We know that such a service would be next to impossible to find. And yet we take for granted our *immune system,* our built-in *specific defense system,* which stalks and eliminates with nearly equal precision almost any type of pathogen that intrudes into the body.

The immune system's response to a threat, called the **immune response,** tremendously increases the inflammatory response, and it provides protection that is carefully targeted against *specific* antigens. Furthermore, the initial exposure to an antigen "primes" the body to react more vigorously to later meetings with the same antigen.

Sometimes referred to as the body's *third line of defense,* the immune system is a functional system that recognizes foreign molecules (antigens) and acts to inactivate or destroy them. Normally it protects us from a wide variety of pathogens, as well as from abnormal body cells. When it fails, malfunctions, or is disabled, some of the most devastating diseases—such as cancer, rheumatoid arthritis, and AIDS—may result.

Although *immunology,* the study of immunity, is a fairly new science, the ancient Greeks knew that once someone had suffered through a certain infectious disease, that person was unlikely to have the same disease again. The basis of this immunity was revealed in the late 1800s when it was shown that animals surviving a serious bacterial infection have "factors" in their blood that protect them from future attacks by the same pathogen. (These factors are now known to be unique proteins, called *antibodies.*) Furthermore, it was demonstrated that if antibody-containing serum from the surviving animals (immune serum) was injected into animals that had not been exposed to the pathogen, those animals would also be protected. These landmark experiments revealed three important aspects of the immune response:

1. **It is antigen specific:** It recognizes and acts against *particular* pathogens or foreign substances.

2. **It is systemic:** Immunity is not restricted to the initial infection site.

3. **It has "memory":** It recognizes and mounts even stronger attacks on previously encountered pathogens.

This was exciting news. But then, in the mid-1900s, it was discovered that injection of serum containing antibodies did not always protect a recipient from diseases the donor had survived. In such cases, however, injection of the donor's lymphocytes *did* provide immunity.

As the pieces began to fall into place, two separate but overlapping arms of immunity were recognized. **Humoral** (hu′mor-al) **immunity,** also called **antibody-mediated immunity,** is provided by antibodies present in the body's "humors," or fluids. When lymphocytes themselves defend the body, the immunity is called **cellular** or **cell-mediated immunity,** because the protective factor is living cells. The cellular arm also has cellular targets—virus-infected tissue cells, cancer cells, and cells of foreign grafts. The lymphocytes act against such targets either *directly,* by lysing the foreign cells, or *indirectly,* by releasing chemicals that enhance the inflammatory response or activate other immune cells. However, before we describe the humoral and cell-mediated responses individually, we will consider the antigens that trigger the activity of the remarkable cells involved in these immune responses.

Antigens

An **antigen** (an′tĭ-jen) **(Ag)** is any substance capable of exciting our immune system and provoking an immune response. Most antigens are large, complex molecules that are not normally present in our bodies. Consequently, as far as our immune system is concerned, they are foreign intruders, or **nonself.** An almost limitless variety of substances can act as antigens, including virtually all foreign proteins, nucleic acids, many large carbohydrates, and some lipids. Of these, proteins are the strongest antigens. Pollen grains and microorganisms such as bacteria, fungi, and virus particles are *antigenic* because their surfaces bear such foreign molecules.

It is also important to remember that our own cells are richly studded with a variety of protein molecules (self-antigens). Somehow, as our immune system develops, it takes an inventory of all these proteins so that, thereafter, they are recognized as self. Although these **self-antigens** do not trigger an immune response in us, they *are* strongly antigenic to other people. This helps explain why our bodies reject cells of transplanted organs or foreign grafts unless special measures (drugs and others) are taken to cripple or stifle the immune response.

As a rule, small molecules are not antigenic. But, when they link up with our own proteins, the immune system may recognize the combination as foreign and mount an attack that is harmful rather than protective. (Such reactions, called allergies, are described later.) In such cases, the troublesome small molecule is called a *hapten* (*haptein* = to grasp), or *incomplete antigen*. Besides certain drugs, chemicals that act as haptens are found in poison ivy and animal dander, and even in some detergents, hair dyes, cosmetics, and other commonly used household and industrial products. Perhaps the most dramatic and familiar example of a drug hapten's provoking an im-

mune response involves the binding of penicillin to blood proteins, which causes a *penicillin reaction* in some people. In such cases, the immune system mounts such a vicious attack that the person's life is endangered. ❑

Cells of the Immune System: An Overview

The crucial cells of the immune system are lymphocytes and macrophages. Lymphocytes exist in two major "flavors." The **B lymphocytes** or **B cells** produce antibodies and oversee humoral immunity, whereas the **T lymphocytes** or **T cells** are nonantibody-producing lymphocytes that constitute the cell-mediated arm of immunity. Unlike the two types of lymphocytes, macrophages do not respond to specific antigens, but instead play an essential role in helping the lymphocytes that do.

Lymphocytes

Like all blood cells, lymphocytes originate from hemocytoblasts in red bone marrow (Figure 12.3). The immature lymphocytes released from the marrow are essentially identical. Whether a given lymphocyte matures into a B cell or a T cell depends on where in the body it becomes **immunocompetent,** that is, capable of responding to a specific antigen by binding to it. T cells arise from lymphocytes that migrate to the thymus (see Figure 12.3), where they undergo a maturation process of 2 to 3 days, directed by thymic hormones (*thymosin* and others). Within the thymus, the immature lymphocytes divide rapidly and their numbers increase enormously, but only those maturing T cells with the sharpest ability to identify *foreign* antigens survive. Lymphocytes capable of binding strongly with *self-antigens* (and of acting against body cells) are vigorously weeded out and destroyed. Thus, the development of tolerance for self-antigens is an essential part of a lymphocyte's "education." This is true not only for T cells but also for B cells. B cells develop immunocompetence in bone marrow, but little is known about the factors that regulate B cell maturation.

Once a lymphocyte is immunocompetent, it will be able to react to one distinct antigen and one only, because *all* the antigen receptors on its external surface are the same. For example, the re-

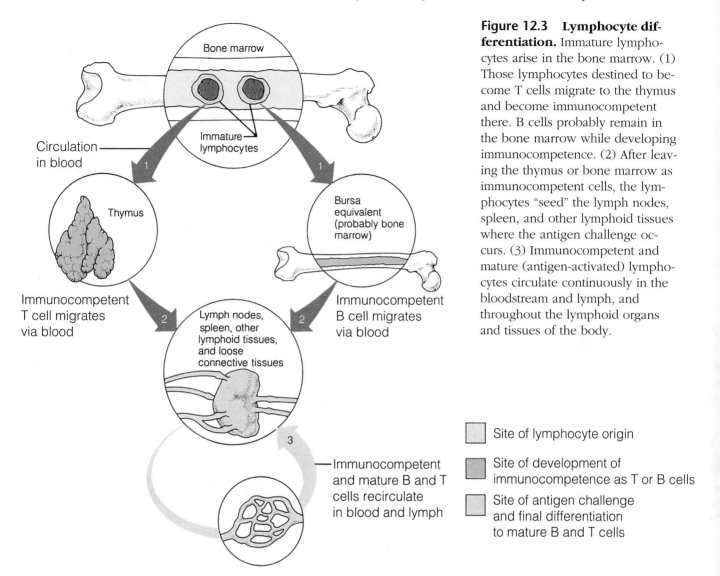

Figure 12.3 Lymphocyte differentiation. Immature lymphocytes arise in the bone marrow. (1) Those lymphocytes destined to become T cells migrate to the thymus and become immunocompetent there. B cells probably remain in the bone marrow while developing immunocompetence. (2) After leaving the thymus or bone marrow as immunocompetent cells, the lymphocytes "seed" the lymph nodes, spleen, and other lymphoid tissues where the antigen challenge occurs. (3) Immunocompetent and mature (antigen-activated) lymphocytes circulate continuously in the bloodstream and lymph, and throughout the lymphoid organs and tissues of the body.

ceptors of one lymphocyte can recognize only a part of the hepatitis A virus, those of another lymphocyte can recognize or bind only to pneumococcus bacteria, and so forth.

Although the details of the maturation process are still beyond our grasp, we know that lymphocytes become immunocompetent *before* meeting the antigens they may later attack. Thus, *it is our genes, not antigens, that determine what specific foreign substances our immune system will be able to recognize and resist*. Only some of the possible antigens our lymphocytes are programmed to resist will ever invade our bodies. Consequently, only some members of our army of immunocompetent cells will be mobilized during our lifetime; the others will be forever idle. As usual, our bodies have done their best to protect us.

After becoming immunocompetent, both T cells and B cells migrate to the lymph nodes and spleen (and loose connective tissues), where their encounters with antigens will occur (Figure 12.3). Then, when the lymphoctyes bind with recognized antigens, they complete their differentiation into fully mature T cells and B cells.

Macrophages

Macrophages, which also become widely distributed throughout the lymphoid organs and connective tissues, arise from monocytes formed in the bone marrow. A major role of macrophages (literally, "big eaters") is to engulf foreign particles and present fragments of these antigens, like signal flags, on their own surfaces, where they can be

Figure 12.4   Clonal selection of a B cell stimulated by antigen binding. The initial meeting stimulates the primary response in which the B cell proliferates rapidly, forming a clone of like cells, most of which become antibody-producing plasma cells. Cells that do not differentiate into plasma cells become memory cells, which respond to subsequent exposures to the same antigen. Should such a meeting occur, the memory cells quickly produce more memory cells and larger numbers of effector plasma cells with the same antigen specificity. Responses generated by memory cells are called secondary responses.

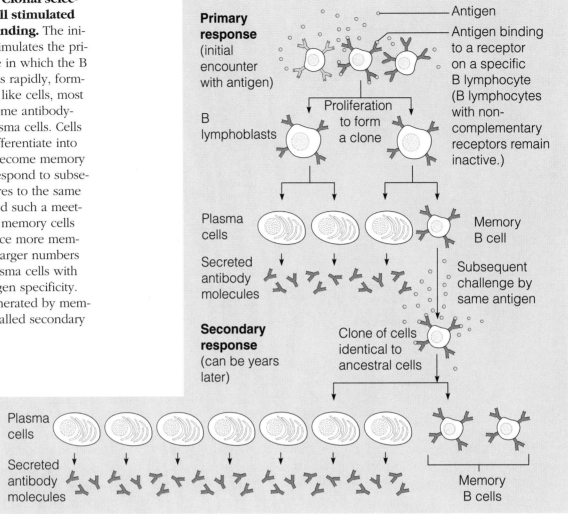

recognized by immunocompetent T cells. Thus, they act as *antigen presenters*. Macrophages also secrete proteins, called *monokines,* that are important in the immune response (see Table 12.4, p. 371). Activated T cells, in turn, release chemicals that cause macrophages to become insatiable phagocytes, or *killer macrophages*. As you will see, interactions between lymphocytes, and between lymphocytes and macrophages, underlie virtually all phases of the immune response.

Macrophages tend to remain fixed in the lymphoid organs (as if waiting for antigens to come to them). But lymphocytes, especially T cells, circulate continuously through the body (Figure 12.3). This makes sense because it greatly increases a lymphocyte's chance of coming into contact with antigens collected by the lymph capillaries from the tissue spaces, as well as with huge numbers of macrophages and other lymphocytes.

To summarize, the immune system is a two-armed defensive system that uses lymphocytes, macrophages, and specific molecules to identify and destroy all substances—both alive and nonliving—that are in the body but are not recognized as being part of the body, or as being self. The immune system's ability to respond to such threats depends on the ability of its cells (1) to recognize foreign substances (antigens) in the body by binding to them, and (2) to communicate with one another so that the system as a whole mounts a response specific to those antigens.

Humoral (Antibody-Mediated) Immune Response

An immunocompetent but as yet immature B lymphocyte is stimulated to complete its development

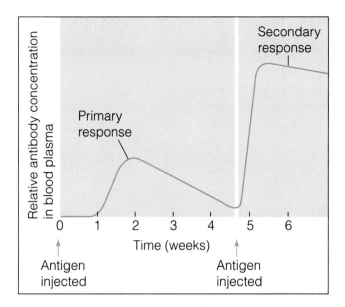

Figure 12.5   Primary and secondary humoral responses to an antigen. In the primary response, there is a gradual rise and then a rapid decline in the level of antibodies in the blood. The secondary response is both more rapid and more intense. Additionally, antibody levels remain high for a much longer time.

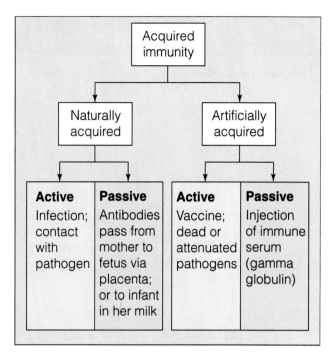

Figure 12.6   Types of acquired immunity. Green boxes signify active types of immunity in which immunological memory is established. Orange boxes signify short-lived passive types of immunity; no immunological memory is established.

into a fully mature B cell when an antigen binds to its surface receptors. This binding event *sensitizes,* or *activates,* the lymphocyte to "switch on" and undergo **clonal selection.** The lymphocyte begins to grow, and then multiplies rapidly to form an army of cells all exactly like itself and bearing the same antigen-specific receptors (see Figure 12.4). The resulting family of identical cells descended from the *same* ancestor cell is called a **clone,** and clone formation represents the **primary humoral response** to that antigen. (As described later, T cells also influence B cell activation.)

Most of the B cell clone members or descendants become **plasma cells.** After an initial lag period, these antibody-producing "factories" swing into action, producing the same highly specific antibodies at an unbelievable rate of about 2000 antibody molecules per second. (The B cells themselves produce only very small amounts of antibodies.) However, this flurry of activity lasts for only 4 or 5 days; then the plasma cells begin to die. Antibody levels in the blood during this primary response peak in about 10 days and then slowly begin to decline (Figure 12.5).

B cell clone members that do not become plasma cells become long-lived **memory cells** ca-

pable of responding to the same antigen at later meetings with it; they are responsible for the immunological "memory" mentioned earlier. These later immune responses, called **secondary responses,** are much faster, more prolonged, and more effective because all the preparations for this attack have already been made. Within hours after recognition of the "old-enemy" antigen, a new army of plasma cells is being generated, and antibodies begin to flood into the bloodstream. Within 2 to 3 days blood antibody levels peak (at much higher levels than seen in the primary response) and their levels remain high for weeks to months. How antibodies protect the body is described shortly.

Active and Passive Humoral Immunity

When your B cells encounter antigens and produce antibodies against them, you are exhibiting **active immunity** (Figure 12.6). Active immunity is (1) *naturally acquired* during bacterial and viral infections, during which we may develop the symptoms of the disease and suffer a little (or a

Table 12.2   Recommended* Immunization Schedule for Children in the U.S.

Recommended age*	Immunizing agent in vaccine
2 months	DPT vaccine (diphtheria toxoid, pertussis [whooping cough] vaccine, and tetanus toxoid); OPV (oral poliomyelitis vaccine)
4 months	DPT vaccine; OPV
6 months	DPT vaccine (OPV optional for areas with high risk of polio exposure)
15 months	DPT vaccine; OPV; MMR vaccine (combined mumps vaccine, measles vaccine, and rubella virus vaccine) or individual mumps, measles, and rubella virus vaccines. Completes primary series of DPT and OPV
18 months	HbCV (Hemophilus influenza type B); conjugate vaccine preferred over the polysaccharide vaccine (HbPV)
4 to 6 years	DPT vaccine; OPV (preferably at or before school entry)
11 to 13 years	MMR (at entrance to middle school or high school); second dose of a two-dose schedule
14 to 16 years	Td booster (tetanus and diphtheria toxoid); repeat every 10 years throughout life

*Recommended by the American Academy of Pediatrics

lot), and (2) *artificially acquired* when we receive **vaccines.** It makes little difference whether the antigen invades the body under its own power or is introduced in the form of a vaccine. The response of the immune system is pretty much the same. Indeed, once it was recognized that secondary responses are so much more vigorous, the race was on to develop vaccines to "prime" the immune response by providing a first meeting with the antigen. Most vaccines contain dead or *attenuated* (living, but extremely weakened) pathogens.

We receive two benefits from vaccines: (1) We are spared most of the signs and symptoms (and discomfort) of the disease which would otherwise occur during the primary response, and (2) the weakened antigens are still able to stimulate antibody production and promote immunologic memory. So-called booster shots, which may intensify the immune response at later meetings with the same antigen, are also available. Vaccines are currently available against microorganisms that cause pneumonia, smallpox, polio, tetanus, diphtheria, measles, and many other diseases. In the United States, many potentially serious childhood diseases have been virtually wiped out by active immuniza-

tion programs. A summary of the currently recommended schedule for administering vaccines to American children is provided in Table 12.2.

Passive immunity is quite different from active immunity, both in the antibody source and in the degree of protection it provides (see Figure 12.6). Instead of being made by your plasma cells, the antibodies are obtained from the serum of an immune human or animal donor. As a result, your B cells are *not* challenged by the antigen, immunologic memory does *not* occur, and the temporary protection provided by the "borrowed antibodies" ends when they naturally degrade in the body.

Passive immunity is conferred *naturally* on a fetus when the mother's antibodies cross the placenta and enter the fetal circulation. For several months after birth, the baby is protected from all the antigens to which the mother has been exposed.

Passive immunity is *artificially* conferred when one receives immune serum or gamma globulin. Gamma globulin is commonly administered after exposure to hepatitis. Other immune sera are used to treat poisonous snake bites (an antivenom), bot-

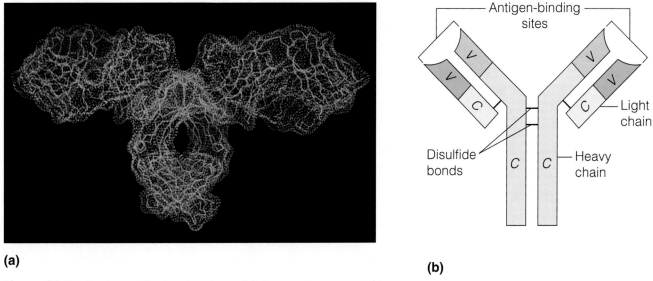

(a)

(b)

Figure 12.7   Basic antibody structure. (**a**) Computer-generated image. (**b**) Diagrammed structure. The basic structure (monomer) of each type of antibody is formed by four amino-acid (polypeptide) chains that are joined by disulfide bonds. Two of the chains are short, light chains; the other two are long, heavy chains. Each chain has a variable (V) region (different in different antibodies) and a constant (C) region (essentially identical in different antibodies of the same class). The variable regions are the antigen-binding sites of the antibody. Hence, each antibody monomer has two antigen-binding sites.

ulism, rabies, and tetanus (an antitoxin) because these diseases will kill a person before active immunity can be established. The donated antibodies provide immediate protection, but their effect is short-lived (2 to 3 weeks). Meanwhile, however, the body's own defenses take over.

In addition to their use to provide passive immunity, antibodies are prepared commercially for use in research, clinical testing for diagnostic purposes, and treating certain cancers. **Monoclonal antibodies** used for such purposes are produced by descendants of a single cell, and are pure antibody preparations that exhibit specificity for one and only one antigen. Besides their use in delivering cancer-fighting drugs to cancerous tissue, monoclonal antibodies are being used for diagnosis of pregnancy, hepatitis, and rabies. They are also used for early diagnosis and to track the extent of cancers hidden deep within the body.

Antibodies

Antibodies, also referred to as **immunoglobulins** (im″mu-no-glob′u-linz), or **Igs,** constitute the *gamma globulin* part of blood proteins. Antibodies are soluble proteins secreted by activated B cells or by their plasma-cell offspring in response to an antigen, and they are capable of binding specifically with that antigen.

Antibodies are formed in response to a huge number of different antigens. Despite their variety, they all have a similar basic anatomy that allows them to be grouped into five Ig classes, each slightly different in structure and function.

Basic Antibody Structure   Regardless of its class, every antibody has a basic structure consisting of four amino-acid (polypeptide) chains linked together by disulfide (sulfur-to-sulfur) bonds (Figure 12.7). Two of the four chains are identical and contain approximately 400 amino acids each; these are the *heavy chains.* The other two chains, the *light chains,* are also identical to each other, but are only about half as long. When the four chains are combined, the antibody molecule formed has two identical halves, each consisting of a heavy and a light chain, and the molecule as a whole is T- or Y-shaped.

Table 12.3   Immunoglobulin Classes

Class	Generalized structure	Where found	Biological function
IgD	Monomer	Virtually always attached to B cell.	Believed to be cell surface receptor of immunocompetent B cell; important in activation of B cell.
IgM	Pentamer (J chain)	Attached to B cell; free in plasma.	When bound to B cell membrane, serves as antigen receptor; first Ig class *released* to plasma by plasma cells during primary response; potent agglutinating agent; fixes complement.
IgG	Monomer	Most abundant antibody in plasma; represents 75% to 85% of circulating antibodies.	Main antibody of both primary and secondary responses; crosses placenta and provides passive immunity to fetus; fixes complement.
IgA	Monomer or dimer (J chain)	Some (monomer) in plasma; dimer in secretions such as saliva, tears, intestinal juice, and milk.	Bathes and protects mucosal surfaces from attachment of pathogens.
IgE	Monomer	Secreted by plasma cells in skin, mucosae of gastrointestinal and respiratory tracts, and tonsils.	Binds to mast cells and basophils, and triggers release of histamine and other chemicals that mediate inflammation and certain allergic responses.

When scientists began investigating antibody structure, they discovered something very peculiar. Each of the four chains forming an antibody had a *variable (V) region* at one end and a much larger *constant (C) region* at the other end. Antibodies responding to different antigens had very different variable regions, but their constant regions were the same or nearly so. This made sense when it was discovered that the variable regions of the heavy and light chains in each arm combine their efforts to form an **antigen-binding site** (Figure 12.7) uniquely shaped to "fit" its specific antigen. Hence, each antibody has two such antigen-binding regions.

The constant regions that form the "stem" of an antibody can be compared to the handle of a key. A key handle has a common function for all keys: It allows you to hold the key and place its tumbler-moving portion into the lock. Similarly, the constant regions of the antibody chains serve common functions in all antibodies: They determine the type of antibody formed (antibody class), how the antibody class will carry out its immune roles in the body, and the cell types or chemicals with which the antibody can bind.

Antibody Classes   There are five major immunoglobulin classes—IgM, IgA, IgD, IgG, and IgE.

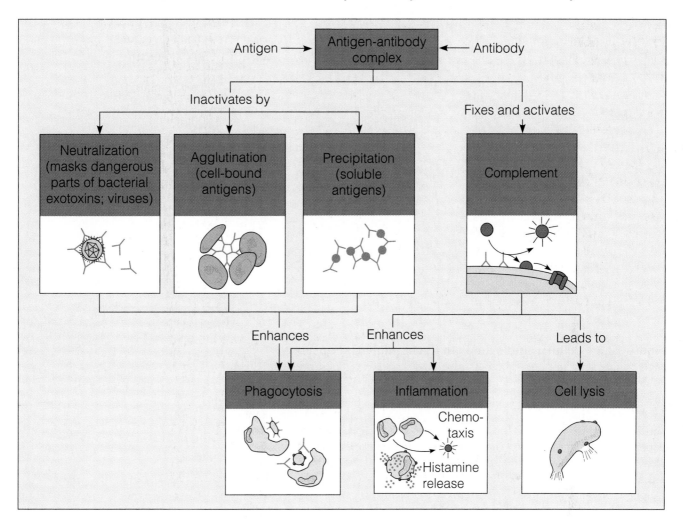

Figure 12.8   Mechanisms of antibody action.

(Remember the woman's name MADGE to recall the five Ig types.) As illustrated in Table 12.3, IgD, IgG, and IgE antibodies have the same basic Y-shaped structure described above, and are referred to as *monomers*. IgA antibodies occur in both monomer and *dimer* (two linked monomers) forms. (Only the dimer form is shown in the table.) Compared to the other antibodies, IgM antibodies are huge. Because they are constructed of five linked monomers, IgM antibodies are called *pentamers* (*penta* = five).

The antibodies of each class have slightly different biological roles and locations in the body. For example, IgG is the most abundant antibody in blood plasma and is the only type that can cross the placental barrier. Hence, the passive immunity that a mother transfers to her fetus is "with the compliments" of her IgG antibodies. Only IgM and IgG can fix complement. The IgA dimer, sometimes called *secretory IgA,* is found mainly in mucus and other secretions that bathe body surfaces. It plays a major role in preventing pathogens from gaining entry into the body. IgE antibodies are the "troublemaker" antibodies involved in allergies. These and other characteristics unique to each of the immunoglobulin classes are summarized in Table 12.3.

Antibody Function   Antibodies inactivate antigens in a number of ways—by complement fixation, neutralization, agglutination, and precipitation (Figure 12.8). Of these, complement fixation and neutralization are most important to body protection.

Complement is the chief antibody ammunition used against cellular antigens, such as bacteria or mismatched red blood cells. As noted earlier, complement is fixed (activated) during nonspecific body defenses. It is also activated very efficiently when it binds to antibodies attached to cellular targets. This triggers events (described earlier) that result in lysis of the foreign cell and release of

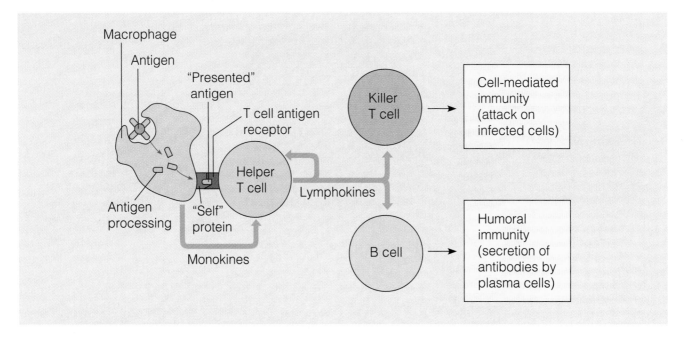

Figure 12.9 Cellular interactions in the immune response.
Macrophages are important both as phagocytes and as antigen presenters.
After they have ingested an antigen, they display parts of it on their surface
membranes, where it can be recognized by a helper T cell bearing receptors
for the same antigen. During the binding process, the T cell binds simultaneously to the antigen and to the macrophage (self) receptor, which leads to
T cell activation and cloning (not illustrated). In addition, the macrophage releases monokines, which enhance T cell activation. Activated helper T cells
release lymphokines, which stimulate proliferation and activity of other
helper T cells and help activate cytotoxic T cells and B cells.

molecules that tremendously enhance the inflammatory process.

Neutralization occurs when antibodies bind
to specific sites on bacterial exotoxins (toxic chemicals secreted by bacteria) or on viruses that can
cause cell injury. In this way, they block the harmful effects of the exotoxin or virus.

Because antibodies have more than one antigen-binding site, they can bind to more than one
antigen at a time; consequently, antigen-antibody
complexes can be cross-linked into large lattices.
When the cross-linking involves cell-bound antigens, the process causes clumping of the foreign
cells, a process called **agglutination.** This type of
antigen-antibody reaction occurs when mismatched blood is transfused (the foreign red blood
cells are clumped) and is the basis of tests used for
blood typing. When the cross-linking process involves soluble antigenic molecules, the resulting
antigen-antibody complexes are so large that they
become insoluble and settle out of solution. This
cross-linking reaction is more precisely called **pre-**

cipitation. There is little question that aggluti-
nated bacteria and immobilized (precipitated) antigen molecules are much more easily captured and
engulfed by the body's phagocytes than are freely
moving antigens.

Cellular (Cell-Mediated) Immune Response

Like B cells, immunocompetent T cells are activated to form a clone by binding with a "recognized" antigen. However, unlike B cells, the T cells
are not able to bind with *free* antigens. Instead, the
antigens must be "presented" by macrophages and
a *double recognition* must occur. The macrophages
engulf the antigens, process them internally, and
then finally display parts of the processed antigens
on their external surface in combination with one
of their own (self) proteins (see Figure 12.9).

Apparently, a T cell must recognize "nonself,"
the antigen fragment presented by the macro-

Table 12.4 Functions of Cells and Molecules Involved in Immunity

Element	Function in the immune response
Cells	
B cell	Lymphocyte that resides in the lymph nodes, spleen, or other lymphoid tissues, where it is induced to replicate by antigen-binding and helper T cell interactions; its progeny (clone members) form plasma cells and memory cells.
Plasma cell	Antibody-producing "machine"; produces huge numbers of the same antibody (immunoglobulin); represents further specialization of B cell clone descendants.
Helper T cell	A *regulatory* T cell that binds with a specific antigen presented by a macrophage; it stimulates the production of other immune cells (killer T cells and B cells) to help fight the invader; acts both directly and indirectly by releasing lymphokines.
Killer T cell	Also called a cytotoxic T cell; activity enhanced by helper T cells; its specialty is killing virus-invaded body cells, as well as body cells that have become cancerous; involved in graft rejection.
Suppressor T cell	Slows or stops the activity of B and T cells once the infection (or attack by foreign cells) has been conquered.
Memory cell	Descendant of an activated B cell or T cell; generated during the initial immune response (primary response); may exist in the body for years thereafter, enabling it to respond quickly and efficiently to subsequent infections or meetings with the same antigen.
Macrophage	Engulfs and digests antigens that it encounters, and presents parts of them on its plasma membrane for recognition by T cells bearing receptors for the same antigen; this function, *antigen presentation,* is essential for normal, cell-mediated responses; also releases chemicals that activate T cells.

phage, and also "self" by coupling with a specific glycoprotein on the macrophage's surface at the same time. Antigen binding alone is not enough to sensitize T cells. They must be "spoon-fed" the antigens by macrophages, and something like a "double handshake" must occur. Although this idea seemed preposterous when it was first suggested, there is no longer any question that **antigen presentation** is a major role of macrophages, and is essential for activation and clonal selection of the T cells. Without macrophage "presenters," the immune response is severely impaired. Chemicals (monokines) released by macrophages also play important roles in the immune response, as shown in Table 12.4.

The different classes of T cell clones, which provide for cell-mediated immunity, are a diverse lot and produce their deadly effects in a variety of ways (see Table 12.4). Some are **killer (cytotoxic) T cells,** cells that specialize in killing virus-infected, cancer, or foreign graft cells. One way they accomplish this is by binding to them and inserting a toxic chemical called *perforin* into the foreign cell's plasma membrane (delivering the so-called kiss of death). Shortly thereafter, the target cell ruptures. By that time, the killer T cell is long gone and is seeking other foreign prey to attack.

Helper T cells are the T cells that act as the "directors" or "managers" of the immune system. Once activated, they circulate through the body, recruiting other cells to fight the invaders. For example, helper T cells interact directly with B cells (that have already attached to antigens), prodding them into more rapid division (clone production) and then, like the "boss" of an assembly line, signaling for antibody formation to begin. They also

Table 12.4	Functions of Cells and Molecules Involved in Immunity (continued)
Element	**Function in the immune response**
Molecules	
Antibody (immunoglobulin)	Protein produced by a B cell or its plasma cell offspring, and released into the body fluids (blood, lymph, saliva, mucus, etc.), where it attaches to antigens causing neutralization, precipitation, or agglutination, which "marks" the antigens for destruction by phagocytes or complement.
Lymphokines	Chemicals released by sensitized T cells: • Macrophage migration inhibiting factor (MIF)—"inhibits" macrophage migration and keeps them in the local area. • Interleukin 2—stimulates T cells and B cells to proliferate. • Helper factors—enhance antibody formation by plasma cells. • Suppressor factors—suppress antibody formation or T cell–mediated immune responses. • Chemotactic factors—attract leukocytes (neutrophils, eosinophils, and basophils) into inflamed area. • Perforin—a cell toxin; causes cell lysis. • Gamma interferon—helps make tissue cells resistant to viral infection; activates macrophages; activates NK cells; enhances maturation of killer T cells.
Monokines	Chemicals released by activated macrophages: • Interleukin 1—stimulates T cells to proliferate and causes fever. • Tumor necrosis factor (TNF)—Like perforin, causes cell killing; attracts granulocytes; activates T cells and macrophages.
Complement	Group of blood-borne proteins activated after binding to antibody-covered antigens; when activated, complement causes lysis of the microorganism and enhances inflammatory response.
Antigen	Substance capable of provoking an immune response; typically a large complex molecule not normally present in the body.

release a variety of chemicals called *lymphokines* (see Table 12.4) that act indirectly to rid the body of antigens by (1) stimulating killer T cells and B cells to grow and divide; (2) attracting other types of protective white blood cells, such as neutrophils, into the area; and (3) enhancing the ability of macrophages to engulf and destroy microorganisms. (Actually, the macrophages are pretty good phagocytes even in the absence of lymphokines, but in their presence they develop an insatiable appetite.) As the released lymphokines summon more and more cells into the battle, the immune response gains momentum, and the antigens are overwhelmed by the sheer numbers of immune elements acting against them.

Another T cell population, the **suppressor T cells,** releases chemicals that suppress the activity of both T and B cells. Suppressor T cells are vital for winding down and finally stopping the immune response after an antigen has been successfully inactivated or destroyed. This helps prevent uncontrolled or unnecessary immune system activity.

Delayed hypersensitivity T cells are effector cells that play a major role in cell-mediated allergies and long-term, or chronic, inflammations. When activated, they release a deluge of lymphokines that promote an intense inflammatory response.

Most of the T cells enlisted to fight in a particular immune response are dead within a few days. However, a few members of each clone are long-lived **memory cells** that remain behind to provide the immunologic memory for each antigen encountered and enable the body to respond quickly to its subsequent invasions.

Organ Transplants and Rejection

For those suffering with end-stage heart or kidney disease, organ transplants are a desirable treatment option. However, organ transplants have had mixed success because the immune system is ever vigilant and rejection is a real problem.

Essentially there are four major types of grafts:

1. **Autografts** are tissue grafts transplanted from one site to another in the same person.

2. **Isografts** are tissue grafts donated by a genetically identical person, the only example being an identical twin.

3. **Allografts** are tissue grafts taken from an unrelated person.

4. **Xenografts** are tissue grafts harvested from a different animal species, such as transplanting a baboon heart into a human being.

Autografts and isografts are ideal donor organs or tissues and are just about always successful given an adequate blood supply and no infection. Xenografts are never successful. The graft type most used is an allograft taken from a recently deceased person.

Before an allograft is even attempted the ABO and other blood group antigens of both the donor and recipient must be determined and must match. Then, the cell membrane antigens of their tissue cells are typed to determine how closely they match. As you might guess, good tissue matches between unrelated people are hard to find; at least a 75 percent match is needed to attempt a graft.

After surgery, to prevent rejection, the patient receives **immunosuppressive therapy,** including one or more of the following: corticosteroids to suppress inflammation, cytotoxic drugs, radiation (X-ray) therapy, and immunosuppressor drugs. Many of these drugs kill rapidly dividing cells (such as activated lymphocytes) and all of them have severe side effects. However, the major problem with immunosuppressive therapy is that while the immune system is suppressed, it cannot protect the body against other foreign agents. Explosive bacterial and viral infection is the most frequent cause of death in these patients.

Disorders of Immunity

The most important disorders of the immune system are allergies, immunodeficiencies, and autoimmune diseases.

Allergies

At first, the immune response was thought to be purely protective. However, it was not long before its dangerous potentials were discovered. **Allergies** or **hypersensitivities** are abnormally vigorous immune responses in which the immune system causes tissue damage as it fights off a perceived "threat" that would otherwise be harmless to the body. The term **allergen** (*allo* = altered; *erg* = reaction) is used to distinguish this type of antigen from those producing essentially normal responses. People rarely die of allergies; they are just miserable with them.

Although there are several different types of allergies, the most common type is **immediate hypersensitivity** (Figure 12.10). This type of response, also called **acute hypersensitivity,** is triggered by the release of a flood of histamine when IgE antibodies bind to *mast cells.* Histamine causes small blood vessels in the area to become dilated and leaky, and is largely to blame for the best-recognized symptoms of allergy: a runny nose, watery eyes, and itching, reddened skin (hives). When the allergen is inhaled, symptoms of asthma appear because smooth muscle in the walls of the bronchioles contracts, constricting the passages and restricting air flow. Over-the-counter (OTC) anti-allergy drugs contain *antihistamines* that counteract these effects. Most of these reactions begin within seconds after contact with the allergen and last about half an hour.

Fortunately, the bodywide or systemic acute allergic response, known as **anaphylactic** (an″ah-fĭ-lak-tik) **shock,** is fairly rare. Anaphylactic shock occurs when the allergen directly enters the blood and circulates rapidly through the body, as might

Sensitization stage

① Antigen (allergen) invades body

② Plasma cells produce large amounts of class IgE antibodies against allergen

③ IgE antibodies attach to mast cells in body tissues (and to circulating basophils)

— Mast cell with fixed IgE antibodies

— IgE

— Granules containing histamine

Subsequent (secondary) responses

④ More of same allergen invades body

⑤ Allergen binding to IgE attached to mast cells triggers degranulation and release of histamine (and other chemicals)

— Antigen

— Mast cell granules release contents after antigen binds with IgE antibodies

— Histamine

⑥ Histamine causes blood vessels to dilate and become leaky, which promotes edema; stimulates release of large amounts of mucus; and causes smooth muscles to contract (if respiratory system is site of allergen entry, asthma may occur)

Outpouring of fluid from capillaries

Release of mucus

Constriction of small respiratory passages (bronchioles)

Figure 12.10   Mechanism of an immediate (acute) hypersensitivity response.

happen with certain bee stings or spider bites. It may also follow an injection of a foreign substance (such as horse serum, penicillin, or other drugs that act as haptens) into susceptible individuals. The mechanism of anaphylactic shock is essentially the same as that of local responses; but when the entire body is involved, the outcome is life-threatening. For example, it is difficult to breathe when smooth muscles of the lung passages contract, and the sudden vasodilation (and fluid loss) that occurs may cause circulatory collapse and death within minutes. Epinephrine is the drug of choice to reverse these histamine-mediated effects.

Delayed hypersensitivities, mediated by *delayed hypersensitivity* and killer T cells, take much longer to appear (1 to 3 days) than any of the acute reactions produced by antibodies. Instead of histamine, the chemicals mediating these reactions are lymphokines released by the activated T cells. Hence, antihistamine drugs are *not* helpful against the delayed types of allergies. Corticosteroid drugs are used to provide relief.

The most familiar examples of delayed hypersensitivity reactions are those classed as *allergic contact dermatitis,* which follow skin contact with poison ivy, some heavy metals (lead, mercury, and others), and certain cosmetic and deodorant chemicals. These agents act as haptens, and after diffusing through the skin and attaching to body proteins, they are perceived as foreign by the immune system. The *Mantoux* and *tine tests,* skin tests for detection of tuberculosis, depend on delayed hypersensitivity reactions. When the tubercle antigens are injected just under (or scratched into) the skin, a small, hard lesion forms if the person has been sensitized to the antigen.

Immunodeficiencies

The **immunodeficiencies** include both congenital and acquired conditions in which the production or function of immune cells or complement is abnormal. The most devastating congenital condition is *severe combined immunodeficiency disease (SCID),* in which there is a marked deficit of both B and T cells. Because T cells are absolutely required for normal operation of *both* arms of the immune response, afflicted children have essentially no protection against pathogens of any type.

Minor infections easily shrugged off by most children are lethal to those with SCID. Bone marrow transplants, which provide normal lymphocyte stem cells, have helped some SCID victims. Without such treatment, the only hope for survival is living behind protective barriers (in a plastic "bubble") that keep out all infectious agents.

Currently, the most important and most devastating of the acquired immunodeficiencies is **acquired immune deficiency syndrome (AIDS).** AIDS, which cripples the immune system by interfering with the activity of helper T cells, is highlighted in the "Closer Look" box on pp. 376–377.

Autoimmune Diseases

Occasionally, the immune system loses its ability to distinguish friend from foe, that is, to tolerate self-antigens, while still recognizing and attacking foreign antigens. When this happens, the body produces antibodies (autoantibodies) and sensitized T cells that attack and damage its own tissues. This puzzling phenomenon is called **autoimmune disease,** because it is one's own immune system that produces the disorder.

Some 5 percent of adults in North America—two-thirds of them women—are afflicted with autoimmune disease. Most common are:

- *Multiple sclerosis (MS),* which destroys the white matter (myelin sheaths) of the brain and spinal cord (see p. 202)

- *Myasthenia gravis,* which impairs communication between nerves and skeletal muscles

- *Graves' disease,* in which the thyroid gland produces excessive amounts of thyroxine

- *Juvenile (type 1) diabetes mellitus,* which destroys pancreatic beta cells, resulting in deficient production of insulin (see pp. 282, 283, and 285)

- *Systemic lupus erythematosus (SLE),* a systemic disease that occurs mainly in young females and particularly affects the kidneys, heart, lungs, and skin

- *Glomerulonephritis,* a severe impairment of kidney function

- *Rheumatoid arthritis (RA),* which systematically destroys joints (see p. 147)

Current therapies include treatment approaches that depress certain aspects of the immune response.

How does the normal state of self-tolerance break down? It appears that one or more of the following may be triggers:

1. Inefficient lymphocyte programming. Instead of being eliminated, self-reactive B or T lymphocytes escape to the rest of the body. This is believed to occur in MS.

2. Appearance of self-proteins in the circulation that were not previously exposed to the immune system during fetal development. Such "hidden" antigens are found in sperm cells, the eye lens, and certain proteins in the thyroid gland. In addition, "new self-antigens" may appear as a result of gene mutations or alterations in self-proteins by hapten attachment or by bacterial or viral damage.

3. Cross-reaction of antibodies produced against foreign antigens with self-antigens. For instance, antibodies produced during an infection caused by streptococcus bacteria are known to cross-react with heart antigens, causing damage to both the heart muscle and its valves, as well as to joints and kidneys. This age-old disease is called *rheumatic fever.* ❑

DEVELOPMENTAL ASPECTS OF BODY DEFENSES

Stem cells of the immune system originate in the spleen and liver during the first month of embryonic development. Later, the bone marrow becomes the predominant source of stem cells (hemocytoblasts), and it persists in this role into adult life. In late fetal life and shortly after birth, the young lymphocytes develop self-tolerance and immunocompetence in their "programming organs" (thymus and bone marrow) and then populate the other lymphoid tissues. Upon meeting "their" antigens, the T and B cells complete their development to fully mature immune cells.

Although the ability of our immune system to recognize foreign substances is determined by our genes, the nervous system may help to control the

A CLOSER LOOK

AIDS: THE PLAGUE OF THE TWENTIETH CENTURY

In October 1347, several ships made port in Sicily and within days all the sailors they carried were dead of bubonic plague. By the end of the fourteenth century, approximately 25 percent of the population of Europe had been wiped out by this "black death." In January 1987, the U.S. Secretary of Health and Human Services warned that acquired immune deficiency syndrome, or AIDS, might be the plague of the twentieth century. These are strong words. Are they true?

Although AIDS was first identified in this country in 1981 among homosexual men and intravenous drug users of both sexes, it had begun afflicting the heterosexual population of Africa several years earlier. AIDS is characterized by severe weight loss, night sweats, swollen lymph nodes, and increasingly frequent infections, including a rare type of pneumonia called *Pneumocystis pneumonia* and the bizarre malignancy *Kaposi's sarcoma,* a cancerlike condition of blood vessels evidenced by purple lesions of the skin. Some AIDS victims develop slurred speech and severe dementia. The course of AIDS is grim, and thus far inescapable, finally ending in complete debilitation and death from cancer or overwhelming infection.

AIDS is caused by a virus transmitted in blood, semen, and vaginal secretions. Most commonly, AIDS enters the body via blood transfusions or blood-contaminated needles and during intimate sexual contact in which the mucosa is torn or where open lesions caused by sexually transmitted diseases allow the virus access to the blood. Although the AIDS virus has also been detected in saliva, it was not believed to be transmitted via that secretion. That conclusion is now being questioned.

The virus, named HIV (human immunodeficiency virus), specifically targets and destroys helper T cells, resulting in depression of cell-mediated immunity. Although antibody levels rise and killer T cells initially mount a vigorous response to viral exposure, in time a profound deficit of normal antibodies develops, and killer T cells become unresponsive to viral cues. The whole immune system is turned topsyturvy. It is now clear that the virus multiplies steadily in the lymph nodes throughout most of the chronic

AIDS viruses attack a T cell. The HIV viruses, artificially colored red in this scanning electron micrograph, surround a T lymphocyte. The virus invades helper T cells, killing them and seriously weakening the immune system of the infected person (14,000×).

asymptomatic period. Symptomatic AIDS appears when the lymph nodes are destroyed. The virus also invades the brain, which could account for the dementia of some AIDS patients. Although there are exceptions, most AIDS victims have died within a few months to 8 years after diagnosis.

The years since 1981 have witnessed an AIDS epidemic in the United States. By the end of 1993, over 339,000 cases had been diagnosed, and over 204,000 had already died. The Centers for Disease Control's grim prophecy is that 333,000 Americans will have died from AIDS complications by the end of 1995. Furthermore, because there is a six-month "window" during which antibodies may develop after exposure to HIV, there are probably 100 asymptomatic carriers of the virus for every diagnosed case. Moreover, the disease has a long incubation period (from a few months to 10 years) between exposure and the appearance of clinical symptoms. Not only has the number of identified cases jumped explosively in the at-risk populations, but the "face of

AIDS" is changing. Victims have begun to include people who do not belong to the original high-risk groups. Before reliable testing of donated blood was available, some people contracted the virus from blood transfusions. Hemophiliacs have been especially vulnerable because the blood factors they need are isolated from pooled blood donations. Although manufacturers began taking measures to kill the virus in 1984, by then an estimated 60 percent of the hemophiliacs in this country were already infected. The virus can also be transmitted from an infected mother to her fetus. Though homosexual men still account for the bulk of cases transmitted by sexual contact, more and more heterosexuals are contracting this disease. Particularly disturbing is the near-epidemic increase in diagnosed cases among teenagers.

Large-city hospitals are seeing more AIDS patients daily, and statistics reporting the number of cases in inner-city ghettos, where intravenous drug use is the chief means of AIDS transmission, are alarming. Currently the drug-abusing community accounts for 25 percent of all AIDS cases. Furthermore, 75 percent of AIDS cases in newborns occur where drug abuse abounds. This shocking revelation prompted the National Research Council to recommend that the government provide sterile needles to intravenous drug users in an attempt to suppress the means of transmission; needle-dispensing programs have been initiated in many cities.

Though the virus is not spread by casual contact, fearful parents are demanding that HIV-infected children be kept out of schools. Because of the high cost of treatment, some employers are requiring AIDS testing before hiring, and insurance companies are denying insurance to those who will not submit to the AIDS test. Many people have stopped donating blood, fearful that they might become infected (they cannot), so that blood supplies needed for surgical patients are deficient.

Although diagnostic tests to identify carriers of the AIDS virus are increasingly sophisticated, no cure has yet been found. Over 100 drugs are now in the Food and Drug Administration pipeline. Vaccines are undergoing clinical trials; but it is unlikely that an approved vaccine will be available soon.

Several antiviral drugs that act by inhibiting the enzymes that the HIV virus needs in order to multiply in the body are now clinically available. The *reverse transcriptase inhibitors,* such as AZT, were early on the scene. AZT has been followed by others, including ddI, ddC, d4T, and 3TC. In late 1995 and early 1996, *protease inhibitors* (saquinavir, ritonavir, and others) were approved. Thus far, it appears that combination therapy using drugs from each class delivers a one–two punch to the HIV virus; that is, combination therapy postpones drug resistance (a problem with mono-drug therapy with AZT alone) and substantially reduces the amount of HIV virus in the blood while boosting the number of helper T cells. Although there is controversy over whether treatment should begin before infected patients have symptoms of full-blown AIDs, the treatment focus is shifting to still-healthy, asymptomatic individuals with low helper T cell counts. Physicians are urging people to be tested for AIDS to enable them to begin treatment early in an attempt to prevent the immunological devastation that precedes the AIDS diagnosis.

A number of nonapproved or "underground" drugs are being prescribed or independently sought out and purchased abroad by AIDS patients. Some of these will prove useless and others will prove to have unforeseen harmful effects; but when life itself is at stake, this may be a small gamble to take. Perhaps, as urged in the media, the best defense is to practice "safer sex" by using condoms or other barriers and knowing one's sexual partner. The only alternative is sexual abstinence.

activity of the immune response. The immune response is definitely impaired in individuals who are under severe stress—for example, in those mourning the death of a beloved family member or friend. Our immune system normally serves us well throughout our lifetime, until old age. But, during the later years, its efficiency begins to wane, as does that of the nonspecific defenses. As a result, the body becomes less able to fight infections and destroy cells that have become cancerous. Additionally, we become more susceptible to both autoimmune and immune-deficiency diseases.

Important Terms

active immunity	**immunoglobulin** (im″mu-no-glob′u-lin)
allergy (hypersensitivity)	**inflammation**
antibody	**killer T cell**
antigen	**macrophage** (mak′ro-faj)
autoimmune disease	**memory cell**
B lymphocyte (B cell)	**neutralization**
cellular immunity	**passive immunity**
clonal selection	**pathogen**
complement fixation	**plasma cell**
helper T cell	**primary response**
humoral (hu′mor-al) **immunity**	**secondary response**
immune response	**suppressor T cell**
immunocompetence	**T lymphocyte (T cell)**
immunodeficiency	

Summary

Nonspecific Body Defenses (pp. 356–361)

1. Surface membranes (skin and mucous membranes) provide mechanical barriers to pathogens. Some produce secretions and/or have structural modifications that enhance their defensive effects: The skin's acidity, lysozyme, mucus, keratin, and ciliated cells are examples.

2. Phagocytes (macrophages and neutrophils) engulf and destroy pathogens that penetrate epithelial barriers. This process is enhanced when the pathogen's surface is altered by attachment of antibodies and/or complement.

3. Natural killer cells are nonimmune cells that act nonspecifically to lyse virus-infected and malignant cells.

4. The inflammatory response prevents spread of harmful agents, disposes of pathogens and dead tissue cells, and promotes healing. Protective leukocytes enter the area; the area is walled off by fibrin; and tissue repair occurs.

5. When complement (a group of plasma proteins) becomes fixed on the membrane of a foreign cell, lysis of the target cell occurs. Complement also enhances phagocytosis and the inflammatory and immune responses.

6. Interferon is a group of proteins synthesized by virus-infected cells and certain immune cells. It prevents viruses from multiplying in other body cells.

7. Fever enhances the fight against infectious microorganisms by increasing metabolism, which speeds up repair processes; and by causing the liver and spleen to store iron and zinc, which are needed for bacterial multiplication.

Specific Body Defenses: The Immune System
(pp. 361–375)

1. The immune system recognizes something as foreign and acts to inactivate or remove it. Immune response is antigen-specific, systemic, and has memory. The two arms of immune response are humoral immunity, mediated by antibodies, and cellular immunity, mediated by living cells (lymphocytes).

2. Antigens
   a. Antigens are large, complex molecules (or parts of them) recognized as foreign by the body. Foreign proteins are the strongest antigens.
   b. Complete antigens provoke an immune response and bind with products of that response (antibodies or sensitized lymphocytes).
   c. Incomplete antigens, or haptens, are small molecules that are unable to cause an immune response by themselves, but do so when they bind to body proteins and the complex is recognized as foreign.

3. Cells of the immune system: An overview
   a. Two main cell populations, lymphocytes and macrophages, provide for immunity.
   b. Lymphocytes arise from hemocytoblasts of bone marrow. T cells develop immunocompetence in the thymus and oversee cell-mediated immunity. B cells develop immunocompetence in bone marrow and provide humoral immunity. Immunocompetent lymphocytes seed lymphoid organs, where antigen challenge occurs, and circulate through blood, lymph, and lymphoid organs.
   c. Immunocompetence is signaled by the appearance of antigen-specific receptors on surfaces of lymphocytes.
   d. Macrophages arise from monocytes produced in bone marrow. They phagocytize pathogens and present parts of the antigens on their surfaces, for recognition by T cells.

4. Humoral (antibody-mediated) immune response
   a. Clonal selection of B cells occurs when antigens bind to their receptors, causing them to proliferate. Most clone members become plasma cells, which secrete antibodies. This is called the *primary* immune response.
   b. Other clone members become memory B cells, capable of mounting a rapid attack against the same antigen in subsequent meetings (*secondary* immune responses). These memory cells provide immunological "memory."
   c. Active humoral immunity is acquired during an infection or via vaccination and provides immunological memory. Passive immunity is conferred when a donor's antibodies are injected into the bloodstream, or when the mother's antibodies cross the placenta. It does not provide immunological memory.
   d. Basic antibody structure
      (1) Antibodies are proteins produced by sensitized B cells or plasma cells in response to an antigen, and are capable of binding with that antigen.
      (2) An antibody is composed of four polypeptide chains (two heavy and two light) that form a Y-shaped molecule.
      (3) Each polypeptide chain has a variable and a constant region. Variable regions form antigen-binding sites, one on each arm of the Y. Constant regions determine antibody function and class.
      (4) Five classes of antibodies exist: IgA, IgG, IgM, IgD, IgE. They differ structurally and functionally.
      (5) Antibody functions include complement fixation, neutralization, precipitation, agglutination.
      (6) Monoclonal antibodies are pure preparations of a single antibody type useful in diagnosis of various infectious disorders and cancer, and in treatment of certain cancers.

5. Cellular (cell-mediated) immune response
   a. T cells are sensitized by binding simultaneously to an antigen and a self-protein displayed on the surface of a macrophage. Clonal selection occurs and clone members differentiate into effector T cells or memory T cells.
   b. There are several different classes of effector T cells. Killer T cells directly attack and lyse infected and cancerous cells. Helper T cells interact directly with B cells bound to antigens. They also liberate lymphokines, chemicals that enhance the killing activity of macrophages, attract other leukocytes, or act as helper factors that stimulate activity of B cells and killer T cells. Delayed hypersensitivity T cells release chemicals that enhance inflammation and promote a delayed allergic reaction. Suppressor T cells terminate the normal immune response by releasing suppressor chemicals.

6. Disorders of immunity
   a. In allergy or hypersensitivity the immune system overreacts to an otherwise harmless antigen and tissue destruction occurs. Immediate (acute) hypersensitivity, as seen in hayfever, hives, and anaphylaxis, is due to IgE antibodies. Delayed hypersensitivity (for example, contact dermatitis) reflects activity of T cells and lym-

phokines, and nonspecific killing by activated macrophages.

b. Immunodeficiencies result from abnormalities in any immune element. Most serious is severe combined immunodeficiency disease (a congenital disease) and AIDS, an acquired immunodeficiency disease caused by a virus that attacks and cripples the helper T cells.

c. Autoimmune disease occurs when the body's self-tolerance breaks down, and antibodies and/or T cells attack the body's own tissues. Most forms of autoimmune disease result from inefficient lymphocyte programming in the fetus, changes in structure of self-antigens or appearance of formerly hidden self-antigens in blood, and cross-reactions with self-antigens and antibodies formed against foreign antigens.

Developmental Aspects of Body Defenses (pp. 375–377)

1. Development of immune response occurs around the time of birth. The thymus gland is the first lymphoid organ to appear in the embryo. Other lymphoid organs remain relatively undeveloped until after birth.

2. The ability of immunocompetent cells to recognize foreign antigens is genetically determined. Stress appears to interfere with normal immune response.

3. Efficiency of immune response wanes in old age, and infections, cancer, immunodeficiencies, and autoimmune diseases become more prevalent.

Review Questions

1. Besides acting as mechanical barriers, the skin and mucosae of the body contribute to body protection in other ways. Cite the common body locations and the importance of mucus, lysozyme, keratin, acid pH, and cilia.

2. What is complement? How does it cause bacterial lysis? What are some of the other roles of complement?

3. Interferons are referred to as antiviral proteins. What stimulates their production, and how do they protect uninfected cells?

4. Define *immune response.*

5. Define *antigen.* What is the difference between a complete antigen and an incomplete antigen (hapten)?

6. Differentiate clearly between humoral and cellular immunity, and between the roles of B lymphocytes and T lymphocytes.

7. Although the immune system has two arms, it has been said, "No T cells, no immunity." How is this so?

8. Define *immunocompetence.* What indicates that a B cell or T cell has developed immunocompetence? Where does the "programming phase" occur in the case of T cells? B cells?

9. Binding of antigens to receptors of immunocompetent lymphocytes leads to clonal selection. Describe the process of clonal selection. What nonlymphocyte cell is a central actor in this process, and what is its function?

10. Name the cell types that would be present in a B cell clone, and give the function of each type.

11. Describe the specific roles of helper, killer, and suppressor T cells in cell-mediated immunity. Which is thought to be disabled in AIDS?

12. Compare and contrast a primary and secondary immune response. Which is more rapid, and why?

13. Describe the structure of an antibody, and explain the importance of its variable and constant regions.

14. Name the five classes of immunoglobulins. Which is most likely to be found attached to a B cell membrane? Which is most abundant in plasma? Which is important in allergic responses? Which is the first Ig to be released during the primary response? Which can cross the placental barrier?

15. How do antibodies help to defend the body?

16. Define *allergy,* and distinguish between immediate types of allergy and delayed allergic reactions relative to cause and consequences.

17. What events can result in the loss of self-tolerance and autoimmune disease?

At the Clinic

1. As an infant receives her first dose of oral polio vaccine, the nurse explains to her parents that the vaccine is a preparation of weakened virus. What type of immunity will the infant develop?

2. Some people with a deficit of IgA exhibit recurrent paranasal sinus and respiratory tract infections. Explain these symptoms.

3. Explain the underlying mechanisms responsible for the cardinal signs of acute inflammation: heat, pain, redness, swelling.

4. Mr. James, an 80-year-old man, is grumbling about having to receive a flu shot every year. Flu viruses have a high mutation rate (change genetically rapidly), which results in the appearance of new proteins on the flu virus's "coat." How does this help explain the need to get a flu shot each year?

13
The Respiratory System

YOUR GOALS

After completing this chapter, you will have a working knowledge of the functions of the respiratory system and will have mastered the objectives listed below.

FUNCTION PREVIEW

■ The respiratory system supplies oxygen to the blood while removing carbon dioxide.

(continued)

The trillions of cells in the body require an abundant and continuous supply of oxygen to carry out their vital functions. We cannot "do without oxygen" for even a little while, as we can without food or water. Furthermore, as cells use oxygen, they give off carbon dioxide, a waste product the body must get rid of.

The *circulatory* and *respiratory systems* share responsibility for supplying the body with oxygen and disposing of carbon dioxide. The respiratory system organs oversee the gas exchanges that occur between the blood and the external environment. The transportation of respiratory gases between the lungs and the tissue cells is accomplished by the circulatory system organs, using blood as the transporting fluid. If either system fails, body cells begin to die from oxygen starvation and accumulation of carbon dioxide.

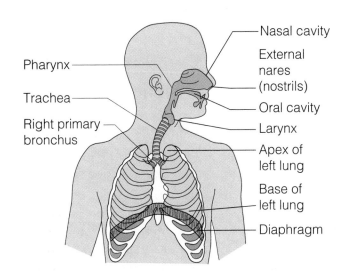

Figure 13.1   The major respiratory organs shown in relation to surrounding structures.

FUNCTIONAL ANATOMY OF THE RESPIRATORY SYSTEM

The organs of the **respiratory system** include the nose, pharynx, larynx, trachea, bronchi and their smaller branches, and the lungs, which contain the *alveoli* (al-ve′o-li), or terminal air sacs. Since gas exchanges with the blood happen only in the alveoli, the other respiratory system structures are really just *conducting passageways* that allow air to

reach the lungs. However, these passageways have another, very important job. They purify, humidify, and warm incoming air. Thus, the air finally reaching the lungs has many fewer irritants (such as dust or bacteria) than when it entered the system, and it resembles the warm, damp air of the tropics. As the respiratory system organs are described in detail next, locate each on Figure 13.1.

The Nose

The **nose,** whether "pug" or "ski-jump" in shape, is the only externally visible part of the respiratory system. During breathing, air enters the nose by passing through the **external nares, or nostrils.** The interior of the nose consists of the **nasal cavity,** divided by a midline **nasal septum.** The *olfactory receptors* for the sense of smell are located in the mucosa in the slitlike superior part of the nasal cavity, just beneath the ethmoid bone. The rest of the mucosa lining the nasal cavity, called the *respiratory mucosa,* rests on a rich network of thin-walled veins which warms the air as it flows past. (Because of the superficial location of these blood vessels, nosebleeds are common and often profuse.) In addition, the sticky mucus produced by the mucosa's glands moistens the air and traps incoming bacteria and other foreign debris. The ciliated cells of the nasal mucosa create a gentle current that moves contaminated mucus posteriorly toward the throat (pharynx), where it is swallowed and digested by stomach juices. We are usually unaware of this important ciliary action, but when the external temperature is extremely cold, these cilia become sluggish, allowing mucus to accumulate in the nasal cavity and to dribble outward through the nostrils. This helps explain why you might have a "runny" nose on a crisp, wintry day.

As shown in Figure 13.2, the lateral walls of the nasal cavity are uneven owing to three mucosa-covered projections or lobes, called **conchae** (kong′ke), which greatly increase the surface area of the mucosa exposed to the air. The conchae also increase the air turbulence in the nasal cavity. As the air swirls through the twists and turns, inhaled particles are deflected onto the mucus-coated surfaces where they are trapped and prevented from reaching the lungs.

The nasal cavity is separated from the oral cavity below by a partition, the **palate** (pal′et). Anteriorly, where the palate is supported by bone, is the **hard palate;** the unsupported posterior part is the **soft palate.**

The genetic defect *cleft palate* (failure of the bones forming the palate to fuse medially) results in breathing difficulty as well as problems with oral cavity functions such as chewing and speaking. ❏

The nasal cavity is surrounded by a ring of **paranasal sinuses** located in the frontal, sphenoid, ethmoid, and maxillary bones. (See Figure 5.10, p. 125.) The sinuses lighten the skull, and they act as resonance chambers for speech. They also produce mucus, which drains into the nasal cavities. The suctioning effect created by nose blowing helps to drain the sinuses. The nasolacrimal ducts, which drain tears from the eyes, also empty into the nasal cavities.

Cold viruses and various allergens can cause *rhinitis* (ri-ni′tis), inflammation of the nasal mucosa. The excessive mucus produced results in nasal congestion and postnasal drip. Because the nasal mucosa is continuous throughout the respiratory tract and extends tentaclelike into the nasolacrimal (tear) ducts and paranasal sinuses, nasal cavity infections often spread to those regions as well. **Sinusitis,** or inflamed sinuses, is difficult to treat and can cause marked changes in voice quality. When the passageways connecting the sinuses to the nasal cavity are blocked with mucus or infectious matter, the air in the sinus cavities is absorbed. The result is a partial vacuum and a *sinus headache* localized over the inflamed area. ❏

Pharynx

The **pharynx** (far′inks) is a muscular passageway about 13 cm (5 inches) long that vaguely resembles a short length of red garden hose. Commonly called the *throat,* the pharynx serves as a common passageway for food and air (Figures 13.1 and 13.2).

Air enters the superior portion, the **nasopharynx** (na″zo-far′inks), from the nasal cavity anteriorly, and then descends through the **oropharynx** (o″ro-far′inks) and **laryngopharynx** (lah-ring″go-far′inks) to enter the **larynx** below. Food enters the mouth, and then travels along with air through the oropharynx and laryngopharynx. But instead of entering the larynx, it is directed into the *esophagus* (ĕ-sof′ah-gus) posteriorly.

The auditory tubes, which drain the middle ear, open into the nasopharynx. Since the mucosae of these two regions are continuous, ear infections such as *otitis media* (o-ti′tis me′de-ah) may follow a sore throat or other types of pharyngeal infections.

Sphenoidal sinus

Pharyngeal tonsil

Opening of auditory tube

Nasopharynx

Soft palate

Uvula

Palatine tonsil

Oropharynx

Laryngopharynx

Vocal fold

Esophagus

Frontal sinus

Cribriform plate of ethmoid bone

Superior concha

Middle concha

Inferior concha

External nares

Hard palate

Tongue

Lingual tonsil

Epiglottis

Hyoid bone

Thyroid cartilage of larynx

Cricoid cartilage of larynx

Thyroid gland

Trachea

Figure 13.2   Basic anatomy of the upper respiratory tract, sagittal section.

Clusters of lymphatic tissue called *tonsils* are also found in the pharynx. The **pharyngeal** (far-rin′je-al) **tonsils,** often called *adenoids,* are located high in the nasopharynx. The **palatine tonsils** are in the oropharynx at the end of the soft palate; the **lingual tonsils** are at the base of the tongue. The role of the tonsils in body protection is described in Chapter 11 (p. 347).

If the pharyngeal tonsils become inflamed and swollen (as during a bacterial infection), they obstruct the nasopharynx and force the person to breathe through the mouth. In mouth breathing, air is not properly moistened, warmed, or filtered before reaching the lungs. Many children seem to have almost continuous *tonsillitis.* Years ago the belief was that, in such cases, the tonsils were more trouble than they were worth, and they were routinely removed. Presently, because of the widespread use of antibiotics, this is no longer necessary (or true). ❑

Larynx

The **larynx** (lar′inks), or *voice box,* routes air and food into the proper channels and plays a role in speech. Located inferior to the pharynx (see Figures 13.1 and 13.2), it is formed by eight rigid hyaline cartilages and a spoon-shaped flap of elastic cartilage, the epiglottis (ep″ĭ-glot′tis). The largest of the hyaline cartilages is the shield-shaped **thyroid cartilage,** which protrudes anteriorly and is commonly called the *Adam's apple.* Sometimes referred to as the "guardian of the airways," the **epiglottis** protects the superior opening of the larynx. When we are not swallowing, the epiglottis does not restrict the passage of air into the lower respiratory passages. When we swallow food or fluids, the situation changes dramatically; the larynx is pulled upward and the epiglottis tips, forming a lid over the opening of the larynx. This routes food into the esophagus, or food tube, posteriorly.

If anything other than air enters the larynx, a *cough reflex* is triggered to expel the substance and prevent it from continuing into the lungs. Because this protective reflex does *not* work when we are unconscious, it is never a good idea to try to give fluids to an unconscious person when attempting to revive him or her.

- Palpate your larynx by placing your hand midway on the anterior surface of your neck. Swallow. Can you feel the larynx rising as you swallow?

Part of the mucous membrane of the larynx forms a pair of folds, called the **vocal folds,** or **true vocal cords,** which vibrate with expelled air. This ability of the vocal folds to vibrate allows us to speak. The slitlike passageway between the vocal folds is the **glottis.**

Trachea

Air entering the **trachea** (tra′ke-ah), or windpipe, from the larynx travels down its length (10–12 cm, or about 4 inches) to the level of the fifth thoracic vertebra, which is approximately midchest (Figure 13.1). The trachea is lined with ciliated mucosa. The cilia beat continuously and in a direction opposite to that of the incoming air. They propel mucus, loaded with dust particles and other debris, away from the lungs to the throat where it can be swallowed or spat out.

Smoking inhibits ciliary activity and ultimately destroys the cilia. Without these cilia, coughing is the only means of preventing mucus from accumulating in the lungs. Smokers with respiratory congestion should avoid medications that inhibit the cough reflex. ❏

The trachea is fairly rigid because its walls are reinforced with C-shaped rings of hyaline cartilage. These rings serve a double purpose. The open parts of the rings abut the esophagus and allow it to expand anteriorly when we swallow a large piece of food. The solid portions support the trachea walls and keep it *patent,* or open, in spite of the pressure changes that occur during breathing.

Since the trachea is the only way air can enter the lungs, tracheal obstruction is life threatening. Many people have suffocated after choking on a piece of food that suddenly closed off the trachea (or the glottis of the larynx). The **Heimlich maneuver,** a procedure in which the air in a person's own lungs is used to "pop out," or expel, an obstructing piece of food, has saved many people from becoming victims of such "café coronaries." The Heimlich maneuver is simple to learn and easy to do. However, it is best learned by demonstration because cracked ribs are a distinct possibility when it is done incorrectly. In some cases of obstructed breathing, an emergency *tracheostomy* (tra′ke-ost′o-me; surgical opening of the trachea) is done to provide an alternate route for air to reach the lungs. Individuals with tracheostomy tubes in place form huge amounts of mucus the first few days because of irritation to the trachea. Thus, they must be suctioned frequently during this time to prevent the mucus from pooling in their lungs. ❏

Primary Bronchi

The right and left **primary bronchi** (brong′ki) are formed by the division of the trachea. Each primary bronchus runs obliquely before it plunges into the medial depression (*hilus*) of the lung on its own side (see Figure 13.1). The right primary bronchus, which is wider, shorter, and straighter than the left, is the more common site for an inhaled foreign object to become lodged. By the time incoming air reaches the bronchi, it is warm, cleansed of most impurities, and well humidified. The smaller subdivisions of the primary bronchi within the lungs are direct routes to the air sacs.

Lungs

The paired **lungs** are fairly large organs. They occupy the entire thoracic cavity except for the most central area, the **mediastinum** (me″de-as-ti′num), which houses the heart, the great blood vessels, bronchi, esophagus, and other organs (see Figure 13.3). The narrow superior portion of each lung, the **apex,** is located just deep to the clavicle. The broad lung area resting on the diaphragm is the **base** (Figure 13.1). Each lung is divided into lobes; the left lung has two lobes and the right lung has three.

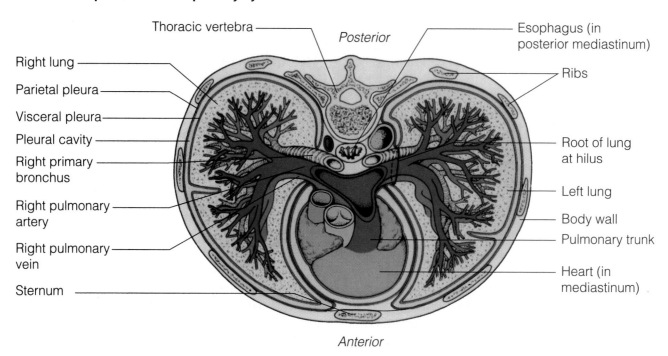

Thoracic vertebra

Posterior

Esophagus (in posterior mediastinum)

Right lung

Parietal pleura

Visceral pleura

Pleural cavity

Right primary bronchus

Right pulmonary artery

Right pulmonary vein

Sternum

Ribs

Root of lung at hilus

Left lung

Body wall

Pulmonary trunk

Heart (in mediastinum)

Anterior

Figure 13.3   Anatomical relationships of organs in the thoracic cavity. Transverse section through the thorax, showing the relationship of the lungs, the pleural membranes, the major organs present in the mediastinum, and the body (thorax).

The surface of each lung is covered with a visceral serosa called the **pulmonary,** or **visceral, pleura** (ploor'ah), and the walls of the thoracic cavity are lined by the **parietal pleura.** The pleural membranes produce a slippery serous secretion, *pleural fluid,* which allows the lungs to glide easily over the thorax wall during breathing movements and causes the two pleural layers to cling tightly together. The pleurae can slide easily from side to side across one another, but they strongly resist being pulled apart. Consequently, the lungs are held tightly to the thorax wall, and the *pleural space* is more of a potential space than an actual one. As described shortly, this condition of tightly adhering pleural membranes is absolutely essential for normal breathing. Figure 13.3 shows the position of the pleura on the lungs and the thorax wall.

Pleurisy (ploo'rĭ-se), inflammation of the pleura, can be caused by decreased secretion of pleural fluid. The pleural surfaces become dry and rough, which results in friction and stabbing pain with each breath. Conversely, the pleurae may produce excessive amounts of fluid, which

Figure 13.4   Resin cast showing the extensive branching of respiratory passages within the lungs (the bronchial or respiratory tree).

(a)

(b)

Figure 13.5   Respiratory zone structures. (a) Diagrammatic view of respiratory zone structures—respiratory bronchioles, alveolar ducts, and alveoli. **(b)** Scanning electron micrograph (SEM) of human lung tissue, showing the final divisions of the respiratory tree (475×).

exerts pressure on the lungs. This type of pleurisy hinders breathing movements, but it is much less painful than the dry rubbing type. ❑

After the primary bronchi enter the lungs, they subdivide into smaller and smaller branches (secondary and tertiary bronchi, and so on), finally ending in the smallest of the conducting passageways, the **bronchioles** (brong′ke-ōlz). Because of this branching and rebranching of the respiratory passageways within the lungs, the network formed is often referred to as the *bronchial* or *respiratory tree* (Figure 13.4). All but the smallest branches have reinforcing cartilage in their walls.

The *terminal bronchioles* lead into the *respiratory zone structures,* even smaller conduits, that

eventually terminate in **alveoli** (al-ve′o-li; *alveol* = small cavity), or air sacs (see Figure 13.5). The **respiratory zone,** which includes the *respiratory bronchioles, alveolar ducts,* alveolar sacs, and alveoli, is the only site of gas exchange. All other respiratory passages are **conducting zone structures** that serve as conduits to and from the respiratory zone. There are millions of the clustered alveoli, which resemble bunches of grapes, and they make up the bulk of the lungs. Consequently, the lungs are mostly air spaces. The balance of the lung tissue, its *stroma,* is elastic connective tissue. Thus, in spite of their relatively large size, the lungs weigh only about 2½ pounds, and they are soft and spongy.

Figure 13.6 **Anatomy of the respiratory membrane.** The respiratory membrane is composed of squamous epithelial cells of the alveoli, the capillary endothelium, and the scant basement membranes between. Surfactant-secreting cells are also shown. Oxygen diffuses from the alveolar air into the pulmonary capillary blood; carbon dioxide diffuses from the pulmonary blood into the alveolus. Neighboring alveoli are connected by small pores.

The Respiratory Membrane

The walls of the alveoli are composed largely of a single, thin layer of squamous epithelial cells. The thinness of their walls is hard to imagine, but a sheet of tissue paper is much thicker. *Alveolar pores* connect neighboring air sacs and provide alternate routes for air to reach alveoli whose feeder bronchioles have been clogged by mucus or otherwise blocked. The external surfaces of the alveoli are covered with a "cobweb" of pulmonary capillaries. Together, the alveolar and capillary walls and their fused basement membranes construct the **respiratory membrane (air-blood barrier),** which has gas (air) flowing past on one side and blood flowing past on the other (see Figure 13.6). The gas exchanges occur by simple diffusion through the respiratory membrane—the oxygen passing from the alveolar air into the capillary blood and the carbon dioxide leaving the blood to enter the gas-filled alveolus. It has been estimated that the total gas exchange surface provided by the alveolar walls of a healthy man is 70 to 80 square meters, or approximately the area of a racquetball court.

The final line of defense for the respiratory system is in the alveoli. Macrophages, sometimes called "dust cells," wander in and out of the alveoli picking up bacteria, carbon particles, and other debris. Also scattered amid the epithelial cells that form most of the alveolar walls are chunky cuboidal cells, which look very different. The cuboidal cells produce a lipid (fat) molecule called *surfactant,* which coats the gas-exposed alveolar surfaces and is very important in lung function (as described on p. 402).

RESPIRATORY PHYSIOLOGY

The major function of the respiratory system is to supply the body with oxygen and to dispose of carbon dioxide. To do this, at least four distinct events, collectively called **respiration,** must occur:

1. **Pulmonary ventilation.** Air must move into and out of the lungs so that the gases in the air sacs (alveoli) of the lungs are continuously changed and refreshed. This process of pulmonary ventilation is commonly called **breathing.**
2. **External respiration.** Gas exchange (oxygen loading and carbon dioxide unloading) between the pulmonary blood and alveoli must take place.
3. **Respiratory gas transport.** Oxygen and carbon dioxide must be transported to and from the lungs and tissue cells of the body via the bloodstream.
4. **Internal respiration.** At systemic capillaries, gas exchanges must be made between the blood and tissue cells.*

Although only the first two processes are the special responsibility of the respiratory system, all four processes are necessary for it to accomplish its goal of gas exchange. Thus, each process is described in turn next.

Mechanics of Breathing

Breathing, or pulmonary ventilation, is a completely mechanical process that depends on volume changes occurring in the thoracic cavity. Here is a rule to keep in mind about the mechanics of breathing: *Volume changes lead to pressure changes, which lead to the flow of gases to equalize the pressure.*

---

*The actual *use* of oxygen and production of carbon dioxide by tissue cells, that is, *cellular respiration,* is the cornerstone of all energy-producing chemical reactions in the body. Cellular respiration, which occurs in all body cells, is discussed in Chapter 14.

A gas, like a liquid, always conforms to the shape of its container. However, unlike liquid, a gas *fills* its container. Therefore, in a large volume, the gas molecules will be far apart and the pressure (created by the gas molecules hitting each other and the walls of the container) will be low. If the volume is reduced, the gas molecules will be closer together and the pressure will rise. Let us see how this relates to the two phases of breathing—**inspiration,** when air is flowing into the lungs, and **expiration,** when air is leaving the lungs.

Inspiration

When the inspiratory muscles, the **diaphragm** and **external intercostals,** contract, the size of the thoracic cavity increases. As the dome-shaped diaphragm contracts, it moves inferiorly and flattens out. As a result, the superior-inferior dimension (height) of the thoracic cavity increases. Contraction of the external intercostals lifts the rib cage and thrusts the sternum forward, which increases the anteroposterior and lateral dimensions of the thorax (Figure 13.7). Since the lungs adhere tightly to the thorax walls (due to the surface tension of the fluid between the pleural membranes), they are stretched to the new, larger size of the thorax. As *intrapulmonary volume* (the volume within the lungs) increases, the gases within the lungs spread out to fill the larger space. The resulting decrease in the gas presssure in the lungs produces a partial vacuum (pressure less than atmospheric pressure), which sucks air into the lungs. Air continues to move into the lungs until the intrapulmonary pressure is equal to atmospheric pressure. This series of events is called *inspiration* (inhalation).

Expiration

Expiration (exhalation) in healthy people is largely a passive process that depends more on the natural elasticity of the lungs than on muscle contraction. As the inspiratory muscles relax and resume their inital resting length, the rib cage descends and the lungs recoil. Thus, both the thoracic and intrapulmonary volumes decrease. As the intrapulmonary volume decreases, the gases inside the lungs are forced more closely together, and the intrapul-

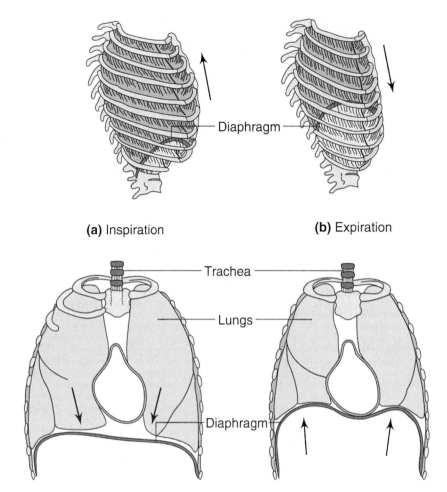

(a) Inspiration   **(b)** Expiration

Trachea

Lungs

Diaphragm

Figure 13.7   Rib cage and diaphragm positions during breathing.
(**a**) At the end of a normal inspiration: Chest is expanded, rib cage is elevated, and diaphragm is depressed and flattened. (**b**) At the end of a normal expiration: Chest is depressed, rib cage is descended, and diaphragm is elevated and dome-shaped.

monary pressure rises to a point higher than atmospheric pressure. This causes the gases to flow out to equalize the pressure inside and outside the lungs. Under normal circumstances, expiration is effortless, but if the respiratory passageways are narrowed by spasms of the bronchioles (as in *asthma*) or clogged with mucus or fluid (as in *chronic bronchitis* or *pneumonia*), expiration becomes an active process. In such cases of *forced expiration,* the internal intercostal muscles are activated to help depress the rib cage, and the abdominal muscles contract and help to force air from the lungs by squeezing the abdominal organs upward against the diaphragm.

The normal pressure within the pleural space *(intrapleural pressure)* is *always* negative, and this is the major factor preventing collapse of the lungs. If for any reason the intrapleural pressure becomes equal to the atmospheric pressure, the lungs immediately recoil completely and collapse.

During *atelectasis* (a″teh-lek′tuh-sis), or lung collapse, the lung is useless for ventilation. This phenomenon is seen when air enters the pleural space through a chest wound, but it may also result from a rupture of the visceral pleura, which allows air to enter the pleural space from the respiratory tract. The presence of air in the intrapleural space is referred to as a *pneumothorax* (nu″mo-tho′rakz). A pneumothorax is reversed by drawing air out of the intrapleural space with chest tubes, which allows the lung to reinflate and resume its normal function. ❏

Table 13.1   Nonrespiratory Air (Gas) Movements

Movement	Mechanism and result
Cough	Taking a deep breath, closing glottis, and forcing air superiorly from lungs against glottis. Then, the glottis opens suddenly and a blast of air rushes upward. Coughs act to clear the lower respiratory passageways.
Sneeze	Similar to a cough, except that expelled air is directed through nasal cavities instead of through oral cavity. The uvula (u'vu-lah), a tag of tissue hanging from the soft palate, becomes depressed and closes oral cavity off from pharynx, routing the air through nasal cavities. Sneezes clear upper respiratory passages.
Crying	Inspiration followed by release of air in a number of short breaths. Primarily an emotionally induced mechanism.
Laughing	Essentially same as crying in terms of the air movements produced. Also an emotionally induced response.
Hiccups	Sudden inspirations resulting from spasms of diaphragm; believed to be initiated by irritation of diaphragm or phrenic nerves, which serve diaphragm. The sound occurs when inspired air hits vocal folds of closed glottis.
Yawn	Very deep inspiration, taken with jaws wide open. Formerly believed to be triggered by need to increase amount of oxygen in blood but this theory is now being questioned; ventilates all alveoli (this is not the case in normal quiet breathing).

Nonrespiratory Air Movements

Many situations other than breathing move air into or out of the lungs and may modify the normal respiratory rhythm. Coughs and sneezes clear the air passages of debris or collected mucus. Laughing and crying reflect our emotions. For the most part, these **nonrespiratory air movements** are a result of reflex activity, but some may be produced voluntarily. Examples of the most common of these movements are given in Table 13.1.

Respiratory Volumes and Capacities

Many factors affect respiratory capacity—for example, a person's size, sex, age, and physical condition. Normal quiet breathing moves approximately 500 ml of air (about a pint) into and out of the lungs with each breath. This respiratory volume is referred to as the **tidal volume (TV).**

As a rule, a person *can* inhale much more air than is taken in during a normal, or tidal, breath.

The amount of air that can be taken in forcibly over the tidal volume is the **inspiratory reserve volume (IRV).** Normally, the inspiratory reserve volume is between 2100 and 3200 ml.

Similarly, after a normal expiration, more air can be exhaled. The amount of air that can be forcibly exhaled after a tidal expiration, the **expiratory reserve volume (ERV),** is approximately 1200 ml.

Even after the most strenuous expiration, about 1200 ml of air still remains in the lungs, and it cannot be voluntarily expelled. This is the **residual volume.** Residual volume air is important, because it allows gas exchange to go on continuously even between breaths and helps to keep the alveoli open (inflated).

The total amount of exchangeable air is typically around 4800 ml in healthy young males, and this respiratory capacity is the **vital capacity (VC).** The vital capacity is the sum of the TV + IRV + ERV. The respiratory volumes are summarized in Figure 13.8.

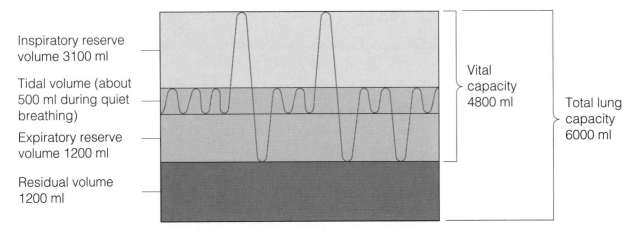

Inspiratory reserve
volume 3100 ml

Tidal volume (about
500 ml during quiet
breathing)

Expiratory reserve
volume 1200 ml

Residual volume
1200 ml

Vital
capacity
4800 ml

Total lung
capacity
6000 ml

Figure 13.8   Idealized tracing of the various respiratory volumes.

Obviously, much of the air that enters the respiratory tract remains in the conducting zone passageways and never reaches the alveoli. This is called the **dead space volume** and during a normal tidal breath, it amounts to about 150 ml (Figure 13.9). The functional volume—air that actually reaches the respiratory zone and contributes to gas exchange—is about 350 ml.

Respiratory capacities are measured with a *spirometer* (spi-rom′ĕ-ter). As a person breathes, the volumes of air exhaled can be read on an indicator, which shows the changes in air volume inside the apparatus. Spirometer testing is useful for evaluating losses in respiratory functioning and in following the course of some respiratory diseases. For example, in pneumonia inspiration is obstructed and the IRV and VC decrease. In emphysema, where expiration is hampered, the ERV is much lower than normal and the residual volume is higher.

Respiratory Sounds

As air flows in and out of the respiratory tree, it produces two recognizable sounds that can be picked up with a stethoscope. **Bronchial sounds** are produced by air rushing through the large respiratory passageways (trachea and bronchi). **Vesicular** (vĕ-sik′u-lar) **breathing sounds** occur as air fills the alveoli. The vesicular sounds are soft, and resemble the sound of a muffled breeze.

Diseased respiratory tissue, mucus, or pus can produce abnormal chest sounds such as *rales* (a rasping sound) and *wheezing* (a whistling sound). ❏

External Respiration, Gas Transport, and Internal Respiration

As explained earlier, **external respiration** is the actual exchange of gases between the alveoli and

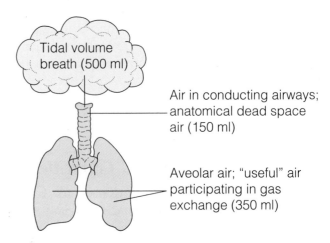

Tidal volume
breath (500 ml)

Air in conducting airways;
anatomical dead space
air (150 ml)

Aveolar air; "useful" air
participating in gas
exchange (350 ml)

Figure 13.9   Tidal volume. The tidal volume is the sum of anatomical dead space air (air occupying the conducting airways that does not participate in gas exchange) and air reaching the alveoli (that participates in gas exchange).

the blood (pulmonary gas exchange) and **internal respiration** is the gas exchange process that occurs between the systemic capillaries and the tissue cells. It is important to remember that all gas exchanges are made according to the laws of diffusion; that is, movement occurs *toward* the area of lower concentration of the diffusing substance. The relative amounts of O_2 and CO_2 in the alveolar tissues, and in the arterial and venous blood are illustrated in Figure 13.10.

External Respiration

During external respiration, dark red blood flowing through the pulmonary circuit is transformed into the scarlet river that is returned to the heart for distribution to the systemic circuit. Although this color change is due to oxygen pickup by hemoglobin in the lungs, carbon dioxide is being unloaded from the blood equally fast. Because body cells continually remove oxygen from blood, there is always more oxygen in the alveoli than there is in the blood. Thus, oxygen tends to move from the air of the alveoli through the alveolar-capillary walls into the more oxygen-poor blood of the pulmonary capillaries. On the other hand, as tissue cells remove oxygen from the blood in the systemic circulation, they release carbon dioxide into the blood. Since the concentration of carbon dioxide is much higher in the pulmonary capillaries than it is in the alveolar air, it will leave the blood to pass into the alveoli and be flushed out of the lungs during expiration. Relatively speaking, blood draining from the lungs into the pulmonary veins is oxygen-rich and carbon dioxide-poor, and is ready to be pumped to the systemic circulation.

Gas Transport in the Blood

Oxygen is transported in the blood in two ways. Most attaches to hemoglobin molecules inside the RBCs to form **oxyhemoglobin** (ok″se-he″moglo′bin), indicated as HbO_2 in Figure 13.11a. A very small amount of oxygen is carried dissolved in the plasma.

Most carbon dioxide is transported in plasma as the **bicarbonate ion** (HCO_3^-). A smaller amount (between 20 and 30 percent of the transported CO_2) is carried inside the RBCs bound to hemoglobin. Carbon dioxide carried inside the

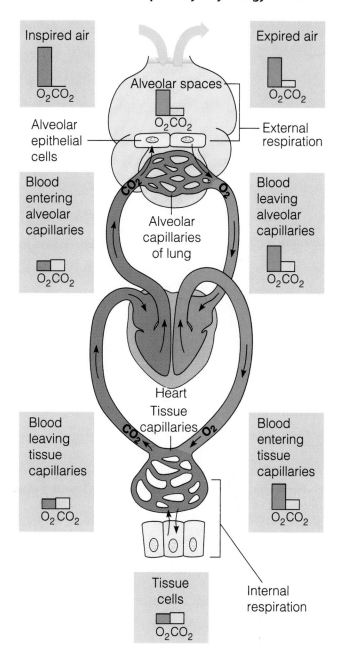

Figure 13.10 Gas exchanges in the body occur according to the laws of diffusion.

RBCs attaches to hemoglobin at a different site than oxygen does, and so it does not interfere in any way with oxygen transport. Before carbon dioxide can diffuse out of the blood into the alveoli, it must first be released from its bicarbonate ion form. For this to occur, bicarbonate ions must combine with hydrogen ions (H^+) to form carbonic

(a)

(b)

Figure 13.11 **Diagrammatic representation of the major means of oxygen (O_2) and carbon dioxide (CO_2) loading and unloading in the body.** (**a**) External respiration in the lungs (pulmonary gas exchange): Oxygen is loaded and carbon dioxide is unloaded. (**b**) Internal respiration in the body tissues (systemic capillary gas exchange): Oxygen is unloaded and carbon dioxide is loaded into the blood. (Note that although the conversion of CO_2 to bicarbonate ion is shown occurring in the plasma, most such conversions occur within the red blood cells.) Additionally, though not illustrated, some CO_2 is carried within red blood cells bound to hemoglobin.

acid (H_2CO_3). Carbonic acid quickly splits to form water and carbon dioxide, and carbon dioxide then diffuses from the blood and enters the alveoli.

Internal Respiration

Internal respiration, the exchange of gases that takes place between the blood and the tissue cells, is opposite to what occurs in the lungs. This process, in which oxygen is unloaded and carbon dioxide is loaded into the blood, is shown in Figure 13.11b. Carbon dioxide diffusing out of tissue cells enters the blood. In the blood, it combines with water to form carbonic acid, which quickly releases the bicarbonate ions. It should be men-

tioned that most conversion of carbon dioxide to bicarbonate ions actually occurs *inside* the RBCs, where a special enzyme (carbonic anhydrase) is available to speed up this reaction. Then the bicarbonate ions diffuse out into plasma where they are transported. At the same time, oxygen is released from hemoglobin, and the oxygen diffuses quickly out of the blood to enter the tissue cells. As a result of these exchanges, venous blood in the systemic circulation is much more oxygen-poor/carbon dioxide-rich than that leaving the lungs.

Impaired oxygen transport: Whatever the cause, inadequate oxygen delivery to body tissues is called **hypoxia** (hi-pok'se-ah). This condition is easy to recognize in fair-skinned people, because their skin and mucosae take on a bluish cast (become cyanotic); in dark-skinned individuals, this color change can be observed only in the mucosae and nailbeds. Hypoxia may be the result of anemia, pulmonary disease, or impaired or blocked blood circulation.

Carbon monoxide poisoning represents a unique type of hypoxia. Carbon monoxide (CO) is an odorless, colorless gas that binds to hemoglobin at the same sites as oxygen and competes vigorously with oxygen for those binding sites. Moreover, since hemoglobin binds to CO more readily than to oxygen, carbon monoxide is a very successful competitor—so much so that it crowds out or displaces oxygen.

Carbon monoxide poisoning is the leading cause of death from fire. It is particularly dangerous because it kills its victims softly and quietly. It does not produce the characteristic signs of hypoxia—cyanosis and respiratory distress. Instead, the victim becomes confused and has a throbbing headache. In rare cases, the skin becomes cherry red (the color of the hemoglobin-CO complex), which is often interpreted as a healthy "blush." Those with CO poisoning are given 100 percent oxygen until the carbon monoxide has been cleared from the body. ❑

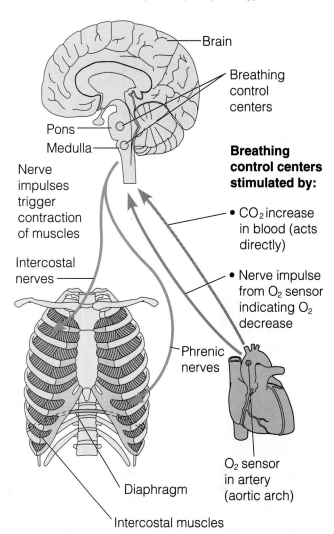

Figure 13.12 **Breathing control centers, sensory inputs, and effector nerves.**

Control of Respiration

Neural Regulation: Setting the Basic Rhythm

Although our tidelike breathing seems so beautifully simple, its control is fairly complex. We will cover only the most basic aspects of the respiratory controls. The activity of the respiratory muscles, the diaphragm and external intercostals, is regulated by nerve impulses transmitted to them from the brain by the **phrenic** and **intercostal nerves.**

The neural centers that control respiratory rhythm and depth are located in the medulla and pons (Figure 13.12). The medulla, which sets the basic rhythm of breathing, contains a **self-exciting inspiratory center,** as well as other respiratory centers. The pons centers appear to smooth out the basic rhythm of inspiration and expiration set by the medulla. Impulses going back and forth between the pons and medulla centers maintain a rate of 12–15 respirations/minute. This normal respiratory rate is referred to as *eupnea* (ūp-ne′ah).

In addition, the bronchioles and alveoli have stretch receptors that respond to extreme overinflation (which might damage the lungs). In the case of overinflation, impulses are sent from the stretch receptors to the medulla by the vagus nerves; soon therafter inspiration ends and expiration occurs.

During exercise, we breathe more deeply and at a faster rate because the brain centers send more rapid impulses to the respiratory muscles. After strenuous exercise, expiration becomes active and the abdominal muscles and any other muscles capable of lifting the ribs are used to aid expiration.

If the medulla centers are completely suppressed (as with an overdose of sleeping pills, morphine, or alcohol), respiration stops completely, and death occurs. ❑

A CLOSER LOOK

DANGEROUS RAPTURES OF THE DEEP

Even the nonswimmers among us are beckoned by photos of scuba divers examining brightly colored coral and other ocean creatures as they glide effortlessly through sparkling blue water. "What a life," we think enviously. But hidden beneath that surface of breathtaking beauty is life-threatening danger for the inexperienced, the careless, and the unlucky. In addition to the significant risk of drowning, scuba divers are exposed to the ever-present dangers of air embolism and decompression sickness.

As a diver descends to greater and greater depths underwater, the pressure on his or her body rises proportionately because of the increasing weight of

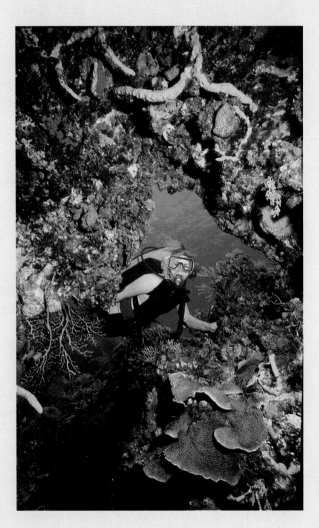

the water. *Scuba* (self-contained underwater breathing apparatus) gear has freed divers from heavy pressurized suits and air lines to the surface because it permits the air pressure (provided by the mixture of compressed air in the tank) to be equalized with the water pressure; that is, air enters the lungs at a higher-than-normal pressure. Descent is not usually a problem, unless the diver dives deeper than 100 feet and remains there for an extended time. Although nitrogen ordinarily has little effect on body functioning, given time and high pressure, so much nitrogen is forced into solution in the blood that it causes *nitrogen narcosis*. Nitrogen is far more soluble in lipids than water, and it tends to concentrate in lipid-rich tissues such as the central nervous system and fat depots. The diver becomes dizzy, giddy, and appears to be intoxicated; for this reason, the condition is sometimes called "rapture of the deep."

Assuming the diver avoids these narcotizing effects, and ascends to the surface gradually, dissolved nitrogen gas can be driven out of the tissues and then eliminated by the lungs without any problems. But if the ascent is rapid, the poorly soluble nitrogen gas appears to "boil" from the tissues. Gas bubbles in blood represent potentially lethal emboli, and those formed within joints, bones, and muscles cause excruciating musculoskeletal pain, commonly called the "bends." Other signs of this *decompression sickness* are mood changes, seizures, itching, numbness, and *chokes*. The syndrome called chokes involves substernal pain, coughing, dyspnea, and, in severe cases, shock. These signs and symptoms usually appear within an hour of surfacing, but they can be delayed as long as 36 hours.

Air emboli may also occur if the diver ascends suddenly and without exhaling. This is common when a diver is faced with some hazard, such as equipment failure or rough waters, and panics. Under such conditions, the alveoli are likely to rupture.

Even with modern scuba gear, divers need to know how to adjust from high underwater pressures to the lower pressures at the surface.

If connections occur between the alveoli and the pulmonary bloodstream, emboli develop as the diver takes the first breath of air at the surface, and the related problems develop within 2 minutes. Since the ascent is typically made head-up, the emboli usually invade the brain, and seizures and unconsciousness are possibilities. Indeed, many scuba-related drownings follow loss of consciousness due to air embolism.

The usual and most effective treatment for decompression sickness is hyperbaric (high pressure) therapy, in which the diver is recompressed and then decompressed . . . slowly. However, the best therapy is to avoid the need for treatment by making sure that the divers are well instructed in the equipment use and possible risks. Assuming that is done, the vacation should be GREAT!

Factors Influencing Respiratory Rate and Depth

Physical Factors Although brain centers set the basic rhythm of breathing, there is no question that physical factors such as talking, coughing, and exercise can modify both the rate and depth of breathing. Some of these factors have already been examined in the earlier discussion of nonrespiratory air movements. Increased body temperature causes an increase in the rate of breathing.

Volition (Conscious Control) We all have consciously controlled our breathing pattern at one time or another. During singing and swallowing, breath control is extremely important, and many of us have held our breath for short periods to swim underwater. However, voluntary control of breathing is limited, and the respiratory centers will simply ignore messages from the cortex (our wishes) when the oxygen supply in the blood is getting low or blood pH is falling. All you need do to prove this is to try to talk normally or to hold your breath after running at breakneck speed for a few minutes. It simply cannot be done. Many toddlers try to manipulate their parents by holding their breath "to death." Even though this threat causes many parents to become anxious, they need not worry because the involuntary controls take over and normal respiration begins again.

Emotional Factors Emotional factors also modify the rate and depth of breathing. Have you ever watched a horror movie with bated (held) breath or been so scared by what you saw that you were nearly panting? Have you ever touched something cold and clammy and gasped? All of these result from reflexes initiated by emotional stimuli acting through centers in the hypothalamus.

Chemical Factors Although many factors can modify respiratory rate and depth, the most important factors are chemical—the levels of carbon dioxide and oxygen in the blood. Increased levels of carbon dioxide and decreased blood pH are the most important stimuli leading to an increase in the rate and depth of breathing. (Actually, an increase in carbon dioxide levels and decreased blood pH are the same thing in this case, because carbon dioxide retention leads to increased levels of carbonic acid that decrease the blood pH.) Changes in carbon dioxide concentrations in the blood seem to act directly on the medulla centers (see Figure 13.12).

Conversely, changes in oxygen concentration in the blood are detected by chemoreceptor regions in the aorta (aortic arch) and carotid artery (carotid body). These, in turn, send impulses to the medulla when blood oxygen levels are dropping. Although every cell in the body must have oxygen to live, it is the body's need to rid itself of carbon dioxide (not to take in oxygen) that is the *most* important stimulus for breathing in a healthy person. Decreases in oxygen levels only become important stimuli when they are dangerously low. However, in people who retain carbon dioxide, as in those with chronic lung diseases like emphysema and chronic bronchitis, increased levels of carbon dioxide are no longer recognized as important by the brain, and dropping oxygen levels become the respiratory stimulus. This is an interesting fact because it explains why such patients are always given low levels of oxygen. If they were given high levels, they would stop breathing because their respiratory stimulus (low oxygen levels) would be gone.

Homeostatic mechanisms of the respiratory system, in healthy individuals, are obvious. As carbon dioxide or other sources of acids begin to accumulate in blood and blood pH starts to drop, you begin to breathe more deeply and more rapidly. This blows off more carbon dioxide and decreases the amount of carbonic acid, which returns blood pH to the normal range. On the other hand, when blood starts to become slightly alkaline, or basic (for whatever reason), breathing slows and becomes more shallow. Slower breathing allows carbon dioxide to accumulate in the blood and brings pH back into normal range. Indeed, control of breathing during rest is aimed primarily at regulating the hydrogen ion concentration in the brain. Extremely slow or shallow breathing *(hypoventilation)* or fast, deep breathing *(hyperventilation)*, can dramatically change the amount of carbonic acid in the blood. Carbonic acid increases greatly during hypoventilation and decreases substantially during hyperventilation. In both situations, the buffering ability of the blood is likely to be overwhelmed, so that acidosis or alkalosis occurs.

Hyperventilation, often brought on by anxiety attacks, frequently leads to brief periods of *apnea* (ap'ne-ah), cessation of breathing, until the carbon dioxide builds up in the blood again. If breathing stops for an extended time, **cyanosis** (si"ah-no'sis), due to insufficient oxygen in the blood, may occur. In addition, the hyperventilating person may get dizzy and faint because the resulting alkalosis causes cerebral blood vessels to constrict. Such attacks can be prevented by having the hyperventilating individual breathe into a paper bag. Because exhaled air contains more carbon dioxide than does atmospheric air, it upsets the normal diffusion gradient that causes CO_2 to be unloaded from the blood and leave the body. As a result, carbon dioxide (and, thus, carbonic acid) levels begin to rise in the blood, ending alkalosis. ❏

RESPIRATORY DISORDERS

The respiratory system is particularly vulnerable to infections because it is open to airborne pathogens. Because many of these inflammatory conditions, such as rhinitis and tonsillitis, have already been considered, we will turn our at-

tention to the most disabling respiratory disorders, the group of diseases collectively referred to as **chronic obstructive pulmonary disease (COPD)** and **lung cancer.** These disorders are "living proof" of cigarette smoking's devastating effects on the body. Long known to promote cardiovascular disease, cigarettes are perhaps even more effective at destroying the lungs. Figure 13.13 contrasts healthy lungs with the blackened, cancer-ridden lungs of a smoker.

Chronic Obstructive Pulmonary Disease (COPD)

The chronic obstructive pulmonary diseases, exemplified by chronic bronchitis and emphysema (em"fĭ-se'mah), are a major cause of death and disability in the United States. These diseases have certain features in common: (1) Patients almost always have a history of smoking; (2) **dyspnea** (disp'ne-ah), difficult or labored breathing often referred to as "air hunger," occurs and becomes progressively more severe; (3) coughing and frequent pulmonary infections are common; and (4) most COPD victims are hypoxic, retain carbon dioxide and have respiratory acidosis, and ultimately develop respiratory failure.

In **emphysema,** the alveoli enlarge as the walls of adjacent chambers break through and chronic inflammation promotes fibrosis of the lungs. As the lungs become less elastic, the airways collapse during expiration and obstruct outflow of air. As a result, these patients must use an incredible amount of energy to exhale, and they are always exhausted. Since air is retained in the lungs, oxygen exchange is surprisingly efficient and cyanosis does not usually appear until late in the disease. Consequently, emphysema sufferers are sometimes referred to as "pink puffers." However, overinflation of the lungs leads to a permanently expanded barrel chest.

In **chronic bronchitis,** the mucosa of the lower respiratory passages becomes severely inflamed and produces excessive amounts of mucus. The pooled mucus impairs ventilation and gas exchange, and dramatically increases the risk of lung infections, including pneumonias. Chronic bronchitis patients are sometimes called "blue bloaters" because hypoxia and carbon dioxide retention occur early in the disease and cyanosis is common.

(a)

(b)

Figure 13.13 **Photographs of healthy lungs of a nonsmoker (a) and cancerous lungs of a smoker (b).**

Lung Cancer

Lung cancer accounts for fully one-third of all cancer deaths in the United States. Its incidence, which is increasing daily, is strongly associated with cigarette smoking (over 90 percent of lung cancer patients were smokers). Lung cancer has a notoriously low cure rate: The overall 5-year survival is about 14 percent. Lung cancer metastasizes (spreads) rapidly and widely, and most cases are not diagnosed until they are well advanced.

Ordinarily, nasal hairs, sticky mucus, and the action of cilia do a fine job of protecting the lungs from chemicals and other irritants, but in smokers, these cleansing devices are overwhelmed and eventually stop functioning. Continuous irritation prompts the production of more mucus, but smoking paralyzes the cilia that clear this mucus and depresses the activity of lung macrophages. However, it is the irritant effects of the free radicals and the other carcinogens (cancer-promoting chemi-

cals) in tobacco smoke that eventually translate into lung cancer (see Figure 13.13b). These carcinogens cause the mucosal cells to proliferate wildly and lose their normal characteristics.

The three most common types of lung cancer are (1) **squamous cell carcinoma** (20 to 40 percent of cases), which arises in the epithelium of the larger bronchi and tends to form masses that hollow out and bleed; (2) **adenocarcinoma** (25 to 35 percent), which originates in the peripheral areas of the lung as solitary nodules that develop from bronchial mucous glands and alveolar cells; and (3) **small cell carcinoma** (10 to 20 percent), also called **oat cell carcinoma,** which consists of lymphocyte-like epithelial cells that originate in the primary bronchi and grow aggressively in cords or small grapelike clusters within the mediastinum.

The most effective treatment for lung cancer is complete resection (removal) of the diseased lung, because it has the greatest potential for prolonging life and cure. However, because of the aggressive spread of most lung cancers, resection is a choice that is open to very few patients. In most cases, radiation therapy and chemotherapy are currently the only options. ❑

DEVELOPMENTAL ASPECTS OF THE RESPIRATORY SYSTEM

In the fetus, the lungs are filled with fluid, and all respiratory exchanges are made by the placenta. At birth, the fluid-filled pathway is drained, and the respiratory passageways fill with air. The alveoli inflate and begin to function in gas exchange, but the lungs are not fully inflated for 2 weeks. The success of this change—that is, from nonfunctional to functional respiration—depends on the presence of **surfactant** (sur-fak′tant), a fatty molecule made by the cuboidal alveolar cells (see Figure 13.6). Surfactant lowers the surface tension of the film of water lining each alveolar sac so that the alveoli do not collapse between each breath. Surfactant is not usually present in large enough amounts to accomplish this function until late in pregnancy; that is, between 28 and 32 weeks.

Infants born prematurely (before week 28) or those in which surfactant production is inadequate for other reasons (as in many infants born to diabetic mothers) have *infant respiratory distress syndrome (IRDS)*. These infants have dyspnea within a few hours after birth and use tremendous amounts of energy just to keep reinflating their alveoli, which collapse after each breath. Although IRDS still accounts for over 20,000 newborn deaths a year, many of these babies survive now because of the current use of equipment that supplies a positive pressure continuously and keeps the alveoli open and working in gas exchange until adequate amounts of surfactant are produced by the maturing lungs.

Important birth defects of the respiratory system include cleft palate and cystic fibrosis. **Cystic fibrosis (CF),** the most common lethal genetic disease in the U.S., strikes one out of every 2000 white children and every day two children die of it. CF causes oversecretion of a thick mucus that clogs the respiratory passages and puts the child at risk for fatal respiratory infections. It affects other secretory processes as well. Most importantly, it impairs food digestion by clogging ducts that deliver pancreatic enzymes and bile to the small intestine.

Also, sweat glands produce an extremely salty perspiration. At the heart of CF is a faulty gene that codes for the CFTR protein, which works as a chloride channel to control the flow of Cl^- in and out of cells. In those with the mutated gene, CFTR is unable to perform its normal role. Consequently, less Cl^- is secreted and less water follows, resulting in the thick mucus typical of CF. Conventional therapy for CF is mucus-dissolving drugs, "clapping" the chest to loosen the thick mucus, and antibiotics to prevent infection. ❑

The respiratory rate is highest in newborn infants, about 40 to 80 respirations/minute. It continues to drop through life: In the infant it is around 30/minute, at five years it is around 25/minute, and in adults it is 12–18/minute. However, the rate often increases again in old age. The lungs continue to mature throughout childhood, and more alveoli are formed until young adulthood. Research has revealed that when smoking is begun during the early teens, the lungs never completely mature, and those additional alveoli are lost forever.

Except for occasional sneezes or coughs (responses to irritants), the respiratory system works so efficiently and smoothly that we are not even aware of it. Most problems that occur are a result of external factors—for example, obstruction of the trachea by a piece of food, or aspiration of food particles or vomitus (which leads to aspiration pneumonia). The common cold blocks the upper respiratory passageways with mucus, and some unfortunate individuals are plagued by **asthma** (caused by factors such as allergy or anxiety), which makes them wheeze and gasp for air as their respiratory passages swell and their bronchioles constrict.

For many years, tuberculosis and pneumonia were the worst killers in the United States. Antibiotics have decreased their lethal threat to a large extent, but they are still dangerous diseases. Newly diagnosed, and frequently drug-resistant, tuberculosis cases in AIDS patients are increasing by leaps and bounds the world over, but *at present* the most damaging and disabling respiratory diseases are still those described above, COPD and lung cancer.

As we age, the chest wall becomes more rigid and the lungs begin to lose their elasticity, resulting in a slowly decreasing ability to ventilate the lungs. Vital capacity decreases by about one-third by the age of 70. In addition, blood oxygen levels decrease and sensitivity to the stimulating effects of carbon dioxide decreases, particularly in a reclin-

SYSTEMS IN SYNC

Homeostatic Relationships Between the *Respiratory System* and Other Body Systems

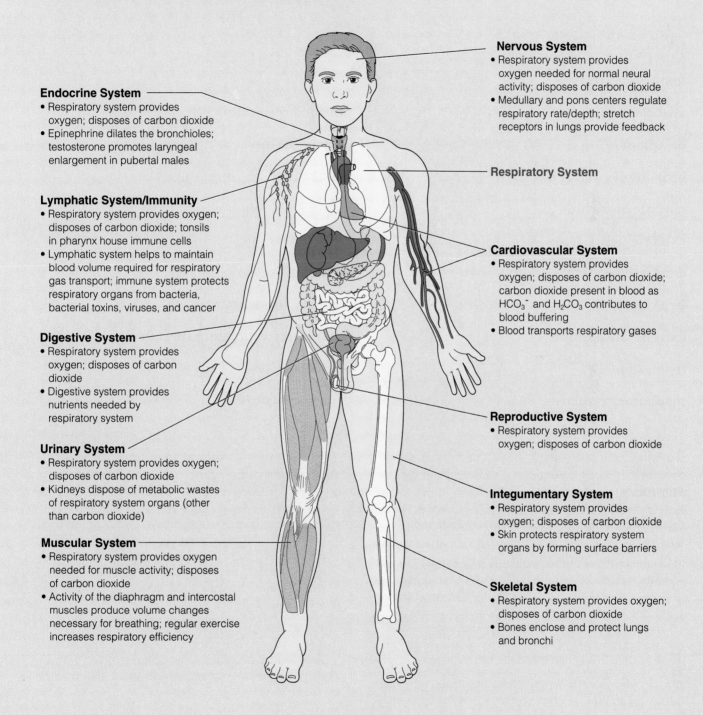

Nervous System
- Respiratory system provides oxygen needed for normal neural activity; disposes of carbon dioxide
- Medullary and pons centers regulate respiratory rate/depth; stretch receptors in lungs provide feedback

Respiratory System

Cardiovascular System
- Respiratory system provides oxygen; disposes of carbon dioxide; carbon dioxide present in blood as HCO_3^- and H_2CO_3 contributes to blood buffering
- Blood transports respiratory gases

Reproductive System
- Respiratory system provides oxygen; disposes of carbon dioxide

Integumentary System
- Respiratory system provides oxygen; disposes of carbon dioxide
- Skin protects respiratory system organs by forming surface barriers

Skeletal System
- Respiratory system provides oxygen; disposes of carbon dioxide
- Bones enclose and protect lungs and bronchi

Endocrine System
- Respiratory system provides oxygen; disposes of carbon dioxide
- Epinephrine dilates the bronchioles; testosterone promotes laryngeal enlargement in pubertal males

Lymphatic System/Immunity
- Respiratory system provides oxygen; disposes of carbon dioxide; tonsils in pharynx house immune cells
- Lymphatic system helps to maintain blood volume required for respiratory gas transport; immune system protects respiratory organs from bacteria, bacterial toxins, viruses, and cancer

Digestive System
- Respiratory system provides oxygen; disposes of carbon dioxide
- Digestive system provides nutrients needed by respiratory system

Urinary System
- Respiratory system provides oxygen; disposes of carbon dioxide
- Kidneys dispose of metabolic wastes of respiratory system organs (other than carbon dioxide)

Muscular System
- Respiratory system provides oxygen needed for muscle activity; disposes of carbon dioxide
- Activity of the diaphragm and intercostal muscles produce volume changes necessary for breathing; regular exercise increases respiratory efficiency

ing or supine position. As a result, many old people tend to become hypoxic during sleep and exhibit *sleep apnea.*

Additionally, many of the respiratory system's protective mechanisms also become less efficient with age. Ciliary activity of the mucosa decreases, and the phagocytes in the lungs become sluggish. The net result is that the elderly population is more at risk for respiratory tract infections, particularly pneumonia and influenza.

Important Terms

alveoli (al-ve′o-li)

bicarbonate ion

bronchioles (brong′ke-ōlz)

chronic bronchitis

COPD

emphysema (em″fĭ-se′mah)

expiration

external respiration

inspiration

inspiratory center

internal respiration

larynx (lar′inks)

nasal cavity

oxyhemoglobin (ok″se-he″mo-glo′bin)

pharynx (far′inks)

pleura (ploor′ah)

primary bronchi (brong′ki)

pulmonary ventilation (breathing)

surfactant (sur-fak′tant)

true vocal cords

Summary

Functional Anatomy of the Respiratory System (pp. 384–390)

1. The nasal cavity, the chamber within the nose, is divided medially by a nasal septum and separated from the oral cavity by the palate. The nasal cavity is lined with a mucosa, which warms, filters, and moistens incoming air. The mucosa also contains receptors for sense of smell. Paranasal sinuses and nasolacrimal ducts drain into the nasal cavity.

2. The pharynx (throat) is a mucosa-lined, muscular tube with three regions—nasopharynx, oropharynx, and laryngopharynx. The nasopharynx functions in respiration only; the others serve both respiratory and digestive functions. The pharynx contains tonsils, which act as part of the body's defense system.

3. The larynx (voice box) is a cartilage structure; most prominent is the thyroid cartilage (Adam's apple).

The larynx connects the pharynx with the trachea below. The laryngeal opening (glottis) is hooded by the epiglottis, which prevents entry of food or drink into respiratory passages when swallowing. The larynx contains the true vocal cords, which produce sounds used in speech.

4. The trachea (windpipe) extends from larynx to primary bronchi. The trachea is a smooth-muscle tube lined with a ciliated mucosa and reinforced with C-shaped cartilage rings, which keep the trachea patent.

5. Right and left primary bronchi result from subdivision of the trachea. Each plunges into the hilus of the lung on its side.

6. The lungs are paired organs flanking the mediastinum in the thoracic cavity. The lungs are covered with visceral pleura; the thorax wall is lined

with parietal pleura. Pleural secretions decrease friction during breathing. The lungs are primarily elastic tissue, plus passageways of the respiratory tree. The smallest passageways end in clusters of alveoli.

7. The conducting zone includes all respiratory passages from the nasal cavity to the terminal bronchioles; they conduct air to and from the lungs. Respiratory bronchioles, alveolar ducts and sacs, and alveoli—which have thin walls through which gas exchanges are made with pulmonary capillary blood—are respiratory zone structures.

Respiratory Physiology *(pp. 391–400)*

1. Mechanics of breathing: Gas travels from high-pressure to low-pressure areas. Pressure outside body is atmospheric pressure; pressure inside lungs is intrapulmonary pressure; pressure in intrapleural space is intrapleural pressure (which is always negative). Movement of air into and out of the lungs is called pulmonary ventilation, or breathing. When inspiratory muscles contract, intrapulmonary volume increases, its pressure decreases, and air rushes in (inspiration). When inspiratory muscles relax, the lungs recoil and air rushes out (expiration). Expansion of the lungs is helped by cohesion between pleurae and by the presence of surfactant in alveoli.

2. Nonrespiratory air movements: Nonrespiratory air movements are voluntary or reflex activities that move air into or out of the lungs. These include coughing, sneezing, laughing, crying, hiccuping, yawning.

3. Respiratory volumes and capacities: Air volumes exchanged during breathing are TV, IRV, ERV, and VC (see p. 393 for values). Residual volume is nonexchangeable respiratory volume and allows gas exchange to go on continually.

4. Respiratory sounds: Bronchial sounds are sounds of air passing through large respiratory passageways. Vesicular breathing sounds occur as air fills alveoli.

5. External respiration, gas transport, and internal respiration: Gases move according to laws of diffusion. Oxygen moves from alveolar air into pulmonary blood. Most oxygen is transported bound to hemoglobin inside RBCs. Carbon dioxide moves from pulmonary blood into alveolar air. Most carbon dioxide is transported as bicarbonate ion in plasma. At body tissues, oxygen moves from blood to the tissues, whereas carbon dioxide moves from the tissues to blood.

6. Control of respiration
   a. Nervous control: Neural centers for control of respiratory rhythm are in the medulla and pons.

   Reflex arcs initiated by stretch receptors in the lungs also play a role in respiration by notifying neural centers of excessive overinflation.
   b. Physical factors: Increased body temperature, exercise, speech, singing, and nonrespiratory air movements modify both rate and depth of breathing.
   c. Volition: To a degree, breathing may be consciously controlled if it does not interfere with homeostasis.
   d. Emotional factors: Some emotional stimuli can modify breathing. Examples are fear, anger, and excitement.
   e. Chemical factors: Changes in blood levels of carbon dioxide are the most important stimuli affecting respiratory rhythm and depth. Carbon dioxide acts directly on the medulla via its effect on reducing blood pH. Rising levels of carbon dioxide in blood result in faster, deeper breathing; falling levels lead to shallow, slow breathing. Hyperventilation may result in apnea and dizziness, due to alkalosis. Oxygen is less important as a respiratory stimulus in normal, healthy people, but it *is* the stimulus for those whose systems have become accustomed to high levels of carbon dioxide.

Respiratory Disorders *(pp. 400–402)*

1. The major respiratory disorders are COPD (emphysema and chronic bronchitis) and lung cancer. A significant cause is cigarette smoking.

2. Emphysema is characterized by permanent enlargement and destruction of alveoli. The lungs lose their elasticity, and expiration becomes an active process.

3. Chronic bronchitis is characterized by excessive mucus production and its pooling in lower respiratory passageways, which severely impairs ventilation and gas exchange. Patients may become cyanotic as a result of chronic hypoxia.

4. Lung cancer is extremely aggressive and metastasizes rapidly. The three most common lung cancers are squamous cell carcinoma, adenocarcinoma, and small cell carcinoma.

Developmental Aspects of the Respiratory System *(pp. 402–404)*

1. Premature infants have problems keeping their lungs inflated due to lack of surfactant in their alveoli. (Surfactant is formed late in pregnancy.)

2. The lungs continue to mature until young adulthood.

3. During youth and middle age, most respiratory system problems are a result of external factors, such as

infections and substances that physically block respiratory passageways.

4. In old age, the thorax becomes more rigid and lungs become less elastic, leading to decreased vital capacity. Protective mechanisms of the respiratory system decrease in effectiveness in elderly persons, predisposing them to more respiratory tract infections.

Review Questions

1. What is the most basic function of respiration?

2. Clearly explain the difference between external and internal respiration.

3. Trace the route of air from the external nares to an alveolus.

4. Why is it important that the trachea is reinforced with cartilage rings? What is the advantage of the fact that the rings are incomplete posteriorly?

5. Where in the respiratory tract is the air filtered, warmed, and moistened?

6. The trachea has cilia and goblet cells that produce mucus. What is the specific protective function of each of these?

7. Which primary bronchus is the more likely site for an inspired object to become lodged? Why?

8. In terms of general health, what is the importance of the fact that the auditory tubes and the sinuses drain into the nasal cavities and nasopharynx?

9. The lungs are mostly passageways and elastic tissue. What is the role of the elastic tissue? of the passageways?

10. What is it about the structure of the alveoli that makes them an ideal site for gas exchange?

11. What do TV, ERV, and VC mean? Which of these values is the largest? Why?

12. Name several nonrespiratory air movements and explain how each differs from normal breathing.

13. The contraction of the diaphragm and the external intercostal muscles begins inspiration. Explain exactly what happens, in terms of volume and pressure changes in the lungs, when these muscles contract.

14. What causes air to flow out of the lungs during expiration?

15. What is the major way that oxygen is transported in the blood? Carbon dioxide?

16. What determines in which direction carbon dioxide and oxygen will diffuse in the lungs? in the tissues?

17. Name the two major brain areas involved in the nervous control of breathing.

18. Name three physical factors that can modify respiratory rate or depth.

19. Name two chemical factors that modify respiratory rate and depth. Which is usually more important?

20. Define *hyperventilation*. If you hyperventilate, do you retain or expel more carbon dioxide? What effect does hyperventilation have on blood pH?

21. Compare and contrast the signs and symptoms of emphysema and chronic bronchitis.

At the Clinic

1. While diapering her 1-year-old boy (who puts virtually everything in his mouth), a mother failed to find one of the small safety pins previously used. Two days later, the boy developed a cough and became feverish. What is likely to have happened to the safety pin, and where (anatomically) would you expect it to be found?

2. Why doesn't Mom have to worry when 3-year-old Johnny threatens to "hold his breath till he dies"?

3. Alvin, a smoker, sees his doctor because he has a persistent cough and becomes short of breath after very little exertion. He has a barrel chest, a red face, and explains that it is difficult for him to exhale but not to inhale. What diagnosis will the doctor make?

4. Mr. Rasputin bumped a bee's nest while making repairs on his roof. As expected, he was promptly stung several times. Since he knew he was allergic to bee stings, he rushed to the hospital. While waiting, he went into a state of shock and had extreme difficulty breathing. Examination showed his larynx to be edematous and a tracheostomy was performed. Why is edema of the larynx likely to obstruct the airway? What is a tracheostomy and what purpose does it serve?

5. As a result of a stroke, Mrs. Minnick's swallowing is uncoordinated. What detrimental effect might this have on her ability to breathe?

14

The Digestive System and Body Metabolism

YOUR GOALS

After completing this chapter, you will have a working knowledge of the functions of the digestive system and body metabolism and will have mastered the objectives listed below.

FUNCTION PREVIEW

■ The digestive system breaks down ingested food to particles small enough to be absorbed into the blood.

■ Metabolism produces energy (ATP) in the cells and accounts for all constructive and degradative cellular activities.

OBJECTIVE CHECKLIST

ANATOMY OF THE DIGESTIVE SYSTEM (pp. 408–419)

❏ Name the organs of the alimentary canal and accessory digestive organs and identify each on an appropriate diagram or model.

❏ Identify the overall function of the digestive system as digestion and absorption of foodstuffs, and describe the general activities of each digestive system organ.

❏ Describe the composition and function(s) of saliva.

❏ Name the deciduous and permanent teeth and describe the basic anatomy of a tooth.

❏ Explain how villi aid digestive processes in the small intestine.

FUNCTIONS OF THE DIGESTIVE SYSTEM (pp. 419–428)

❏ Describe the mechanisms of swallowing, vomiting, and defecation.

❏ Describe how foodstuffs in the digestive tract are mixed and moved along the tract.

❏ Describe the function of local hormones in the digestive process.

❏ List the major enzymes or enzyme groups produced by the digestive organs or accessory glands and name the foodstuffs on which they act.

(continued)

❏ Name the end products of protein, fat, and carbohydrate digestion.

❏ State the function of bile in the digestive process.

NUTRITION AND METABOLISM (pp. 428–442)

❏ Define *nutrient, essential nutrient,* and *calorie.*

❏ List the six major nutrient categories. Note important dietary sources and the principal cellular uses of each.

❏ Define *enzyme, metabolism, anabolism,* and *catabolism.*

❏ Describe the metabolic roles of the liver.

❏ Recognize the sources of carbohydrates, fats, and proteins and their uses in cell metabolism.

❏ Explain the importance of energy balance in the body and indicate consequences of energy imbalance.

❏ List several factors that influence metabolic rate, and note the effect of each.

❏ Describe how body temperature is regulated.

DEVELOPMENTAL ASPECTS OF THE DIGESTIVE SYSTEM AND METABOLISM (pp. 442–443)

❏ Name important congenital disorders of the digestive system and significant inborn errors of metabolism.

❏ Describe the effect of aging on the digestive system.

Children have a special fascination with the workings of the digestive system: They relish crunching a potato chip, delight in making "mustaches" with milk, and giggle when their stomach "growls." As adults, we know that a healthy digestive system is essential for good health, because it converts food into the raw materials that build and fuel our body's cells. Specifically, the digestive system takes in food, breaks it down into nutrient molecules and absorbs them into the bloodstream, and then rids the body of the indigestible remains.

ANATOMY OF THE DIGESTIVE SYSTEM

The organs of the digestive system can be separated into two main groups: those forming the *alimentary* (al″ĕ-men′tar-e; *aliment* = nourish) *canal,* and the *accessory digestive organs* (see Figure 14.1). The alimentary canal **digests** food—breaks it down into smaller fragments—and **absorbs** the digested fragments through its lining into the blood. The accessory organs (teeth, tongue, and several large digestive glands) assist the process of digestive breakdown in various ways.

Organs of the Alimentary Canal

The **alimentary canal,** also called the **gastrointestinal (GI) tract,** is a continuous, coiled, hollow, muscular tube that winds through the ventral body cavity and is open at both ends. Its organs are the *mouth, pharynx, esophagus, stomach, small intestine,* and *large intestine.* The large intestine leads to the terminal opening, or *anus.* In a cadaver, the alimentary canal is approximately 9 m (about 30 feet) long, but in a living person, it is considerably

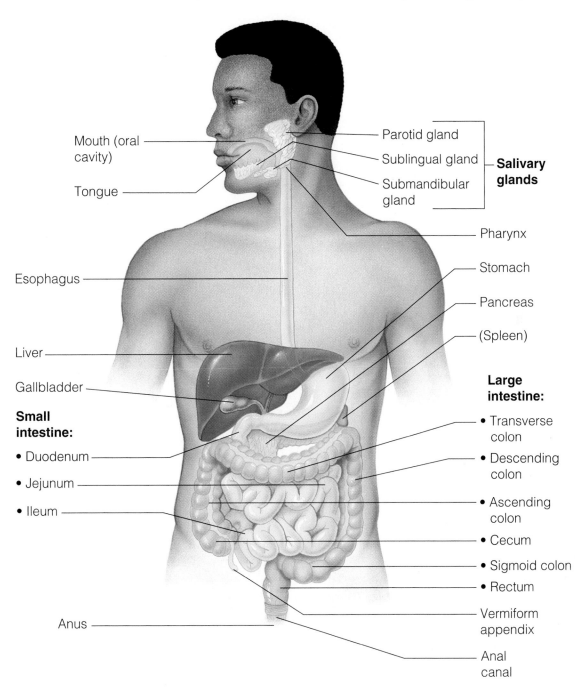

Mouth (oral cavity)

Tongue

Parotid gland

Sublingual gland

Submandibular gland

Salivary glands

Pharynx

Stomach

Pancreas

(Spleen)

Esophagus

Liver

Gallbladder

Small intestine:

• Duodenum

• Jejunum

• Ileum

Large intestine:

• Transverse colon

• Descending colon

• Ascending colon

• Cecum

• Sigmoid colon

• Rectum

• Vermiform appendix

• Anal canal

Anus

Figure 14.1   The human digestive system: Alimentary canal and accessory organs. (Liver and gallbladder are reflected superiorly and to the right side of the body.)

shorter because of its relatively constant muscle tone. Food material within this tube is technically outside the body, because it has contact only with cells lining the tract and is open to the external environment at both ends. As each organ of the alimentary canal is described next, find it on Figure 14.1.

Mouth

Food enters the digestive tract through the **mouth,** or **oral cavity,** a mucous membrane-lined cavity. The **lips (labia)** protect its anterior opening, the **cheeks** form its lateral walls, the **hard palate** forms its anterior roof, and the **soft palate** forms its posterior roof. The **uvula** (u′vu-lah) is a fleshy

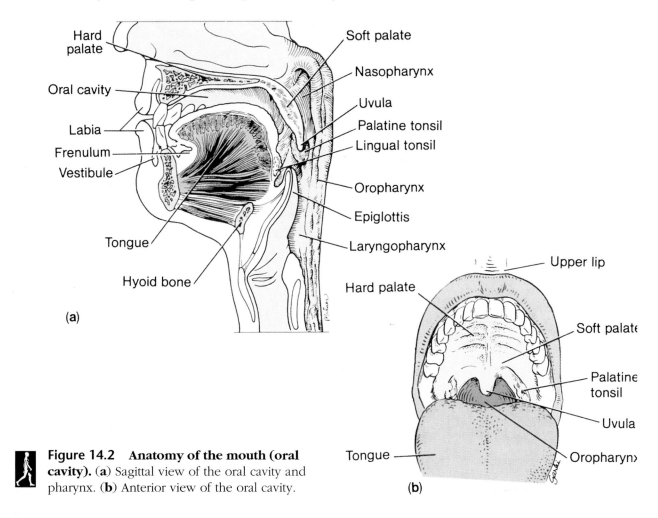

Figure 14.2 **Anatomy of the mouth (oral cavity).** (**a**) Sagittal view of the oral cavity and pharynx. (**b**) Anterior view of the oral cavity.

fingerlike projection of the soft palate, which extends downward from its posterior edge. The space between the lips and cheeks externally and the teeth and gums internally is the **vestibule.** The area contained by the teeth is the **oral cavity proper** (Figure 14.2). The muscular **tongue** occupies the floor of the mouth. The tongue has several bony attachments—two of these are to the hyoid bone and the styloid processes of the skull. The **frenulum** (fren'u-lum), a fold of mucous membrane, secures the tongue to the floor of the mouth and limits its posterior movements (see Figure 14.2a).

Children born with an extremely short frenulum are often referred to as "tongue-tied" because distorted speech results when tongue movement is restricted. This congenital condition can be corrected surgically by cutting the frenulum. ❑

At the posterior end of the oral cavity are paired masses of lymphatic tissue, the *palatine tonsils.* The *lingual* (ling'gwal) *tonsil* covers the base of the tongue just beyond. The tonsils, along with other lymphatic tissues, are part of the body's

defense system. When the tonsils become inflamed and enlarge, they partially block the entrance into the throat (pharynx), making swallowing difficult and painful.

As food enters the mouth, it is mixed with saliva and *masticated* (chewed). The cheeks and closed lips hold the food between the teeth during chewing. The nimble tongue continually mixes food with saliva during chewing and initiates swallowing. Thus, the breakdown of food begins before the food has even left the mouth. As noted in Chapter 8, *papillae* containing taste buds, or taste receptors, are found on the tongue surface. And so, besides its food manipulating function, the tongue allows us to enjoy and appreciate the food as it is eaten.

Pharynx

From the mouth, food passes posteriorly into the *oropharynx* and *laryngopharynx,* both of which are common passageways for food, fluids, and air.

Visceral peritoneum

Nerve

Blood vessel

Mesentery

Lumen

Submucosal gland

Lymph nodule

Villi (in small intestine only)

Intrinsic nerve plexuses:
Myenteric nerve plexus

Submucosal nerve plexus

Mucosa:
Surface epithelium

Lamina propria

Muscular layer

Submucosa

Muscularis externa:
Circular muscle layer

Longitudinal muscle layer

Serosa (visceral peritoneum)

Figure 14.3   Basic structure of the alimentary canal wall.

As explained in Chapter 13, the pharynx is subdivided into the *nasopharynx,* part of the respiratory passageway; the **oropharynx,** posterior to the oral cavity; and **laryngopharynx,** which is continuous with the esophagus below.

The walls of the pharynx contain two skeletal muscle layers. The cells of the inner layer run longitudinally; those of the outer layer (the constrictor muscles) run around the wall in a circular fashion. Alternating contractions of these two muscle layers propel food through the pharynx into the esophagus below. This propelling mechanism, called *peristalsis* (per″i-stal′sis), is described later.

Esophagus

The **esophagus,** (ĕ-sof′ah-gus), or *gullet,* runs from the pharynx through the diaphragm to the stomach. About 25 cm (10 inches) long, it is essen-

tially a passageway that conducts food (by peristalsis) to the stomach.

The walls of the alimentary canal organs from the esophagus to the large intestine are made up of the same four basic tissue layers or tunics (Figure 14.3):

1. The **mucosa** is the innermost layer, a moist membrane that lines the cavity or lumen of the organ. It consists primarily of a surface epithelium, plus a small amount of connective tissue (lamina propria) and a scanty smooth muscle layer. Beyond the esophagus, which has a friction-resisting stratified squamous epithelium, the epithelium is mostly simple columnar.

2. The **submucosa** is found just beneath the mucosa. It is a soft connective tissue layer containing blood vessels, nerve endings, lymph nodules, and lymphatic vessels.

(a)

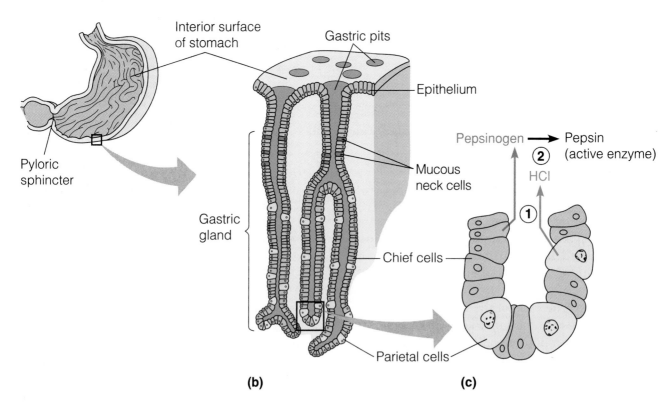

(b) **(c)**

Figure 14.4 **Anatomy of the stomach.** (**a**) Gross internal anatomy (frontal section). Pancreatic, cystic, and hepatic ducts are also shown. (**b**) Enlarged view of gastric pits and glands (longitudinal section). (**c**) Schematic showing the sequence of events from the production of pepsinogen by the chief cells (1) to its activation (to pepsin) by HCl secreted by the parietal cells (2).

3. The **muscularis externa** is a muscle layer typically made up of an inner *circular layer* and an outer *longitudinal layer* of smooth muscle cells.

4. The **serosa** is the outermost layer of the wall. It consists of a single layer of flat serous fluid-producing cells, the **visceral peritoneum** (per″i-to-ne′um). The visceral peritoneum is continuous with the slick, slippery **parietal peritoneum,** which lines the abdominopelvic cavity by way of a membrane extension, the **mesentery** (mes′en-ter″e). These relationships are illustrated in Figure 14.5 on p. 414.

The alimentary canal wall contains two important *nerve plexuses*—the **submucosal nerve plexus** and the **myenteric nerve plexus.** These networks of nerve fibers are actually part of the autonomic nervous system. They help regulate the mobility and secretory activity of GI tract organs.

Stomach

The C-shaped **stomach** (Figure 14.4) is on the left side of the abdominal cavity, nearly hidden by the liver and diaphragm. Different regions of the stomach have been named. The *cardiac region* (named for its position near the heart) surrounds the **cardioesophageal** (kar″de-o-ĕ-sof″ah-je′al) **sphincter** through which food enters the stomach from the esophagus. The *fundus* is the expanded part of the stomach lateral to the cardiac region. The *body* is the midportion, and the funnel-shaped *pylorus* (pi-lo′rus) is the terminal part of the stomach. The pylorus is continuous with the small intestine through the **pyloric sphincter,** or **valve.** The stomach is approximately 25 cm (10 inches) long, but its diameter depends on how much food it contains. When it is full, it can hold about 4 liters (1 gallon) of food. When it is empty, it collapses inward on itself, and its mucosa is thrown into large folds called **rugae** (roo′ge; *ruga* = wrinkle, fold). The convex lateral surface of the stomach is the **greater curvature;** its concave medial surface is the **lesser curvature.**

The **lesser omentum** (o-men′tum), a double layer of peritoneum, extends from the liver to the lesser curvature. The **greater omentum,** another extension of the peritoneum, drapes downward and covers the abdominal organs like a lacy apron before attaching to the posterior body wall (see

Figure 14.5). The greater omentum is riddled with fat which helps to insulate, cushion, and protect the abdominal organs, and has large collections of lymph nodules containing macrophages and defensive cells of the immune system.

When the peritoneum is infected, a condition called *peritonitis* (per″i-to-ni′tis), the peritoneal membranes tend to stick together around the infection site. This helps to seal off and localize many intraperitoneal infections (at least initially), providing time for macrophages in the lymphatic tissue to mount an attack. ❏

The stomach acts as a temporary "storage tank" for food as well as a site for food breakdown. Besides the usual longitudinal and circular layers, its wall contains a third obliquely arranged layer of muscle in the muscularis externa. This muscle arrangement allows the stomach not only to move food along the tract but also to churn, mix, and pummel the food, physically breaking it down to smaller fragments. In addition, chemical breakdown of proteins begins in the stomach. The mucosa of the stomach is a simple columnar epithelium that produces large amounts of mucus. This otherwise smooth lining is dotted with millions of deep *gastric pits,* which lead into *gastric glands* that secrete the solution called **gastric juice.** For example, some stomach cells produce *intrinsic factor,* a substance needed for the absorption of vitamin B_{12} from the small intestine. The **chief cells** produce protein-digesting enzymes, mostly **pepsinogens,** and the **parietal cells** produce corrosive hydrochloric acid, which makes the stomach contents acidic and activates the enzymes (Figure 14.4b). The *mucous neck cells* produce a sticky alkaline mucus, which clings to the stomach mucosa and protects the stomach wall itself from being damaged by the acid and digested by the enzymes. Still other cells produce local hormones, such as *gastrin,* that are important to the digestive activities of the stomach (see Table 14.2 on p. 427).

Most digestive activity occurs in the pyloric region of the stomach. After food has been processed in the stomach, it resembles heavy cream and is called **chyme** (kīm). The chyme enters the small intestine through the pyloric sphincter.

Small Intestine

The **small intestine** is the body's major digestive organ. Within its twisted passageways, usable food

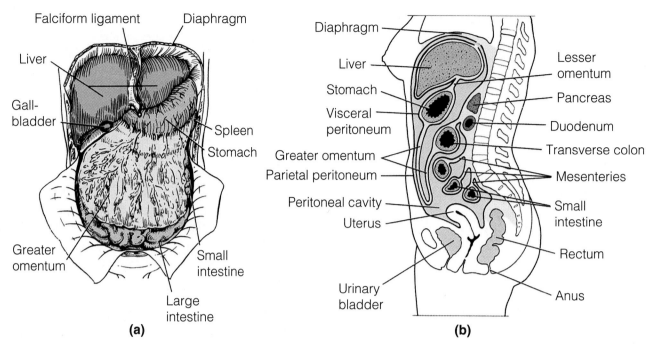

Falciform ligament Diaphragm
Liver
Gall-
bladder
Spleen
Stomach
Greater
omentum
Small
intestine
Large
intestine

(a)

Diaphragm
Liver
Stomach
Visceral
peritoneum
Greater omentum
Parietal peritoneum
Peritoneal cavity
Uterus
Urinary
bladder
Lesser
omentum
Pancreas
Duodenum
Transverse colon
Mesenteries
Small
intestine
Rectum
Anus

(b)

Figure 14.5   Peritoneal attachments of the abdominal organs. (**a**) Anterior view; the greater omentum is shown in its normal position covering the abdominal viscera. (**b**) Sagittal view of the abdominopelvic cavity of a female.

is finally prepared for its journey into the cells of the body. The small intestine is a convoluted tube extending from the pyloric sphincter to the **ileocecal** (il″e-o-se′kal) **valve.** It is the longest section of the alimentary tube, with an average length of 2 m (6 feet) in a living person. The small intestine hangs in sausagelike coils in the abdominal cavity, suspended from the posterior abdominal wall by the fan-shaped mesentery (see Figure 14.5). The large intestine encircles and frames it in the abdominal cavity.

The small intestine has three subdivisions. The **duodenum** (du″o-de′num; "twelve finger widths long"), which curves around the head of the pancreas, is about 25 cm (10 inches) long; the **jejunum** (je-joo′num) is about 2.5 m (8 feet) long and extends from the duodenum to the ileum. The **ileum** (il′e-um), about 3.6 m (12 feet) long, is the terminal part of the small intestine (see Figure 14.1). It joins the large intestine at the ileocecal valve.

Chemical digestion of foods begins in earnest in the small intestine. The small intestine is able to process only a small amount of food at one time. The pyloric sphincter (literally, "gatekeeper") controls food movement into the small intestine from

the stomach and prevents the small intestine from being overwhelmed. Enzymes, produced by the intestinal cells and (more importantly) by the pancreas and ducted into the duodenum through the **pancreatic duct,** complete the chemical breakdown of foods in the small intestine. Bile (formed by the liver) also enters the duodenum through the **bile duct** in the same area (see Figure 14.4a). The pancreatic and bile ducts join at the duodenum to form the flasklike *hepatopancreatic ampulla* (hepah″to-pan-kre-a′tik am-pu′lah), literally, the "liver-pancreatic enlargement." From there, the bile and pancreatic juice travel through the *duodenal papilla* and enter the duodenum together.

Nearly all food absorption occurs in the small intestine. The small intestine is well suited for its function. Its wall has three structures that increase the absorptive surface tremendously—microvilli, villi, and circular folds (see Figure 14.6). **Microvilli** (mi″kro-vih′li) are tiny projections of the plasma membrane of the mucosa cells that give the cell surface a fuzzy appearance, sometimes referred to as the **brush border. Villi** are fingerlike projections of the mucosa that give it a velvety appearance and feel, much like the soft nap of a Turkish towel. Within each villus is a rich capillary

Figure 14.6 **Structural modifications of the small intestine.** (**a**) Several circular folds (plicae circulares), seen on the inner surface of the small intestine. (**b**) Enlargement of one villus extension of the circular fold. (**c**) Enlargement of an absorptive cell to show microvilli.

bed and a modified lymphatic capillary called a **lacteal.** The digested foodstuffs are absorbed through the mucosa cells into both the capillaries and the lacteal, as discussed later in this chapter. **Circular folds,** also called **plicae circulares** (pli′se ser-ku-la′res), are deep folds of both mucosa and submucosa layers. Unlike the rugae folds of the stomach, the circular folds do not disappear when food fills the small intestine. All these structural modifications, which increase the surface area, decrease in number toward the end of the small intestine. On the other hand, local collections of lymphatic tissue (called **Peyer's patches**) found in the submucosa increase toward the end of the small intestine. This reflects the fact that the remaining (undigested) food residue in the intestine contains huge numbers of bacteria, which must be prevented from entering the bloodstream if at all possible.

Large Intestine

The **large intestine** is about 1.5 m (5 feet) long and extends from the ileocecal valve to the anus (see Figure 14.7). Its major functions are to dry out the indigestible food residue by absorbing water and to eliminate these residues from the body as feces. It frames the small intestine on three sides and has the following subdivisions: **cecum** (se′kum), **appendix, colon, rectum,** and **anal canal.** The saclike cecum is the first part of the large intestine. Hanging from the cecum is the wormlike appendix, a potential trouble spot. Since it is usually twisted, it is an ideal location for bacteria to accumulate and multiply. Inflammation of the appendix, *appendicitis,* is the usual result. The colon is divided into several distinct regions. The **ascending colon** travels up the right side of the abdominal cavity and makes a turn, the *right colic*

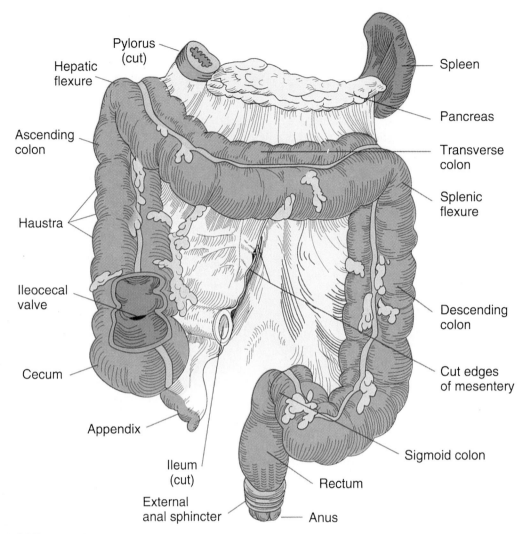

Figure 14.7   The large intestine. A section of the cecum is removed to show the ileocecal valve.

or *hepatic flexure,* to travel across the abdominal cavity as the **transverse colon.** It then turns again at the *left colic* or *splenic flexure* and continues down the left side as the **descending colon,** to enter the pelvis where it becomes the S-shaped **sigmoid** (sig′moid) **colon.** The sigmoid colon, rectum, and anal canal lie in the pelvis. The anal canal ends at the **anus** (a′nus), which opens to the exterior. The anal canal has an external *voluntary sphincter* composed of skeletal muscle and an internal *involuntary sphincter* formed by smooth muscle. These sphincters, which act rather like purse strings to open and close the anus, are ordinarily closed except during defecation, when feces are eliminated from the body.

Because most nutrient absorption has occurred before the large intestine is reached, no villi are seen in the large intestine, but there are tremen-

dous numbers of *goblet cells* in its mucosa that produce mucus. The mucus acts as a lubricant to ease the passage of feces to the end of the digestive tract.

In the large intestine, the longitudinal muscle layer of the muscularis externa is reduced to three bands of muscle. Since these muscle bands usually display some degree of tone (are partially contracted), they cause the wall to pucker into small pocketlike sacs called **haustra** (haws′trah).

Accessory Digestive Organs

Pancreas

The **pancreas** is a soft, pink, triangular gland that extends across the abdomen from the spleen to the duodenum (see Figures 14.1 and 14.4a). Most of the pancreas lies posterior to the parietal peritoneum; hence its location is referred to as *retroperitoneal.*

The pancreas produces a wide spectrum of enzymes (described later) that break down all categories of digestible foods. The pancreatic enzymes are secreted into the duodenum in an alkaline fluid, which neutralizes the acidic chyme coming in from the stomach. The pancreas also has an endocrine function; it produces the hormones insulin and glucagon, as explained in Chapter 9.

Liver and Gallbladder

The **liver** is the largest gland in the body. It is located under the diaphragm, more to the right side of the body (see Figure 14.1). As described earlier, the liver overlies and almost completely covers the stomach. The liver has four lobes and is suspended from the diaphragm and abdominal wall by a delicate mesentery cord, the **falciform** (fal′si-form) **ligament** (see Figure 14.5a).

There is no question that the liver is one of the body's most important organs. It has many metabolic and regulatory roles; however, its digestive function is to produce **bile.** Bile leaves the liver through the **common hepatic duct** and enters the duodenum through the bile duct (see Figures 14.1 and 14.4a).

Bile is a yellow-to-green, watery solution containing bile salts, bile pigments (chiefly bilirubin, a breakdown product of hemoglobin), cholesterol, phospholipids, and a variety of electrolytes. Of these components, only the bile salts (derived from cholesterol) and phospholipids aid the digestive process. Bile does not contain enzymes, but its bile salts *emulsify* fats by physically breaking large fat globules into smaller ones, thus providing more surface area for the fat-digesting enzymes to work on.

The **gallbladder** is a small, thin-walled green sac that snuggles in a shallow fossa in the inferior surface of the liver (see Figures 14.1 and 14.4a). When food digestion is not occurring, bile backs up the **cystic duct** and enters the gallbladder to be stored. While being stored in the gallbladder, bile is concentrated by the removal of water. Later, when fatty food enters the duodenum, a hormonal stimulus prompts the gallbladder to contract and spurt out stored bile, making it available to the duodenum.

If bile is stored in the gallbladder for too long or too much water is removed, the cholesterol it contains may crystallize, forming *gallstones.* Since gallstones tend to be quite sharp, agonizing pain may occur when the gallbladder contracts (the typical *gallbladder attack*).

Blockage of the common hepatic or bile ducts (for example, by wedged gallstones) prevents bile from entering the small intestine, and it begins to accumulate and eventually backs up into the liver. This exerts pressure on the liver cells, and bile salts and bile pigments begins to enter the bloodstream. As the bile pigments circulate through the body, the tissues become yellow, or *jaundiced.* Blockage of the ducts is just one cause of jaundice. More often it results from actual liver problems such as *hepatitis* (an inflammation of the liver) or *cirrhosis* (sir-ro′sis), a chronic inflammatory condition in which the liver is severely damaged and becomes hard and fibrous. Hepatitis is most often due to viral infection resulting from drinking contaminated water or transmitted in blood via transfusion or contaminated needles. Cirrhosis is almost guaranteed when one drinks alcoholic beverages in excess for many years, and is a common consequence of severe hepatitis. ❑

Salivary Glands

Three pairs of **salivary glands** empty their secretions into the mouth. The large **parotid** (pah-rot′id) **glands** lie anterior to the ears. *Mumps,* a common childhood disease, is an inflammation of the parotid glands. If you look at the location of the parotid glands in Figure 14.1, you can readily understand why people with mumps complain that it hurts to open their mouth or chew.

The **submandibular glands** and the small **sublingual** (sub-ling′gwal) **glands** empty their secretions into the floor of the mouth through tiny ducts. The product of the salivary glands, *saliva,* is a mixture of mucus and serous fluids. The mucus moistens and helps to bind food together into a mass called a *bolus* (bo′lus), which makes chewing and swallowing easier. The clear serous portion contains an enzyme, **salivary amylase** (am′ĭ-lās), which begins the process of starch digestion in the mouth. Saliva also contains substances such as lysozyme and antibodies (IgA) that inhibit bacteria; therefore, it has a protective function as well. Last but not least, saliva dissolves food chemicals so they can be tasted.

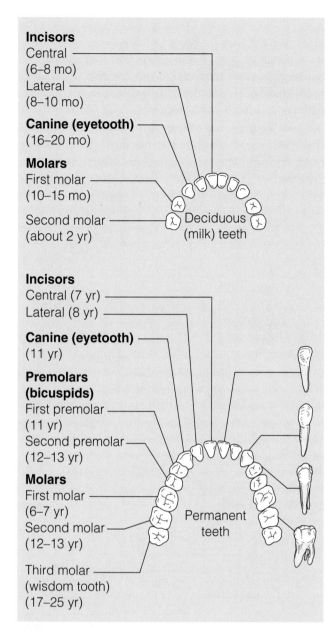

Incisors
Central
(6–8 mo)
Lateral
(8–10 mo)

Canine (eyetooth)
(16–20 mo)

Molars
First molar
(10–15 mo)

Second molar
(about 2 yr)

Deciduous
(milk) teeth

Incisors
Central (7 yr)
Lateral (8 yr)

Canine (eyetooth)
(11 yr)

**Premolars
(bicuspids)**
First premolar
(11 yr)
Second premolar
(12–13 yr)

Molars
First molar
(6–7 yr)
Second molar
(12–13 yr)

Third molar
(wisdom tooth)
(17–25 yr)

Permanent
teeth

Figure 14.8   Human deciduous and permanent teeth. Approximate time of tooth eruption is shown in parentheses. Since the same number and arrangement of teeth exist in both upper and lower jaws, only the lower jaw is shown in each case. The shapes of individual teeth are shown on the right.

Teeth

The role of the teeth in food processing needs little introduction. We **masticate,** or **chew,** by opening and closing our jaws and moving them from side to side while continually using our tongue to move the food between our teeth. In the process, the teeth tear and grind the food, breaking it down into smaller fragments.

Ordinarily, by the age of 21, two sets of teeth have been formed (Figure 14.8). The first set is the **deciduous** (de-sid'u-us) teeth, also called **baby teeth** or **milk teeth.** The deciduous teeth begin to erupt around six months, and a baby has a full set (20 teeth) by the age of two years. The first teeth to appear are the lower central incisors, an event that is usually anxiously awaited by the child's parents.

As the second set of teeth, the deeper **permanent teeth,** enlarge and develop, the roots of the milk teeth are reabsorbed, and between the ages of 6 and 12 years they loosen and fall out. All of the permanent teeth but the third molars have erupted by the end of adolescence. The third molars, also called *wisdom teeth,* emerge later, between the ages of 17 and 25. Although there are 32 permanent teeth in a full set, the wisdom teeth often fail to erupt; sometimes they are completely absent.

When teeth remain embedded in the jawbone, they are said to be *impacted.* Impacted teeth exert pressure and cause a good deal of pain and must be removed surgically. Wisdom teeth are the most commonly impacted. ❑

The teeth are classified according to shape and function as incisors, canines, premolars, and molars (see Figure 14.8). The chisel-shaped **incisors** are adapted for cutting, the fanglike **canines** (eyeteeth) for tearing or piercing. The **premolars** (bicuspids) and **molars** have broad crowns with rounded cusps (tips) and are best suited for grinding.

A tooth consists of two major regions, the **crown** and the **root,** as shown in Figure 14.9. The enamel-covered crown is the exposed part of the tooth above the **gingiva** (jin-ji'vah), or **gum. Enamel** is the hardest substance in the body and is fairly brittle because it is heavily mineralized with calcium salts. The portion of the tooth embedded in the jawbone is the root; the root and crown are connected by the tooth region called the **neck.** The outer surface of the root is covered by a substance called **cementum,** which attaches the tooth to the **periodontal** (per"e-o-don'tal) **membrane (ligament).** This ligament holds the tooth in place in the bony jaw. **Dentin,** a bonelike material, underlies the enamel and forms the bulk of the tooth. It surrounds a central **pulp cavity,** which contains a number of structures (connective tissue, blood vessels, and nerve fibers) collectively called **pulp.** Pulp supplies nutrients to the tooth tissues and provides for tooth sensations. Where the pulp cavity extends into the root, it becomes

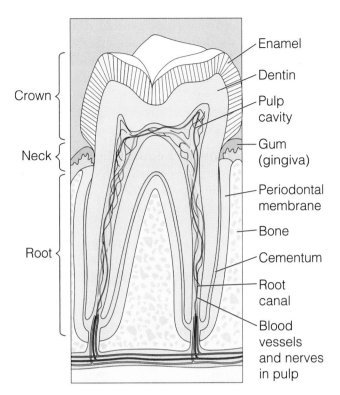

Figure 14.9 **Longitudinal section of a molar.**

the **root canal,** which provides a route for blood vessels, nerves, and other pulp structures to enter the pulp cavity of the tooth.

FUNCTIONS OF THE DIGESTIVE SYSTEM

Overview of Gastrointestinal Processes and Controls

The major functions of the digestive tract are usually summarized in two words—*digestion* and *absorption.* However, many of its specific activities, such as smooth muscle activity, and certain regulatory events, are not really covered by either term. To describe digestive system processes a little more accurately, we really have to consider a few more functional terms. The essential activities of the GI tract include the following six processes, summarized in Figure 14.10.

1. **Ingestion**—Food must be placed into the mouth before it can be acted on. This is an active, voluntary process called ingestion.

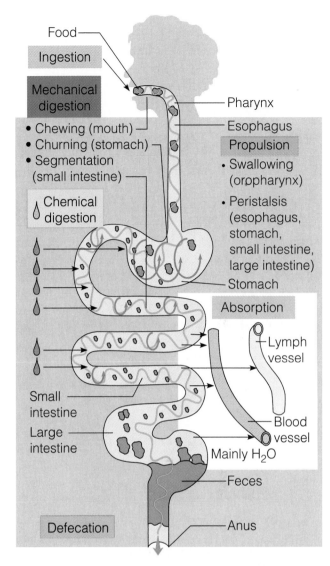

Figure 14.10 **Schematic summary of gastrointestinal tract activities.** Gastrointestinal tract activities include ingestion, mechanical digestion, chemical (enzymatic) digestion, propulsion, absorption, and defecation. Sites of chemical digestion are also sites that produce enzymes or that receive enzymes or other secretions made by accessory organs outside the alimentary canal. The mucosa of virtually the entire GI tract secretes mucus, which protects and lubricates.

2. **Propulsion**—If foods are to be processed by more than one digestive organ (and indeed they are), they must be propelled from one organ to the next. Swallowing is one example of food movement that depends largely on the propulsive process called **peristalsis.** Peristalsis is involuntary and involves alternating waves of contraction and relaxation of the muscles in the organ wall (Figure 14.11a). The

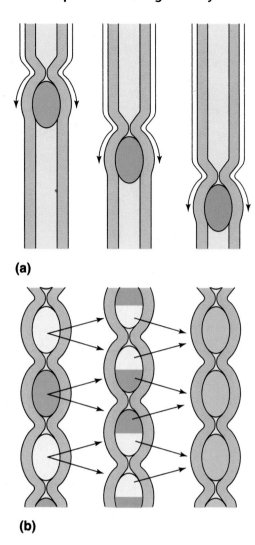

(a)

(b)

Figure 14.11 Peristaltic and segmental movements of the digestive tract. (**a**) In peristalsis, adjacent or neighboring segments of the intestine (or other alimentary canal organs) alternately contract and relax, which moves food distally along the tract. (**b**) In segmentation, single segments of the intestine alternately contract and relax. Because active segments are separated by inactive ones, the food is moved forward and then backward. Thus the food is mixed rather than simply propelled along the tract.

net effect is to squeeze the food along the tract. Although **segmentation** (Figure 14.11b) may help to propel foodstuffs through the small intestine, it normally only moves food back and forth across the internal wall of the organ, serving to mix it with the digestive juices. Thus, segmentation is more an example of mechanical digestion than of propulsion.

3. **Food breakdown: mechanical digestion**— Mixing of food in the mouth by the tongue, churning of food in the stomach, and segmentation in the small intestine are all examples of physical processes contributing to mechanical digestion. Mechanical digestion prepares food for further degradation by enzymes.

4. **Food breakdown: chemical digestion**—The sequence of steps in which large food molecules are broken down to their building blocks by enzymes (protein molecules that act as catalysts) is called chemical digestion. (Recall from Chapter 2 that these reactions are called *hydrolysis* reactions, because a water molecule is added to each bond to be broken [see p. 37].)

Since each of the major food groups has very different building blocks, it is worth taking a little time to review these chemical units, which were first introduced in Chapter 2. The building blocks, or units, of *carbohydrate* foods are *monosaccharides* (mon″o-sak′ah-rīdz), or simple sugars. We need to remember only three of these that are common in our diet—*glucose, fructose,* and *galactose.* Glucose is by far the most important, and when we talk about blood sugar levels, glucose is the "sugar" being referred to. Fructose is the most abundant sugar in fruits, and galactose is found in milk. Essentially, the only carbohydrates that our digestive system is able to digest, or to break down to simple sugars, are *sucrose* (table sugar), *lactose* (milk sugar), *maltose* (malt sugar), and *starch.* Sucrose, maltose, and lactose are referred to as *disaccharides,* or double sugars, because each consists of two simple sugars linked together. Starch is a *polysaccharide* (literally, "many sugars") formed of hundreds of glucose units. Although we eat foods containing other polysaccharides, such as *cellulose,* we do not have enzymes capable of breaking them down. The indigestible polysaccharides do not provide us with any nutrients, but they help move the foodstuffs along the gastrointestinal tract by providing bulk, or fiber, in our diet.

Proteins are digested to their building blocks, which are amino (ah-me′no) acids. Intermediate products of protein digestion are polypeptides and peptides. When lipids (fats) are digested, they yield two different types of building blocks—fatty acids and an alcohol called glycerol (glis*er-ol). The chemical breakdown of carbohydrates, proteins, and fats is summarized in Table 14.1 on page 422 and will be described in more detail shortly.

5. **Absorption**—The transport of digested end products from the lumen of the GI tract to the blood or lymph is absorption. For absorption to occur, the digested foods must first enter the mucosal cells by active or passive transport processes. The small intestine is the major absorptive site.

6. **Defecation**—Defecation is the elimination of indigestible substances from the body via the anus in the form of feces (fe'sēz).

Some of these processes are the job of a single organ. For example, only the mouth ingests, and only the large intestine defecates. But most digestive system activities occur bit by bit as food is moved along the tract. Thus, in one sense, the digestive tract can be viewed as a "disassembly line" in which food is carried from one stage of its processing to the next, and its nutrients are made available to the cells in the body en route.

A point that has been stressed throughout this book has been the drive of the body to maintain a constant internal environment, particularly in terms of homeostasis of the blood, which comes into intimate contact with all body cells. The digestive system, however, creates an optimal environment for its functioning in the lumen (cavity) of the alimentary canal, an area that is actually *outside* the body. Conditions in that lumen are controlled so that digestive processes occur efficiently. Digestive activity is mostly controlled by reflexes via the parasympathetic division of the autonomic nervous system. (Recall from Chapter 7 that this is the "resting-digesting" arm.) The sensors (mechanoreceptors, chemoreceptors) involved in these reflexes are located in the walls of the alimentary canal organs and respond to a number of stimuli, the most important being (1) stretch of the organ by food in its lumen, (2) pH of the contents, and (3) presence of certain breakdown products of digestion. When these receptors are activated, they start reflexes that activate or inhibit (1) the glands that secrete digestive juices into the lumen or hormones into the blood, and (2) the smooth muscles of the muscularis that mix and propel the foods along the tract.

Now that we have summarized some points that apply to the function of the digestive organs as a group, we are ready to look at their special capabilities.

Activities Occurring in the Mouth, Pharynx, and Esophagus

Food Ingestion and Breakdown

Once food has been placed in the mouth, both mechanical and chemical digestion begin. First the food is *physically* broken down into smaller particles by chewing. Then, as the food is mixed with saliva, salivary amylase begins the chemical digestion of starch, breaking it down into maltose (see Table 14.1). The next time you eat a piece of bread, chew it for a few minutes before swallowing it. You will notice that it will begin to taste sweet as the sugars are released.

Saliva is normally secreted continuously to keep the mouth moist, but when food enters the mouth, much larger amounts of saliva pour out. However, the simple pressure of anything put in the mouth and chewed, such as rubber bands or sugarless gum, will also stimulate the release of saliva. Some emotional stimuli can also cause salivation. For example, the mere thought of a hot fudge sundae will make many a mouth water. All these reflexes, though initiated by different stimuli, are brought about by parasympathetic fibers in the fifth and ninth cranial nerves.

Essentially no food absorption occurs in the mouth. (However, some drugs such as nitroglycerine are absorbed easily through the oral mucosa.) The pharynx and esophagus have no digestive function; they simply provide passageways to carry food to the next processing site, the stomach.

Food Propulsion—Swallowing and Peristalsis

In order for food to be sent on its way from the mouth, it must first be swallowed. **Deglutition** (deg"loo-tish'un), or **swallowing,** is a complicated process that involves the coordinated activity of several structures (tongue, soft palate, pharynx,

Table 14.1 Flowchart of Digestion and Absorption of Foodstuffs

(a) Carbohydrates: sequence and sites of chemical digestion

Foodstuff	Enzymes and source	Site of action

Lactose   Starch   Sucrose

| | Salivary amylase | Mouth |
| | Pancreatic amylase | Small intestine |

Dextrins and oligosaccharides

| | Intestinal (brush border) enzymes: dextrinase and glucoamylase | Small intestine |

Lactose   Maltose   Sucrose

| | Intestinal (brush border) enzymes: lactase, maltase, and sucrase | Small intestine |

Galactose   Glucose   Fructose

Absorption: Monosaccharides (glucose, galactose, and fructose) enter the capillaries of the villi and are transported to the liver via the hepatic portal vein.

(b) Proteins: sequence and sites of chemical digestion

Foodstuff	Enzymes and source	Site of action

Proteins

| | Pepsin (gastric glands) in the presence of HCl | Stomach |

Large polypeptides

| | Pancreatic enzymes: trypsin, chymotrypsin, carboxypeptidase | Small intestine |

Small polypeptides, small peptides

| | Intestinal (brush border) enzymes: aminopeptidase, carboxypeptidase, and dipeptidase | Small intestine |

Amino acids (some dipeptides and tripeptides)

Absorption: Amino acids enter the capillaries of the villi and are transported to the liver via the hepatic portal vein.

(c) Lipids: sequence and sites of chemical digestion

Foodstuff	Enzymes and source	Site of action

Unemulsified fats

| | Emulsified by detergent action of bile salts from liver | Small intestine |
| | Pancreatic lipase | Small intestine |

Monoglycerides   Glycerol and
and fatty acids   fatty acids

Absorption: Absorbed primarily into the lacteals of the villi (glycerol and short-chain fatty acids are absorbed into the capillary blood in the villi), and are transported to the liver via the systemic circulation (hepatic artery), which receives the lymphatic flow from the thoracic duct or via the hepatic portal vein.

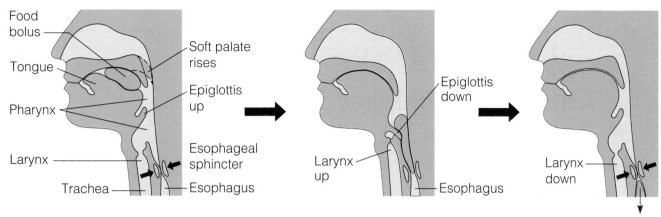

Figure 14.12 **Swallowing.** Left to right: The tongue pushes the food bolus posteriorly and against the soft palate. The soft palate rises to close off the nasal passages as the bolus enters the pharynx. The larynx rises so that the epiglottis covers its opening as peristalsis carries the food through the pharynx and the esophagus to the stomach. The esophageal sphincter is relaxed to allow food passage and then contracts again as the larynx and epiglottis return to their former positions.

and esophagus). It has two major phases. The first phase, the voluntary **buccal phase,** occurs in the mouth. Once the food has been chewed and well mixed with saliva, the bolus (food mass) is forced into the pharynx by the tongue. As food enters the pharynx, it passes out of our control and into the realm of reflex activity.

The second phase, the involuntary **pharyngeal-esophageal phase,** transports food through the pharynx and esophagus. The parasympathetic division of the autonomic nervous system (primarily the vagus nerves) controls this phase and promotes the mobility of the digestive organs from this point on. All routes that the food might take except the desired route distal into the digestive tract are blocked off. The tongue blocks off the mouth, and the soft palate closes off the nasal passages. The larynx rises so that its opening (into the respiratory passageways) is covered by the flaplike epiglottis. Food is moved through the pharynx and then into the esophagus below by wavelike peristaltic contractions of their muscular walls—first the longitudinal muscles contract, and then the circular muscles contract. The events of the swallowing process are illustrated in Figure 14.12.

If we try to talk while swallowing, our routing mechanisms may be "short-circuited," and food may manage to enter the respiratory passages. This triggers still another protective reflex—coughing—during which air rushes upward from the lungs in an attempt to expel the food.

Once food reaches the end of the esophagus, it presses against the cardioesophageal sphincter, causing it to open, and the food enters the stomach. The movement of food through the pharynx and esophagus is so automatic that a person can swallow, and food will reach the stomach even if he is standing on his head. Gravity plays no part in the transport of food once it has left the mouth, which explains why astronauts (in zero gravity of outer space) can still swallow and get nourishment.

Activities of the Stomach

Food Breakdown

Secretion of **gastric juice** is regulated by both neural and hormonal factors. The sight, smell, and taste of food stimulate parasympathetic nervous system reflexes, which increase the secretion of gastric juice by the stomach glands. In addition, the presence of food and a falling pH in the stomach stimulates the stomach cells to release the hormone **gastrin.** Gastrin prods the stomach glands to produce still more of the protein-digesting enzymes (pepsinogens), mucus, and hydrochloric acid. Under normal conditions, 2 to 3 liters of gastric juice are produced every day.

Hydrochloric acid makes the stomach contents very acid. This is (somewhat) dangerous since

A CLOSER LOOK

PEPTIC ULCERS: "SOMETHING IS EATING AT ME"

Archie, a 53-year-old factory worker, began to experience a gnawing pain in his upper abdomen an hour or two after each meal. At first, he blamed the quality of his home cooking, but he experienced the same symptoms after eating at the factory cafeteria or at restaurants. Archie always responded to stress by drinking and smoking heavily, and his abdominal pain became markedly worse during a hectic week when he worked 15 hours overtime on the assembly line. Finally, after 2 months of increasingly severe pain, Archie consulted a physician and was told that he had a peptic ulcer.

Peptic ulcers affect one of every eight Americans. A peptic ulcer is a craterlike erosion in the mucosa of any part of the alimentary canal that is exposed to the secretions of the stomach. Hydrochloric acid and pepsin cause this damage; people whose stomachs fail to secrete these substances never develop peptic ulcers. A few peptic ulcers occur in the lower esophagus, following the regurgitation of stomach contents, but most (98 percent) occur in the pyloric part of the stomach (gastric ulcers) or the first part of the duodenum (duodenal ulcers). Duodenal ulcers are about three times more common than gastric ulcers.

A peptic ulcer in the mucosa of the stomach is visible at the arrow.

In the past, more men than women developed peptic ulcers, but the incidence is now nearly equal for both sexes. Peptic ulcers may appear at any age, but they develop most frequently between the ages of 50 and 70 years. After developing, they tend to recur—healing, then flaring up periodically—for the rest of one's life if not treated.

Gastric and duodenal ulcers may produce a gnawing or burning pain in the epigastric region of the abdomen. This pain often appears 1 to 3 hours after a meal (or causes one to awaken at night) and is relieved by eating. Other potential symptoms include loss of appetite, burping, nausea, and vomiting. Not all people with ulcers experience the above symptoms, however, and many exhibit no symptoms at all.

Despite years of intensive study, the cause of peptic ulcers remains incompletely understood. For over a century, it has been "common knowledge" that stress causes ulcers, and the stereotypical ulcer patient has been the overworked business executive. Recent studies have not been able to demonstrate such a link between stress and ulcers, however, and many researchers now doubt the validity of this claim. Nonetheless, a stressful lifestyle does seem to *aggravate* existing ulcers. Recent studies indicate that ulcers are associated with a strain of acid-resistant bacteria *(Helicobacter pylori)* that inhabit the stomach of 40 percent of healthy people and 70 to 90 percent of those with ulcers. The tendency to develop ulcers is known to be inherited to a certain extent.

The anatomy of a peptic ulcer is shown in the figure in this box. It is a round, sharply defined crater in the mucosa. Typical ulcers are 1 to 4 cm in diameter. The base of the ulcer contains dead tissue cells, granulation tissue, and scar tissue. Eroded blood vessels may sometimes be seen there as well.

Peptic ulcers can produce serious complications. In about 20 percent of cases, eroded blood vessels bleed into the alimentary canal, causing vomiting of blood and blood in the feces. In such cases, anemia may result from a severe loss of blood. In 5 to 10 percent of ulcer patients, scarring within the stomach obstructs the pyloric opening, blocking digestion. About 5 percent of peptic ulcers *perforate,* allowing the contents of the stomach or duodenum to leak into the peritoneal cavity. This can cause either peritonitis or the digestion and destruction of the nearby pancreas. A perforated ulcer is a life-threatening condition.

In spite of these potential complications, most peptic ulcers heal readily and respond well to treatment. The first steps in treatment are to avoid smoking, alcohol, and aspirin, all of which aggravate ulcers. Antacid drugs are often suggested to neutralize the stomach acids. In ulcers found to be colonized by *Helicobacter,* the goal is to kill the embedded bacteria. Triple drug therapy (a combination of antibiotics, traditional antiulcer drugs, and bismuth [the active ingredient in Pepto-Bismol]) effectively promotes healing and prevents recurrence. In other cases, drugs that inhibit HCl secretion (H_2 blockers) such as cimetidine (Tagamet) and ranitidine (Zantac) are the therapy of choice. In rare, persistent cases, surgery to cut the vagus nerves to the stomach or to remove part of the stomach may be necessary. Both of these treatments cure the ulcer by decreasing the production of acid and pepsin.

both hydrochloric acid and the protein-digesting enzymes have the ability to digest the stomach itself, causing *ulcers* (see the "Closer Look" box beginning at left). However, as long as enough mucus is made, the stomach is "safe" and will remain unharmed.

Occasionally, the cardioesophageal sphincter fails to close tightly and gastric juice backs up into the esophagus, which has little mucus protection. This results in a characteristic pain known as *heartburn,* which uncorrected leads to inflammation of the esophagus (*esophagitis* [ĕ-sof′ah-ji′tis]) and perhaps even to ulceration of the esophagus. A common cause is a *hiatal hernia,* a structural abnormality in which the superior part of the stomach protrudes slightly above the diaphragm. Since the diaphragm no longer reinforces the cardioesophageal sphincter, which is a weak sphincter to begin with, gastric juice flows into the unprotected esophagus. Conservative treatment involves restricting food intake after the evening meal, taking antacids, and sleeping with the head elevated. ❑

The extremely acid environment that hydrochloric acid provides *is* necessary, because it activates *pepsinogen* to *pepsin,* the active protein-digesting enzyme. *Rennin,* the second protein-digesting enzyme produced by the stomach, works primarily on milk protein and converts it to a substance that looks like sour milk. Many mothers mistakenly think that when their infants spit up a curdy substance after having their bottle that the milk has soured in their stomach. Rennin is produced in large amounts in infants, but it is not believed to be produced in adults.

Other than beginning protein digestion, little chemical digestion occurs in the stomach. With the exception of aspirin and alcohol (which seem somehow to have a "special pass"), virtually no absorption occurs through the stomach walls.

As food enters and fills the stomach, its wall begins to stretch (at the same time the gastric juices are being secreted, as just described). Then the three muscle layers of the stomach wall become active. They compress and pummel the food, breaking it apart physically, all the while continually mixing the food with the enzyme-containing gastric juice so that the semifluid chyme is formed. Perhaps it would help if you visualize this process as being like the preparation of a cake mix in which the floury mixture is continually folded on itself and mixed with the liquid until it reaches uniform texture.

Food Propulsion

Once the food has been well mixed, a rippling peristalsis begins in the lower half of the stomach, and the contractions increase in force as the pyloric valve is approached. The pylorus of the stomach, which holds about 30 ml of chyme, acts like a meter that allows only liquids and very small particles to pass through the pyloric sphincter (see Figure 14.4a). Because the pyloric sphincter is barely open, each contraction of the stomach muscle "spits" or squirts 3 ml or less of chyme into the small intestine. Since the contraction also *closes* the valve, the rest (about 27 ml) is propelled backward into the stomach for more mixing. When the duodenum is filled with chyme and its wall is stretched, a nervous reflex, the *enterogastric* (en″ter-o-gas′trik) *reflex,* occurs. This reflex "puts the brakes on" gastric activity and slows the emptying of the stomach by inhibiting the vagus nerves and tightening the pyloric sphincter, thus allowing time for intestinal processing to catch up. Generally, it takes about 4 hours for the stomach to empty completely after eating a well-balanced meal, and 6 hours or more if the meal has a high fat content.

Local irritation of the stomach, such as occurs with bacterial food poisoning, may activate the *emetic* (ĕ-met′ik) *center* in the brain (medulla). The emetic center, in turn, causes **vomiting.** Vomiting is essentially a reverse peristalsis occurring in the stomach (and perhaps the small intestine), accompanied by contraction of the abdominal muscles and the diaphragm, which increases the pressure on the abdominal organs. The emetic center may also be activated through other pathways; disturbance of the equilibrium apparatus of the inner ear during a boat ride on rough water is one example. ❑

Activities of the Small Intestine

Food Breakdown and Absorption

Food reaching the small intestine is only partially digested. Carbohydrate and protein digestion have been started, but virtually no fats have been digested up to this point. Here the process of chemical food digestion is accelerated as the food now takes a rather wild 3- to 6-hour journey through the looping coils and twists of the small intestine. By the time the food reaches the end of the small intestine, digestion is complete and nearly all food absorption has occurred.

The microvilli of small intestine cells bear a few important enzymes, the so-called **brush border enzymes** that break down double sugars into simple sugars and complete protein digestion (see Table 14.1). Intestinal juice itself is relatively enzyme-poor, and protective mucus is probably the most important intestinal gland secretion. However, foods entering the small intestine are literally deluged with enzyme-rich **pancreatic juice** ducted in from the pancreas.

Pancreatic juice contains enzymes that: (1) along with brush border enzymes, complete the digestion of starch *(pancreatic amylase);* (2) carry out about half of protein digestion (via the action of *trypsin, chymotrypsin, carboxypeptidase,* and others); (3) are totally responsible for fat digestion since the pancreas is the only source of *lipases,* and (4) digest nucleic acids *(nucleases).*

In addition to enzymes, pancreatic juice contains a rich supply of bicarbonate, which makes it very basic (about pH 8). When pancreatic juice reaches the small intestine, it neutralizes the acid chyme coming in from the stomach and provides the proper environment for activation and activity of intestinal and pancreatic digestive enzymes.

Pancreatitis (pan″kre-ah-ti′tis) is a rare but extremely serious inflammation of the pancreas that results from activation of pancreatic enzymes in the pancreatic duct. Since pancreatic enzymes break down all categories of biological molecules, the pancreatic tissue and duct are digested. This painful condition can lead to nutritional deficiencies, because pancreatic enzymes are essential to digestion in the small intestine. ❑

The release of pancreatic juice into the duodenum is stimulated by both the vagus nerves and local hormones. When chyme enters the small intestine, it stimulates the mucosa cells to produce several hormones (see Table 14.2). Two of these hormones, **secretin** (se-kre′tin) and **cholecystokinin** (ko″le-sis″to-kin′in) **(CCK),** influence the release of pancreatic juice and bile.

The hormones enter the blood and circulate to their target organs, the pancreas, liver, and gallbladder. Both hormones work together to stimulate the pancreas to release its enzyme- and bicarbonate-rich product. In addition, secretin causes the liver to increase its output of bile. Cholecystokinin causes the gallbladder to contract and release stored bile into the bile duct so that bile and pancreatic juice enter the small intestine together. As mentioned before, bile is not an enzyme. Instead, it acts like a detergent to emulsify, or mechanically break down, large fat globules into thousands of tiny ones, providing a much greater surface area for the pancreatic lipases to work on. Bile is also necessary for absorption of fats and fat-soluble vitamins (K, D, and A) from the intestinal tract.

If *either* bile or pancreatic juice are absent, essentially no fat digestion or absorption goes on, and fatty, bulky stools are the result. In such cases, blood-clotting problems also occur because the liver needs vitamin K to make prothrombin, one of the clotting factors. ❑

Absorption of water and of the end products of digestion occurs all along the length of the small intestine. Most substances are absorbed through

Table 14.2		Hormones and Hormonelike Products That Act in Digestion	
Hormone	**Source**	**Stimulus for secretion**	**Action**
Gastrin	Stomach	Food in stomach (chemical stimulus)	Stimulates release of gastric juice; stimulates mobility of small intestine; relaxes ileocecal valve.
Histamine	Stomach	Food in stomach	Activates parietal cells to secrete hydrochloric acid.
Somatostatin	Stomach	Food in stomach	Inhibits secretion of gastric juice and pancreatic juice; inhibits emptying of stomach and gallbladder.
Secretin	Duodenum	Acidic chyme and partially digested foods in duodenum	Increases output of pancreatic juice rich in bicarbonate ions; increases bile output by liver; inhibits gastric mobility and gastric gland secretion.
Cholecystokinin (CCK)	Duodenum	Fatty chyme in duodenum	Increases output of enzyme-rich pancreatic juice; stimulates gallbladder to expel stored bile; relaxes sphincter of duodenal papilla to allow bile and pancreatic juice to enter the duodenum.
Gastric inhibitory peptide (GIP)	Duodenum	Fatty chyme in duodenum	Inhibits gastric mobility and secretion of gastric juice.

the intestinal cell plasma membranes by the process of *active transport*. Then they enter the capillary bed in the villus to be transported in the blood to the liver via the hepatic portal vein. The exception seems to be lipids, or fats, which are absorbed passively by the process of *diffusion*. Lipid breakdown products enter both the capillary bed and the lacteal in the villus and are carried to the liver by both blood and lymphatic fluids.

At the end of the ileum, all that remains is some water, indigestible food materials (plant fibers such as cellulose), and large amounts of bacteria. This debris enters the large intestine through the ileocecal valve. The complete process of food digestion and absorption is summarized in Table 14.1 (p. 422).

Food Propulsion

As mentioned previously, *peristalsis* is the major means of propelling food through the digestive tract. It involves waves of contraction that move along the length of the intestine, followed by waves of relaxation. The net effect is that the food is moved through the small intestine much in the same way that toothpaste is squeezed from a tube. Rhythmic segmental movements produce local constrictions of the intestine (see Figure 14.11b). Although it was once presumed that these movements were only important to mix the chyme with the digestive juices, we now know that they also play a role in propelling food through the intestine.

Activities of the Large Intestine

Food Breakdown and Absorption

What is finally delivered to the large intestine contains few nutrients, but that residue still has 12 to 24 hours more to spend there. The colon itself produces no digestive enzymes. However, the "resident" bacteria that live within its lumen metabolize some of the remaining nutrients, releasing gases (methane and hydrogen sulfide) that contribute to the odor of feces. About 500 ml of gas (flatus) is produced each day, much more when certain carbohydrate-rich foods (such as beans) are eaten.

Bacteria residing in the large intestine also make some vitamins (vitamin K and some B vitamins). Absorption by the large intestine is limited to the absorption of these vitamins, some ions, and most of the remaining water. **Feces,** the more or less solid product delivered to the rectum, contain undigested food residues, mucus, millions of bacteria, and just enough water to allow their smooth passage.

Propulsion of the Residue and Defecation

Peristalsis and mass movements are the two major types of propulsive movements occurring in the large intestine. Colon peristalsis is sluggish and, compared to the mass movements, probably contributes very little to propulsion. **Mass movements** are long, slow-moving but powerful contractile waves that move over large areas of the colon three or four times daily and force the contents toward the rectum. Typically, they occur during or just after eating, when food begins to fill the stomach and small intestine. Bulk, or fiber, in the diet increases the strength of colon contractions and softens the stool, allowing the colon to act as a well-oiled machine.

When the diet lacks bulk, the colon narrows and its circular muscles contract more powerfully, which increases the pressure on its walls. This encourages formation of *diverticula* (di"ver-tik'u-lah), in which the mucosa protrudes through the colon walls, a condition called *diverticulosis. Diverticulitis,* a condition in which the diverticula become inflamed, can be life-threatening if ruptures occur. ❏

The rectum is generally empty, but when feces are forced into it by mass movements and its wall is stretched, the **defecation reflex** is initiated. The defecation reflex is a spinal (sacral region) reflex that causes the walls of the sigmoid colon and the rectum to contract and the anal sphincters to relax. As the feces are forced through the anal canal, messages reach the brain giving us time to make a decision as to whether the external voluntary sphincter should remain open or be constricted to stop passage of feces. If it is not convenient, defecation (or moving the bowels) can be delayed temporarily. Within a few seconds, the reflex contrac-

tions end and the rectal walls relax. With the next mass movement, the defecation reflex is initiated again.

Watery stools, or *diarrhea* (di"ah-re'ah), result from any condition that rushes food residue through the large intestine before that organ has had sufficient time to absorb the water (as in irritation of the colon by bacteria). Because fluids and ions are lost from the body, prolonged diarrhea may result in dehydration and electrolyte imbalance. On the other hand, when food residue remains in the large intestine for extended periods, too much water is absorbed, and the stool becomes hard and difficult to pass. This condition, called *constipation*, may result from lack of fiber in the diet, poor bowel habits ("failing to heed the call"), and laxative abuse. ❏

NUTRITION AND METABOLISM

Although it seems at times that people can be divided into two camps—those who live to eat and those who eat to live—we all recognize the vital importance of food for life. It has been said that "you are what you eat," and this is true in that part of the food we eat is converted to our living flesh. In other words, a certain fraction of nutrients is used to build cellular molecules and structures and to replace worn-out parts. However, most foods are used as metabolic fuels. That is, they are oxidized and transformed into **ATP,** the chemical energy form needed by body cells to drive their many activities. The energy value of foods is measured in units called **kilocalories (kcal)** or "large calories (C)," the units conscientiously counted by dieters.

We have just considered how foods are digested and absorbed. But what happens to these foods once they enter the blood? Why do we need bread, meat, and fresh vegetables? Why does everything we eat seem to turn to fat? We will try to answer these questions in this section.

Nutrition

A **nutrient** is a substance in food that is used by the body to promote normal growth, maintenance, and repair. The nutrients divide neatly into six categories. The *major nutrients*—carbohydrates, lipids, and proteins—make up the bulk of what we

Table 14.3 Five Basic Food Groups and Some of Their Major Nutrients			
		Major nutrients supplied in significant amounts:	
Group	**Example foods**	**By all in group**	**By only some in group**
Fruits	Apples, bananas, dates, oranges, tomatoes	Carbohydrate Water	Vitamins: A, C, folic acid Minerals: iron, potassium Fiber
Vegetables	Broccoli, cabbage, green beans, lettuce, potatoes	Carbohydrate Water	Vitamins: A, C, E, K, and B vitamins except B_{12} Minerals: calcium, magnesium, iodine, manganese, phosphorus Fiber
Grain products (preferably whole grain; otherwise, enriched or fortified)	Breads, rolls, bagels Cereals, dry and cooked Pasta Rice, other grains Tortillas, pancakes, waffles Crackers Popcorn	Carbohydrate Protein Vitamins: thiamin (B_1), niacin	Water Fiber Minerals: iron, magnesium, selenium
Milk products	Milk, yogurt Cheese Ice cream, ice milk, frozen yogurt	Protein Fat Vitamins: riboflavin, B_{12} Minerals: calcium, phosphorus Water	Carbohydrate Vitamins: A, D
Meats and meat alternates	Meat, fish, poultry Eggs Seeds Nuts, nut butters Soybeans, tofu Other legumes (peas and beans)	Protein Vitamins: niacin, B_6 Minerals: iron, zinc	Carbohydrate Fat Vitamins: B_{12}, thiamin (B_1) Water Fiber

Source: Christian, Janet, and Janet Greger. *Nutrition for Living,* 3rd ed. Redwood City, CA: Benjamin/Cummings, 1991.

eat. Vitamins and minerals, while equally crucial for health, are required in minute amounts. Water, which accounts for about 60 percent of the volume of the food we eat, is also considered to be a major nutrient. However, because its importance as a solvent and in many other aspects of body functioning is described in Chapter 2, only the other five classes of nutrients will be considered here.

Most foods offer a combination of nutrients. For example, a bowl of cream of mushroom soup contains all of the major nutrients plus some vitamins and minerals. A diet consisting of foods selected from each of the five food groups (see Table 14.3), that is, grains, fruits, vegetables, meats and fish, and milk products, normally guarantees adequate amounts of all of the needed nutrients.

Dietary Sources of the Major Nutrients

Carbohydrates Except for milk sugar (lactose) and small amounts of glycogen in meats, all the **carbohydrates**—sugars and starches—we ingest are derived from plants. Sugars come mainly from fruits, sugar cane, and milk. The polysaccharide starch is found in grains, legumes, and root vegetables. The polysaccharide cellulose, which is plentiful in most vegetables, is not digested by humans, but it provides roughage, or fiber, which increases the bulk of the stool and aids defecation.

Lipids Although we also ingest cholesterol and phospholipids, most dietary **lipids** are neutral fats—triglycerides. We eat saturated fats in animal products such as meat and dairy foods and in a few plant products such as coconut. Unsaturated fats are present in seeds, nuts, and most vegetable oils. Major sources of cholesterol are egg yolk, meats, and milk products.

Proteins Animal products contain the highest-quality **proteins,** molecules that are basically amino acid polymers. Eggs, milk, and most meat proteins are *complete proteins* that meet all of the body's amino acid requirements for tissue maintenance and growth. Legumes (beans and peas), nuts, and cereals are also protein-rich, but their proteins are nutritionally incomplete because they are low in one or more of the essential amino acids. As you can see, strict vegetarians must carefully plan their diets to obtain all the essential amino acids and prevent protein malnutrition. Cereal grains and legumes when ingested together provide all the essential amino acids, and some variety of this combination is found in the diets of all cultures (most obviously in the rice and beans seen on nearly every plate in a Mexican restaurant).

Vitamins **Vitamins** are organic nutrients of various forms that the body requires in small amounts. Although vitamins are found in all major food groups, no one food contains all the required vitamins. Thus, a balanced diet is the best way to ensure a full vitamin complement, particularly since certain vitamins (A, C, and E) appear to have anti-cancer effects. Diets rich in broccoli, cabbage, and brussels sprouts (all good sources of vitamins A and C) appear to reduce cancer risk. However, controversy abounds concerning the ability of vitamins to work wonders.

Most vitamins function as **coenzymes** (or parts of coenzymes); that is, they act with an enzyme to accomplish a particular type of catalysis.

Minerals The body requires adequate supplies of seven **minerals** (calcium, phosphorus, potassium, sulfur, sodium, chloride, and magnesium) and trace amounts of about a dozen others.

Fats and sugars have practically no minerals, and cereals and grains are poor sources. The most mineral-rich foods are vegetables, legumes, milk, and some meats.

The main uses of the major nutrients in the body are discussed in the section on metabolism. Appendix C details some important roles of vitamins and minerals in the body.

Metabolism

Metabolism (mĕ-tab′o-lizm; *metabol* = change) is a broad term referring to all chemical reactions that are necessary to maintain life. It involves **catabolism** (kah-tab′o-lizm), in which substances are broken down to simpler substances, and **anabolism** (ah-nab′o-lizm), in which larger molecules or structures are built from smaller ones. During catabolism, energy is released and captured to make ATP, the energy-rich molecule used to energize all cellular activities, including catabolic reactions (Figure 14.13).

Not all foodstuffs are treated in the same way by body cells. For example, carbohydrates, particularly glucose, are usually broken down to make ATP. Fats are used to build cell membranes, make myelin sheaths, and insulate the body with a fatty cushion. They are also used as the body's main energy fuel for making ATP when there are inadequate carbohydrates in the diet. Proteins tend to be carefully conserved (even hoarded) by the body cells. This is easy to understand when you recognize that proteins are the major structural materials used for building cell structures.

Carbohydrate, Fat, and Protein Metabolism in Body Cells

Carbohydrate Metabolism

Just as an oil furnace uses oil (its fuel) to produce heat, the cells of the body use **carbohydrates** as

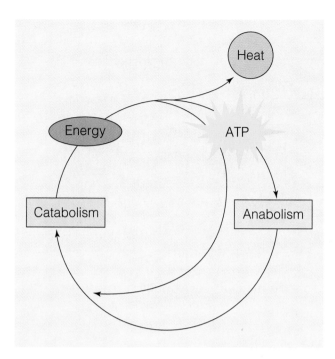

Figure 14.13   Central position of ATP in metabolism (anabolic and catabolic reactions).

trons) are eventually combined with oxygen to form water. These oxygen-using events are referred to collectively as **cellular respiration.** The events of the three main metabolic pathways that are involved in cellular respiration—**glycolysis, the Krebs cycle,** and the **electron transport chain**—are shown schematically in Figure 14.15, but we need not concern ourselves with further details of these pathways here.

Homeostasis of blood glucose levels is critically important. If there are excessively high levels of glucose in the blood (*hyperglycemia* [hi″per-gli-se′me-ah]), some of the excess is stored in body cells (particularly liver and muscle cells) as glycogen. If blood glucose levels are still too high, excesses are converted to fat. There is no question that eating large amounts of empty-calorie foods such as candy and other sugary sweets causes a rapid deposit of fat in the body's adipose tissues. When blood glucose levels are too low *(hypoglycemia),* the liver breaks down stored glycogen and releases glucose to the blood for cellular use. These various fates of carbohydrates are shown in Figure 14.16a.

Fat Metabolism

As described shortly, the liver handles most lipid, or fat, metabolism that goes on in the body. The liver cells use some **fats** to make ATP for their own use, some to synthesize lipoproteins, thromboplastin (a clotting protein), and cholesterol, and then release the rest to the blood in the form of relatively small, fat-breakdown products. Body cells remove the fat products and cholesterol from the blood and build them into their membranes or steroid hormones as needed. Fats are also used to form myelin sheaths of neurons (see Chapter 7) and fatty cushions around body organs. In addition, stored fats are the body's most concentrated source of energy. (Catabolism of 1 gram of fat

their preferred fuel to produce cellular energy (ATP). **Glucose,** also known as **blood sugar,** is the major breakdown product of carbohydrate digestion. Glucose is also the major fuel used for making ATP in most body cells. The liver is an exception; it routinely uses fats as well, thus saving glucose for other body cells. Essentially, glucose is broken apart piece by piece and some of the chemical energy released when its bonds are broken is captured and used to bind phosphate to ADP molecules to make ATP.

The overall reaction is summed up simply in Figure 14.14. Basically, the carbon atoms released leave the cells as carbon dioxide and the hydrogen atoms removed (which contain energy-rich elec-

$$C_6H_{12}O_6 \quad + \quad 6 \ O_2 \quad \longrightarrow \quad 6 \ CO_2 \quad + \quad 6 \ H_2O \quad + \quad ATPs$$

Glucose   Oxygen   Carbon   Water   Energy
                gas   dioxide

Figure 14.14   Summary equation for cellular respiration.

Figure 14.15   An overview of sites of ATP formation during cellular respiration. Glycolysis occurs outside the mitochondria in the cytosol. The Krebs cycle and the electron transport chain reactions occur within the mitochondria. (1) During glycolysis, hydrogen atoms containing high-energy electrons are removed as each glucose molecule is broken down to two molecules of pyruvic acid. (2) The pyruvic acid enters the mitochondrial matrix, where Krebs cycle enzymes remove more hydrogen and decompose the pyruvic acid to carbon dioxide. During glycolysis and the Krebs cycle, small amounts of ATP are formed. (3) Chemical energy from glycolysis and the Krebs cycle, in the form of hydrogen atoms containing energy-rich electrons, is then transferred to the electron transport chain, which is built into the membrane of the cristae. The electron transport chain carries out oxidative phosphorylation, which accounts for most of the ATP generated by cellular respiration and finally unites the removed hydrogen with oxygen to form water.

yields twice as much energy as the breakdown of 1 gram of carbohydrate or protein.)

For fat products to be used for ATP synthesis, they must first be broken down to acetic acid. Within the mitochondria, the acetic acid (like the pyruvic acid product of carbohydrates) is then completely oxidized and carbon dioxide, water, and ATP are formed. When there is not enough glucose to fuel the needs of the cells for energy, larger amounts of fats are used to produce ATP. Under such conditions, fat oxidation is fast but incomplete and some of the intermediate products such as acetoacetic acid and acetone begin to accumulate in the blood. These cause the blood to become acidic (a condition called *acidosis,* or *ketosis*), and the breath takes on a fruity odor as ace-

tone diffuses from the lungs. Ketosis is a common consequence of "no-carbohydrate" diets, uncontrolled diabetes mellitus, and starvation in which the body is forced to rely almost totally on fats to fuel its energy needs. While fats are an important energy source, cholesterol is *never* used as a cellular fuel. Its importance lies in the functional molecules and in the structures it helps to form.

Excess fats are stored in fat depots such as the hips, abdomen, breasts, and subcutaneous tissues. Although the fat in subcutaneous tissue is important as insulation for the deeper body organs, excessive amounts restrict movement and place greater demands on the circulatory system. The metabolism and uses of fats are shown in Figure 14.16b.

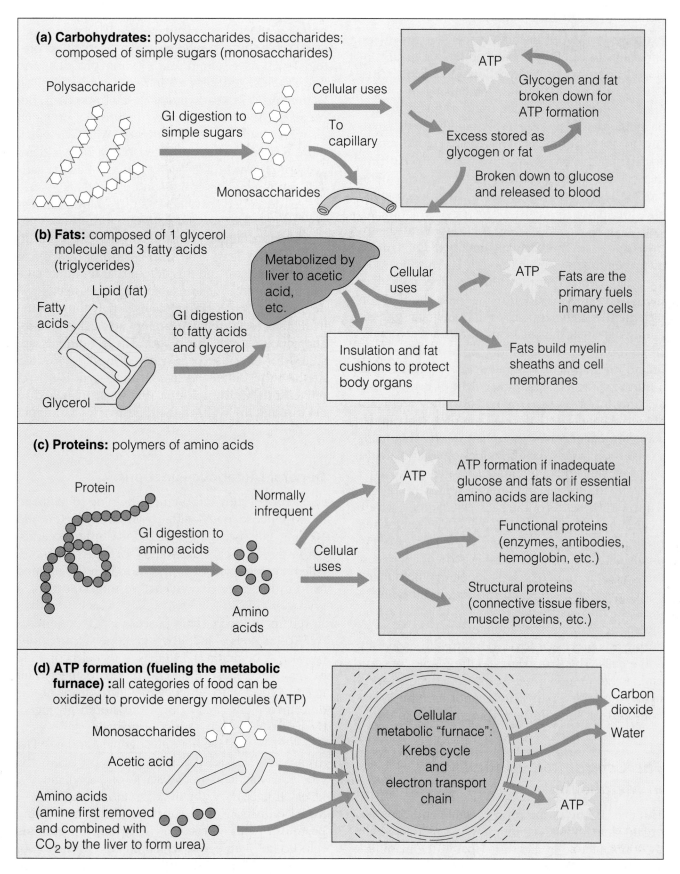

Figure 14.16   Metabolism by body cells. (**a**) Carbohydrate metabolism.
(**b**) Fat metabolism. (**c**) Protein metabolism. (**d**) ATP formation.

Protein Metabolism

Proteins make up the bulk of cellular structures, and they are carefully conserved by body cells. Ingested proteins are broken down to amino acids. Once the liver has finished processing the blood draining the digestive tract, and has taken its "fill" of amino acids, the remaining amino acids circulate to the body cells. The cells remove amino acids from the blood and use them to build proteins, both for their own use (enzymes, membranes, mitotic spindle proteins, muscle proteins) and for export (mucus, hormones, and others). Cells take few chances with their amino acid supply. They use ATP to actively transport amino acids into their interior even though (in many cases) they may contain more of those amino acids than there is in the blood flowing past them. Even though this may appear to be "cellular greed," there is an important reason for this active uptake of amino acids. Cells cannot build their proteins unless *all* the needed amino acids, which number around 20, are present. Since eight of these amino acids cannot be made by the cells, they are available to the cells only through the diet; such amino acids are called *essential amino acids.* This helps explain the avid accumulation of amino acids, which ensures that all amino acids needed will be available for present and (at least some) future protein-building needs of the cells (see Figure 14.16c).

Amino acids are used to make ATP only when proteins are overabundant and/or when carbohydrates and fats are not available. When it is necessary to oxidize amino acids for energy, their amine groups are removed as **ammonia,** and the rest of the molecule enters the Krebs cycle pathways in the mitochondria. The ammonia that is released during this process is toxic to body cells, especially nerve cells. The liver comes to the rescue by combining the ammonia with carbon dioxide to form **urea** (u-re′ah). Urea, which is not harmful to the body cells, is then flushed from the body in urine.

The Central Role of the Liver in Metabolism

The liver is one of the most versatile and complex organs in the body. Without it we would die within 24 hours. Its role in digestion (that is, the manufacture of bile) is important to the digestive process to be sure, but it is only one of the many functions of liver cells. The liver cells detoxify drugs and alcohol, degrade hormones, make many substances vital to the body as a whole (cholesterol, blood proteins such as albumin and clotting proteins, and lipoproteins), and play a central role in metabolism as they process nearly every class of nutrient. Because of the liver's key roles, nature has provided us with a surplus of liver tissue. We have much more than we need, and even if part of it is damaged or removed, it is one of the few body organs that can regenerate rapidly and easily.

As described in Chapter 11, a unique circulation, the *hepatic portal circulation,* brings nutrient-rich blood draining from the digestive viscera directly to the liver. The liver is the body's major metabolic organ, and this detour that nutrients take through the liver ensures that the liver's needs will be met first. As blood circulates slowly through the liver, liver cells remove amino acids, fatty acids, and glucose from the blood. These nutrients are stored for later use or processed in various ways. At the same time, the liver's phagocytic cells remove and destroy bacteria that have managed to get through the walls of the digestive tract and into the blood.

General Metabolic Functions

The liver is vitally important in helping to maintain blood glucose levels within normal range (around 100 mg glucose/100 ml of blood). After a carbohydrate-rich meal, thousands of glucose molecules are removed from the blood and combined to form the large polysaccharide molecules called *glycogen* (gli′ko-jen) *molecules,* which are then stored in the liver. This process is **glycogenesis** (gli″ko-jen′ĕ-sis), literally, "glycogen formation" (*genesis* = beginning). Later, as body cells continue to remove glucose from the blood to meet their needs, blood glucose levels begin to drop. At this time, liver cells break down the stored glycogen, by a process called **glycogenolysis** (gli″ko-jen-ol′ĭ-sis), which means "glycogen splitting." The liver cells then release glucose bit by bit to the blood to maintain homeostasis of blood glucose levels. If necessary, the liver can also make glucose from noncarbohydrate substances such as fats and proteins. This process is **gluconeogenesis** (glu″ko-ne″o-jen′ĕ-sis), which means "formation of new sugar" (see Figure 14.17). As described in Chapter 9, hormones such as thyroxine, insulin,

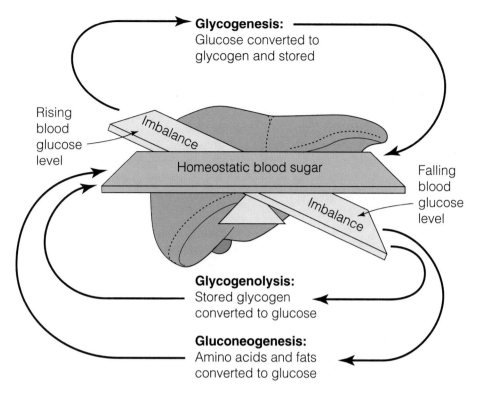

Figure 14.17   Metabolic events occurring in the liver as blood glucose levels rise and fall. When the blood glucose level is rising, the liver removes glucose from the blood and stores it as glycogen (glycogenesis). When the blood glucose level falls, the liver breaks down stored glycogen (glycogenolysis) and makes new glucose from amino acids and fats (gluconeogenesis). The glucose is then released to the blood to restore homeostasis of blood sugar.

and glucagon are vitally important in controlling the blood sugar levels and the handling of glucose in all body cells.

Some of the fats and fatty acids picked up by the liver cells are oxidized for energy (to make ATP) for use by the liver cells themselves. The rest are broken down to simpler substances such as *acetic acid* and *acetoacetic acid* (two acetic acids linked together) and released into the blood, or stored as fat reserves in the liver. The liver also makes cholesterol and secretes cholesterol's breakdown products in bile.

All blood proteins made by the liver are built from the amino acids its cells pick up from the blood. The completed proteins are then released back into the blood to travel throughout the circulation. *Albumin,* the most abundant protein in blood, holds fluids in the bloodstream. When insufficient albumin is present in blood, fluid leaves the bloodstream and accumulates in the tissue spaces, causing edema. The role of the protective

clotting proteins made by the liver was discussed in Chapter 10. Liver cells also synthesize nonessential amino acids and, as mentioned earlier, detoxify ammonia (produced when amino acids are oxidized for energy) by converting it to urea.

Nutrients not needed by the liver cells, as well as the products of liver metabolism, are released into the blood and drain from the liver in the hepatic vein to enter the systemic circulation, where they become available to other body cells.

Cholesterol Metabolism and Transport

Although it's a very important lipid in the diet, cholesterol is not used as an energy fuel. Instead, it serves as the structural basis of steroid hormones and vitamin D, and is a major building block of plasma membranes. Because we hear so much about "cutting down our cholesterol intake" in the media, it is always surprising to learn that only about 15 percent of blood cholesterol comes from the diet. The other 85 percent or so is made by the liver. Cholesterol is lost from the body when it is broken down and secreted in bile salts, which eventually leave the body in feces.

Because of the important role they play in fat and cholesterol transport, the lipoproteins, one class of proteins made by the liver and known by the buzzwords *HDLs* and *LDLs,* deserve a bit more attention.

Since fatty acids, fats, and cholesterol are insoluble in water, they cannot circulate freely in the bloodstream. Instead they are transported bound to the small lipid-protein complexes called lipoproteins. Although the entire story is pretty complex, the important thing to know is that the **low-density lipoproteins,** or **LDLs,** transport cholesterol and other lipids *to* body cells where they are used in various ways. If large amounts of LDLs are circulating, the chance that fatty substances will be deposited on the arterial walls, initiating atherosclerosis, is high. Because of this possibility, the LDLs are unkindly tagged as "bad lipoproteins." By contrast, the lipoproteins that transport cholesterol *from* the tissue cells (or arteries) to the liver for its disposal in bile are **high-density lipoproteins,** or **HDLs.** High HDL levels are considered "good" because the cholesterol is destined to be broken down and eliminated from the body. Obviously both LDLs and HDLs are "good and necessary"; it is their relative ratio in the blood that determines whether or not potentially lethal cholesterol deposits are likely to be laid down in the artery walls. In general, aerobic exercise, a diet low in saturated fats and cholesterol, and abstaining from smoking and drinking coffee all appear to favor a desirable HDL/LDL ratio.

Body Energy Balance

When any fuel is burned, it consumes oxygen and liberates heat. The "burning" of food fuels by body cells is no exception. As stated in Chapter 2, energy cannot be created or destroyed—it can only be converted from one form to another. If we apply this principle to cell metabolism, it means that a dynamic balance exists between the body's energy intake and its energy output:

Energy intake = total energy output
(heat + work + energy storage)

Energy intake is the energy liberated during food oxidation—that is, during the reactions of glycolysis, the Krebs cycle, and the electron transport chain. **Energy output** includes the energy we immediately lose as heat (about 60 percent of the total), plus that used to do work (driven by ATP), plus energy that is stored in the form of fat or glycogen. Energy storage is important only during periods of growth and during net fat deposit.

Regulation of Food Intake

When energy intake and energy outflow are balanced, body weight remains stable. When they are not, weight is either gained or lost. Since body weight in most people is surprisingly stable, mechanisms that control food intake or heat production or both must exist.

But how is food intake controlled? That is a difficult question, and one that is still unanswered. Researchers believe that several factors—such as rising or falling blood levels of nutrients (glucose and amino acids), hormones (insulin and glucagon), or body temperature (rising temperature is inhibitory), and psychological factors—have an effect on eating behavior through feedback signals to the brain. Indeed, psychological factors are believed to be a very important cause of obesity. However, even when psychological factors *are* the underlying cause of obesity, individuals do *not* continue to gain weight endlessly. It seems that their feeding controls still operate, but act to maintain total body energy content at higher-than-normal levels.

Metabolic Rate and Body Heat Production

Basal Metabolic Rate When nutrients are broken down to produce cellular energy (ATP), they yield different amounts of energy. As mentioned earlier, the energy value of foods is measured in a unit called the *kilocalorie (kcal)*. In general, carbohydrates and proteins each yield 4 kcal/gram, and fats yield 9 kcal/gram when they are broken down for energy production. Most meals, and even many individual foods, are mixtures of carbohydrates, fats, and proteins. To determine the caloric value of a meal, we must know how many grams of each type of foodstuff it contains. For most of us, this is a difficult chore indeed, but approximations can

Table 14.4	Factors Determining the Basal Metabolic Rate (BMR)	
Factor	**Variation**	**Effect on BMR**
Surface area	Large surface area in relation to body volume, as in thin, small individuals	Increased
	Small surface area in relation to body volume, as in large, heavy individuals	Decreased
Sex	Male	Increased
	Female	Decreased
Thyroxine production	Increased	Increased
	Decreased	Decreased
Age	Young, rapid growth	Increased
	Aging, elderly	Decreased
Strong emotions (anger or fear) and infections		Increased

easily be made with the help of a simple, calorie-values guide available in most drugstores.

The amount of energy used by the body is also measured in kilocalories. The **basal metabolic rate (BMR)** is the amount of heat produced by the body per unit of time when it is under basal conditions—that is, at rest. It reflects the energy supply a person's body needs just to perform essential life activities such as breathing, maintaining the heartbeat, and kidney function. An average 70-kg (154-pound) adult has a BMR of about 60 to 72 kcal/hour.

Many factors influence BMR, including surface area and gender. As shown in Table 14.4, small, thin males tend to have a higher BMR than large, obese females. Age is also important; children and adolescents require large amounts of energy for growth and have relatively high BMRs. In old age, the BMR decreases dramatically as the muscles begin to atrophy.

The amount of **thyroxine** produced by the thyroid gland is probably the most important factor in determining a person's BMR; hence, thyroxine has been dubbed the "metabolic hormone." The more thyroxine produced, the higher the oxygen consumption and ATP use, and the higher the metabolic rate. In the past, most BMR tests were done to determine whether sufficient thyroxine was being made. Today, thyroid activity is more easily assessed by blood tests.

Hyperthyroidism causes a host of effects due to the excessive metabolic rate it produces. The body catabolizes stored fats and tissue proteins and, despite increased hunger and food intake, the person often loses weight. Bones weaken and body muscles, including the heart, atrophy. In contrast, *hypothyroidism* results in slowed metabolism, obesity, and diminished thought processes. ❑

Total Metabolic Rate   When we are active, more glucose must be oxidized to provide energy for the additional activities. Digesting food and even modest physical activity increase the body's caloric requirements dramatically. These additional fuel requirements are above and beyond the energy required to maintain the body in the basal state. **Total metabolic rate (TMR)** refers to the total amount of kilocalories the body must consume to fuel all ongoing activities. Muscular work is the major body activity that increases the TMR. Even

A CLOSER LOOK

OBESITY: MAGICAL SOLUTION WANTED

How fat is too fat? The bathroom scale is an inaccurate guide because body weight tells nothing of body composition. A skilled dancer with dense bones and well-developed muscles may weigh several pounds more than an inactive person of the same relative size. Most experts agree that a person is obese when he or she is 20 percent heavier than the "ideal weight" published in insurance company tables (which, by the way, are artificially low). What is really needed is a measure of body fat, because the most common view of obesity is that it is a condition of excessive triglyceride storage. Although we bewail our inability to rid ourselves of fat, the real problem is that we keep refilling the storehouses by consuming too many calories. A body fat content of 18 to 22 percent of body weight (males and females, respectively) is considered normal for adults. Anything over that is defined as obesity.

However it's defined, obesity is a perplexing and poorly understood disease. The term *disease* is appropriate because all forms of obesity involve some imbalance in food intake control mechanisms. Despite its well-known adverse effects on health (the obese have a higher incidence of arteriosclerosis, hypertension, coronary artery disease, and diabetes mellitus), it is the most common health problem in the United States. About 50 percent of adults and 20 percent of teenagers are obese. U.S. kids are not only getting fatter but, because they are opting for home video games and nachos instead of touch football and apples, their general cardiovascular fitness is declining as well.

As if this were not enough, the social stigma and economic disadvantages of obesity are legendary. A fat person pays higher insurance premiums, is discriminated against in the job market, has fewer clothing choices, and is frequently humiliated during both childhood and adulthood.

With all its attendant problems, it's a pretty fair bet that few people choose to be obese. So what causes obesity? Let's look at three of the more recent theories.

SET POINT THEORY

Large numbers of fat cells deliver signals that stimulate overeating. Some believe that overeating behaviors develop early in life (the "clean your plate" syndrome) and set the stage for adult obesity by increasing the number of fat cells formed during childhood. During adulthood, increases in adipose tissue mass occur by depositing more fat in the existing cells. Thus, the more cells there are, the more fat can be stored.

Besides the signals delivered by blood-borne nutrients or so-called satiety chemicals (hormones and others), researchers have found hints that the fat cells themselves may stimulate overeating. Supporting this idea is the observation that when yo-yo dieters lose weight, their metabolic rate falls sharply. But when they subsequently gain weight, their metabolic rate increases like a furnace being stoked. Each successive weight loss occurs more slowly, but lost weight is regained three times as fast. Thus, it appears that people, like laboratory animals subjected to alternating "feasts and fasts," become increasingly food efficient, and their metabolic rates adjust to prevent any deviation from their weight "set point." The instrument used to solve the problem—dieting (again and again)—becomes self-defeating.

As shown in this photograph of two obese men, excess weight in males is typically carried in the abdomen.

FUEL EFFICIENCY THEORY

Obese people are more fuel efficient and more effective "fat storers." Although it is often assumed that obese people eat more than other people, this is not necessarily true—many actually eat less than those of normal weight.

Fat, the nutrient, is the obese person's worst enemy. Fats pack more wallop per calorie (are more fattening) than proteins or carbohydrates because very little energy is used in their processing. For example, when someone ingests 100 excess fat calories, only 3 calories are "burned" and the rest (97) are stored.

These facts apply to everyone, but when you are obese the picture is even bleaker. For example, fat cells of overweight people "sprout" more alpha receptors, and their lipoprotein lipase enzyme, which unloads fat from the blood (usually to fat cells), is exceptionally efficient.

GENETIC THEORY

Morbid obesity is the fate of those inheriting two obesity genes. However, a true genetic predisposition for "fatness" appears to account for only about 5 percent of the U.S. obese. These people, given excess calories, will always deposit them as fat, as opposed to those who lay down more muscle with some of the excess calories.

FALSE AND RISKY CURES

Rumors and poor choices for dealing with obesity abound. Some of the most unfortunate strategies used for coping with obesity are listed here.

- *"Water pills."* Diuretics, which prompt the kidneys to excrete more water, are touted as a means of losing weight. At best, these may cause a few pounds of weight loss for a few hours; they can also cause serious electrolyte imbalance and dehydration.

- *Diet pills.* Some obese people use amphetamines (best known are Dexedrine and Benzedrine) to reduce appetite. These work, but only temporarily (until tolerance develops) and can cause a dangerous dependency. Furthermore, diet aids that provide fiber to prevent absorption of nutrients can cause serious malnutrition.

- *Fad diets.* Many magazines print at least one new diet regimen yearly, and diet products sell well. However, many of these diets are nutritionally unhealthy. For example, some of the liquid high-protein diets contain such poor-quality (incomplete) protein that they are actually dangerous. (The worst are those that contain collagen protein instead of a milk or soybean protein source.)

- *Surgery.* Sometimes sheer desperation prompts surgical solutions, such as having the jaws wired shut or the stomach stapled, intestinal bypass surgery, lipectomy (removal of fat by suction), and biliopancreatic diversion (BPD). BPD "rearranges" the digestive tract: Two-thirds of the stomach is removed; the small intestine is cut in half and one 8-foot long portion is sutured into the stomach opening. Since pancreatic juice and bile are diverted away from this "new intestine," fewer nutrients (and no fats) are digested and absorbed. Although the patients can eat anything they want without gaining weight, BPD is major surgery and carries all of its risks.

Unfortunately, there is no magical solution for obesity. The only way for most of us to lose weight is to take in fewer fat calories and increase physical activity to increase muscle mass (muscle consumes more energy at rest than does fat). This prescription stresses the fact that there are two sides to the energy balance equation—caloric intake and energy output—and recognizes the unique nature of fat handling in the body. Furthermore, low activity levels actually stimulate eating, while physical exercise depresses food intake and increases metabolic rate not only during activity but also for some time after. Let's not despair though—there's a glimmer of hope on the research horizon. Clinical studies of drugs that influence the hypothalamus "appestat" (*appe*tite thermo*stat*) appear to be bearing fruit for those who starve themselves and still don't lose weight. This so-called *phen-fen treatment* (after bits of the generic names of the drugs used) may provide the help that they need to achieve their desired weight. However, this approach is yet to be proven completely safe, as there are hints of lung damage in some experimental clients.

slight increases in the activity of skeletal muscles cause remarkable leaps in metabolic rate. When a well-trained athlete exercises vigorously for several minutes the TMR may increase to 15 to 20 times normal, and it remains elevated for several hours afterward.

When the total amount of calories consumed is equal to the TMR, homeostasis is maintained, and our weight remains constant. However, if we eat more than we need to sustain our activities, excess calories appear in the form of fat deposits. Conversely, if we are extremely active and do not properly feed the "metabolic furnace," we begin to break down fat reserves and even tissue proteins to satisfy our TMR. This principle is used in every good weight-loss diet. (The total calories needed are calculated on the basis of body size and age. Then, 20 percent or more of the requirements are cut from the daily diet.) If the dieting person exercises regularly, weight drops off even more quickly because the TMR increases above the person's former rate.

Body Temperature Regulation

Although we have been emphasizing that foods are "burned" to produce ATP, remember that ATP is not the only product of cell catabolism. Most of the energy released as foods are oxidized escapes as heat. Less than 40 percent of available food energy is actually captured to form ATP. The heat released warms the tissues and, more importantly, the blood, which circulates to all body tissues, keeping them at homeostatic temperatures, which allows metabolism to occur efficiently.

Body temperature reflects the balance between heat production and heat loss. The body's thermostat is in the *hypothalamus* of the brain. Through autonomic nervous system pathways, the hypothalamus continually regulates body temperature around a constant setpoint of 35.6° to 37.8°C (96° to 100°F) by initiating heat-loss or heat-promoting mechanisms (see Figure 14.18).

Heat-Promoting Mechanisms When the environmental temperature is cold (or the temperature of circulating blood falls), body heat must be conserved (increased). Short-term means of accomplishing this are **vasoconstriction** of blood vessels of the skin and **shivering.**

When the skin vasculature constricts, the skin is temporarily bypassed by the blood; blood is rerouted to the deeper, more vital, body organs. When this happens, the temperature of the exposed skin drops to that of the external environment.

Restriction of blood delivery to the skin is no problem for brief periods of time. But if it is extended, the skin cells, deprived of oxygen and nutrients, begin to die. This condition, called *frostbite,* is extremely serious. ❏

When the *core* body temperature (the temperature of the deep organs) drops to the point beyond which simple constriction of skin capillaries can handle the situation, shivering begins. Shivering, involuntary shudderlike contractions of the voluntary muscles, is very effective in increasing the body temperature, because skeletal muscle activity produces large amounts of heat.

Extremely low body temperature resulting from prolonged exposure to cold is *hypothermia.* In hypothermia, the individual's vital signs (respiratory rate, blood pressure, heart rate) decrease. The person becomes drowsy and oddly comfortable, even though previously he or she felt extremely cold. Uncorrected, the situation progresses to coma and finally death as metabolic processes grind to a stop. ❏

Heat Loss Mechanisms Just as the body must be protected from becoming too cold, it must also be protected from excessively high temperatures. Most heat loss occurs through the skin via **radiation or evaporation.** When body temperature increases above what is desirable, the blood vessels serving the skin dilate and capillary beds in the skin become flushed with warm blood. As a result, heat radiates from the skin surface. However, if the external environment is as hot as or hotter than the body, heat cannot be lost by radiation and the only means of getting rid of excess heat is by the evaporation of perspiration off the skin surface. This is an efficient means of body-heat loss as long as the air is dry. If it is very humid, evaporation occurs at a much slower rate. In such cases, our heat-liberating mechanisms cannot work well, and we feel miserable and irritable. This is why the hot, humid days of August are often called the "dog days."

When normal heat loss processes become ineffective, the *hyperthermia,* or elevated body

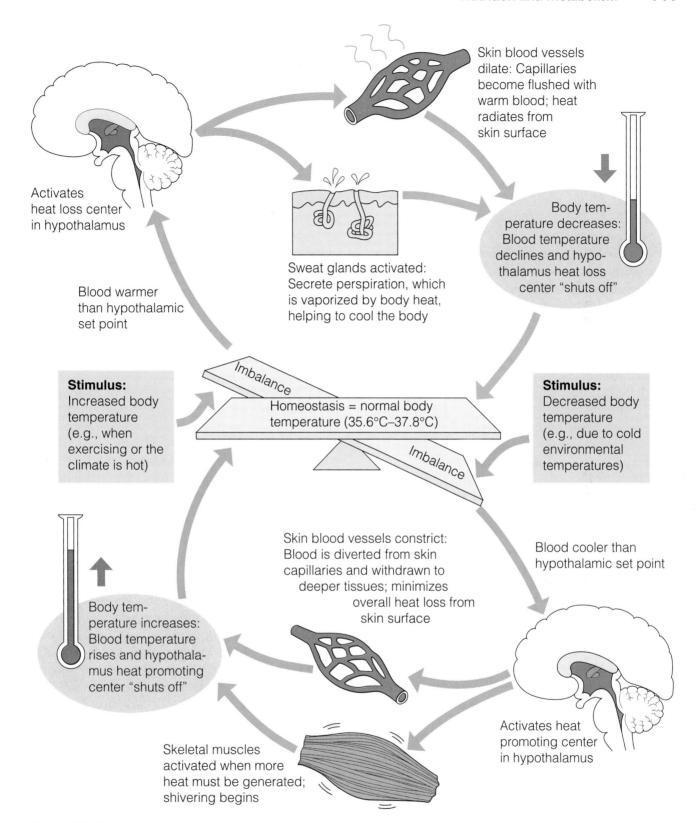

Figure 14.18   **Mechanisms of body temperature regulation.**

temperature, that results depresses the hypothalamus. As a result, a vicious positive-feedback cycle occurs: Soaring body temperature increases the metabolic rate, which in turn increases heat production. The skin becomes hot and dry and, as the temperature continues to spiral upward, permanent brain damage becomes a distinct possibility. This condition, called **heat stroke,** can be fatal unless rapid corrective measures are taken immediately (immersion in cool water and administration of fluids).

Heat exhaustion is the term used to describe the heat-associated collapse of an individual during or following vigorous physical activity. Heat exhaustion results from excessive loss of body fluids (dehydration) and is evidenced by low blood pressure, a rapid heartbeat, and cool, clammy skin. In contrast to heat stroke, heat-loss mechanisms still operate in heat exhaustion. ❑

Fever is *controlled hyperthermia*. Most often, it results from infection somewhere in the body, but it may be caused by other conditions (cancer, allergic reactions, and CNS injuries). Macrophages, white blood cells, and injured tissue cells release chemical substances called *pyrogens* (pi'ro-jenz) that act directly on the hypothalamus, causing its thermostat to be set to a higher temperature (*pyro* = fire). After the thermostat resetting, heat-promoting mechanisms are initiated. Because of vasoconstriction, the skin becomes cool and shivering begins to generate heat. This situation, called "the chills," is a sure sign that body temperature is rising. Body temperature is allowed to rise until it reaches the new setting. Then, it is maintained at the "fever setting" until natural body defense processes or antibiotics reverse the disease process. At that point, the thermostat is reset again to a lower (or normal) level, causing heat-loss mechanisms to swing into action—the individual begins to sweat, and the skin becomes flushed and warm. Physicians have long recognized that these signs signaled a turn for the better in their patients, and said that the patient had "passed the crisis" because body temperature was falling.

As explained in Chapter 12, fever, by increasing the metabolic rate, helps speed the various healing processes, and it also appears to inhibit bacterial growth. The danger of fever is that if the body thermostat is set too high, body proteins may be denatured and permanent brain damage may occur.

DEVELOPMENTAL ASPECTS OF THE DIGESTIVE SYSTEM AND METABOLISM

The very young embryo is flat and pancake-shaped. However, it soon folds to form a cylindrical body, and its internal cavity becomes the cavity of the alimentary canal. By the fifth week of development, the alimentary canal is a continuous tube-like structure extending from the mouth to the anus. Shortly after, the digestive glands (salivary glands, liver, and pancreas) bud out from the mucosa of the alimentary tube. These glands retain their connections (ducts) and can easily empty their secretions into the alimentary canal to promote its digestive functions.

The digestive system is susceptible to many congenital defects that interfere with feeding. The most common is the *cleft palate/cleft lip* defect. Of the two, cleft palate is more serious because the child is unable to suck properly. Another relatively common congenital defect is a *tracheoesophageal fistula* (tra″ke-o-ĕ-sof′ah-je-al fis′tu-lah). In this condition, there is a connection between the esophagus and the trachea. In addition, the esophagus often (but not always) ends in a blind sac and does not connect to the stomach. The baby chokes, drools, and becomes cyanotic during feedings, because food is entering the respiratory passageways. All three defects can be corrected surgically.

There are many types of inborn errors of metabolism (genetically based problems that interfere with metabolism), but perhaps the two most common are *cystic fibrosis* and *phenylketonuria* (fen″il-ke′to-nu″re-ah) *(PKU)*. Cystic fibrosis primarily affects the lungs, but it also significantly impairs the activity of the pancreas. In cystic fibrosis, huge amounts of mucus are produced, which block the passages of involved organs. Blockage of the pancreatic duct prevents pancreatic fluid from reaching the small intestine. As a result, fats and fat-soluble vitamins are not digested or absorbed and bulky, fat-laden stools result. This condition is usually handled by administering pancreatic enzymes with meals.

PKU involves an inability of tissue cells to use phenylalanine (fen″il-al′ah-nin), an amino acid present in all protein foods. In such cases, brain damage and retardation occur unless a special diet low in phenylalanine is prescribed. ❑

The developing infant receives all its nutrients through the placenta, and at least at this period of life, obtaining and processing nutrients is no problem (assuming the mother is adequately nourished). Obtaining nutrition is the most important activity of the newborn baby and several reflexes present at this time help in this activity. For example, the *rooting reflex* helps the infant find the nipple (mother's or bottle), and the *sucking reflex* helps him or her to hold on to the nipple and swallow. The stomach of a newborn infant is very small, so feeding must be frequent (every 3 to 4 hours). Peristalsis is rather inefficient at this time, and vomiting is not at all unusual.

Teething begins around age 6 months and continues until about the age of 2. During this interval, the infant progresses to more and more solid foods and usually is eating an adult diet by toddlerhood. Appetite decreases in the elementary school-age child and then increases again during the rapid growth of adolescence. (Parents of adolescents usually bewail their high grocery bills!)

All through childhood and into adulthood, the digestive system operates with relatively few problems unless there are abnormal interferences such as contaminated food or extremely spicy or irritating foods (which may cause inflammation of the gastrointestinal tract, or *gastroenteritis* [gas″tro-en-ter-i′tis]). Inflammation of the appendix, *appendicitis,* is particularly common in teenagers for some unknown reason. Between middle age and early old age, the metabolic rate decreases by 5 to 8 percent in every 10-year period. This is the time of life when the weight seems to creep up, and obesity often becomes a fact of life. To maintain desired weight, we must be aware of this gradual change and be prepared to reduce caloric intake. Two distinctly middle age digestive problems are *ulcers* (see the "Closer Look" box on p. 424) and *gallbladder problems* (inflammation of the gallbladder or gallstones).

During old age, activity of the GI tract declines. Fewer digestive juices are produced, and peristalsis slows. Taste and smell become less acute and periodontal disease often develops. Many elderly individuals live alone or on a reduced income. These factors, along with increasing physical disability, tend to make eating less appealing, and nutrition is inadequate in many of our elderly citizens. Diverticulosis and cancer of the gastrointestinal tract are fairly common problems in the elderly. Cancer of the stomach and colon rarely have early signs, and often progress to an inoperable stage (that is, spread to distant parts of the body as well) before the individual seeks medical attention. However, when detected early, both diseases are treatable, and it has been suggested that diets high in plant fiber and low in fat might help to decrease the incidence of colon cancer. Additionally, since most colorectal cancers derive from initially benign mucosal tumors called *polyps,* and the incidence of polyp formation increases with age, a yearly colon examination should be a health priority in everyone over the age of 50.

Important Terms

alimentary (al″ ĕ-men′tar-e) **canal**	**digestion**
basal metabolic rate (BMR)	**esophagus** (ĕ-sof′ah-gus)
bile	**feces** (fe′sēz)
carbohydrates	**gallbladder**
cellular respiration	**gastric juice**
chyme (kīm)	**lacteal**
defecation	**large intestine**

SYSTEMS IN SYNC

Homeostatic Relationships Between the Digestive System and Other Body Systems

Nervous System
- Digestive system provides nutrients for normal neural functioning
- Neural controls of digestive function; in general, parasympathetic fibers accelerate and sympathetic fibers inhibit digestive activity; reflex and voluntary controls of defecation

Respiratory System
- Digestive system provides nutrients for energy metabolism, growth, and repair
- Respiratory system provides oxygen and carries away carbon dioxide produced by digestive system organs

Cardiovascular System
- Digestive system provides nutrients to heart and blood vessels; absorbs iron needed for hemoglobin synthesis; absorbs water necessary for normal blood volume
- Cardiovascular system transports nutrients absorbed by alimentary canal to all tissues of the body; distributes hormones of the digestive tract

Reproductive System
- Digestive system provides nutrients for energy fuel, growth, and repair and extra nutrition needed to support fetal growth

Integumentary System
- Digestive system provides nutrients for energy fuel, growth, and repair; supplies fats that provide insulation in the dermal and subcutaneous tissues
- The skin synthesizes vitamin D needed for calcium absorption from the intestine; protects by enclosure

Skeletal System
- Digestive system provides nutrients for energy fuel, growth, and repair; absorbs calcium needed for bone salts
- Skeletal system protects some digestive organs by bone; cavities store some nutrients (e.g., calcium, fats)

Endocrine System
- Liver removes hormones from blood ending their activity; digestive system provides nutrients needed for energy fuel, growth, and repair; pancreas has hormone-producing cells
- Local hormones help regulate digestive function

Lymphatic System/Immunity
- Digestive system provides nutrients for normal functioning; HCl of stomach provides nonspecific protection against bacteria
- Lacteals drain fatty lymph from digestive tract organs and convey it to blood; Peyer's patches and lymphoid tissue in mesentery house macrophages and immune cells that protect digestive tract organs against infection

Digestive System

Urinary System
- Digestive system provides nutrients for energy fuel, growth, and repair; excretes some bilirubin produced by the liver
- Kidneys transform vitamin D to its active form, which is needed for calcium absorption

Muscular System
- Digestive system provides nutrients for energy fuel, growth, and repair; liver removes lactic acid, resulting from muscle activity, from the blood
- Skeletal muscle activity increases motility of GI tract

lipids

lipoproteins

liver

mesentery (mes'en-ter"e)

metabolism

minerals

mucosa

nutrient

oral cavity

pancreas

pancreatic juice

peristalsis (per"ĭ-stal'sis)

peritoneum (per"i-to-ne'um)

proteins

rectum

salivary glands

segmentation

small intestine

stomach

villi

vitamins

Summary

Anatomy of the Digestive System (pp. 408–419)

1. The digestive system consists of the alimentary canal (a hollow tube extending from mouth to anus) and several accessory digestive organs. The wall of the alimentary canal has four main tissue layers—mucosa, submucosa, muscularis externa, serosa. The serosa (visceral peritoneum) is continuous with the parietal peritoneum, which lines the abdominal cavity wall.

2. Organs of the alimentary canal:
   a. The mouth or oral cavity contains teeth and tongue and is bounded by lips, cheeks, and palate. Tonsils guard its posterior margin.
   b. The pharynx is a muscular tube that provides a passageway for food and air.
   c. The esophagus is a muscular tube that completes the passageway from the pharynx to the stomach.
   d. The stomach is a C-shaped organ located on the left side of the abdomen beneath the diaphragm. Food enters it through the cardioesophageal sphincter and leaves it to enter the small intestine through the pyloric sphincter. The stomach has a third oblique layer of muscle in its wall that allows it to perform mixing or churning movements. Gastric glands produce hydrochloric acid, pepsin, rennin, mucus, and

intrinsic factor. Mucus protects the stomach itself from being digested.
   e. The tubelike small intestine is suspended from posterior body wall by the mesentery. Its subdivisions are the duodenum, jejunum, and ileum. Food digestion and absorption are completed here. Pancreatic juice and bile enter the duodenum through a sphincter at the distal end of the common bile duct. Microvilli, villi, and circular folds increase the surface area of small intestine for enhanced absorption.
   f. The large intestine frames the small intestine. Subdivisions are cecum; appendix; ascending, transverse, and descending colon; sigmoid colon; rectum; anal canal. The large intestine delivers undigested food residue (feces) to the body exterior.

3. Accessory organs duct substances into the alimentary tube.
   a. The pancreas is a soft gland lying in the mesentery between the stomach and small intestine. Pancreatic juice contains enzymes (that digest all categories of foods) in an alkaline fluid.
   b. The liver is a four-lobed organ overlying the stomach. Its digestive function is to produce bile, which it ducts into the small intestine.
   c. The gallbladder is a muscular sac that stores and concentrates bile. When fat digestion is not on-

going, continuously made bile backs up the cystic duct and enters the gallbladder.

d. Salivary glands (three pairs—parotid, submandibular, sublingual) secrete saliva into the oral cavity. Saliva contains mucus and serous fluids. The serous component contains salivary amylase.

4. Two sets of teeth are formed. The first set consists of 20 deciduous teeth that begin to appear at 6 months and are lost by 12 years. Permanent teeth (32) begin to replace deciduous teeth around 7 years. A typical tooth consists of crown covered with enamel and root covered with cementum. The bulk of the tooth is bonelike dentin. The pulp cavity contains blood vessels and nerves.

Functions of the Digestive System *(pp. 419–428)*

1. Foods must be broken down to their building blocks to be absorbed. Building blocks of carbohydrates are simple sugars, or monosaccharides; building blocks of proteins are amino acids; building blocks of fats, or lipids, are fatty acids and glycerol.

2. Both mechanical (chewing) and chemical food breakdown begin in the mouth. Saliva contains mucus, which helps bind food together into a bolus, and salivary amylase, which begins the chemical breakdown of starch. Saliva is secreted in response to food in mouth, mechanical pressure, and psychic stimuli. Essentially no food absorption occurs in the mouth.

3. Swallowing has two phases: The buccal phase is voluntary; the tongue pushes the bolus into the pharynx. The involuntary pharyngeal–esophageal phase involves closing off of nasal and respiratory passages and conduction of food to the stomach by peristalsis.

4. When food enters the stomach, gastric secretion is stimulated by vagus nerves and by gastrin (a local hormone). Hydrochloric acid activates the protein-digesting enzyme pepsin, and chemical digestion of proteins begins. Food is also mechanically broken down by churning activity of stomach muscles. Movement of chyme into the small intestine is controlled by the enterogastric reflex.

5. Chemical digestion of fats, proteins, and carbohydrates is completed in the small intestine by intestinal enzymes and, more importantly, pancreatic enzymes. Alkaline pancreatic juice neutralizes acid chyme and provides the proper environment for operation of enzymes. Both pancreatic juice (the only source of lipases) and bile (formed by the liver) are necessary for normal fat breakdown and absorption. Bile acts as a fat emulsifier. Secretin and cholecystokinin, hormones produced by the small intestine, stimulate release of bile and pancreatic juice. Segmental movements mix foods; peristaltic movements move foodstuffs along small intestine. Most nutrient absorption occurs by active transport into capillary blood of the villi. Fats are absorbed by diffusion into both capillary blood and lacteals in the villi.

6. The large intestine receives bacteria-laden indigestible food residue. Activities of the large intestine are absorption of water, salts, and of vitamins made by resident bacteria. When feces are delivered to the rectum by peristalsis and mass peristalsis, the defecation reflex is initiated.

Nutrition and Metabolism *(pp. 428–442)*

1. Metabolism includes all chemical breakdown (catabolic) and building (anabolic) reactions needed to maintain life.

2. Carbohydrates, most importantly glucose, are the body's major energy fuel. As glucose is oxidized, carbon dioxide, water, and ATP are formed. The sequential pathways of glucose catabolism are glycolysis, which occurs in the cytosol, and the Krebs cycle and electron transport chain (in the mitochondria). During hyperglycemia, glucose is stored as glycogen or converted to fat. In hypoglycemia, glycogenolysis, gluconeogenesis, and fat breakdown occur to restore normal blood glucose levels.

3. Fats insulate the body, protect organs, build some cell structures (membranes and myelin sheaths), and provide reserve energy. When carbohydrates are not available, more fats are oxidized to produce ATP. Excessive fat breakdown causes blood to become acidic. Excess dietary fat is stored in subcutaneous tissue and other fat depots.

4. Proteins form the bulk of cell structure and most functional molecules. They are carefully conserved by body cells. Amino acids are actively taken up from blood by tissue cells; those that cannot be made by body cells are called *essential* amino acids. Amino acids are oxidized to form ATP mainly when other fuel sources are not available. Ammonia, released as amino acids are catabolized, is detoxified by liver cells that combine it with carbon dioxide to form urea.

5. The liver is the body's key metabolic organ. Its cells remove nutrients from hepatic portal blood. It performs glycogenesis, glycogenolysis, and gluconeogenesis to maintain homeostasis of blood glucose levels. Its cells make blood proteins and other substances and release them to blood. Fats are burned by liver cells to provide some of their energy (ATP); excesses are stored or released to blood in simpler forms that can be used by other tissue cells. Phagocytic cells remove bacteria from hepatic portal blood. Most cholesterol is made by the liver; cholesterol breakdown products are secreted in bile. Fats and cholesterol are transported in the blood by lipoproteins. LDLs transport cholesterol to body cells; HDLs carry cholesterol to the liver for degradation. Cholesterol is used to make functional molecules and for some structural purposes; it is not used for energy.

6. A dynamic balance exists between energy intake and total energy output (heat + work + energy storage). Interference with this balance results in obesity or malnutrition leading to body wasting.

7. When the three major types of foods are oxidized for energy, they yield different amounts of energy. Carbohydrates and proteins yield 4 kcal/gram; fats yield 9 kcal/gram. Basal metabolic rate (BMR) is the total amount of energy used by the body when one is in a basal state. Age, sex, body surface area, and amount of thyroxine produced influence BMR.

8. Total metabolic rate (TMR) is number of calories used by the body to accomplish all ongoing daily activities. It increases dramatically as muscle activity increases. When TMR equals total caloric intake, weight remains constant.

9. As foods are catabolized to form ATP, more than 60 percent of energy released escapes as heat, warming the body. The hypothalamus initiates heat-loss processes (radiation of heat from skin and evaporation of sweat) or heat-promoting processes (vasoconstriction of skin blood vessels and shivering) as necessary to maintain body temperature within normal limits. Fever (hyperthermia) represents body temperature regulated at higher than normal levels.

Developmental Aspects of the Digestive System and Metabolism (pp. 442–443)

1. The alimentary tract forms as a hollow tube. Accessory glands form as outpocketings from this tube.

2. Common congenital defects include cleft palate, cleft lip, and tracheoesophageal fistula, all of which interfere with normal nutrition. Common inborn errors of metabolism are phenylketonuria (PKU) and cystic fibrosis.

3. Various inflammatory conditions plague the digestive system through life. Appendicitis is common in adolescents, gastroenteritis and food poisoning can occur at any time (given the proper irritating factors), ulcers and gallbladder problems increase in middle age. Obesity and diabetes mellitus are bothersome during later middle age.

4. Efficiency of all digestive system processes decreases in the elderly. Gastrointestinal cancers, such as stomach and colon cancer, appear with increasing frequency in an aging population.

Review Questions

1. Make a simple line drawing of the organs of the alimentary tube and label each organ.

2. Add three labels to your drawing—salivary glands, liver, and pancreas—and use arrows to show where each of these organs empties its secretion into the alimentary tube.

3. Name the layers of the alimentary tube wall from the lumen out.

4. What is the mesentery? The peritoneum?

5. Name the subdivisions of the small intestine in an anterior-to-posterior direction. Do the same for the subdivisions of the large intestine.

6. The digestive system has many structural modifications. Describe the structure and function of villi.

7. What is the normal number of permanent teeth? of deciduous teeth? What substance covers the tooth crown? What substance makes up the bulk of a tooth? What is pulp, and where is it?

8. Name the three pairs of salivary glands. Name two functions of saliva.

9. Assume you have been chewing a piece of bread for 5 or 6 minutes. How would you expect its taste to change during this time? Why?

10. Name two regions of the digestive tract where mechanical food breakdown occurs, and explain how it is accomplished in those regions.

11. Name the organ where protein digestion begins.

12. Why is it necessary for the stomach contents to be so acidic? How does the stomach protect itself from being digested?

13. Only one organ produces enzymes capable of digesting all groups of foodstuffs. What organ is this?

14. Explain why fatty stools result from the absence of bile and/or pancreatic juice.

15. Define *emulsify*.

16. What is the function of gastrin? of secretin?

17. Describe the two phases of swallowing.

18. How do segmental and peristaltic movements differ?

19. What are the end products of protein digestion? of fat digestion? of carbohydrate digestion?

20. Where does most nutrient absorption occur?

21. What substances are absorbed in the large intestine?

22. What is the composition of feces?

23. Define *defecation reflex, constipation,* and *diarrhea*.

24. Define *metabolism, anabolism,* and *catabolism*.

25. Define *gluconeogenesis, glycogenolysis,* and *glycogenesis*.

26. Which food group is most important as a fuel source (that is, for catabolism and ATP production)? Which is most important for building cell structures?

27. What is the harmful result when excessive amounts of fats are burned to produce ATP? Name two conditions that might lead to this result.

28. Define *BMR* and name two factors that are important in determining an individual's BMR.

29. If your total caloric intake exceeds your TMR, what can you expect to happen?

30. How many calories are produced when 1 gram of carbohydrate is oxidized? 1 gram of protein? 1 gram of fat? If you just ate 100 grams of food that was 20 percent protein, 30 percent carbohydrate, and 10 percent fat, how many calories did you consume?

31. Some of the energy released as foods are oxidized is captured to make ATP. What happens to the rest of it?

32. Where is the body's thermostat?

33. Name two ways in which heat is lost from the body. Name two ways in which heat is retained or generated.

34. What is fever? What does it indicate?

35. Name three digestive system problems common to middle-aged adults. Name one common to teenagers. Name three common in elderly persons.

At the Clinic

1. After chopping wood for about 2 hours on a hot but breezy afternoon, John stumbled into the house and then fainted. His T-shirt was wringing wet with perspiration and his pulse was faint and rapid. Was he suffering from heat stroke or heat exhaustion? Explain your reasoning and note what you should do to help John recover.

2. Harry is hospitalized with bacterial pneumonia. When you visit him, his teeth are chattering, his skin is cool and clammy to the touch, and he complains of feeling cold even though his room is quite warm. Explain his symptoms.

3. A young woman is put through an extensive battery of tests to determine the cause of her "stomach pains." She is diagnosed with gastric ulcers. An antihistamine drug is prescribed and she is sent home. What is the mechanism of her medication? What life-threatening problems can result from a poorly managed ulcer? Why did the clinic doctor warn the woman not to take aspirin?

4. Continuing from the previous question, the woman's ulcer got worse. She started complaining of back pain. The physician discovered that the back pain occurred because the pancreas was now damaged. Use logic to deduce how a perforating gastric ulcer could come to damage the pancreas.

5. Benny, a 6-year-old child who is allergic to milk, has extremely bowed legs. What condition do you suspect and what is the connection to not drinking milk?

15
The Urinary System

YOUR GOALS

After completing this chapter, you will have a working knowledge of the functions of the urinary system and will have mastered the objectives listed below.

FUNCTION PREVIEW

■ The urinary system rids the body of nitrogenous wastes while regulating water, electrolyte, and acid-base balance of the blood.

(continued)

The kidneys, which maintain the purity and constancy of our internal fluids, are perfect examples of homeostatic organs. Much like sanitation workers who keep a city's water supply drinkable and dispose of its waste, the kidneys are usually unappreciated until there is a malfunction and "internal garbage" piles up. Every day, the kidneys filter gallons of fluid from the bloodstream. They then process this filtrate, allowing wastes and excess ions to leave the body in urine, while returning needed substances to the blood in just the right proportions. Although the lungs and the skin also play roles in excretion, the kidneys bear the major responsibility for eliminating nitrogenous (nitrogen-containing) wastes, toxins, and drugs from the body.

Disposing of wastes and excess ions is only one part of the work of the kidneys. As they perform these excretory functions, they also regulate the volume and chemical makeup of the blood, so that the proper balance between water and salts and between acids and bases is maintained. Frankly, this would be tricky work for a chemical engineer, but the kidneys do it efficiently most of the time.

The kidneys have other regulatory functions as well: By producing the enzyme *renin* (rĕ′nin), they help regulate blood pressure, and their hormone *erythropoietin* stimulates red blood cell production in bone marrow (see Chapter 10). Kidney cells also convert vitamin D to its active form.

The kidneys alone perform the functions just described and manufacture urine in the process. The other organs of the **urinary system**—the paired ureters and the single urinary bladder and urethra (see Figure 15.1)—provide temporary storage reservoirs for urine or serve as transportation channels to carry it from one body region to another.

KIDNEYS

Location and Structure

Although many believe that the kidneys are located in the lower back, this is *not* their location. Instead, these small, dark red organs with a kidney-bean shape lie against the dorsal body wall in a *retroperitoneal* position (beneath the parietal peritoneum) in the *superior* lumbar region. The kidneys extend from T_{12} to L_3; thus they receive some protection from the lower part of the rib cage. Because it is crowded by the liver, the right kidney is positioned slightly lower than the left. An adult kidney is about 12 cm (5 inches) long, 6 cm (2.5 inches) wide, and 3 cm (1 inch) thick, about the size of a large bar of soap. It is convex laterally and has a medial indentation called the *hilus.* Several structures, including the ureters, the renal blood vessels, and nerves, enter or exit the kidney at the hilus (see Figure 15.1). Atop each kidney is an *adrenal gland,* which is part of the endocrine system and is a distinctly separate organ functionally.

A fibrous, transparent **renal capsule** encloses each kidney and gives a fresh kidney a glistening appearance. In a living person, a fatty mass, the **adipose capsule,** surrounds each kidney and helps hold it in place against the muscles of the trunk wall.

The fat surrounding the kidneys is extremely important in holding them in their normal body position. If the amount of fatty tissue dwindles (as with rapid weight loss), the kidneys may drop to a lower position, a condition called *ptosis* (to′sis; "a fall"). Ptosis creates problems if the ureters, which drain urine from the kidneys, become kinked. When this happens, urine that can

Figure 15.1 Anterior view of the urinary organs. (Most unrelated abdominal organs have been removed.)

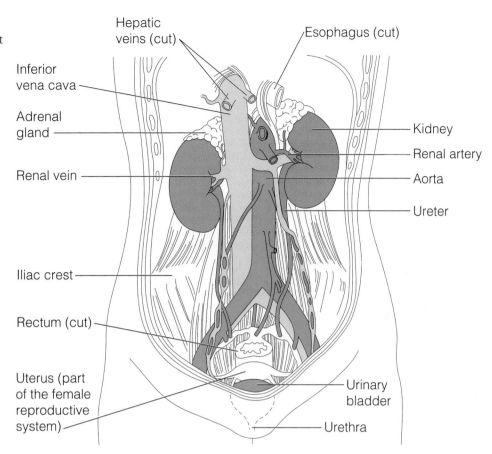

Hepatic veins (cut)

Esophagus (cut)

Inferior vena cava

Adrenal gland

Renal vein

Iliac crest

Rectum (cut)

Uterus (part of the female reproductive system)

Kidney

Renal artery

Aorta

Ureter

Urinary bladder

Urethra

no longer pass through the ureters backs up and exerts pressure on the kidney tissue. This condition, called *hydronephrosis* (hi″dro-nĕ-fro′sis), can severely damage the kidney. ❑

When a kidney is cut lengthwise, three distinct regions become apparent, as can be seen in Figure 15.2. The outer region, which is light in color, is the **renal cortex.** (The word *cortex* comes from the Latin word meaning "bark.") Deep to the cortex is a darker reddish-brown area, the **renal medulla.** The medulla has many basically triangular regions with a striped appearance, the **medullary** (med′u-lar″e) **pyramids.** The broader base of each pyramid faces toward the cortex; its tip, the **apex,** points toward the inner region of the kidney. The pyramids are separated by extensions of cortical tissue, the **renal columns.**

Medial to the hilus is a flat, basinlike cavity, the **renal pelvis.** As Figure 15.2b shows, the pelvis is continuous with the ureter leaving the hilus. Extensions of the pelvis, **calyces** (kal′ĭ-sēz), form cup-shaped areas that enclose the tips of the pyramids. The calyces collect urine, which continu-

Renal pelvis

Renal cortex

Medullary pyramid (in medulla)

Apex of pyramid

Renal column

Calyx

(a)

Figure 15.2 Internal anatomy of the kidney. (a) Photograph of a coronally sectioned kidney.

➤

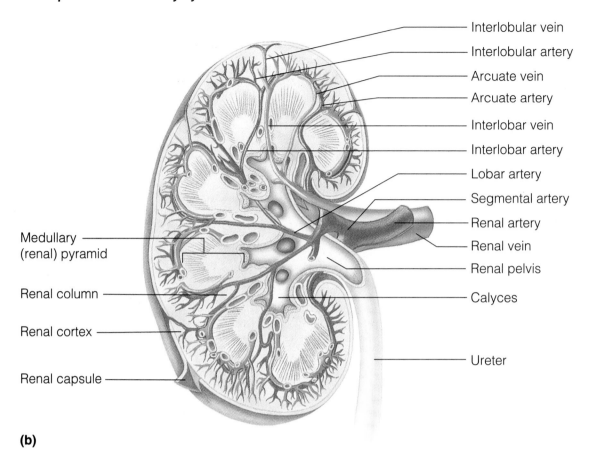

Interlobular vein

Interlobular artery

Arcuate vein

Arcuate artery

Interlobar vein

Interlobar artery

Lobar artery

Segmental artery

Renal artery

Renal vein

Renal pelvis

Calyces

Ureter

Medullary (renal) pyramid

Renal column

Renal cortex

Renal capsule

(b)

Figure 15.2 (continued) **Internal anatomy of the kidney. (b)** Diagrammatic view of a coronally sectioned kidney, illustrating major blood vessels.

ously drains from the tips of the pyramids into the renal pelvis. Urine then flows from the pelvis into the ureter, which transports it to the bladder for temporary storage.

Blood Supply

The kidneys continuously cleanse the blood and adjust its composition, so it is not surprising that they have a very rich blood supply (see Figure 15.2b). Approximately one-quarter of the total blood supply of the body passes through the kidneys each minute. The arterial supply of each kidney is the **renal artery.** As the renal artery approaches the hilus, it divides into **segmental arteries.** Once inside the pelvis, the segmental arteries break up into **lobar arteries,** each of which gives off several branches called **interlobar arteries,** which travel through the renal columns to reach the cortex. At the medulla-cortex junction, interlobar arteries give off the **arcuate** (ar′ku-at)

arteries, which curve over the medullary pyramids. Small **interlobular arteries** then branch off the arcuate arteries and run outward to supply the cortex tissue. Venous blood draining from the kidney flows through veins that trace the pathway of the arterial supply but in a reverse direction—**interlobular veins** to **arcuate veins** to **interlobar veins** to the **renal vein** that emerges from the kidney hilus. (There are no lobar or segmental veins.)

Nephrons and Urine Formation

Nephrons

Each kidney contains over a million tiny structures called **nephrons** (nef′ronz). Nephrons are the structural and functional units of the kidneys and, as such, they are responsible for forming the urine product. Figure 15.3 shows the anatomy and relative positioning of nephrons in each kidney.

Figure 15.3   Structure of the nephron.
(**a**) Wedge-shaped section of kidney tissue indicating the positioning of nephrons in the kidney. (**b**) Detailed anatomy of a nephron and its associated blood supply. Part of the distal convoluted tubule and afferent arteriole have been sectioned to reveal the location of the juxtaglomerular apparatus.

Each nephron consists of two main structures: a **glomerulus** (glo-mer′u-lus), which is a knot of capillaries, and a **renal tubule.** The closed end of the renal tubule is enlarged and cup-shaped and completely surrounds the glomerulus. This portion of the renal tubule is called the **glomerular** (*glom* = little ball), or **Bowman's, capsule.** The inner layer of the capsule is made up of highly modified octopuslike cells called *podocytes* (pod′o-sītz). Podocytes have long branching processes that intertwine with one another and cling to the glomer-

ulus. Since openings or slits exist between their extensions, the podocytes form a porous, or holey, membrane around the glomerulus.

The rest of the tubule is about 3 cm (approximately 1.25 inches) long. As it extends from the glomerular capsule, it coils and twists before forming a hairpin loop, and then again becomes coiled and twisted before entering a collecting tubule called the collecting duct. These different regions of the tubule have specific names; in order from the glomerular capsule they are the **proximal convoluted tubule,** the **loop of Henle** (hen′le), and the **distal convoluted tubule** (see Figure 15.3). The lumen surfaces (surface exposed to the filtrate) of the tubule cells in the proximal convoluted tubules are covered with dense microvilli, which increases their surface area tremendously.

Microvilli also occur on the tubule cells in other parts of the tubule but in much reduced numbers.

Most nephrons are called **cortical nephrons** because they are located almost entirely within the cortex. In a few cases, the nephrons are called **juxtamedullary nephrons** because they are situated close to the cortex-medulla junction and their loops of Henle dip deep into the medulla (Figure 15.3a). The **collecting ducts,** each of which receives urine from many nephrons, run downward through the medullary pyramids, giving them their striped appearance. They deliver the final urine product into the calyces and renal pelvis.

Each and every nephron is associated with two capillary beds—the glomerulus (mentioned earlier) and the *peritubular* (per″ĭ-tu′bu-lar) *capillary bed.* The glomerulus is both fed and drained by *arterioles.* The **afferent arteriole,** which arises from an *interlobular artery,* is the "feeder vessel," and the **efferent arteriole** receives blood that has passed through the glomerulus. The glomerulus, specialized for filtration, is unlike any other capillary bed in the entire body. Because it is both fed *and* drained by arterioles, which are high-resistance vessels, and the afferent arteriole has a larger diameter than the efferent, blood pressure in the glomerulus is extraordinarily high. This extremely high pressure forces fluid and solutes (smaller than proteins) out of the blood into the glomerular capsule. Most of this filtrate (99 percent) is eventually reclaimed by the renal tubule cells and returned to the blood in the peritubular capillary beds.

The second capillary bed, the **peritubular capillaries,** arises from the efferent arteriole that drains the glomerulus. Unlike the high-pressure glomerulus, these capillaries are low-pressure, porous vessels that are adapted for absorption instead of filtration. They cling closely to the whole length of the renal tubule, where they are in an ideal position to receive solutes and water from the tubule cells as these substances are reabsorbed from the filtrate percolating through the tubule. The peritubular capillaries drain into interlobular veins leaving the cortex.

Urine Formation

Urine formation is a result of three processes—*filtration, reabsorption,* and *secretion.* Each of these processes is illustrated in Figure 15.4 and described in more detail next.

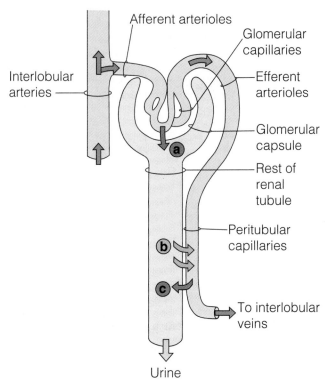

Key:

→ **Filtration:** Water and solutes smaller than
a proteins are forced through the capillary walls and pores of the glomerular capsule into the renal tubule.

→ **Reabsorption:** Water, glucose, amino acids,
b and needed ions are transported out of the filtrate into the tubule cells and then enter the capillary blood.

→ **Secretion:** H⁺, K⁺, creatinine, and drugs are
c removed from the peritubular blood and secreted by the tubule cells into the filtrate.

Figure 15.4   The kidney depicted schematically as a single, large, uncoiled nephron. A kidney actually has millions of nephrons acting in parallel. The three processes by which the kidneys adjust the composition of plasma are (**a**) filtration, (**b**) reabsorption, and (**c**) secretion.

Filtration   As just described, the glomerulus acts as a filter. **Filtration** is a nonselective, passive process. The filtrate that is formed is essentially blood plasma without blood proteins. Both proteins and blood cells are normally too large to pass through the filtration membrane, and when either of these appear in the urine, it is a pretty fair bet that there is some problem with the glomerular filters. As long as the systemic blood pressure is normal, filtrate will be formed. If arterial blood pressure drops too low, the glomerular pressure becomes inadequate to force substances out of the blood into the tubules, and filtrate formation stops. An abnormally low urinary output is called *oliguria* (ol″i-gu′re-ah) if it is between 100 and 400 ml/day, and *anuria* (ah-nu′re-ah) if urinary output is less than 100 ml/day. Low urinary output usually indicates that glomerular blood pressure is too low to cause filtration, but anuria may also result from transfusion reactions and acute inflammation, or crush injuries of the kidneys. ❑

Reabsorption Besides wastes and excess ions that must be removed from the blood, the filtrate contains many useful substances (including water, glucose, amino acids, and ions), which must be reclaimed from the filtrate and returned to the blood. **Tubular reabsorption** begins as soon as the filtrate enters the proximal convoluted tubule. The tubule cells are "transporters," taking up needed substances from the filtrate and then passing them out their posterior aspect into the extracellular space, from which they are absorbed into peritubular capillary blood. Some reabsorption is done passively (for example, water passes by osmosis), but the reabsorption of most substances depends on active transport processes, which use membrane carriers and are very selective. There is an abundance of carriers for substances that need to be retained, and few or no carriers for substances of no use to the body. This is why certain substances (for example, glucose and amino acids) are usually entirely removed from the filtrate and nitrogenous waste products (such as urea, creatinine, and uric acid) are poorly reabsorbed. Various ions are reabsorbed or allowed to go out in the urine, according to what is needed at a particular time to maintain the proper pH and electrolyte composition of the blood. Most reabsorption occurs in the proximal convoluted tubules, but under certain conditions the distal convoluted tubule (and even the collecting duct) is also active.

Secretion **Tubular secretion** is essentially reabsorption in reverse. Some substances, such as hydrogen and potassium ions and creatinine, move from the blood of the peritubular capillaries through the tubule cells or from the tubule cells themselves into the filtrate to be eliminated in urine. This process seems to be important for getting rid of substances not already in the filtrate, such as certain drugs, or as an additional means for controlling blood pH.

Control of Blood Composition by the Kidneys

Blood composition depends on three major factors: diet, cellular metabolism, and urine output. In 24 hours, the marvelously complex kidneys filter approximately 150 to 180 liters of blood plasma through their glomeruli into the tubules, which process the filtrate by taking substances out of it (reabsorption) and adding substances to it (secretion). In the same 24 hours, only about 1.0 to 1.8 liters of urine are produced. Obviously, urine and filtrate are quite different. Filtrate contains everything that blood plasma does (except proteins), but by the time it reaches the collecting ducts, the filtrate has lost most of its water and just about all of its nutrients and necessary ions. What remains, **urine,** contains most of the waste and unneeded substances. Assuming we are healthy, our kidneys can keep our blood composition fairly constant despite wide variations in diet and cell activity.

In general, the kidneys have four major roles to play, which help keep the blood composition relatively constant. These are (1) excretion of nitrogen-containing wastes, maintaining (2) water and (3) electrolyte balance of the blood, and (4) ensuring proper blood pH. Each of these roles is discussed briefly next.

Excretion of Nitrogen-Containing Wastes

Urea, uric acid, and creatinine are the most important **nitrogenous** (nitrogen-containing) **wastes** found in blood. **Urea** (u-re′ah), which is formed by the liver when amino acids are used to produce energy, is an end product of protein breakdown. **Uric acid** is released when nucleic acids are metabolized, and **creatinine** (kre-at′ĭ-nin) is associated with creatine (kre′ah-tin) metabolism in muscle tissue. Because tubule cells have few membrane carriers to reabsorb these substances, they tend to remain in the filtrate and are found in high concentrations in urine. In addition, creatinine is actively secreted into the filtrate.

Figure 15.5   The major fluid compartments of the body. Approximate values are noted for a 70-kg (154-pound) male.

Maintaining Water and Electrolyte Balance of Blood

If you are a healthy young adult, water probably accounts for half or more of your body weight—50 percent in females and about 60 percent in males. These differences reflect the fact that females have relatively less muscle and a larger amount of body fat (and of all body tissues, fat contains the least water). Babies, with little fat and low bone mass, are about 75 percent water, but total body water content declines through life and accounts for only about 45 percent of body weight in old age. The importance of water to functioning of the body and its cells is described in Chapter 2. That information will not be repeated here except to say that water is the universal body solvent within which all solutes (including the very important electrolytes) are dissolved.

Water occupies three main locations within the body, referred to as *fluid compartments* (Figure 15.5). About two-thirds of body fluid, the so-called **intracellular fluid,** is contained within the living cells. The balance, called **extracellular fluid (ECF),** includes all body fluids located outside the cells. Although ECF most importantly includes blood plasma and interstitial (or tissue) fluid, it also accounts for cerebrospinal and serous fluids, the humors of the eye, lymph, and others.

Water certainly accounts for nearly the entire volume of body fluids, regardless of type, and all body fluids are similar, but there is more to *fluid balance* than just water. The types and amounts of solutes in body fluids, especially electrolytes such as sodium, potassium, and calcium ions, are also vitally important to overall body homeostasis and **water** and **electrolyte balance** are tightly linked as the kidneys continuously process the blood. (Recall from Chapter 2 that electrolytes are charged particles [ions] that conduct an electrical current in an aqueous solution.) Very small changes in solute concentrations in the various fluid compartments cause water to move from one compartment to another. Not only does this alter blood volume and blood pressure, but it also can severely impair the activity of irritable cells like nerve and muscle cells. For example, a deficit of sodium ions (Na^+) in the ECF results in water loss from the bloodstream into the tissue spaces (edema) and muscular weakness.

If the body is to remain properly hydrated, we cannot lose more water than we take in. Most water intake is a result of fluids and foods we ingest in our diet. However, a small amount (about 10 percent) is produced during cellular metabolism, as explained in Chapter 14 and indicated in Figure 15.6. There are several routes for water to leave the body. Some water vaporizes out of the lungs, some is lost in perspiration, and some leaves the body in

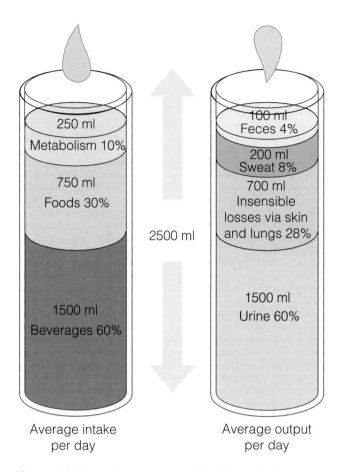

250 ml Metabolism 10%	100 ml Feces 4%
750 ml Foods 30%	200 ml Sweat 8%
	700 ml Insensible losses via skin and lungs 28%
1500 ml Beverages 60%	1500 ml Urine 60%
Average intake per day	Average output per day

2500 ml

Figure 15.6   Major sources of body water (water intake) and routes of water loss from the body (water output). When intake and output are in balance, the body is adequately hydrated.

the stool. The job of the kidneys is like that of a juggler. If large amounts of water are lost in other ways, they compensate by putting out less urine to conserve body water. On the other hand, when water intake is excessive, the kidneys excrete generous amounts of urine and the anguish of a too-full bladder becomes very real.

Likewise, the proper concentrations of the various electrolytes must be present in both intracellular and extracellular fluids. Most electrolytes enter the body in foods and "hard" (mineral-rich) water. Although very small amounts are lost in perspiration and in feces, the major factor regulating the electrolyte composition of body fluids is the kidneys. Just how the kidneys accomplish their balancing act is explained in more detail next.

Reabsorption of water and electrolytes by the kidneys is regulated primarily by hormones. When

blood volume drops for any reason (for example, due to hemorrhage or excessive water loss through sweating or diarrhea), arterial blood pressure drops, which in turn decreases the amount of filtrate formed by the kidneys. In addition, highly sensitive cells in the hypothalamus called *osmoreceptors* (oz″mo-re-sep′torz) react to the change in blood composition (that is, less water and more solutes) by becoming very irritable. The result is that nerve impulses are sent to the posterior pituitary (Figure 15.7), which then releases **antidiuretic hormone (ADH).** (The term *antidiuretic* is derived from **diuresis** [di″u-re′sis], which means "flow of urine from the kidney," and *anti,* which means "against.") As one might guess, this hormone prevents excessive water loss in the urine. ADH travels in the blood to its main target, the kidney's collecting ducts, where it causes the duct cells to reabsorb more water. As more water is returned to the bloodstream, blood volume and blood pressure increase to normal levels, and only a small amount of very concentrated urine is formed. ADH is released more or less continuously unless the solute concentration of the blood drops too low. When this happens, the osmoreceptors become "quiet," and excess water is allowed to leave the body in the urine.

When ADH is *not* released (perhaps because of injury or destruction of the hypothalamus or posterior pituitary gland), huge amounts of very dilute urine (up to 25 liters/day) flush from the body day after day. This condition, *diabetes insipidus* (in-sip′ĭ-dus), can lead to severe dehydration and electrolyte imbalances. Affected individuals are always thirsty and have to drink fluids almost continuously to maintain normal fluid balance. ❑

A second hormone that helps to regulate blood composition and blood volume by acting on the kidney is **aldosterone** (al″dos′ter-on). Aldosterone is the major factor regulating sodium ion content of the ECF and in the process helps regulate the concentration of other ions (Cl^-, K^+, Mg^{2+}) as well. Sodium ion is the electrolyte most responsible for osmotic water flows. When too little sodium is in the blood, the blood becomes too dilute. Consequently, water leaves the bloodstream and flows out into the tissue spaces, causing edema and possible circulatory collapse. But whether aldosterone is present or not, about 80 percent of the sodium in the filtrate is reabsorbed

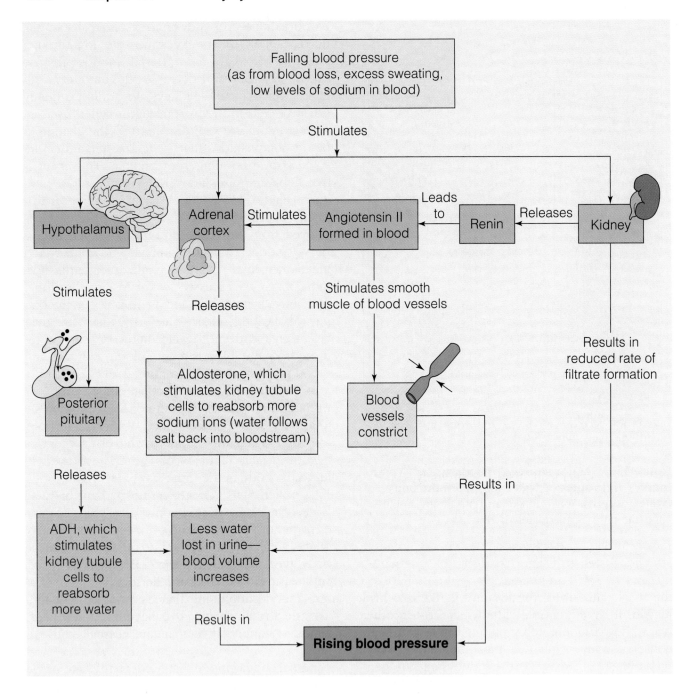

**Figure 15.7   Factors helping to maintain normal blood volume and
blood pressure by regulating sodium and water reabsorption by the
kidneys.**

in the proximal convoluted tubules of the kidneys.
When aldosterone concentrations are high, most of
the remaining Na$^+$ (actually sodium chloride because chloride ions follow) is actively reabsorbed
in the distal convoluted tubules and the collecting
ducts. Generally speaking, for each sodium ion re-

absorbed, a potassium ion is secreted into the filtrate. Thus, as the sodium content of the blood increases, potassium concentration decreases, bringing these two ions back to their normal balance in
the blood. Still another effect of aldosterone is to
increase water reabsorption by the tubule cells, be-

cause as sodium is reclaimed, water follows it passively back into the blood. A little rule to keep in mind here is: *Water follows salt.*

Recall that aldosterone is produced by the adrenal cortex. Although rising potassium levels or falling sodium levels in the ECF directly stimulate the adrenal cells to release aldosterone, the most important trigger for aldosterone release is the **renin-angiotensin mechanism** (see Figure 15.7) mediated by the *juxtaglomerular (JG) apparatus* of the renal tubules. The juxtaglomerular apparatus (see Figure 15.3b) consists of a complex of modified smooth muscle cells in the afferent arteriole plus some modified epithelial cells forming part of the distal convoluted tubule. The naming of this cell cluster reflects its location close by *(juxta)* the glomerulus. When the cells of the JG apparatus are stimulated by low blood pressure in the afferent arteriole or changes in solute content of the filtrate, they respond by releasing the enzyme **renin** into the blood. Renin catalyzes the series of reactions that produce angiotensin II, which in turn acts directly on the blood vessels to cause vasoconstriction and on the adrenal cortical cells to promote aldosterone release. As a result, blood volume and blood pressure increase (see Figure 15.7). The renin-angiotensin mechanism is extremely important for regulating blood pressure.

People with Addison's disease (hypoaldosteronism) have *polyuria* (excrete large volumes of urine) and lose tremendous amounts of salt and water to urine. As long as adequate amounts of salt and fluids are ingested, people with this condition can avoid problems, but they are constantly teetering on the brink of dehydration. ❑

Maintaining Acid-Base Balance of Blood

For the cells of the body to function properly, blood pH must be maintained between 7.35 and 7.45, a very narrow range. Whenever the pH of arterial blood rises above 7.45, a person is said to have **alkalosis** (al″kah-lo′sis). A drop in arterial pH to below 7.35 results in **acidosis** (as″ĭ-do′sis). Because a pH of 7.0 is neutral, 7.35 is not acidic, chemically speaking; however, it represents a higher-than-optimal H^+ concentration for the functioning of most body cells. Therefore, any arterial pH between 7.35 and 7.0 is called **physiological acidosis.**

Although small amounts of acidic substances enter the body via ingested foods, most hydrogen ions originate as by-products of cellular metabolism, which continuously adds substances to the blood that tend to disturb its **acid-base balance.** Many different acids are produced (for example, phosphoric acid, lactic acid, and many types of fatty acids). In addition, carbon dioxide, which is released during energy production, forms carbonic acid. Ammonia and other basic substances are also released to the blood as cells go about their usual "business." Although blood buffers can temporarily "tie up" excess acids and bases, and the lungs have the chief responsibility for eliminating carbon dioxide from the body, the kidneys assume most of the load for maintaining acid-base balance of the blood. Before describing how the kidneys function in acid-base balance, let's take a look at how each of our other two pH-controlling systems, blood buffers and the respiratory system, works.

Blood Buffers   Chemical buffers are systems of one or two molecules that act to prevent dramatic changes in hydrogen ion (H^+) concentration when a strong acid or strong base is added. They do this by binding to hydrogen ions whenever the pH drops and by releasing hydrogen ions when the pH rises. Since the chemical buffers act within a fraction of a second, they are the first line of defense in resisting pH changes.

To better understand how a chemical buffer system works, let's review the definitions of strong and weak acids and bases. Recall that acids are proton (H^+) donors, and that the acidity of a solution reflects only the *free* hydrogen ions, not those still bound to anions. *Strong acids* dissociate completely and liberate all their H^+ in water. Consequently they can cause large changes in pH. By contrast, *weak acids* like carbonic acid dissociate only partially and so have a much slighter effect on a solution's pH (Figure 15.8). However, weak acids are very effective at preventing pH changes since they are forced to dissociate and release more H^+ when the pH rises over the desirable pH range. This feature allows them to play a very important role in the chemical buffer systems.

Also recall that bases are proton or hydrogen ion acceptors. *Strong bases* like hydroxides dissociate easily in water and quickly tie up H^+, but *weak bases* like bicarbonate ion (HCO_3^-) and ammonia (NH_3) are slower to accept H^+. However, as pH

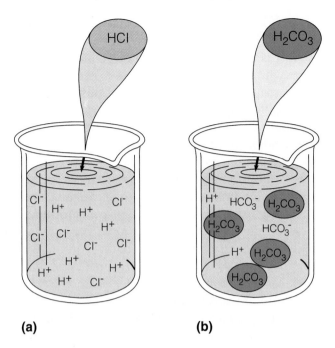

(a) **(b)**

Figure 15.8 Comparison of dissociation of strong and weak acids. (a) When HCl, a strong acid, is added to water it dissociates completely into its ions (H^+ and Cl^-). **(b)** By contrast, dissociation of H_2CO_3, a weak acid, is very incomplete, and some molecules of H_2CO_3 remain undissociated (symbols shown in green circles) in solution.

drops, the weak bases become "stronger" and begin to tie up more hydrogen ions. Thus, like weak acids, they are valuable members of the chemical buffer systems.

The three major chemical buffer systems of the body are the *bicarbonate, phosphate,* and *protein buffer systems,* each of which helps to maintain the pH in one or more of the fluid compartments. They all work together and anything that causes a shift in H^+ concentration in one compartment also causes changes in the others. Thus, drifts in pH are resisted by the entire buffering system. Since all three systems operate in a similar way, examining just one, the bicarbonate buffer system which is so important in preventing changes in blood pH, should be sufficient.

The **bicarbonate buffer system** is a mixture of *carbonic acid* (H_2CO_3) and its salt, *sodium bicarbonate* ($NaHCO_3$). Since carbonic acid is a weak acid, it does not dissociate much in neutral or acidic solutions. Thus, when a strong acid, such as hydrochloric acid (HCl) is added, most of the carbonic acid remains intact. However, the *bicar-*

bonate ions (HCO_3^-) of the salt act as bases to tie up the H^+ released by the stronger acid, forming more carbonic acid:

$$HCl + NaHCO_3 \longrightarrow H_2CO_3 + NaCl$$

strong acid   weak base   weak acid   salt

Because the strong acid is (effectively) changed to a weak one, it lowers the pH of the solution only very slightly.

Similarly, if a strong base like sodium hydroxide (NaOH) is added to a solution containing the bicarbonate buffer system, $NaHCO_3$ will not dissociate further under such alkaline conditions. However, carbonic acid will be forced to dissociate further by the presence of the strong base—liberating more H^+ to bind with the OH^- released by NaOH.

$$NaOH + H_2CO_3 \longrightarrow NaHCO_3 + H_2O$$

strong base   weak acid   weak base   water

The net result is replacement of a strong base by a weak one, so that the pH of the solution rises very little.

Respiratory System Controls As described in Chapter 13, the respiratory system eliminates carbon dioxide from the blood while it "loads" oxygen into the blood. Remember that when carbon dioxide (CO_2) enters the blood from the tissue cells, it is converted to bicarbonate ion (HCO_3^-) for transport in the plasma as shown by the equation:

$$CO_2 + H_2O \underset{}{\overset{\text{Carbonic anhydrase}}{\rightleftharpoons}} H_2CO_3 \rightleftharpoons H^+ + HCO_3^-$$

carbon dioxide   water   carbonic acid   hydrogen ion   bicarbonate ion

The double-headed arrows reveal that an increase in carbon dioxide pushes the reaction to the right, producing more carbonic acid. Likewise, an increase in hydrogen ions pushes the equation to the left, producing more carbonic acid. In healthy people, carbon dioxide is expelled from the lungs at the same rate as it is formed in the tissues. Thus the H^+ released when carbon dioxide is loaded into the blood is not allowed to accumulate because it is tied up in water when CO_2 is unloaded in the lungs. Thus, under normal conditions, the hydrogen ions produced by carbon dioxide transport have essentially no effect on blood pH. However, when CO_2 accumulates in the blood (for example, during restricted breathing) or more H^+ is released to the blood by metabolic processes, the chemoreceptors in the respiratory control centers of the brain (or in peripheral blood vessels) are activated. As a result, breathing rate and depth increase and the excess H^+ is "blown off" as more carbon dioxide is removed from the blood.

On the other hand, when blood pH begins to rise (alkalosis), the respiratory center is depressed. The respiratory rate and depth fall, allowing carbon dioxide (hence, H^+) to accumulate in the blood. Again blood pH is restored to the normal range. Generally, these respiratory system corrections of blood pH (via regulation of CO_2 content of the blood) are accomplished within a minute or so.

Renal Mechanisms Chemical buffers can tie up excess acids or bases temporarily, but they cannot eliminate them from the body. And while the lungs can dispose of carbonic acid by eliminating carbon dioxide, only the kidneys can rid the body of other acids generated during metabolism. Additionally, only the kidneys have the power to regulate blood levels of alkaline substances. Thus, although the kidneys act slowly and require hours or days to bring about changes in blood pH, they are the most potent of the mechanisms for regulating blood pH.

The most important means by which the kidneys maintain acid-base balance of the blood are by (1) excreting bicarbonate ions and (2) conserving (reabsorbing) or generating new bicarbonate ions. Look back again at the equation showing how the carbonic acid–bicarbonate buffer system operates. Notice that losing a HCO_3^- from the body has the same effect as gaining a H^+ because it pushes the equation to the right (i.e., it leaves a free hydrogen ion). By the same token, reabsorbing or generating new HCO_3^- is the same as losing H^+ because it tends to combine with a hydrogen ion and pushes the equation to the left. Renal mechanisms undertake these adjustments: As blood pH rises, bicarbonate ions are excreted and hydrogen ions are retained by the tubule cells. Conversely, when blood pH falls, bicarbonate is reabsorbed and hydrogen ions are secreted. Urine pH varies from 4.5 to 8.0, which reflects the ability of the renal tubules to excrete basic or acid ions to maintain blood pH homeostasis.

Characteristics of Urine

Freshly voided urine is generally clear and pale to deep yellow. The normal yellow color is due to *urochrome* (u'ro-krōm), a pigment that results from the body's destruction of hemoglobin. The more solutes there are in the urine, the deeper yellow its color; on the other hand, dilute urine is a pale, straw color. At times, urine may have an abnormal color. This might be a result of eating certain foods (beets, for example) or the presence of bile or blood in the urine.

When formed, urine is sterile and its odor is slightly aromatic. If it is allowed to stand, it takes on an ammonia odor caused by bacterial action on the urine solutes. Some drugs, vegetables (such as asparagus), and various diseases (such as diabetes mellitus) alter the usual odor of urine.

Urine pH is usually slightly acid (around 6), but changes in body metabolism and certain types of food may cause it to be much more acidic or basic. For example, a diet that contains large amounts of protein (eggs and cheese) and whole-wheat products causes urine to become quite acid; thus, such foods are called *acid-ash foods*. Conversely, a vegetarian diet is called an *alkaline-ash diet*, because it causes urine to become quite basic as the kidneys excrete the excess bases. Bacterial infection of the urinary tract also may cause the urine to be quite alkaline.

Since urine is water plus solutes, urine weighs more, or is more dense, than distilled water. The term used to compare how *much* heavier urine is than distilled water is **specific gravity.** Whereas the specific gravity of pure water is 1.0, the specific gravity of urine usually ranges from 1.001 to 1.035 (dilute to concentrated urine, respectively). Urine is generally dilute (that is, it has a low specific gravity) when a person drinks excessive fluids, uses diuretics (drugs that increase urine output), or has chronic renal failure (a condition in which the kidney loses its ability to concentrate urine). Conditions that produce urine with a high specific gravity include inadequate fluid intake, fever, and a kidney inflammation called *pyelonephritis* (pi''ĕ-lo-nĕ-fri'tis).

A CLOSER LOOK

RENAL FAILURE AND THE ARTIFICIAL KIDNEY

Like master chemists, the kidneys continually maintain the purity of our internal environment. Without their continual efforts, body fluids quickly become contaminated with nitrogen-containing wastes, blood pH tumbles into the acidic range, and *uremia* sets in, totally disrupting life processes. Signs and symptoms of uncontrolled uremia include diarrhea, vomiting, labored breathing, irregular heartbeat, convulsions, coma, and finally death.

Although not common, renal failure may occur when the number of functioning nephrons becomes too low to carry out the normal kidney functions. Possible causes of renal failure include:

■ Repeated damaging infections of the kidneys

■ Physical trauma to the kidneys (crush injury and others)

■ Chemical poisoning of the tubule cells by heavy metals (mercury or lead) or organic solvents (dry-cleaning fluids, paint thinner)

■ Inadequate blood delivery to the tubule cells (as sometimes happens with arteriosclerosis)

In renal failure, filtrate formation decreases or stops completely. Because toxic wastes accumulate quickly in the blood when the tubule cells are not working, *dialysis* by an artificial kidney must be performed to cleanse the blood while the kidneys are shut down. In hemodialysis, which uses an "artificial kidney" apparatus (see illustration), the patient's blood is passed through a membrane tubing that is permeable only to selected substances, and the tubing is immersed in a bathing solution that differs slightly from normal "cleansed" plasma. As blood circulates through the tubing, substances such as nitrogenous wastes and K^+ present in the blood (but not in the bath) diffuse out of the blood through the tubing into the surrounding solution, and substances to be added to the blood (mainly buffers for H^+) move from the bathing solution into the blood. In this way, needed substances are retained in the blood or added to it, while

Blood pump

Arterial blood line (to apparatus)

Dialyzing (bathing) solution

Membrane tubing containing patient's blood

Bubble trap

Venous blood line (from apparatus)

Compressed air — Fresh dialyzing solution — Constant-temperature bath

Used dialyzing solution

wastes and ion excesses are removed. Hemodialysis is routinely done three times weekly and each session takes 4 to 8 hours.

A less efficient but more convenient procedure for patients who are not hospitalized is *continuous ambulatory peritoneal dialysis (CAPD)*. CAPD uses the patient's own peritoneal membrane as the dialyzing membrane. Fluid that is equal in chemical content to normal plasma and interstitial fluid is introduced into the patient's peritoneal cavity with a catheter and left to equilibrate there for 15 to 60 minutes. Then the dialysis fluid is retrieved from the peritoneal cavity and replaced with fresh dialysis fluid. The procedure is repeated until the patient's blood chemistry reaches normal. Because some ambulatory patients may be inattentive to cloudy or bloody dialysis fluid, infection is more common in CAPD than in hemodialysis.

When renal damage is nonreversible, as in chronic, slowly progressing renal failure, the kidneys become totally unable to process plasma or concentrate urine, and a kidney transplant is the only answer. Unhappily, the signs and symptoms of this life-threatening problem become obvious only after about 75 percent of renal function has been lost. The end-stage of renal failure, uremia, occurs when about 90 percent of the nephrons have been lost.

Solutes normally found in urine include sodium and potassium ions, urea, uric acid, creatinine, ammonia, bicarbonate ions, and various other ions depending on blood composition. With certain diseases, urine composition can change dramatically, and the presence of abnormal substances in urine is often helpful in diagnosing the problem. This is why a routine urinalysis should always be part of any good physical examination.

Substances *not* normally found in urine are glucose, blood proteins (primarily albumin), red blood cells, hemoglobin, white blood cells (pus), and bile. Names and possible causes of conditions in which abnormal urinary constituents and volumes might be seen are given in Table 15.1.

Table 15.1 Abnormal Urinary Constituents

Substance	Name of Condition	Possible Causes
Glucose	Glycosuria (gli″ko-su′re-ah)	Nonpathologic: Excessive intake of sugary foods Pathologic: Diabetes mellitus
Proteins	Proteinuria (pro″te-ĭ-nu′re-ah) (also called albuminuria)	Nonpathologic: Physical exertion, pregnancy Pathologic: Glomerulonephritis, hypertension
Pus (WBCs and bacteria)	Pyuria (pi-u′re-ah)	Urinary tract infection
RBCs	Hematuria (he″mah-tu′re-ah)	Bleeding in the urinary tract (due to trauma, kidney stones, infection)
Hemoglobin	Hemoglobinuria (he″mo-glo-bĭ-nu′re-ah)	Various: Transfusion reaction, hemolytic anemia
Bile pigment	Bilirubinuria (bil″ĭ-roo-bĭ-nu′re-ah)	Liver disease (hepatitis)

URETERS, URINARY BLADDER, AND URETHRA

Ureters

The **ureters** (u-re′terz) are slender tubes each 25 to 30 cm (10 to 12 inches) long and 6 mm (¼ inch) in diameter. Each ureter runs behind the peritoneum from the hilus of a kidney to the posterior aspect of the bladder, which it enters at a slight angle (see Figure 15.1). The superior end of each ureter is continuous with the pelvis of the kidney, and its mucosa lining is continuous with that lining the renal pelvis and the bladder below.

Essentially, the ureters are passageways that carry urine from the kidneys to the bladder. Although it might appear that urine could simply drain to the bladder below by gravity, the ureters *do* play an active role in urine transport. Smooth muscle layers in their walls contract to propel urine into the bladder by peristalsis. Once urine has entered the bladder, it is prevented from flowing back into the ureters by small valvelike folds of bladder mucosa that flap over the ureter openings.

When urine becomes extremely concentrated, solutes such as uric acid salts form crystals that precipitate in the renal pelvis. These crystals are called **renal calculi** (kal′kyoo-li; *calculus* = little stone), or kidney stones. Excruciating pain that radiates to the flank occurs when the ureter walls close in on the sharp calculi as they are being eased through the ureter by peristalsis, or when the calculi become wedged in a ureter. Frequent bacterial infections of the urinary tract, urinary retention, and alkaline urine all favor calculi formation. Surgery has been the treatment of choice, but a newer noninvasive procedure *(lithotripsy)* that uses ultrasound waves to shatter the calculi is becoming more popular. The pulverized, sandlike remnants of the calculi are painlessly eliminated in the urine. ◻

Urinary Bladder

The **urinary bladder** is a smooth, collapsible, muscular sac that stores urine temporarily. It is located retroperitoneally in the pelvis just posterior to the pubic symphysis. If the interior of the bladder is scanned, three openings are seen—the two ureter openings and the single opening of the *urethra* (u-re′thrah), which drains the bladder (Figure 15.9). The smooth triangular region of the bladder base outlined by these three openings is called the **trigone** (tri′gon). The trigone is important clinically because infections tend to persist in this region. In males, the *prostate gland* (part of the male reproductive system) surrounds the neck of the bladder where it empties into the urethra.

The bladder wall contains three layers of smooth muscle, collectively called the *detrusor muscle,* and its mucosa is a special type of epithelium, *transitional epithelium* (see p. 78). Both of these structural features make the bladder uniquely suited for its function of urine storage. When the bladder is empty, it is collapsed, 5 to 7.5 cm (2 to 3 inches) long at most, and its walls are thick and thrown into folds. As urine accumulates, the bladder expands and rises in the abdominal cavity. Its muscular wall stretches and the transitional epithelial layer thins, allowing the bladder to store more urine without substantially increasing its internal pressure. A moderately full bladder is about 12.5 cm (5 inches) long and holds about 500 ml (1 pint) of urine, but it is capable of holding more than twice that amount. When the bladder is really distended or stretched by urine, it becomes firm and pear-shaped and may be felt just above the pubic symphysis. Although urine is formed continuously by the kidneys, it is usually stored in the bladder until its release is convenient.

Urethra

The **urethra** is a thin-walled tube that carries urine by peristalsis from the bladder to the outside of the body. At the bladder-urethra junction, a thickening of the smooth muscle forms the **internal urethral sphincter,** an involuntary sphincter that keeps the urethra closed when urine is not being passed. A second sphincter, the **external urethral sphincter,** is fashioned by skeletal muscle as the urethra passes through the pelvic floor. This sphincter is voluntarily controlled.

The length and relative function of the urethra differ in the two sexes. In females, it is about 3 to 4 cm (1½ inches) long and its external orifice, or opening, lies anteriorly to the vaginal opening. Its function is to conduct urine to the body exterior.

Since the female urinary orifice is so close to the anal opening and feces contain a good

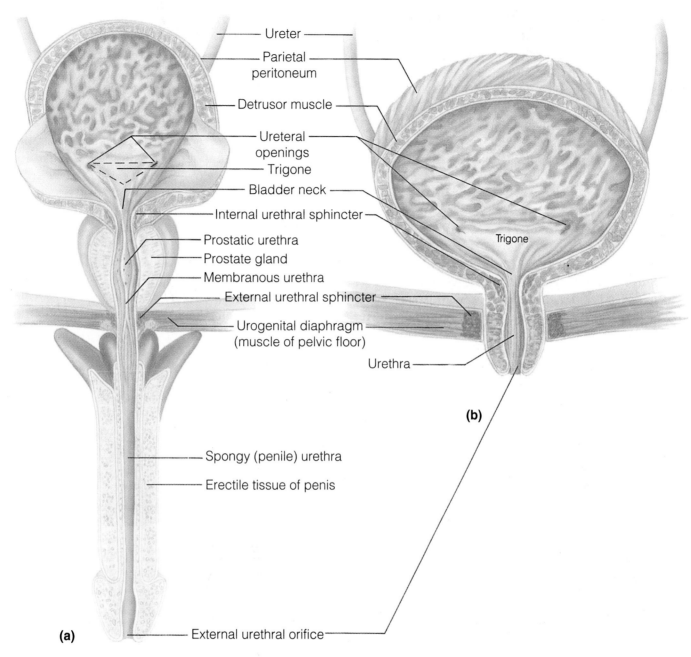

Figure 15.9   Structure of the urinary bladder and urethra. The anterior wall of the bladder has been reflected or omitted to reveal the position of the trigone. (**a**) The bladder and urethra of the male. The urethra of the male is substantially longer than that of the female and has three regions: prostatic, membranous, and penile. (**b**) The bladder and urethra of the female.

deal of bacteria, improper toileting habits (that is, wiping from back to front rather than from front to back) can easily carry bacteria into the urethra. Moreover, since the mucosa of the urethra is continuous with that of the rest of the urinary tract organs, an inflammation of the urethra, or *urethritis* (u″re-thri′tis), can easily ascend the tract to cause

bladder inflammation *(cystitis)* or even kidney inflammation *(pyelitis* or *pyelonephritis)*. Symptoms of urinary tract infection include *dysuria* (painful urination), urinary *urgency* and *frequency*, fever, and sometimes cloudy or blood-tinged urine. When the kidneys are involved, back pain and a severe headache often occur. ❏

In males, the urethra is approximately 20 cm (8 inches) long and has three named regions, the *prostatic, membranous,* and *spongy,* or *penile, urethrae.* It opens at the tip of the penis after traveling down its length. The urethra of the male has a double function. It carries urine out of the body, and it provides the passageway through which sperm is ejected from the body. Thus, in males, the urethra is part of both the urinary and reproductive systems.

Micturition

Micturition (mik″tu-rish′un), or **voiding,** is the act of emptying the bladder. Two sphincters or valves, the internal urethral sphincter (more superiorly located) and the external urethral sphincter (more inferiorly located) control the flow of urine from the bladder. Ordinarily, the bladder continues to collect urine until about 200 ml have accumulated. At about this point, stretching of the bladder wall activates stretch receptors. Impulses transmitted to the sacral region of the spinal cord, and then back to the bladder via the *pelvic splanchnic nerves,* cause the bladder to go into reflex contractions. As the contractions become stronger, stored urine is forced past the internal sphincter (the smooth muscle, involuntary sphincter) into the upper part of the urethra. It is then that a person feels the urge to void. Because the lower external sphincter is skeletal muscle and is voluntarily controlled, we can choose to keep it closed and postpone bladder emptying temporarily. On the other hand, if it is convenient, the external sphincter can be relaxed so that urine is flushed from the body. When one chooses not to void, the reflex contractions of the bladder will stop within a minute or so, and urine will continue to accumulate in the bladder. After 200–300 ml more have been collected, the micturition reflex occurs again. Eventually, micturition occurs whether one wills it or not.

Incontinence (in-kon′tǐ-nens) occurs when we are unable to voluntarily control the external sphincter. Incontinence is normal in children 2 years old or younger, because they have not yet gained control over their voluntary sphincter. It may also occur in older children who sleep so soundly they are not awakened by the stimulus. However, after the toddler years, incontinence is usually a result of emotional problems, pressure (as in pregnancy), or nervous system problems (stroke or spinal cord injury).

Urinary retention is essentially the opposite of incontinence. It is a condition in which the bladder is unable to expel its contained urine. There are various causes for urinary retention, but it often occurs after surgery in which general anesthesia has been given. It appears that it takes a little time for the smooth muscles to regain their activity. Another cause of urinary retention, occurring primarily in elderly men, is enlargement, or *hypertrophy,* of the prostate gland, which surrounds the neck of the bladder. As it enlarges, it narrows the urethra, making it very difficult to void. When urinary retention is prolonged, a slender rubber drainage tube called a *catheter* (kath′ĭ-ter) must be inserted through the urethra to drain the urine and prevent bladder trauma from excessive stretching. ❑

DEVELOPMENTAL ASPECTS OF THE URINARY SYSTEM

When you trace the development of the kidneys in a young embryo, it almost seems as if they can't "make up their mind" about whether to come or go. The first tubule system forms and then begins to degenerate as a second, lower set appears. The second set, in turn, degenerates as a third set makes its appearance. This third set develops into the functional kidneys, which are excreting urine by the third month of fetal life. It is important to remember that the fetal kidneys do not work nearly as hard as they will after birth, because exchanges with the mother's blood through the placenta allow her system to clear many of the undesirable substances from the fetal blood.

There are many congenital abnormalities of this system. Two of the most common are polycystic kidney and hypospadias.

Polycystic (pol″e-sis′tik) *kidney* is a degenerative condition that appears to run in families. In this disease, one or both kidneys are enlarged and have many blisterlike sacs (cysts) containing urine. These cysts interfere with renal function by obstructing urine drainage. Currently, not too much can be done for this condition except to prevent further kidney damage by avoiding infection. Renal failure is the eventual outcome but kidney transplants improve chances for survival.

SYSTEMS IN SYNC

Homeostatic Relationships Between the Urinary System and Other Body Systems

Endocrine System
- Kidneys dispose of nitrogenous wastes; maintain fluid, electrolyte, and acid-base balance of blood; produce the hormone erythropoietin; renal regulation of Na^+ and water balance essential for blood pressure homeostasis; Na^+, K^+, and Ca^{2+} regulation help maintain normal heart function
- ADH, aldosterone, ANF, and other hormones help regulate renal reabsorption of water and electrolytes

Lymphatic System/Immunity
- Kidneys dispose of nitrogenous wastes; maintain fluid, electrolyte, and acid-base balance of blood
- By returning leaked plasma fluid to cardiovascular system, lymphatic vessels help assure normal systemic blood pressure needed for kidney function; immune cells protect urinary organs from infection, cancer, and other foreign substances

Digestive System
- Kidneys dispose of nitrogenous wastes; maintain fluid, electrolyte, and acid-base balance of blood; also, metabolize vitamin D to the active form needed for calcium absorption
- Digestive organs provide nutrients needed for kidney cell health; liver synthesizes most urea, a nitrogenous waste that must be excreted by the kidneys

Urinary System

Muscular System
- Kidneys dispose of nitrogenous wastes; maintain fluid, electrolyte, and acid-base balance of blood; renal regulation of Na^+, K^+, and Ca^{2+} content in ECF crucial for muscle activity
- Muscles of pelvic diaphragm and external urethral sphincter function in voluntary control of micturition; creatinine is a nitrogenous waste product that must be excreted by the kidneys

Nervous System
- Kidneys dispose of nitrogenous wastes; maintain fluid, electrolyte, and acid-base balance of blood; renal regulation of Na^+, K^+, and Ca^{2+} content in ECF essential for normal neural function
- Neural controls involved in micturition; sympathetic nervous system activity triggers the renin-angiotensin mechanism

Respiratory System
- Kidneys dispose of nitrogenous wastes; maintain fluid, electrolyte, and acid-base balance of blood
- Respiratory system provides oxygen required by kidney cells; disposes of carbon dioxide; cells in the lungs convert angiotensin I to angiotensin II

Cardiovascular System
- Kidneys dispose of nitrogenous wastes; maintain fluid, electrolyte, and acid-base balance of blood; renal regulation of Na^+ and water balance essential for blood pressure homeostasis. Na^+, K^+, and Ca^{2+} regulation help maintain normal heart function
- Systemic arterial blood pressure is the driving force for glomerular filtration; heart secretes atrial natriuretic factor; transports nutrients, oxygen, etc. to urinary organs

Reproductive System
- Kidneys dispose of nitrogenous wastes; maintain fluid, electrolyte, and acid-base balance of blood

Integumentary System
- Kidneys dispose of nitrogenous wastes; maintain fluid, electrolyte, and acid-base balance of blood
- Skin provides external protective barrier; serves as site for water loss (via perspiration); vitamin D synthesis site

Skeletal System
- Kidneys dispose of nitrogenous wastes; maintain fluid, electrolyte, and acid-base balance of blood
- Bones of rib cage provide some protection to kidneys

Hypospadias (hi″po-spa′de-as) is a condition found in male babies only. It occurs when the urethral orifice is located on the ventral surface of the penis. Corrective surgery is generally done when the child is around 12 months old. ❑

Because the bladder is very small and the kidneys are unable to concentrate urine for the first 2 months, a newborn baby voids from 5 to 40 times per day depending on fluid intake. By 2 months, the infant is voiding approximately 400 ml/day, and the amount steadily increases until adolescence, when adult urine output (about 1500 ml/day) is achieved.

Control of the voluntary urethral sphincter goes hand in hand with nervous system development. By 15 months, most toddlers are aware when they have voided, and by 18 months they can hold urine in their bladder for about 2 hours, which is the first sign that toilet training (for voiding) can begin. Daytime control usually occurs well before nighttime control is achieved. It is generally unrealistic to expect that complete nighttime control will occur before the child is 4 years old.

During childhood and through late middle age, most urinary system problems are infectious, or inflammatory, conditions. Many types of bacteria may invade the urinary tract to cause urethritis, cystitis, or pyelonephritis. *Escherichia coli* (esh″er-i′ke-ah ko′li) are normal residents of the digestive tract and generally cause no problems there. But these bacteria act as pathogens (disease-causers) in the sterile environment of the urinary tract. Bacteria and viruses responsible for *sexually transmitted diseases (STDs)*, which are primarily reproductive tract infections, may also invade and cause inflammation in the urinary tract, which leads to the clogging of some of its ducts.

Childhood streptococcal (strep″to-kok′al) infections, such as strep throat and scarlet fever, may cause inflammatory damage to the kidneys if the original infections are not treated promptly and properly. A common sequel to untreated childhood strep infections is *glomerulonephritis* (glo-mer″ŭ-lo-ne-fri′tis), in which the glomerular filters become clogged with antigen-antibody complexes resulting from the strep infection. ❑

As we age, there is a progressive decline in kidney function. By age 70, the rate of filtrate formation is only about half that of the middle-aged adult. This is believed to result from impaired renal circulation due to atherosclerosis, which affects the entire circulatory system of the aging person. In addition to a decrease in the number of functional nephrons, the tubule cells become less efficient in their ability to concentrate urine.

Another consequence of aging is bladder shrinkage and loss of bladder tone, causing many elderly individuals to experience *urgency* (a feeling that it is necessary to void) and *frequency* (frequent voiding of small amounts of urine). *Nocturia* (nok-tu′re-ah), the need to get up during the night to urinate, plagues almost two-thirds of this population. In many, incontinence is the final outcome of the aging process. This loss of control is a tremendous blow to the pride of many aging people. Urinary retention is another common problem; most often it is a result of hypertrophy of the prostate gland in males. Some of the problems of incontinence and retention can be avoided by a regular regimen of activity that keeps the body as a whole in optimum condition and promotes alertness to elimination signals.

Important Terms

acid-base balance

diuresis (di″u-re′sis)

electrolyte balance

filtration

glomerular (Bowman's) capsule

glomerulus (glo-mer′u-lus)

micturition (mik″tu-rish′un)

nephrons (nef′ronz)

nitrogenous wastes

peritubular (per″ĭ-tu′bu-lar) **capillaries**

reabsorption

renal cortex

renal medulla

renal pelvis

renal tubule

secretion

ureters (u-re′terz)

urethra (u-re′thrah)

urinary bladder

water balance

Summary

Kidneys (pp. 450–464)

1. The paired kidneys are retroperitoneal in the superior lumbar region. Each kidney has a medial indentation (hilus) where renal artery, renal vein, and ureter are seen. Each kidney is enclosed in a tough fibrous capsule. A fatty cushion holds the kidneys against the trunk wall.

2. A longitudinal section of a kidney reveals an outer cortex, deeper medulla, and medial pelvis. Extensions of the pelvis (calyces) surround the tips of medullary pyramids and collect urine draining from them.

3. The renal artery, which enters the kidney, breaks up into segmental, lobar, and then interlobar arteries that travel outward through medulla. Interlobar arteries split into arcuate arteries that branch to produce interlobular arteries, which serve the cortex.

4. Nephrons are structural and functional units of the kidneys. Each consists of a glomerulus and a renal tubule. Subdivisions of the renal tubule (from the glomerulus) are glomerular capsule, proximal convoluted tubule, loop of Henle, and distal convoluted tubule. A second (peritubular) capillary bed is also associated with each nephron.

5. Nephron functions include filtration, reabsorption, and secretion. Filtrate formation is the role of the high-pressure glomerulus. Filtrate is essentially plasma without blood proteins. In reabsorption, done by tubule cells, needed substances are removed from filtrate (amino acids, glucose, water, some ions) and returned to blood. The tubule cells also secrete additional substances into filtrate. Secretion is important to rid the body of drugs and excess ions and to maintain acid-base balance of blood.

6. Blood composition depends on diet, cellular metabolism, and urinary output. To maintain blood composition, the kidneys must:
   a. Allow nitrogen-containing wastes (urea, ammonia, creatinine, uric acid) to go out in the urine.

   b. Maintain water and electrolyte balance by absorbing more or less water and reclaiming ions in response to hormonal signals. ADH increases water reabsorption and conserves body water. Aldosterone increases reabsorption of sodium and water, and decreases potassium reabsorption

   c. Maintain acid-base balance by actively secreting bicarbonate ions (and retaining H^+) and by absorbing bicarbonate ions (and secreting H^+). Chemical buffers tye up excess H^+ or bases, temporarily; respiratory centers modify blood pH by retaining CO_2 (decreases pH) or by eliminating more CO_2 from the blood (increases blood pH). Only the kidney can remove metabolic acids and excess bases from the body.

7. Urine is clear, yellow, and usually slightly acid, but its pH value varies widely. Substances normally found in urine are nitrogenous wastes, water, various ions (always sodium and potassium). Substances normally absent from urine include glucose, blood proteins, blood, pus (WBCs), bile.

Ureters, Urinary Bladder, and Urethra (pp. 464–466)

1. The ureters are slender tubes running from each kidney to the bladder. They conduct urine by peristalsis from kidney to bladder.

2. The bladder is a muscular sac posterior to the pubic symphysis. It has two inlets (ureters) and one outlet (urethra). In males, the prostate gland surrounds its outlet. The function of the bladder is to store urine.

3. The urethra is a tube that leads urine from the bladder to body exterior. In females, it is 3–4 cm long and conducts only urine. In males, it is 20 cm long and conducts both urine and sperm. Internal sphincter of smooth muscle is at the bladder-urethra junction. External sphincter, of skeletal muscle, is located more inferiorly.

4. Micturition is emptying of the bladder. The micturition reflex causes the involuntary internal sphincter to open when stretch receptors in the bladder wall

are stimulated. Since the external sphincter is voluntarily controlled, micturition can ordinarily be temporarily delayed. Incontinence is inability to control micturition.

Developmental Aspects of the Urinary System
(pp. 466–468)

1. The kidneys begin to develop in the first few weeks of embryonic life and are excreting urine by the third month.

2. Common congenital abnormalities include polycystic kidney and hypospadias.

3. Common urinary system problems in children and young to middle-aged adults are infections caused by fecal microorganisms, sexually transmitted disease-causing microorganisms, and streptococcus.

4. Renal failure is an uncommon, but extremely serious, problem in which kidneys are unable to concentrate urine, and dialysis must be done to maintain chemical homeostasis of blood.

5. With age, filtration rate decreases and tubule cells become less efficient at concentrating urine, leading to urgency, frequency, and incontinence. Urinary retention is another common problem.

Review Questions

1. Name the organs of the urinary system and describe the general function of each organ.

2. Describe the location of the kidneys in the body.

3. Make a diagram of a longitudinal section of a kidney. Identify and label the cortex, medulla, medullary pyramids, renal columns, and pelvis.

4. Name the structural and functional unit of the kidney.

5. Trace the pathway a uric acid molecule takes from a glomerulus to the urethra. Name every gross or microscopic structure it passes through on its journey.

6. What is the function of the glomerulus? What two functions do the renal tubules perform?

7. Besides ridding the body of wastes formed during cell metabolism, the kidney adjusts blood chemistry in other ways. What are these three other ways?

8. Explain the difference between filtrate and urine.

9. How does aldosterone modify the chemical composition of urine?

10. What hormone's name means "against urine flow"? What condition happens if it is not secreted?

11. Name three substances normally found in blood that are not normally found in urine. Name the condition when each named substance *is* found in urine.

12. Why is urinalysis a routine part of any good physical examination?

13. Define *micturition* and then describe the micturition reflex.

14. What sometimes happens when urine becomes too concentrated or remains too long in the bladder?

15. Define *incontinence*.

16. How is the female urethra different from that of the male in structure and function?

17. Why is cystitis more common in females?

18. What type of problem most commonly affects the urinary system organs?

19. Describe the changes that occur in kidney and bladder function in old age.

At the Clinic

1. A 55-year-old woman is awakened by an excruciating pain that radiates from her right abdomen to her flank on the same side. The pain is not continuous, but recurs every 3 to 4 minutes. Diagnose this patient's problem and cite factors that might favor its occurrence. Explain why this woman's pain comes in "waves."

2. Mitchell's parents bring him to the clinic because his "wee-wee" (urine) is tinged with blood. Two days before he was diagnosed with strep throat. His face and hands are swollen. What is the probable cause of Mitchell's current kidney problem?

3. A young woman has come to the clinic with dysuria and frequent urination. What is the most likely diagnosis?

4. A patient exhibits excessive thirst and polyuria, with excessive blood levels of sodium. Will you check the pituitary or the adrenal glands? Do you suspect hyposecretion or hypersecretion, and of what hormone? What is the name of this condition?

5. BJ has been diagnosed with a hypersecretory tumor of a parathyroid gland. What effect will this have on his renal tubule function?

16
The Reproductive System

YOUR GOALS

After completing this chapter, you will have a working knowledge of the functions of the reproductive system and will have mastered the objectives below.

FUNCTION PREVIEW

■ The reproductive system ensures continuity of the species by producing offspring.

OBJECTIVE CHECKLIST

ANATOMY OF THE MALE REPRODUCTIVE SYSTEM (pp. 473–476)

❑ Discuss the common purpose of the reproductive system organs.

❑ When provided with a model or diagram, identify the organs of the male reproductive system and discuss the general function of each.

❑ Name the endocrine and exocrine products of the testes.

❑ Discuss the composition of semen and name the glands that produce it.

❑ Trace the pathway followed by a sperm from the testis to the body exterior.

❑ Define *erection, ejaculation,* and *circumcision.*

MALE REPRODUCTIVE FUNCTIONS (pp. 476–478)

❑ Define *meiosis* and *spermatogenesis.*

❑ Describe the structure of a sperm and relate its structure to its function.

❑ Describe the effect of FSH and LH on testis functioning.

ANATOMY OF THE FEMALE REPRODUCTIVE SYSTEM (pp. 478–482)

❑ When provided with an appropriate model or diagram, identify the organs of the female reproductive system and discuss the general function of each.

❑ Describe the functions of the vesicular follicle and corpus luteum of the ovary.

❑ Define *endometrium, myometrium,* and *ovulation.*

❑ Explain the location of the following regions of the female uterus: cervix, fundus, body.

(continued)

Most organ systems of the body function almost continuously to maintain the well-being of the individual. The **reproductive system,** however, appears to "slumber" until puberty. The **primary sex organs,** or **gonads** (go′nadz; "seeds"), are the *testes* in males and the *ovaries* in females. The gonads produce sex cells, or **gametes** (gam′ēts; "spouses"), and secrete sex hormones. The remaining reproductive system structures are **accessory reproductive organs.** Although male and female reproductive organs are quite different, their joint purpose is to produce offspring.

The reproductive role of the male is to manufacture male gametes called **sperm** and deliver them to the female reproductive tract. The female, in turn, produces female gametes, called **ova,** or *eggs.* If the time is suitable, the sperm and egg fuse to produce a fertilized egg, which is the first cell of a new individual. Once fertilization has occurred, the female uterus provides a protective environment in which the *embryo,* later called the *fetus,* develops until birth.

The sex hormones play vital roles both in the development and function of the reproductive organs and in sexual behavior and drives. These gonadal hormones also influence the growth and development of many other organs and tissues of the body.

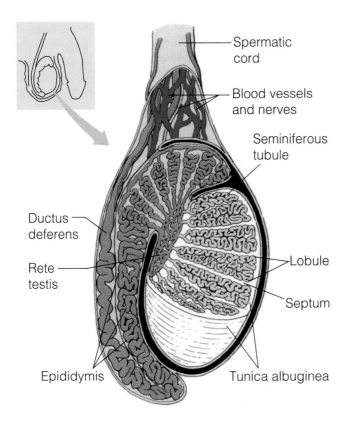

Spermatic cord

Blood vessels and nerves

Seminiferous tubule

Ductus deferens

Rete testis

Lobule

Septum

Epididymis

Tunica albuginea

Figure 16.1 **Sagittal section of the testis and associated epididymis.**

ANATOMY OF THE MALE REPRODUCTIVE SYSTEM

As already noted, the primary reproductive organs of the male are the **testes** (tes'tēz) or *male gonads,* which have both an exocrine (sperm-producing) function and an endocrine (testosterone-producing) function. The accessory reproductive structures are ducts or glands that aid in the delivery of sperm to the body exterior or to the female reproductive tract.

Testes

Each olive-sized testis is approximately 4 cm (1½ inches) long and 2.5 cm (1 inch) wide. A fibrous connective tissue capsule, the *tunica albuginea* (tu'nĭ-kah al"bu-jin'e-ah; "white coat") surrounds each testis. Extensions of this capsule *(septa)*

plunge into the testis and divide it into a large number of lobules. Each lobule contains one to four tightly coiled **seminiferous** (sem"in-if'er-us) **tubules,** the actual "sperm-forming factories" (see Figure 16.1). Seminiferous tubules of each lobe empty sperm into another set of tubules, the *rete* (re'te) *testis,* located at one side of the testis. Sperm travel through the rete testis to enter the first part of the duct system, the *epididymis* (ep"ĭ-did'ĭ-mis), which hugs the external surface of the testis.

Lying in the soft connective tissue surrounding the seminiferous tubules are the **interstitial** (in"ter-stish'al) **cells,** functionally distinct cells that produce androgens—most importantly, *testosterone.* Thus, the sperm-producing and hormone-producing functions of the testes are carried out by completely different cell populations.

Duct System

The accessory organs forming the male duct system, which transports sperm from the body, are the epididymis, ductus deferens, and urethra (see Figure 16.2).

Epididymis

The comma-shaped **epididymis** is a highly coiled tube about 6 m (20 feet) long that caps the superior part of the testis and then runs down its posterolateral side. The epididymis is the first part of the male duct system and provides a temporary storage site for the immature sperm that enter it from the testis. While the sperm make their way along the snaking course of the epididymis (a trip that takes about 20 days), they complete their maturation process, gaining the abiity to swim. When a male is sexually stimulated, the walls of the epididymis contract to expel the sperm into the next part of the duct system, the ductus deferens.

Ductus Deferens

The **ductus deferens** (duk'tus def'er-enz; "carrying away"), or **vas deferens,** runs upward from the epididymis through the inguinal canal into the pelvic cavity and arches over the superior aspect of the bladder. This tube is enclosed, along with blood vessels and nerves, in a connective tissue sheath called the **spermatic cord** (see Figure

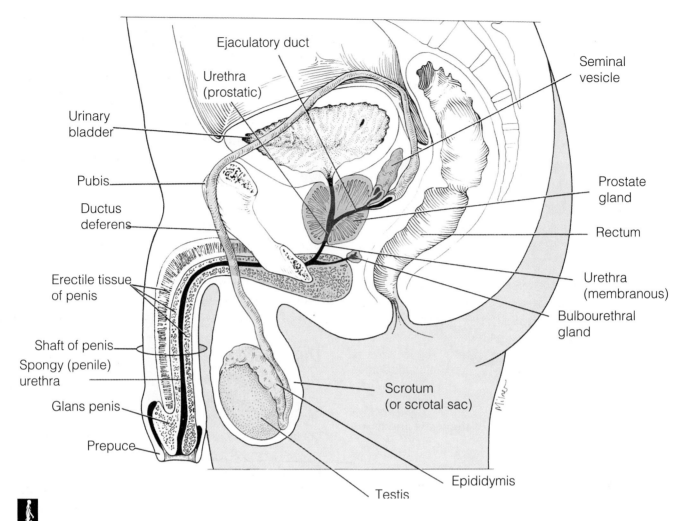

Figure 16.2 **Reproductive organs of the male (sagittal view).**

16.1). The end of the ductus deferens empties into the **ejaculatory** (e-jak′u-lah-to″re) **duct,** which passes through the prostate gland to merge with the urethra. The main function of the ductus deferens is to propel live sperm from their storage sites, the epididymis and distal part of the ductus deferens, into the urethra. At the moment of ejaculation (*ejac* = to shoot forth), the thick layers of smooth muscle in its walls create peristaltic waves that rapidly squeeze the sperm forward.

As Figure 16.2 illustrates, part of the ductus deferens lies in the scrotal sac, which hangs outside the body cavity. Some men voluntarily opt to take full responsibility for birth control by having a **vasectomy** (vah-sek′to-me). In this relatively minor operation, the surgeon makes a small incision into the scrotum and then cuts through or cauterizes the ductus deferens. Sperm are still produced,

but they can no longer reach the body exterior and eventually deteriorate and are reabsorbed. A man is sterile after this procedure, but because testosterone is still produced, the secondary sex characteristics are retained.

Urethra

The **urethra,** which extends from the base of the urinary bladder to the tip of the penis, is the terminal part of the male duct system. It has three named regions: (1) the **prostatic urethra** surrounded by the prostate gland; (2) the **membranous urethra** spanning the distance from the prostatic urethra to the penis; and (3) the **spongy (penile) urethra** running within the length of the penis. As mentioned in Chapter 15, the male urethra carries both urine and sperm to the body exterior; thus, it serves two masters, the urinary and reproductive systems. However, urine and sperm never pass at the same time. When ejaculation occurs and sperm enters the prostatic urethra from the ejaculatory ducts, the bladder sphincter constricts. This event not only prevents the passage of urine into the urethra, but also prevents sperm from entering the urinary bladder.

Accessory Glands and Semen

The accessory glands include the paired seminal vesicles, the single prostate gland, and the bulbourethral (bul-bo-u-re'thral) glands (see Figure 16.2). These glands produce the bulk of *semen* (se'men), the sperm-containing fluid that is propelled out of the male's reproductive tract during **ejaculation.**

Seminal Vesicles

The **seminal** (sem'ĭ-nul) **vesicles,** located at the base of the bladder, produce about 60 percent of the fluid volume of semen. Their thick, yellowish secretion is rich in sugar (fructose), vitamin C, prostaglandins, and other substances, which nourish and activate the sperm passing through the tract. The duct of each seminal vesicle joins that of the ductus deferens on the same side to form the ejaculatory duct (see Figure 16.2). Thus, sperm and seminal fluid enter the urethra together during ejaculation.

Prostate Gland

The **prostate gland** is a single gland about the size and shape of a chestnut (see Figures 16.2 and 15.9, p. 465). It encircles the upper (prostatic) part of the urethra just below the bladder. Prostate gland secretion is a milky fluid that plays a role in activating sperm. During ejaculation, it enters the urethra through several small ducts. Since the prostate is located immediately anterior to the rectum, its size and texture can be palpated (felt) by digital (finger) examination through the anterior rectal wall.

The prostate gland has a reputation as a health destroyer. Hypertrophy of the prostate gland, which affects nearly every elderly male, strangles the urethra. This troublesome condition makes urination difficult and enhances the risk of bladder infections *(cystitis)* and kidney damage. Treatment is usually surgical. Inflammation of the prostate is the single most common reason for a man to consult a urologist, and prostatic cancer is the third most prevalent type of cancer in men. As a rule, prostatic cancer is a slow-growing, hidden condition, but it can also be a swift and deadly killer. ❑

Bulbourethral Glands

The **bulbourethral glands** are tiny, pea-sized glands inferior to the prostate gland. They produce a thick, clear mucus, which drains into the penile urethra. This secretion is the first to pass down the urethra when a man becomes sexually excited. It is believed to cleanse the urethra of traces of acidic urine, and it serves as a lubricant during sexual intercourse.

Semen

Semen is a milky white, somewhat sticky mixture of sperm and accessory gland secretions. The liquid provides a transport medium and nutrients and contains chemicals that protect the sperm and aid their movement. Mature sperm cells are streamlined cellular "missiles" containing little cytoplasm or stored nutrients. The fructose in the seminal vesicle secretion provides essentially all of their energy fuel. The relative alkalinity of semen as a whole (pH 7.2–7.6) helps neutralize the acid environment (pH 3.5–4) of the female's vagina, protecting the delicate sperm and enhancing their motility. Sperm are very sluggish under acidic conditions (below pH 6). Semen also contains a chemical that inhibits bacterial multiplication.

Semen also dilutes sperm; without such dilution, sperm motility is severely impaired. The amount of semen propelled out of the male duct system during ejaculation is relatively small, only 2 to 5 ml (about a teaspoonful), but there are between 50 and 100 million sperm in each milliliter.

Male infertility may be caused by obstructions of the duct system, hormonal imbalances, and many other factors. One of the first series of tests done when a couple has been unable to conceive is *semen analysis.* Factors analyzed include sperm count, motility and morphology (shape and maturity), and semen volume, pH, and fructose content. A sperm count lower than 20 million per milliliter makes impregnation improbable. ❑

External Genitalia

The **external genitalia** (jen"i-tal'e-ah) of the male include the scrotum and the penis (see Figure 16.2). The **scrotum** (skro'tum; "pouch") is a divided sac of skin that hangs outside the abdominal cavity, between the legs and at the root of the penis. Under normal conditions, the scrotum hangs loosely from its attachments, providing the testes with a temperature that is below body temperature. This is a rather exposed location for a man's testes, which contain his entire genetic heritage, but apparently viable sperm cannot be produced at normal body temperature. The scrotum, which provides a temperature about 3°C lower, is necessary for the production of healthy sperm. When the external temperature is very cold, the scrotum becomes heavily wrinkled as it pulls the testes closer to the warmth of the body wall. Thus, changes in scrotal surface area can maintain a temperature that favors viable sperm production.

The **penis** (pe'nis; "tail") is designed to deliver sperm into the female reproductive tract. The skin-covered penis consists of a **shaft,** which ends in an enlarged tip, the **glans penis.** The skin covering the penis is loose and it folds downward to form a cuff of skin, the **prepuce** (pre'pus), or **foreskin,** around the proximal end of the glans. Frequently, the foreskin is surgically removed shortly after birth, by a procedure called *circumcision.*

Internally, the spongy urethra (see Figure 16.2) is surrounded by three elongated areas of *erectile tissue,* a spongy tissue that fills with blood during sexual excitement. This causes the penis to enlarge and become rigid. This event, called **erection,** helps the penis serve as a penetrating organ to deliver the semen into the female's reproductive tract.

MALE REPRODUCTIVE FUNCTIONS

The chief roles of males in the reproductive process are to produce sperm and the hormone testosterone. These processes are described next.

Spermatogenesis

Sperm production, or **spermatogenesis** (sper"mah-to-jen'ĕ-sis), begins during puberty and continues throughout life. Every day a man makes millions of sperm. Since only one sperm fertilizes an egg, it seems that nature has made sure that the human species will not be endangered for lack of sperm.

Sperm formation occurs in the seminiferous tubules of the testis, as noted earlier. As shown in Figure 16.3, the process is begun by primitive stem cells called **spermatogonia** (sper"mah-to-go'ne-ah), found in the outer edge, or periphery, of each tubule. Spermatogonia go through rapid mitotic divisions to build up the stem cell line. Before puberty, all such divisions simply produce more stem cells. During puberty, however, *follicle-stimulating hormone (FSH)* is secreted in increasing amounts by the anterior pituitary gland and, from this time on, each division of a spermatogonium produces one stem cell (a type A daughter cell) and another cell called a type B daughter cell. The type A cell remains at the tubule periphery to maintain the stem cell population. The type B cell gets pushed toward the tubule lumen where it becomes a **primary spermatocyte,** destined to undergo **meiosis** (mi-o'sis) and form four sperm. Meiosis is a special type of nuclear division that differs from mitosis (described in Chapter 3) in two major ways. Meiosis consists of two successive divisions of the nucleus (called meiosis I and II) and results in four (instead of two) daughter cells. In spermatogenesis, the daughter cells are called **spermatids** (sper'mah-tidz). Spermatids have only half as much genetic material as other body cells. In humans, this is 23 chromosomes (or the so-called *n* number of chromosomes) rather than the usual 46 ($2n$). Then, when the sperm and the egg (which also has 23 chromosomes) unite, the normal $2n$ number of 46 chromosomes is reestablished.

As meiosis occurs, the dividing cells (primary and then secondary spermatocytes) are pushed toward the lumen of the tubule. Thus, the progress of meiosis can be followed from the tubule periphery to the lumen. The spermatids, which are the products of meiosis, are *not* functional sperm. They are nonmotile cells and have too much excess baggage to function well in reproduction. They must still undergo further changes in which their excess cytoplasm is stripped away and a tail is formed (see Figure 16.3). In this last stage of

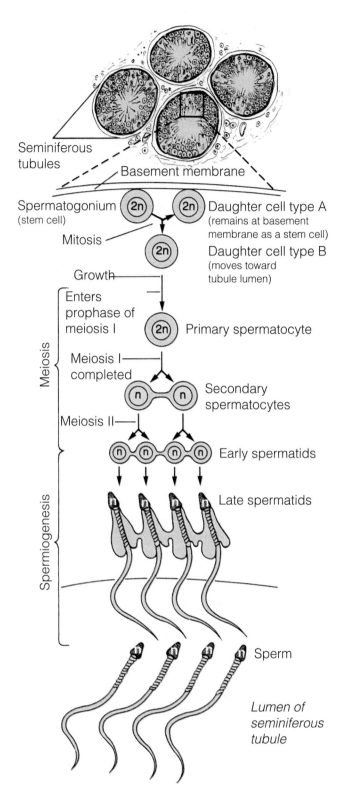

Seminiferous tubules

Basement membrane

Spermatogonium (stem cell) ⟨2n⟩ ⟨2n⟩ Daughter cell type A (remains at basement membrane as a stem cell)

Mitosis

⟨2n⟩ Daughter cell type B (moves toward tubule lumen)

Growth

Meiosis

Enters prophase of meiosis I

⟨2n⟩ Primary spermatocyte

Meiosis I completed

⟨n⟩ ⟨n⟩ Secondary spermatocytes

Meiosis II

⟨n⟩ ⟨n⟩ ⟨n⟩ ⟨n⟩ Early spermatids

Spermiogenesis

Late spermatids

Sperm

Lumen of seminiferous tubule

Figure 16.3 Spermatogenesis. Flowchart showing the relative position of the spermatogenic cells in the wall of the seminiferous tubule. Although the stem cells and the primary spermatocytes have the same number of chromosomes (46, designated as 2n) as other body cells, the products of meiosis (spermatids and sperm) have only half as many (23, designated as n).

sperm development, called **spermiogenesis** (sper″me-o-gen′ĕ-sis), all the excess cytoplasm is sloughed off, and what remains is compacted into the three regions of the mature sperm—the *head, midpiece,* and *tail* (Figure 16.4). The mature sperm is a greatly streamlined cell equipped with a high rate of metabolism and a means of propelling itself, enabling it to move long distances in a short time to get to the egg. It is a prime example of the fit between form and function.

The sperm head contains DNA, the genetic material. Essentially, it *is* the nucleus of the spermatid. Anterior to the nucleus is the helmetlike **acrosome** (ak′ro-sōm), which is similar to a large lysosome. When a sperm comes into close contact with an egg (or more precisely, an *oocyte*), the acrosomal membrane breaks down and releases enzymes that help the sperm penetrate the egg. Filaments, which form the tail, arise from centrioles in the midpiece. Mitochondria, wrapped tightly around these filaments, provide the ATP needed for the whiplike movements of the tail that propel the sperm.

The entire process of spermatogenesis, from the formation of a primary spermatocyte to release of immature sperm in the tubule lumen, takes 64 to 72 days. Sperm in the lumen are unable to "swim" and incapable of fertilizing an egg. They are moved by peristalsis through the tubules of the testes into the epididymis. There they undergo further maturation, which results in increased motility and fertilizing power.

Environmental threats can alter the normal process of sperm formation. For example, some common antibiotics such as penicillin and tetracycline may suppress sperm formation. Radiation, lead, certain pesticides, marijuana, tobacco, and excessive alcohol can cause production of abnormal sperm (two-headed, multiple-tailed, and so on). ❑

Testosterone Production

As noted earlier, the interstitial cells produce **testosterone** (tes-tos′tĕ-rōn), the most important hormonal product of the testes. During puberty, interstitial cells are activated by *luteinizing hormone (LH),* also called *interstitial cell–stimulating hormone (ICSH),* which is released by the anterior pituitary gland. From this time on, testosterone is

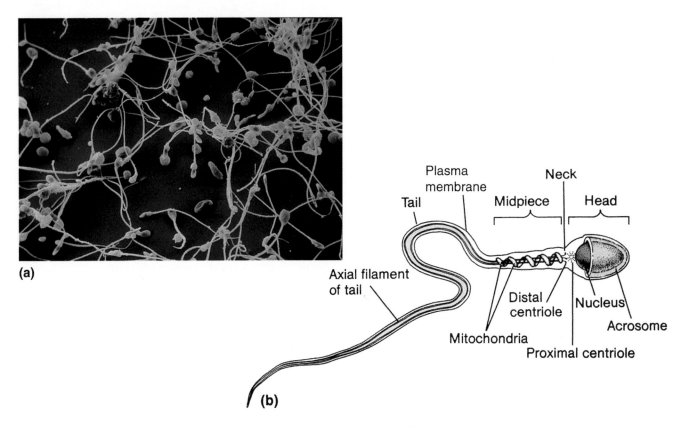

(a)

(b)

Figure 16.4   Structure of a sperm. (a) Scanning electron micrograph of mature sperm (430×). **(b)** Diagrammatic view of a sperm.

produced continuously (more or less) for the rest of a man's life. The rising blood level of testosterone in the young male stimulates his reproductive organs to develop to their adult size, underlies the sex drive, and causes the secondary male sex characteristics to appear. **Secondary sex characteristics** typical of males include:

- Deepening of the voice due to enlargement of the larynx

- Increased hair growth all over the body, and particularly in the axillary and pubic regions and the face (the beard)

- Enlargement of skeletal muscles to produce the heavier muscle mass typical of the male physique

- Increased heaviness of the skeleton due to thickening of the bones

Because testosterone is responsible for the appearance of these typical masculine characteristics, it is often referred to as the "masculinizing" hormone.

If testosterone is not produced, the secondary sex characteristics never appear in the young man, and his other reproductive organs remain childlike. This is *sexual infantilism*. Castration of the adult male (or the inability of his interstitial cells to produce testosterone) results in a decrease in the size and function of his reproductive organs as well as a decrease in his sex drive. Sterility also occurs because testosterone is necessary for the final stages of sperm production. ❏

ANATOMY OF THE FEMALE REPRODUCTIVE SYSTEM

The reproductive role of the female is much more complex than that of the male. Not only must she produce the female gametes (ova), but her body must also nurture and protect a developing fetus during 9 months of pregnancy. **Ovaries** are the primary reproductive organs of a female. Like the

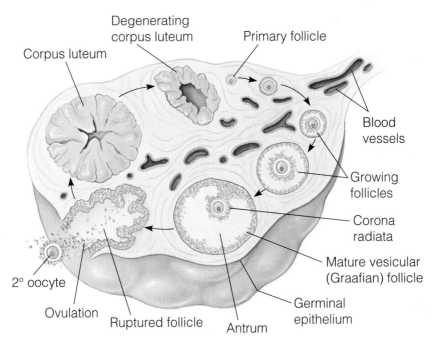

Figure 16.5   Diagrammatic view of a human ovary.

testes of a male, ovaries produce both an exocrine product (the eggs, or *ova*) and endocrine products (estrogens and progesterone). The other organs of the female reproductive system serve as accessory structures to transport, nurture, or otherwise serve the needs of the reproductive cells and/or the developing fetus.

Ovaries

The paired *ovaries* (o′vah-rēz) are pretty much the size and shape of almonds. An internal view of an ovary reveals many tiny saclike structures called **ovarian follicles** (Figure 16.5). Each follicle consists of an immature egg, called an **oocyte** (o′o-sīt), surrounded by one or more layers of very different cells called **follicle cells.** As a developing egg within a follicle begins to ripen or mature, the follicle enlarges and develops a fluid-filled central region called an *antrum*. At this stage, the follicle, called a **vesicular,** or **Graafian** (graf′e-an), **follicle,** is mature and the developing egg is ready to be ejected from the ovary, an event called **ovulation.** After ovulation, the ruptured follicle is transformed into a very different-looking structure called a **corpus luteum** (kor′pus lu′te-um; "yellow body"), which eventually degenerates. Ovula-

tion generally occurs every 28 days, but it can occur more or less frequently in some women. In older women, the surfaces of the ovaries are scarred and pitted, which attests to the fact that many eggs have been released.

The ovaries are secured to the lateral walls of the pelvis by the *suspensory ligaments*. They flank the uterus laterally and anchor to it medially by the *ovarian ligaments* (see Figure 16.6). In between, they are enclosed and held in place by a fold of peritoneum, the *broad ligament*.

Duct System

The uterine tubes, uterus, and vagina form the duct system of the female reproductive tract (Figure 16.6).

Uterine (Fallopian) Tubes

The **uterine** (u′ter-in), or **fallopian** (fal-lo′pe-an), **tubes** form the initial part of the duct system. They receive the ovulated oocyte and provide a site where fertilization can occur. Each of the uterine tubes is about 10 cm (4 inches) long and extends medially from an ovary to empty into the superior region of the uterus. Like the ovaries, the uterine

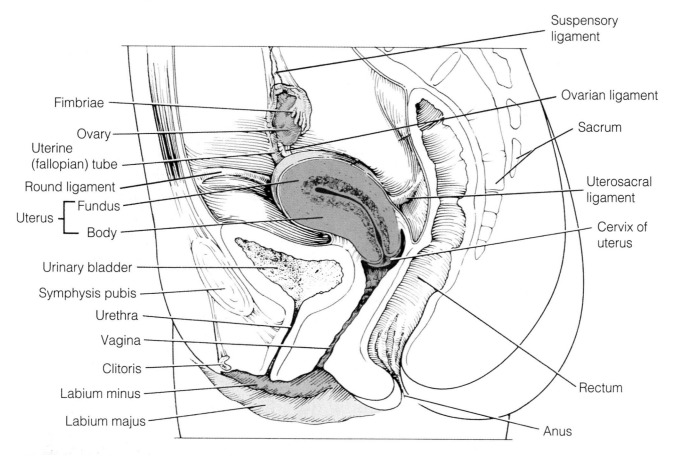

Figure 16.6   Sagittal section of the human female reproductive system. (The plural of labium minus and majus is labia minora and majora, respectively.)

tubes are enclosed and supported by the broad ligament. Unlike the male duct system, which is continuous with the tubule system of the testes, there is little or no actual contact between the uterine tubes and the ovaries. The distal end of each uterine tube is expanded and has fingerlike projections called **fimbriae** (fim′bre-e), which partially surround the ovary. As an oocyte is expelled from an ovary during ovulation, the waving fimbriae create fluid currents that act to carry the oocyte into the uterine tube where it begins its journey toward the uterus. (Obviously, however, many potential eggs are lost in the peritoneal cavity.) The oocyte is carried toward the uterus by a combination of peristalsis and the rhythmic beating of *cilia.* Because the journey to the uterus takes 3 to 4 days and the oocyte is viable for up to 24 hours after ovulation, the usual site of fertilization is the uterine tube. To reach the oocyte, the sperm must swim upward through the vagina and uterus to reach the uterine tubes. This must be a difficult journey. Because

they must swim against the downward current created by the cilia, it is rather like swimming against the tide!

The fact that the uterine tubes are not continuous distally with the ovaries places women at risk for infections spreading into the peritoneal cavity from other parts of the reproductive tract. *Gonorrhea* (gon″o-re′ah) bacteria sometimes infect the peritoneal cavity in this way, causing an extremely severe inflammation called *pelvic inflammatory disease (PID).* Unless treated promptly, PID can cause scarring and closure of the narrow uterine tubes, which is one of the major causes of female infertility. ❑

Uterus

The **uterus** (u′ter-us; "womb"), located in the pelvis between the urinary bladder and rectum, is a hollow organ that functions to receive, retain, and nourish a fertilized egg. In a woman who has never been pregnant, it is about the size and shape of a pear. (During pregnancy, the uterus increases tremendously in size to accommodate the growing fetus and can be felt well above the umbilicus during the later part of pregnancy.) The uterus is suspended in the pelvis by the broad ligament and anchored anteriorly and posteriorly by the *round* and *uterosacral ligaments,* respectively (see Figure 16.6).

The major portion of the uterus is referred to as the **body.** Its superior rounded region above the entrance of the uterine tubes is the **fundus,** and its narrow outlet, which protrudes into the vagina below, is the **cervix.**

The wall of the uterus is thick and composed of three layers. The inner layer or mucosa is the **endometrium** (en-do-me′tre-um). If fertilization occurs, the fertilized egg (actually the young embryo by the time it reaches the uterus) burrows into the endometrium of the uterus (this process is called **implantation**) and resides there for the rest of its development. When a woman is not pregnant, the endometrial lining sloughs off periodically, usually about every 28 days, in response to changes in the levels of ovarian hormones in the blood. This process, called *menses,* is discussed on p. 484.

Cancer of the cervix is common among women between the ages of 30 and 50. Risk factors include frequent cervical inflammation, sexually transmitted diseases, multiple pregnancies, and an active sex life with many partners. A yearly *Pap smear* is the single most important diagnostic test for detecting this slow-growing cancer. ❏

The **myometrium** (mi-o-me′tre-um) is the bulky middle layer of the uterus. It is composed of interlacing bundles of smooth muscle. The myometrium plays an active role during the delivery of a baby, when it contracts rhythmically to force the baby out of the mother's body. The outermost serous layer of the uterus is the **epimetrium** (ep-ĭ-me′tre-um), or the visceral peritoneum.

Vagina

The **vagina** (vah-ji′nah) is a thin-walled tube 8 to 10 cm (3 to 4 inches) long. It lies between the bladder and rectum and extends from the cervix to the body exterior. Often called the *birth canal,* the vagina provides a passageway for the delivery of an infant and for the menstrual flow to leave the

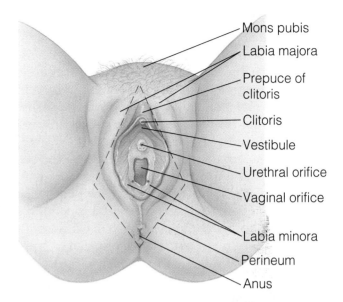

Figure 16.7   External genitalia of the human female.

body. Since it receives the penis (and semen) during sexual intercourse, it is the female organ of copulation.

The distal end of the vagina is partially closed by a thin fold of the mucosa called the **hymen** (hi′men). The hymen is very vascular and tends to bleed when it is ruptured during the first sexual intercourse. However, its durability varies. In some females, it is torn during a sports activity, tampon insertion, or pelvic examination. Occasionally, it is so tough that it must be ruptured surgically if intercourse is to occur.

External Genitalia

The female reproductive structures that are located external to the vagina are the **external genitalia** (Figure 16.7). The external genitalia, also called the **vulva,** include the mons pubis, labia, clitoris, urethral and vaginal orifices, and greater vestibular glands.

The **mons pubis** ("mountain on the pubis") is a fatty, rounded area overlying the pubic symphysis. After puberty, this area is covered with pubic hair. Running posteriorly from the mons pubis are two elongated hair-covered skin folds, the **labia majora** (la′be-ah ma-jo′ra), which enclose two delicate hair-free folds, the **labia minora.** The labia majora enclose a region called the **vestibule,** which contains the external openings of the urethra,* followed posteriorly by that of the vagina. A pair of mucus-producing glands, the **greater vestibular glands,** flank the vagina, one on each side. Their secretion lubricates the distal end of the vagina during intercourse. (These glands are not shown in Figure 16.7.)

Just anterior to the vestibule is the **clitoris** (kli′to-ris; "hill"), a small, protruding structure that corresponds to the male penis. Like the penis, it is hooded by a prepuce and is composed of sensitive erectile tissue that becomes swollen with blood during sexual excitement. The clitoris differs from the penis in that it lacks a reproductive duct. The diamond-shaped region between the anterior end of the labial folds, the anus posteriorly, and the ischial tuberosities laterally is the **perineum** (per″ĭ-ne′um).

FEMALE REPRODUCTIVE FUNCTIONS AND CYCLES

Oogenesis and the Ovarian Cycle

As described earlier, sperm production in males begins at puberty and generally continues throughout life. The situation is quite different in females. The total supply of eggs that a female can release is already determined by the time she is born. In addition, a female's reproductive ability (that is, her ability to release eggs) usually begins during puberty and ends in her 50s or before. The period in which a woman's reproductive capability gradually declines and then finally ends is called *menopause.* (Menopause is described in more detail on p. 498.)

The special kind of cell division (meiosis) that occurs in the male testes to produce sperm also occurs in the female ovaries. But in this case, female gametes, or sex cells, are produced, and the process is called **oogenesis** (o″o-jen′ĕ-sis), literally,

"the beginning of an egg." This process is shown in Figure 16.8 and described in more detail next.

In the developing female fetus, **oogonia** (o″o-go′ne-ah), the female stem cells, multiply rapidly to increase their number, and then their daughter cells, **primary oocytes,** push into the ovary connective tissue where they become surrounded by a single layer of cells to form the primary follicles. By birth, the oogonia no longer exist, and a female's lifetime supply of primary oocytes (approximately 700,000 of them) is already in place in the ovarian follicles, awaiting the chance to undergo meiosis to produce functional eggs. Since the primary oocytes remain in this state of suspended animation all through childhood, their wait is a long one—10 to 14 years at the very least.

At puberty, the anterior pituitary gland begins to release *follicle-stimulating hormone (FSH),* which stimulates a small number of primary follicles to grow and mature each month, and ovulation begins to occur each month. These cyclic changes that occur monthly in the ovary constitute the **ovarian cycle.** Since the reproductive life of a female is at best about 45 years (from the age of 11 to approximately 55) and there is typically only one ovulation per month, only 400 to 500 ova out of a potential of 700,000 are released during a woman's lifetime. Again, nature has provided us with a generous oversupply of sex cells.

As a follicle prodded by FSH grows larger, it accumulates fluid in the central chamber called the *antrum,* and the primary oocyte it contains begins meiosis and undergoes the first meiotic division to produce two cells that are very dissimilar in size (see Figure 16.8). The larger cell is a **secondary oocyte** and the other, very tiny cell is a **polar body.** By the time a follicle has ripened to the mature (vesicular follicle) stage, it contains a secondary oocyte and protrudes like an angry boil from the external surface of the ovary. Follicle development to this stage takes about 14 days, and ovulation (of a secondary oocyte) occurs just about that time in response to the burstlike release of a second anterior pituitary hormone, *luteinizing hormone (LH).* As shown in Figure 16.8, the ovulated secondary oocyte is still surrounded by its follicle-cell capsule, now called the *corona radiata* ("radiating crown"). Some women experience a twinge of abdominal pain in the lower abdomen when ovulation occurs. This phenomenon, called *mittelschmerz* (mit′el-shmārts; German for "middle

---

*The male urethra carries both urine and semen, but the female urethra has no reproductive function; it is strictly a passageway for urine.

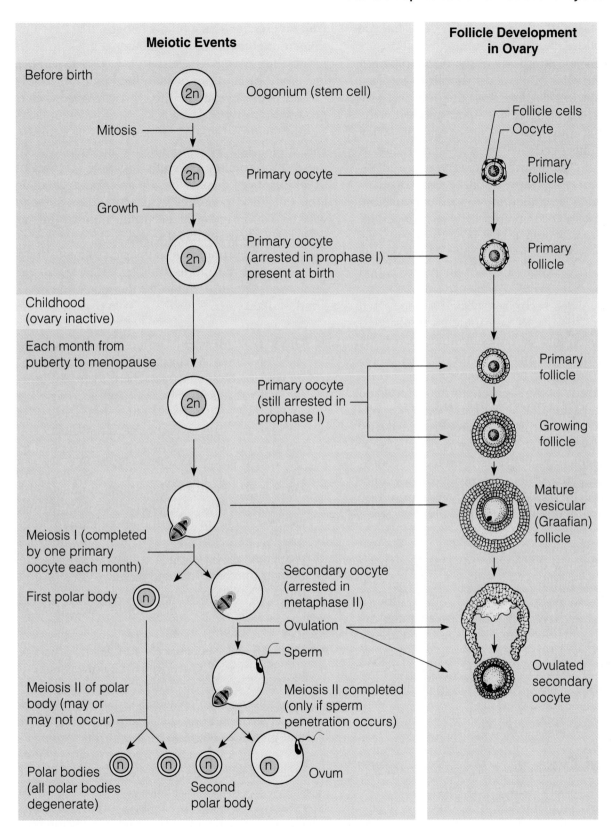

Figure 16.8   Events of oogenesis. Left, flowchart of meiotic events. Right, correlation with follicle development and ovulation in the ovary.

pain"), is thought to be caused by the intense stretching of the ovarian wall during ovulation.

Generally speaking, one of the developing follicles outstrips the others each month to become the dominant follicle. Just how this follicle is selected or selects itself is not understood, but the follicle that is at the proper stage of maturity when the LH stimulus occurs will rupture and release its oocyte into the peritoneal cavity. The mature follicles that are not ovulated soon become overripe and deteriorate. In addition to triggering ovulation, LH also causes the ruptured follicle to change into a very different glandular structure, the *corpus luteum*. (Both the maturing follicles and the corpus luteum produce hormones, as will be described later.)

If the ovulated secondary oocyte is penetrated by a sperm, its nucleus will undergo the second meiotic division that produces another polar body and the **ovum nucleus.** Once the ovum nucleus has been formed, its 23 chromosomes are combined with those of the sperm to form the fertilized egg, which is the first cell of the yet-to-be offspring. However, if the secondary oocyte is not penetrated by a sperm, it simply deteriorates without ever completing meiosis to form the functional egg. Although meiosis in males results in four functional sperm, meiosis in females yields only one functional ovum and three tiny polar bodies. Since the polar bodies have essentially no cytoplasm, they deteriorate and die quickly.

Another major difference between males and females concerns the size and structure of their sex cells. Sperm are tiny and equipped with tails for locomotion. They have little nutrient-containing cytoplasm; thus, the nutrients in seminal fluid are vital to their survival. In contrast, the egg is a large, nonmotile cell, well stocked with nutrient reserves that nourish the developing embryo until it can take up residence in the uterus.

Menstrual Cycle

Although the uterus is the receptacle in which the young embryo implants and develops, it is receptive to implantation only for a very short period each month. Not surprisingly, this brief interval coincides exactly to the time when a fertilized egg would begin to implant, approximately 7 days after ovulation. The events of the **menstrual,** or **uter-**ine, cycle are the cyclic changes that the endometrium, or mucosa, of the uterus goes through month after month as it responds to changes in the levels of ovarian hormones in the blood.

Since the cyclic production of estrogens and progesterone by the ovaries is, in turn, regulated by the anterior pituitary gonadotropic hormones, FSH and LH, it is important to understand how these "hormonal pieces" fit together. Generally speaking, both female cycles are about 28 days long (a period commonly called a *lunar month*), with ovulation typically occurring midway in the cycles, on or about day 14. Figure 16.9 illustrates the events occurring both in the ovary (the ovarian cycle) and in the uterus (menstrual cycle) at the same time. The three stages of the menstrual cycle are described next.

- **Days 1–5: Menses.** During this interval, the thick endometrial lining of the uterus is sloughing off, or becoming detached, from the uterine wall. This is accompanied by bleeding for 3 to 5 days. The detached tissues and blood pass through the vagina as the menstrual flow. The average blood loss during this period is 50 to 150 ml (or about ¼ to ½ cup).

- **Days 6–14: Proliferative stage.** Stimulated by rising estrogen levels produced by the growing follicles of the ovaries, the endometrium is repaired, glands are formed in it, and the endometrial blood supply is increased. The endometrium once again becomes velvety and thick. (Ovulation occurs in the ovary at the end of this stage, in response to the sudden surge of LH in the blood.)

- **Days 15–28: Secretory stage.** Rising levels of progesterone production by the corpus luteum of the ovary act on the estrogen-primed endometrium and increase its blood supply even more. Progesterone also causes the endometrial glands to increase in size and to begin secreting nutrients into the uterine cavity. These nutrients will sustain a developing embryo (if one is present) until it has implanted. If fertilization does occur, the embryo produces a hormone very similar to LH, which causes the corpus luteum to continue producing its hormones. If fertilization does not occur, the corpus luteum begins to degenerate toward the end of this period as LH blood levels decline.

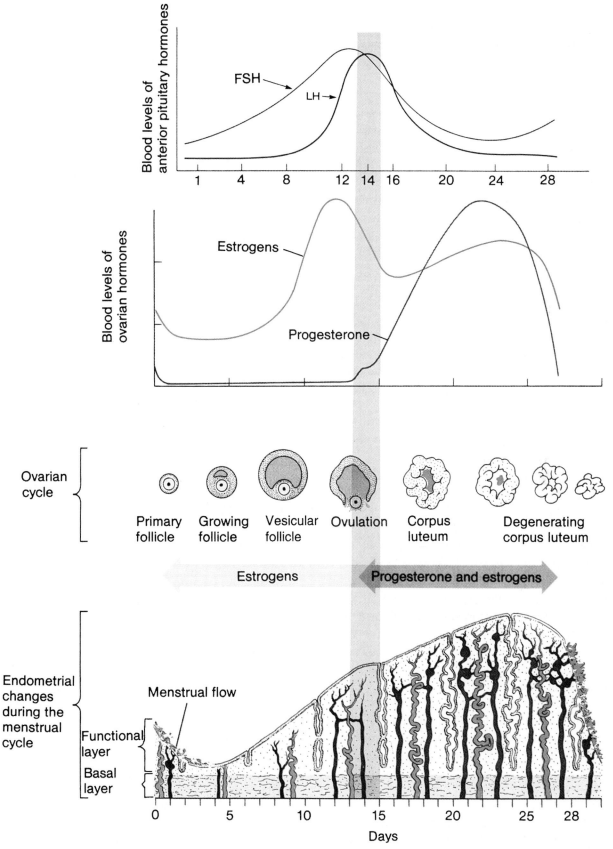

Figure 16.9   Hormonal interactions of the female cycles. Relative levels of anterior pituitary gonadotropins correlated with follicular and hormonal changes of the ovary. Menstrual cycle also depicted.

Lack of ovarian hormones in the blood causes the blood vessels supplying the endometrium to go into spasms and kink. When deprived of oxygen and nutrients, the endometrial cells begin to die, which sets the stage for menses to begin again on day 28.

Although this explanation assumes a classic 28-day cycle, the length of the menstrual cycle is quite variable. It can be as short as 21 days or as long as 40 days. Only one interval is fairly constant in all females; the time from ovulation to the beginning of menses is almost always 14 or 15 days.

Hormone Production by the Ovaries

As the ovaries become active at puberty and start to produce ova, production of ovarian hormones also begins. The follicle cells of the growing and mature follicles produce **estrogens,*** which cause the appearance of the *secondary sex characteristics* in the young woman. Such changes include:

- Enlargement of the accessory organs of the female reproductive system (uterine tubes, uterus, vagina, external genitals)
- Development of the breasts
- Appearance of axillary and pubic hair
- Increased deposits of fat beneath the skin in general, and particularly in the hips and breasts
- Widening and lightening of the pelvis
- Onset of menses, or the menstrual cycle

The second ovarian hormone, **progesterone,** is produced by a special glandular structure of the ovaries, the *corpus luteum* (see Figure 16.5). As mentioned earlier, after ovulation occurs the ruptured follicle is converted to the corpus luteum, which looks and acts completely different from the growing and mature follicle. Once formed, the corpus luteum produces progesterone (and some estrogen) as long as LH is still present in the blood.

Generally speaking, the corpus luteum has stopped producing hormones by 10 to 14 days after ovulation. Except for working with estrogen to establish the menstrual cycle, progesterone does not contribute to the appearance of the secondary sex characteristics. Its other major effects are exerted during pregnancy, when it helps maintain the pregnancy and prepare the breasts for milk production. (However, the source of progesterone during pregnancy is the placenta, not the ovaries.)

MAMMARY GLANDS

The **mammary glands** are present in both sexes, but they normally function only in females. Since the biological role of the mammary glands is to produce milk to nourish a newborn baby, they are actually important only when reproduction has already been accomplished. Stimulation by female sex hormones, especially estrogens, causes the female mammary glands to increase in size at puberty.

Developmentally, the mammary glands are modified *sweat glands* that are actually part of the skin. Each mammary gland is contained within a rounded skin-covered breast anterior to the pectoral muscles of the thorax. Slightly below the center of each breast is a pigmented area, the **areola** (ah-re′o-lah), which surrounds a central protruding **nipple** (Figure 16.10).

Internally, each mammary gland consists of 15 to 25 *lobes*, which radiate around the nipple. The lobes are padded and separated from each other by connective tissue and fat. Within each lobe are smaller chambers called *lobules*, which contain clusters of **alveolar glands** that produce the milk when a woman is *lactating* (producing milk). The alveolar glands of each lobule pass the milk into the **lactiferous** (lak-tif′er-us) **ducts,** which open to the outside at the nipple.

Cancer of the breast is a leading cause of death in American women. One woman in 8 will develop this condition. Breast cancer is often signaled by a change in skin texture, puckering, or leakage from the nipple. Early detection by breast self-examination and mammography is unquestionably the best way to increase one's chances of

---

*Although the ovaries produce several different estrogens, the most important are *estradiol, estrone,* and *estriol.* Of these, estradiol is the most abundant and is most responsible for mediating estrogenic effects.

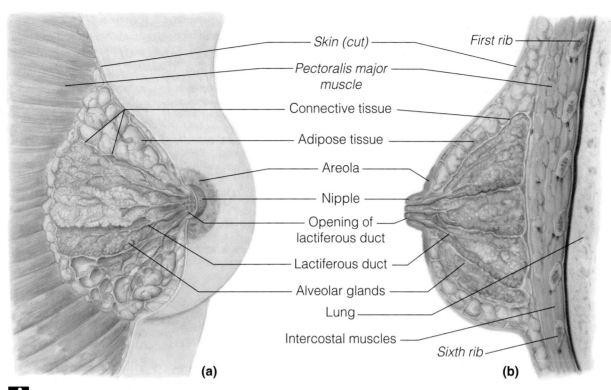

Figure 16.10   Female mammary glands. (a) Anterior view.
(b) Sagittal section.

surviving breast cancer. Since most breast lumps are discovered by women themselves in routine monthly breast exams, this simple examination should be a priority in every woman's life. Currently the American Cancer Society recommends scheduling **mammography**—X-ray examination that detects breast cancers too small to feel (less than 1 cm)—every 2 years for women between 40 and 49 years old and yearly thereafter (Figure 16.11). ❑

Figure 16.11   Mammograms. (a) Photograph of woman undergoing mammography. **(b)** Normal breast. **(c)** Breast with tumor.

SURVEY OF PREGNANCY AND EMBRYONIC DEVELOPMENT

Because the birth of a baby is such a familiar event, we tend to lose sight of the wonder of this accomplishment. In every instance it begins with a single cell, the fertilized egg, and ends with an extremely complex human being consisting of trillions of cells. The development of an embryo is very complex, and the details of this process can fill a good-sized book. Our intention here is simply to outline the important events of pregnancy and embryonic development.

Accomplishing Fertilization

Before fertilization can occur, the sperm must reach the ovulated secondary oocyte. The oocyte is viable for 12 to 24 hours after it is cast out of the ovary, and sperm generally retain their fertilizing power within the female reproductive tract for 12 to 48 hours after ejaculation. Some "super sperm," however, are viable for 72 hours. Consequently, for fertilization to occur, sexual intercourse must occur no more than 72 hours before ovulation and no later than 24 hours after, at which point the oocyte is approximately one-third of the way down the length of the uterine tube. Remember that sperm are motile cells that can propel themselves by lashing movements of their tails. If sperm are deposited in a female's vagina at the approximate time of ovulation, they are attracted to the oocyte by chemicals which act as "homing devices," allowing them to locate the oocyte. It takes 1 to 2 hours for sperm to complete the journey up the female duct system to the end of the uterine tubes, and if an oocyte is en route in the tube, fertilization is a distinct possibility.

When the swarming sperm reach the oocyte, hundreds of their acrosomes rupture, releasing enzymes that break down the "cement" that holds the follicle cells of the corona radiata together around the oocyte. Once a path has been cleared and a single sperm makes contact with the oocyte's membrane receptors, its head (nucleus) is pulled into the oocyte cytoplasm. This is one case that does not bear out the adage, "The early bird catches the worm." A sperm that comes along later,

after virtually hundreds of sperm have undergone acrosomal reactions to expose the oocyte membrane, is in the best position to be *the* fertilizing sperm. Once a single sperm has penetrated the oocyte, the oocyte nucleus completes the second meiotic division, forming the ovum and a polar body.

After sperm entry, changes occur in the fertilized egg to prevent other sperm from gaining entry. Of the millions of sperm ejaculated by a male, only *one* can penetrate an oocyte. **Fertilization** occurs at the moment the genetic material of a sperm combines with that of an ovum to form a fertilized egg, or **zygote** (zi'gōt). The zygote represents the first cell of the new individual.

Events of Embryonic and Fetal Development

As the zygote journeys down the uterine tube (propelled along by peristalsis and cilia), it begins to undergo rapid mitotic cell divisions—forming first two cells, then four, and so on. This early stage of embryonic development, called **cleavage,** is shown in Figure 16.12. Since there is not much time for cell growth between divisions, the daughter cells become smaller and smaller. Cleavage provides a large number of cells to serve as building blocks for constructing the **embryo** (developmental stage until the ninth week). Consider for a moment how difficult it would be to construct a building from one huge block of granite. If you now consider how much easier your task would be if you could use hundreds of brick-size granite blocks, you will quickly grasp the importance of cleavage. By the time the developing embryo reaches the uterus (about 3 days after ovulation, or on day 17 of the woman's cycle), it is a tiny ball of 16 cells that looks like a microscopic raspberry. The uterine endometrium is still not fully prepared to receive the embryo at this point, so the embryo floats free in the uterine cavity, temporarily using the uterine secretions for nutrition. While still unattached, the embryo continues to develop until it has about 100 cells, and then it hollows out to form a ball-like structure, a **blastocyst** (blas'to-sist) or **chorionic** (ko"re-on'ik) **vesicle.** At the same time, it secretes an LH-like hormone called **human chorionic gonadotropin (hCG),** which prods

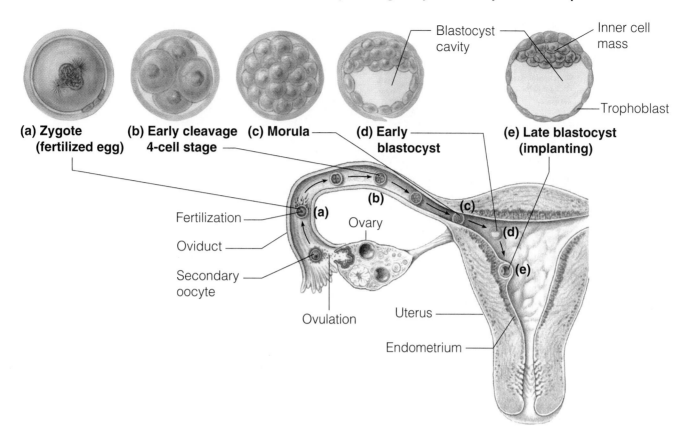

Figure 16.12 **Cleavage is a rapid series of mitotic divisions that begins with the zygote and ends with the blastocyst.** The zygote begins to divide about 24 hours after fertilization, and continues to divide rapidly (undergo cleavage) as it travels down the uterine tube. Three to four days after ovulation, the preembryo reaches the uterus and floats freely for two to three days, nourished by secretions of the endometrial glands.

At the late blastocyst stage, the embryo is implanting into the endometrium; this begins at about day 7 after ovulation. (**a**) Zygote (fertilized egg). (**b**) Four-cell stage. (**c**) Morula, a solid ball of blastomeres. (**d**) Early blastocyst; the morula hollows out, fills with fluid, and "hatches" from the zona pellucida. (**e**) Late blastocyst, composed of an outer sphere of trophoblast cells and an off-center cell cluster called the inner cell mass.

the corpus luteum of the ovary to continue producing its hormones. (If this were not the case, the endometrium would be sloughing off shortly in menses.)

The blastocyst has two important functional areas: the *trophoblast,* which forms the large fluid-filled sphere, and the *inner cell mass,* a small cluster of cells displaced to one side. By day 7 after ovulation, the blastocyst has attached to the endometrium and has eroded the lining away in a small area, embedding itself in the thick velvety mucosa. All of this is occurring even while devel-

opment is continuing and the three primary germ layers are being formed from the inner cell mass (Figure 16.13). The *primary germ layers* are the **ectoderm** (which gives rise to the nervous system and the epidermis of the skin), the **endoderm** (which forms mucosae and associated glands), and the **mesoderm** (which gives rise to virtually everything else). Implantation has usually been completed and the uterine mucosa has grown over the burrowed-in embryo by day 14 after ovulation—the day the woman would ordinarily be expecting to start menses. After it is securely implanted, the

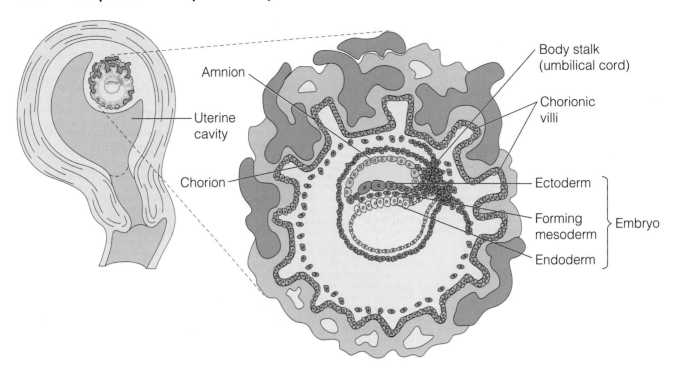

Figure 16.13 **Embryo of approximately 22 days.** Embryonic membranes present.

trophoblast part of the blastocyst develops elaborate projections, called **chorionic villi,** which cooperate with the tissues of the mother's uterus to form the **placenta** (plah-sen′tah). Once the placenta has formed, the platelike embryonic body, now surrounded by a fluid-filled sac called the **amnion** (am′ne-on), is attached to the placenta by a blood vessel-containing stalk of tissue, the **umbilical cord** (Figure 16.13). (The special features of the umbilical blood vessels and fetal circulation are discussed on p. 333.) Generally by the third week, the placenta is functioning to deliver nutrients and oxygen to and remove wastes from the embryonic blood. All exchanges are made through the placental barrier. By the end of the second month of pregnancy, the placenta has also become an endocrine organ and is producing estrogen, progesterone, and other hormones that help to maintain the pregnancy. At this time, the corpus luteum of the ovary becomes inactive.

By the eighth week of embryonic development, all the groundwork has been completed. All the organ systems have been laid down, at least in rudimentary form, and the embryo looks distinctly human. Beginning the ninth week of development,

the embryo is referred to as a **fetus.** From this point on, the major activities are growth and organ specialization, accompanied by changes in body proportions. During the fetal period, the developing fetus grows from a crown-to-rump length of about 3 cm (slightly more than 1 inch) and a weight of approximately 1 g (0.03 ounce) to about 36 cm (14 inches) and 2.7–4.1 kg (6–10 pounds) or more. (Total body length at birth is about 55 cm, or 22 inches.) As you might expect with such tremendous growth, the changes in fetal appearance are quite dramatic (Figure 16.14). The most significant of these changes are summarized in Table 16.1. By approximately 270 days after fertilization (the end of the tenth lunar month), the fetus is said to be "full-term" and is ready to be born.

Effects of Pregnancy on the Mother

Pregnancy (the period from conception to the birth of her baby) can be a difficult time for the mother. Not only are there obvious anatomical changes, but striking changes occur in her physiology as well.

Figure 16.14   Photographs of developing fetuses. The major events of fetal development are growth and tissue specialization. All the organ systems are laid down, at least in rudimentary form, during the embryonic period of development. (**a**) Fetus of 14 weeks, approximately 10 cm long. (**b**) Fetus of 20 weeks, approximately 19 cm long. By birth the fetus is typically 35 cm in length crown to rump.

Anatomical Changes

The ability of the uterus to enlarge during pregnancy is nothing less than remarkable. Starting as a fist-sized organ, the uterus grows to occupy most of the pelvic cavity by 16 weeks. As pregnancy continues, the uterus pushes higher and higher into the abdominal cavity. As birth nears, the uterus reaches the level of the xiphoid process and occupies the bulk of the abdominal cavity. The crowded abdominal organs press superiorly against the diaphragm, which intrudes on the thoracic cavity. As a result, the ribs flare, causing the thorax to widen.

The increasing bulkiness of the abdomen changes the woman's center of gravity, and many women develop an accentuated lumbar curvature (lordosis), often accompanied by backaches, during the last few months of pregnancy. Placental production of the hormone **relaxin** causes pelvic ligaments and the pubic symphysis to relax, widen, and become more flexible. This increased motility eases birth passage, but it may also result in a waddling gait during pregnancy.

Obviously, good maternal nutrition is necessary throughout pregnancy if the developing fetus is to have all the building materials (proteins, calcium, iron, and the like) it needs to form its tissues and organs. The old expression, "A pregnant woman is eating for two," has encouraged many women to eat *twice* the amount of food actually needed during pregnancy, which, of course, leads to excessive weight gain. Actually, a pregnant woman needs only about 300 additional calories daily to sustain proper fetal growth. The emphasis should be on high-quality food, not just more food.

Since many potentially harmful substances can cross through the placental barrier into the fetal blood, the pregnant woman should be very much aware of what she is taking into her body. Substances that may cause life-threatening birth defects (and even fetal death) include alcohol, nicotine, and many types of drugs (anticoagulants, antihypertensives, sedatives, and some antibiotics). Maternal infections, particularly German measles, may also cause severe fetal damage. Termination of a pregnancy by loss of a fetus during the first 20 weeks of pregnancy is called *abortion*. ❑

Physiological Changes

Gastrointestinal System   Many women suffer nausea, commonly called *morning sickness*, during the first few months of pregnancy, until their system adjusts to the elevated levels of progesterone and estrogens. *Heartburn* is common because the esophagus is displaced and the stomach is crowded by the growing uterus, which favors reflux of stomach acid into the esophagus. Another problem is constipation, because motility of the digestive tract declines during pregnancy.

Urinary System   The kidneys have the additional burden of disposing of fetal metabolic wastes, and they produce more urine during pregnancy. Because the uterus compresses the bladder, urination becomes more frequent, more urgent, and sometimes uncontrollable. (The last condition is called *stress incontinence*.)

Table 16.1   Development of the Human Fetus

Time	Changes/Accomplishments
8 weeks (end of embryonic period) *8 weeks*	Head nearly as large as body; all major brain regions present Liver disproportionately large and begins to form blood cells Limbs present; though initially webbed, fingers and toes are free by the end of this interval Bone formation begun Heart has been pumping blood since the fourth week All body systems present in at least rudimentary form Approximate crown-to-rump length: 30 mm (3 cm; 1.2 inches); weight: 1 gram (0.03 ounces)
9–12 weeks (third month) *12 weeks*	Head still dominant, but body elongating; brain continues to enlarge Facial features present in crude form Walls of hollow visceral organs gaining smooth muscle Blood cell formation begins in bone marrow Bone formation accelerating Sex readily detected from the genitals Approximate crown-to-rump length at end of interval: 90 mm (9 cm)
13–16 weeks (fourth month) *16 weeks*	General sensory organs are present; eyes and ears assume characteristic position and shape; blinking of eyes and sucking motions of lips occur Face looks human and body beginning to outgrow head Kidneys attain typical structure Most bones are distinct and joint cavities apparent Approximate crown-to-rump length at end of interval: 140 mm (14 cm)
17–20 weeks (fifth month)	Vernix caseosa (fatty secretions of sebaceous glands) covers body; silklike hair (lanugo) covers skin Fetal position (body flexed anteriorly) assumed because of space restrictions Limbs achieve near-final proportions Quickening occurs (mother feels spontaneous muscular activity of fetus) Approximate crown-to-rump length at end of interval: 190 mm (19 cm)
21–30 weeks (sixth and seventh months) *At birth*	Substantial increase in weight (may survive if born prematurely at 27–28 weeks, but hypothalamus still too immature to regulate body temperature and surfactant production by the lungs is still inadequate) Myelination of spinal cord begins; eyes are open Skin is wrinkled and red; fingernails and toenails are present Body is lean and well-proportioned Bone marrow becomes sole site of blood cell formation Testes enter scrotum in seventh month (in males) Approximate crown-to-rump length at end of interval: 280 mm (28 cm)
30–40 weeks (term) (eighth and ninth months)	Skin whitish pink; fat laid down in subcutaneous tissue Approximate crown-to-rump length at end of interval: 350–400 mm (35–40 cm; 14–16 inches) weight: 2.7–4.1 kg (6–10 pounds)

Respiratory System   The nasal mucosa responds to estrogen by becoming swollen and congested; thus, nasal stuffiness and occasional nosebleeds may occur. Vital capacity and respiratory rate increase during pregnancy, but residual volume declines, and many women exhibit *dyspnea* (difficult breathing) during the later stages of pregnancy.

Cardiovascular System   Perhaps the most dramatic physiological changes occur in the cardiovascular system. Total body water rises and blood volume increases by 25 to 40 percent to accommodate the additional needs of the fetus. The rise in blood volume also acts as a safeguard against blood loss during birth. Blood pressure and pulse typically rise and increase cardiac output by 20 to 40 percent; this helps propel the greater blood volume around the body. Because the uterus presses on the pelvic blood vessels, venous return from the lower limbs may be impaired somewhat, resulting in varicose veins.

Childbirth

Childbirth, also called **parturition** (par″tu-rish′un; "bringing forth young"), is the culmination of pregnancy. It usually occurs within 15 days of the calculated due date (which is 280 days from the last menstrual period). The series of events that expel the infant from the uterus are referred to collectively as **labor.**

Initiation of Labor

The precise trigger for labor is not clear, but several events appear to be interlocked in this process. During the last few weeks of pregnancy, estrogens reach their highest levels in the mother's blood. This has two important consequences: It causes the myometrium to form abundant *oxytocin* receptors (so that it becomes more sensitive to the hormone oxytocin), and it interferes with progesterone's quieting influence on the uterine muscle. As a result, weak, irregular uterine contractions begin to occur. These contractions, called *Braxton Hicks contractions*, have caused many women to go to the hospital, only to be told that they were in **false labor** and sent home.

As birth nears, two more chemical signals cooperate to convert these false labor pains into the real thing. Certain cells of the fetus begin to pro-

(4) Hypothalamus sends efferent impulses to posterior pituitary, where oxytocin is stored

(5) Posterior pituitary releases oxytocin to blood; oxytocin targets mother's uterine muscle

(6) Uterus responds by contracting more vigorously

(1) Baby moves deeper into mother's birth canal

(3) Afferent impulses to hypothalamus

(2) Pressoreceptors in cervix of uterus excited

Positive feedback mechanism continues to cycle until interrupted by birth of baby

Figure 16.15   The positive feedback mechanism by which oxytocin promotes labor contractions during birth.

duce oxytocin, which in turn stimulates the placenta to release *prostaglandins.* Both hormones stimulate more frequent and powerful contractions of the uterus. At this point, the increasing emotional and physical stresses activate the mother's hypothalamus, which signals for oxytocin release by the posterior pituitary. The combined effects of rising levels of oxytocin and prostaglandins initiate the rhythmic, expulsive contractions of true labor. Once the hypothalamus is involved, a positive feedback mechanism is propelled into action: Stronger contractions cause the release of more oxytocin, which causes even more vigorous contractions forcing the baby ever deeper into the mother's pelvis, and so on (Figure 16.15).

(Text continues on p. 496)

A CLOSER LOOK

CONTRACEPTION: TO BE OR NOT TO BE

In a society such as ours, where many women opt for professional careers or must work for economic reasons, **contraception** (*contra* = against; *cept* = taking), or *birth control,* is often seen as a necessity. Thus far much of the burden for birth control has fallen on women's shoulders, and most birth control products are female-directed.

The key to birth control is dependability. As shown by the red arrows in the accompanying flowchart, the birth control techniques and products currently available have many sites of action for blocking the reproductive process. Let's examine the relative advantages of a few of these current methods more closely.

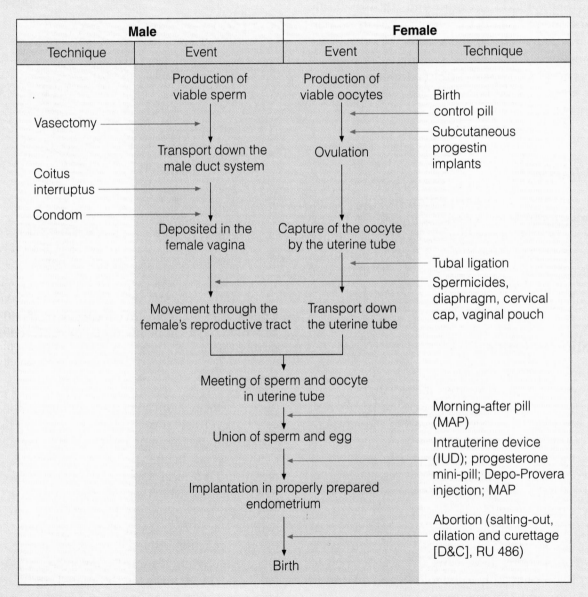

Flowchart of the events that must occur to produce a baby. Techniques or products that interfere with the process are indicated by red arrows at the site of interference; they act to prevent the next step.

The most-used contraceptive product in the United States is the *birth control pill*, or simply, "the pill," a preparation containing tiny amounts of estrogens and progestins (progesterone-like hormones) that is taken daily, except for the last 5 days of the 28-day cycle. The pill tricks the hypothalamic-pituitary control system and "lulls it to sleep," because the relatively constant blood levels of ovarian hormones make it appear that the woman is pregnant (both estrogen and progesterone are produced throughout pregnancy). Ovarian follicles do not mature and ovulation ceases. The endometrium is sloughed off when the pill is discontinued each month, but menstrual flow is much reduced. However, since hormonal balance in the body is precisely controlled, some women simply cannot tolerate the changes caused by the pill—they become nauseated and/or hypertensive. For a while, the pill was suspected of increasing the incidence of breast and uterine cancer. Its influence on breast cancer is still a question. However, it appears that the new very-low-dose preparations may actually help protect against some forms of cancer (ovarian and endometrial) and have also reduced the incidence of serious cardiovascular side effects, such as strokes, heart attacks, and blood clots, that occurred (rarely) with earlier forms of the pill. Presently, the pill is one of the most widely used drugs in the world; well over 50 million women use these drugs to prevent pregnancy. Its incidence of failure is less than six pregnancies per 100 women per year.

Morning-after pills (MAPs) are one of the best-kept secrets in the U.S.—most teenagers and women in their 30s and 40s have never heard of them. Taken within 72 hours of unprotected intercourse, these estrogen-progesterone combination pills "mess up" the normal hormonal signals so much that fertilization is prevented altogether or a fertilized egg is prevented from implanting. Within the past 4 years, Planned Parenthood Clinics have begun to publicize the availability of MAPs; however, since the therapy is not FDA approved, it cannot be dispensed in federally supported clinics.

Other approaches include a device that is implanted just under the skin and injections of synthetic progesterone. The Norplant implant, 6 tiny silicone rods that release progestin over a 5-year period, was approved late in 1990 and quickly gained users. Its failure rate (0.05%) is even lower than that of the pill, but there have been some problems with rod removal. *Depo-Provera*, a synthetic progesterone developed 35 years before, was finally approved for use as an injectable contraceptive in October 1992. Administered in a 150-mg dose once every three months, Depo-Provera's failure rate is only 0.4%.

For several years, the second most used contraceptive method was the *intrauterine device (IUD)*, a plastic or metal device inserted into the uterus that prevented implantation of the fertilized egg. The failure rate of the IUD was nearly as low as that of the pill. However, most IUDs have been taken off the market in the U.S. because of problems with occasional contraceptive failure, uterine perforation, or PID.

Some methods, such as *tubal ligation* or *vasectomy* (cutting or cauterizing the uterine tubes or ductus deferens, respectively) are nearly foolproof and are the choice of approximately 33 percent of couples of childbearing age in the United States. Both procedures can be done in the physician's office. These changes are usually permanent, however, making them unpopular with individuals who still plan to have children but want to choose the time.

Coitus interruptus, or withdrawal of the penis just before ejaculation, is simply against nature, and control of ejaculation is never assured. *Rhythm* or *fertility awareness methods* depend on recognizing the period of ovulation or fertility and avoiding intercourse during those intervals. This may be accomplished by (1) recording daily basal body temperatures (body temperature drops slightly [0.2° to 0.6°F] immediately prior to ovulation and then rises slightly [0.2° to 0.6°F] after ovulation) or (2) recording changes in the consistency of vaginal mucus (the mucus first becomes sticky and then clear and stringy, much like egg white, during the fertile period). Rhythm techniques require accurate record-keeping for several cycles before they can be used with confidence. *Barrier methods*, such as diaphragms, cervical caps, condoms, spermicidal foams, gels, and sponges, are quite effective, especially when some agent is used by both partners. But many people avoid them because they can reduce the spontaneity of sexual encounters.

RU 486, the so-called *abortion pill* developed in France, may well be the next "big seller" in the U.S. When taken during the first 7 weeks of pregnancy in conjunction with a tiny amount of prostaglandin to induce uterine contractions, it induces miscarriage by blocking progesterone's quieting effect on the uterus. RU 486 has a 96 to 98 percent success rate with virtually no side effects, but its proposed approval by the FDA has become a topic of bitter controversy among pro-choice and pro-life groups in the U.S.

This summary has barely touched on the large number of experimental birth control drugs now awaiting clinical trials; and other methods are sure to be developed in the near future. In the final analysis, however, the only 100 percent effective means of birth control is the age-old one—*total abstinence.*

Since both oxytocin and prostaglandins are needed to initiate labor in humans, anything that interferes with production of either of these hormones will hinder the onset of labor. For example, antiprostaglandin drugs such as aspirin and ibuprofen can inhibit labor at the early stages, and such drugs are used occasionally to prevent premature births.

Stages of Labor

The process of labor is commonly divided into three stages. These stages are described next.

Stage 1: Dilation Stage   The **dilation stage** is the time from the appearance of true contractions until the cervix is fully dilated by the baby's head (about 10 cm in diameter). As labor starts, regular but weak uterine contractions begin in the upper part of the uterus and move downward toward the vagina. Gradually, the contractions become more vigorous and more rapid, and as the infant's head is forced against the cervix with each contraction, the cervix begins to soften, becomes thinner (effaces), and dilates. Eventually, the amnion ruptures, releasing the amniotic fluid, an event commonly referred to as "breaking the water." The dilation stage is the longest part of labor and usually lasts for 6 to 12 hours (sometimes considerably more).

Stage 2: Expulsion Stage   The **expulsion stage** is the period from full dilation to delivery of the infant. In this stage, the infant passes through the cervix and vagina to the outside of the body. During this stage, a mother experiencing natural childbirth (that is, undergoing labor without local anesthesia) has an increasing urge to push or bear down with the abdominal muscles. Although this phase can take as long as 2 hours, it is typically 50 minutes in a first birth and around 20 minutes in subsequent births.

When the infant is in the usual head-first (*vertex*) position, the skull (its largest diameter) acts as a wedge to dilate the cervix. The head-first presentation also allows the baby to be suctioned free of mucus and to breathe even before it has completely exited from the birth canal. Once the head has been delivered, the rest of the baby's body is delivered much more easily. After birth, the umbilical cord is clamped and cut. In *breech* (buttocks-first) presentations and other nonvertex presenta-

tions, these advantages are lost and delivery is much more difficult, often requiring the use of forceps.

During an extremely prolonged or difficult stage 2, a condition called *dystocia* (dis-to′se-ah) may occur. In dystocia, oxygen delivery to the infant is inadequate leading to fetal brain damage (resulting in cerebral palsy or epilepsy) and decreased viability of the infant. To prevent these outcomes, a *cesarean* (se-zayr′e-an) *section,* also called a *C-section* may be performed. A C-section is delivery of the infant through a surgical incision made through the abdominal and uterine walls. ❑

Stage 3: Placental Stage   The **placental stage,** or the delivery of the placenta, is usually accomplished within 15 minutes after birth of the infant. The strong uterine contractions that continue after birth compress uterine blood vessels, limit bleeding, and cause the placenta to detach from the uterine wall. The placenta and its attached fetal membranes, collectively called the **afterbirth,** are then easily removed by a slight tug on the umbilical cord. It is very important that all placental fragments be removed to prevent continued uterine bleeding after birth (*postpartum bleeding).*

DEVELOPMENTAL ASPECTS OF THE REPRODUCTIVE SYSTEM

Although the sex of an individual is determined at the time of fertilization (males have X and Y sex chromosomes and females have XX), the gonads do not begin to form until about the eighth week of embryonic development. Prior to this time, the embryonic reproductive structures of males and females are identical and are said to be in the *indifferent stage.* After the gonads have formed, development of the accessory structures and external genitalia begins. Whether male or female structures will form depends entirely on whether testosterone is present or absent. The usual case is that, once formed, the embryonic testes produce testosterone, and the development of the male duct system and external genitalia follows. When testosterone is not produced, as is the case in female embryos that form ovaries, the female ducts and external genitalia result.

on a cyclic basis. The female egg (ovum) is formed only if sperm penetrates the secondary oocyte. In females, meiosis produces only one functional ovum (plus three nonfunctional polar bodies) as opposed to the four functional sperm per meiosis produced by males.

2. Hormone production: Estrogens are produced by ovarian follicles in response to FSH. Progesterone, produced in response to LH, is the main hormonal product of the corpus luteum. Estrogens stimulate development of female secondary sex characteristics.

3. The menstrual cycle concerns changes in the endometrium in response to changes in blood levels of ovarian hormones. The phases are:
   a. Menses. Endometrium sloughs off and bleeding occurs. Ovarian hormones are at their lowest levels.
   b. Proliferative phase. Endometrium is repaired, thickens, and becomes well vascularized in response to increasing levels of estrogens.
   c. Secretory phase. Endometrial glands begin to secrete nutrients, and lining becomes more vascular in response to increasing levels of progesterone.

   If fertilization does not occur, the phases are repeated about every 28 days.

Mammary Glands *(pp. 486–487)*

1. Mammary glands are milk-producing glands found in the breasts. They produce milk after birth of a baby in response to hormonal stimulation.

Survey of Pregnancy and Embryonic Development *(pp. 488–496)*

1. An oocyte is fertilizable for up to 24 hours; sperm are viable within the female reproductive tract for up to 72 hours. Hundreds of sperm must release their acrosomal enzymes to break down the egg's corona radiata.

2. Following sperm penetration, the secondary oocyte completes meiosis II. Then ovum and sperm nuclei fuse (fertilization), forming a zygote.

3. If fertilization occurs, embryonic development begins immediately. Cleavage, a rapid series of mitotic divisions without intervening growth, begins with the zygote and ends with a blastocyst.

4. By day 14 after ovulation, the young embryo (blastocyst) has implanted in the endometrium, and the placenta is being formed. hCG released by the blastocyst maintains hormone production of the corpus luteum, preventing menses, until the placenta assumes its endocrine role.

5. The placenta serves respiratory, nutritive, and excretory needs of the embryo and produces hormones of pregnancy.

6. All major organ systems have been laid down by 8 weeks, and at 9 weeks the embryo is called a *fetus*. Growth and tissue/organ specialization are the major events of the fetal period.

7. A pregnant woman has increased respiratory, circulatory, and urinary demands placed on her system by the developing fetus. Good nutrition is necessary to produce a healthy baby.

8. Childbirth (parturition) includes a series of events called *labor*. It is initiated by several factors, but most importantly by rising levels of oxytocin and prostaglandins, which promote vigorous uterine contractions. The three stages of labor are dilation, expulsion, placental.

Developmental Aspects of the Reproductive System *(pp. 496–498)*

1. Reproductive system structures of males and females are identical during early development. Gonads develop by the eighth week. The presence or absence of testosterone determines whether male or female accessory reproductive organs are formed.

2. Important congenital defects result from abnormal separation of sex chromosomes during sex cell formation.

3. The reproductive system is inactive during childhood. Reproductive organs mature and become functional for childbearing at puberty.

4. Common reproductive problems during young adulthood are infections of the reproductive tract. Neoplasms of breast and cervix are major threats to females. Prostatic cancer is the most common reproductive system cancer seen in males.

5. During menopause, female reproductive capabilities end, and reproductive organs begin to atrophy. Hot flashes and mood changes may occur. Reproductive capacity does not appear to decline significantly in aging males.

Review Questions

1. What are the primary sex organs, or gonads, of males? What are their two major functions?

2. Name the organs forming the male duct system, from the male gonad to the body exterior.

3. What is the function of seminal fluid? Name the three types of glands that help produce it.

4. The penis contains erectile tissue that becomes engorged with blood during sexual excitement. What term is used to describe this event?

5. Define *ejaculation*.

6. Why are the male gonads not found in the abdominal cavity? Where are they found?

7. How does enlargement of the prostate gland interfere with a male's reproductive function?

8. What structures in the testes form the sex cells? When does spermatogenesis begin? What causes it to begin?

9. The process of spermatogenesis actually forms cells called spermatids. How are the spermatids converted to functional sperm?

10. Testosterone causes the male secondary sex characteristics to appear at puberty. Name three examples of male secondary sex characteristics.

11. Name the female gonad and describe its two major functions.

12. Name the structures of the female duct system and describe the important functions of each.

13. Since the uterine tubes are not continuous with the ovaries, how can you explain the fact that all ovulated "eggs" do not end up in the female's peritoneal cavity?

14. What anterior pituitary hormones cause follicle development and ovulation to occur in the ovary? What is a follicle? What is ovulation?

15. The female cell that is ovulated is not a mature sex cell (ovum). When or under what conditions does it become mature?

16. What hormone can be called the "feminizing" hormone? What ovarian structures produce this hormone? Name the second hormone produced by the ovary.

17. List and describe the events of the menstrual cycle. Why is the menstrual cycle so important?

18. Define *menopause*. What does this mean to a female?

19. What is the role of the mammary glands?

20. Define *fertilization*. Where does fertilization usually occur? Describe the process of implantation.

21. What are the functions of the placenta?

22. How is a pregnant woman's body functioning altered by her pregnancy?

23. Compare the effects of aging on the male and female reproductive systems.

At the Clinic

1. A pregnant woman in substantial pain called her doctor and explained (between sobs) that she was about to have her baby "right away." The doctor calmed her and asked how she had come to that conclusion. She said that her water had broken and that her husband could see the baby's head. (a) Was she right to believe that birth was imminent? If so, what stage of labor was she in? (b) Do you think she had time to make it to the hospital 60 miles away? Why or why not?

2. Lucy had both her left ovary and her right uterine tube removed surgically at age 17 because of a cyst and a tumor in these organs. Now, at age 32, she remains healthy and is expecting her second child. How could Lucy conceive a child with just one ovary and one uterine tube, widely separated on opposite sides of the pelvis like this?

3. Adolf, a 68-year-old man, has trouble urinating and is given a rectal exam. What is Adolf's most probable condition, and what is the purpose of the rectal exam?

4. Some anatomy students were saying that the bulbourethral glands of males act like city workers who come around and clear parked cars from the street before a parade. What did they mean by this analogy?

5. A young woman visited her doctor for a premarital gynecological examination and requested information on contraception. She plans to have a large family but "not just yet." She reveals that her faith prevents her from using any drugs or mechanical means of birth control. What are the only options for this young woman?

6. Mrs. Monica is experiencing Braxton Hicks contractions in the sixth month of her pregnancy. Why did her doctor prescribe ibuprofen (with what mechanism does this drug interfere)?

Word Roots, Prefixes, and Suffixes

Word element	Meaning	Word element	Meaning	Word element	Meaning

WORD ROOTS

Circulatory system

Word element	Meaning
cardio	heart
angio, vaso	vessel
hem, hema, hemato	blood
vena, phlebo	vein
arteria	artery
lympho	lymph
thrombo	clot (of blood)
embolus	moving clot

Digestive system

Word element	Meaning
bucca	cheek
os, stomato	mouth
gingiva	gum
glossa	tongue
pharyngo	pharynx
esophago	esophagus
gastro	stomach
hepato	liver
cholecyst	gallbladder
pancreas	pancreas
entero	intestines
duodeno	duodenum
jejuno	jejunum
ileo	ileum
appendeco	appendix
colo	colon
recto	rectum
ano, procto	anus

Skeletal system

Word element	Meaning
skeleto	skeleton

Respiratory system

Word element	Meaning
naso, rhino	nose
tonsillo	tonsil
laryngo	larynx
tracheo	trachea
bronchus, broncho	bronchus (pl. bronchi)
pulmo, pneuma, pneum	lung (sac with air)

Nervous system

Word element	Meaning
neuro	nerve
cerebro	brain; cerebrum
oculo, ophthalmo	eye
oto	ear
psych, psycho	mind

Urinary system

Word element	Meaning
urethro	urethra
cysto	bladder
uretero	ureter
reni, reno, nephro	kidney
pyelo	pelvis of kidney
uro	urine

Female reproductive system

Word element	Meaning
vulvo	vulva
perineo	perineum
labio	labium (pl. labia)
vagino, colpo	vagina
cervico	cervix
utero	womb; uterus
tubo, salpingo	fallopian tube
ovario, oophoro	ovary

Male reproductive system

Word element	Meaning
orchido	testes

Regions of the body

Word element	Meaning
crani, cephalo	head
cervico, tracheo	neck
thoraco	chest
abdomino	abdomen
dorsum	back

Tissues

Word element	Meaning
cutis, dermato	skin
lipo	fat
musculo, myo	muscle
osteo	bone
myelo	marrow
chondro	cartilage

Miscellaneous

Word element	Meaning
cyto	cell
gene	formation, origin
gram	tracing or mark
graph	writing, description
kinesis	motion
lapar	through the abdominal wall
meter	measure
oligo	small, few
phobia	fear
photo	light
pyo	pus
roentgen	X ray
scope	instrument for visual examination

PREFIXES

Word element	Meaning	Word element	Meaning	Word element	Meaning
a, an, ar	without or not	adeno	glandular	amyl	starch
ab	away from	aero	air	ante	before, forward
acro	extremities	ambi	around, on both sides	anti	against, counteracting
ad	toward, to				

continued

Word Roots, Prefixes, and Suffixes (continued)

Word element	Meaning	Word element	Meaning	Word element	Meaning
PREFIXES (continued)					
bi	double	homo	same	path	disease
bili	bile	hydro	water	ped	child, foot
bio	life	hygro	moisture	per	by, through
bis	two	hyper	too much, high	peri	around
brachio	arm	hypo	under, decreased	pharyngo	pharynx
brady	slow			phlebo	vein
broncho	bronchus (pl. bronchi)	hyster	uterus	photo	light
		ileo	ileum	phren	diaphragm, mind
cardio	heart	in	in, within, into		
cervico	neck	inter	between	pneumo	air, lungs
chole	gall or bile	intra	within	pod	foot
cholecysto	gallbladder	intro	in, within, into	poly	many, much
circum	around	juxta	near, close to	post	after
co	together	lapar	abdomen	pre	before
contra	against, opposite	laryngo	larynx	proct	rectum
		latero	side	pseudo	false
costo	ribs	leuk	white	psych	mind
cysto	bladder	macro	large, big	pyel	pelvis of the kidney
cyto	cell	mal	bad, poor		
demi	half	mast	breast	pyo	pus
derma	skin	medio	middle	pyro	fever, heat
dis	from	mega, megalo	large, great	quadri	four
dorso	back	meno	menses	radio	radiation
dys	abnormal, difficult	mono	single	re	back, again
		multi	many	reno	kidney
electro	electric	myelo	bone marrow, spinal cord	retro	backward
en	into, in, within			rhin	nose
encephal	brain	myo	muscle	sacro	sacrum
entero	intestine	neo	new	salpingo	fallopian tube
equi	equal	nephro	kidney	sarco	flesh
eryth	red	neuro	nerve	sclero	hard, hardening
ex	out, out of, away from	nitro	nitrogen	semi	half
		noct	night	sex	six
extra	outside of, in addition to	non	not	skeleto	skeleton
		ob	against, in front of	steno	narrowing, constriction
ferro	iron				
fibro	fiber	oculo	eye	sub	under
fore	before, in front of	odonto	tooth	super	above, excess
		ophthalmo	eye	supra	above
gastro	stomach	ortho	straight, normal	syn	together
glosso	tongue	os	mouth, bone	tachy	fast
glyco	sugar	osteo	bone	thyro	thyroid, gland
hemi	half	oto	ear	trache	trachea
hemo	blood	pan	all	trans	across, over
hepa, hepato	liver	para	beside, accessory to	tri	three
histo	tissue			ultra	beyond

Word Roots, Prefixes, and Suffixes (continued)

Word element	Meaning	Word element	Meaning	Word element	Meaning
PREFIXES (continued)					
un	not, back reversal	uretero	ureter	uro	urine, urinary organs
uni	one	urethro	urethra	vaso	vessel
SUFFIXES					
able	able to	graph	writing	otomy	to incise
algia	pain	ism	condition	pexy	fixation
cele	tumor, swelling	itis	inflammation	phage	ingesting
centesis	surgical puncture to remove fluid	ize	to treat	phobia	fear
		lith	stone, calculus	plasty	plastic surgery
		lithiasis	presence of stones	plegia	paralysis
cide	killing, destructive	lysis	disintegration	rhage	to burst forth
cule	little	megaly	enlargement	rhea	excessive discharge
cyte	cell	meter	instrument that measures	rhexis	rupture
ectasia	dilating, stretching	oid	likeness, resemblance	scope	lighted instrument for visual examination
ectomy	excision, surgical removal of	oma	tumor		
		opathy	disease of	scopy	to examine visually
emia	blood	orrhaphy	surgical repair	stomy	to form an opening
esis	action	osis	disease, condition of		
form	shaped like				
genesis, genetic	formation, origin	ostomy	to form an opening or outlet	tomy	incision into
				uria	urine
gram	tracing, mark				

Sources: Courtesy of Margaret Ling, Director of Vocational Nursing, Santa Rosa Junior College, Santa Rosa, Calif.; Kozier, B., and G. Erb: *Fundamentals of Nursing: Concepts and Procedures*, 3d ed., Addison-Wesley, 1987. Used by permission.

Periodic Table of the Elements

Legend: Atomic number / Symbol / Element / Atomic weight

Example: 1 / H / Hydrogen / 1.00794

IA 1	IIA 2	IIIB 3	IVB 4	VB 5	VIB 6	VIIB 7	VIIIB 8	VIIIB 9	VIIIB 10	IB 11	IIB 12	IIIA 13	IVA 14	VA 15	VIA 16	VIIA 17	VIIIA 18
1 H Hydrogen 1.008																	2 He Helium 4.003
3 Li Lithium 6.941	4 Be Beryllium 9.012											5 B Boron 10.81	6 C Carbon 12.01	7 N Nitrogen 14.01	8 O Oxygen 16.00	9 F Fluorine 19.00	10 Ne Neon 20.18
11 Na Sodium 22.99	12 Mg Magnesium 24.31											13 Al Aluminum 26.98	14 Si Silicon 28.09	15 P Phosphorus 30.97	16 S Sulfur 32.06	17 Cl Chlorine 35.45	18 Ar Argon 39.95
19 K Potassium 39.10	20 Ca Calcium 40.08	21 Sc Scandium 44.96	22 Ti Titanium 47.88	23 V Vanadium 50.94	24 Cr Chromium 52.00	25 Mn Manganese 54.94	26 Fe Iron 55.85	27 Co Cobalt 58.93	28 Ni Nickel 58.70	29 Cu Copper 63.55	30 Zn Zinc 65.38	31 Ga Gallium 69.72	32 Ge Germanium 72.59	33 As Arsenic 74.92	34 Se Selenium 78.96	35 Br Bromine 79.90	36 Kr Krypton 83.80
37 Rb Rubidium 85.47	38 Sr Strontium 87.62	39 Y Yttrium 88.91	40 Zr Zirconium 91.22	41 Nb Niobium 92.91	42 Mo Molybdenum 95.94	43 Tc Technetium (98)	44 Ru Ruthenium 101.1	45 Rh Rhodium 102.9	46 Pd Palladium 106.4	47 Ag Silver 107.9	48 Cd Cadmium 112.4	49 In Indium 114.8	50 Sn Tin 118.7	51 Sb Antimony 121.8	52 Te Tellurium 127.6	53 I Iodine 126.9	54 Xe Xenon 131.3
55 Cs Cesium 132.9	56 Ba Barium 137.3	57 La* Lanthanum 138.9	72 Hf Hafnium 178.5	73 Ta Tantalum 180.9	74 W Tungsten 183.9	75 Re Rhenium 186.2	76 Os Osmium 190.2	77 Ir Iridium 192.2	78 Pt Platinum 195.1	79 Au Gold 197.0	80 Hg Mercury 200.6	81 Tl Thallium 204.4	82 Pb Lead 207.2	83 Bi Bismuth 209.0	84 Po Polonium (209)	85 At Astatine (210)	86 Rn Radon (222)
87 Fr Francium (223)	88 Ra Radium (226.0)	89 Ac** Actinium (227)	104 Rf Rutherfordium (261)	105 Ha Hahnium (262)	106 Unh Unnilhexium (263)	107 Uns Unnilseptium (262)	108 Uno Unniloctium (265)	109 Une Unnilennium (266)									

*Lanthanides

58 Ce Cerium 140.1	59 Pr Praseodymium 140.9	60 Nd Neodymium 144.2	61 Pm Promethium (145)	62 Sm Samarium 150.4	63 Eu Europium 152.0	64 Gd Gadolinium 157.3	65 Tb Terbium 158.9	66 Dy Dysprosium 162.5	67 Ho Holmium 164.9	68 Er Erbium 167.3	69 Tm Thulium 168.9	70 Yb Ytterbium 173.0	71 Lu Lutetium 175.0

**Actinides

90 Th Thorium 232.0	91 Pa Protactinium (231)	92 U Uranium 238.0	93 Np Neptunium (237)	94 Pu Plutonium (244)	95 Am Americium (243)	96 Cm Curium (247)	97 Bk Berkelium (247)	98 Cf Californium (251)	99 Es Einsteinium (252)	100 Fm Fermium (257)	101 Md Mendelevium (258)	102 No Nobelium (259)	103 Lr Lawrencium (260)

The periodic table arranges elements according to atomic number and atomic weight into horizontal rows called *periods* and eighteen vertical columns called *groups* or *families*. (Elements 110 and 111 were discovered late in 1994 and element 112 was detected in February of 1995. As of preparation of this edition, these three elements had not been named and, hence, are not included in this version of the periodic table.) The elements in the groups are classified as being in either A or B classes.

Elements of each group of the A series have similar chemical and physical properties. This reflects the fact that members of a particular group have the same number of valence shell electrons, which is indicated by the number of the group. For example, group IA elements have one valence shell electron, group IIA elements have two, and group VA elements have five.

In contrast, as you progress across a period from left to right, the properties of the elements change in discrete steps, varying gradually from the very metallic properties of groups IA and IIA elements to the nonmetallic properties seen in group VIIA (chlorine and others), and finally to the inert elements (noble gases) in group VIIIA. This change reflects the continual increase in the number of valence shell electrons seen in elements (from left to right) within a period.

Class B elements are referred to as *transition elements*. All transition elements are metals, and in most cases they have one or two valence shell electrons. (In these elements, some electrons occupy more distant electron shells before the deeper shells are filled.)

The 24 elements found in the human body are listed in Table 2.1 on p. 24 with their atomic symbol, percentage of body mass, and role.

Appendix C

Key Information About Vitamins and Many Essential Minerals

Vitamin	Major dietary sources	Major functions	Signs of severe, prolonged deficiency	Signs of extreme excess
Fat-Soluble				
A	Fat-containing and fortified dairy products; liver; provitamin carotene in orange and deep green fruits and vegetables.	Vitamin A is a component of rhodopsin; carotenoids can serve as antioxidants; still under intense study.	Night blindness; keratinization of epithelial tissues including the cornea of the eye (xerophthalmia) causing permanent blindness; dry, scaling skin; increased susceptibility to infection.	Damage to liver, bone; headache, irritability, vomiting, hair loss, blurred vision; some fetal defects; yellowed skin.
D	Fortified and full-fat dairy products, egg yolk (diet often not as important as sunlight exposure).	Promotes absorption and use of calcium and phosphorus.	Rickets (bone deformities) in children; osteomalacia (bone softening) in adults.	Calcium deposition in soft tissues leading to cerebral, cardiovascular, and kidney damage.
E	Vegetable oils and their products; nuts, seeds.	Antioxidant to prevent cell membrane damage; still under intense study.	Possible anemia and neurological effects.	Generally nontoxic; may worsen clotting defect in vitamin K deficiency.
K	Green vegetables; tea.	Aids in formation of certain proteins, especially those for blood clotting.	Defective blood coagulation causing severe bleeding on injury.	Liver damage and anemia from high doses of the synthetic form menadione.
Water-Soluble				
Thiamin (B$_1$)	Pork, legumes, peanuts, enriched or whole-grain products.	Coenzyme used in energy metabolism.	Nerve changes, sometimes edema, heart failure; beriberi.	Generally nontoxic, but repeated injections may cause shock reaction.
Riboflavin (B$_2$)	Dairy products, meats, eggs, enriched grain products, green leafy vegetables.	Coenzyme used in energy metabolism.	Skin lesions.	Generally nontoxic.
Niacin	Nuts, meats; provitamin tryptophan in most proteins.	Coenzyme used in energy metabolism.	Pellagra (multiple vitamin deficiencies including niacin).	Flushing of face, neck, hands; potential liver damage.
B$_6$	High-protein foods in general.	Coenzyme used in amino-acid metabolism.	Nervous, skin, and muscular disorders; anemia.	Unstable gait, numb feet, poor coordination.
Folic acid	Green vegetables, orange juice, nuts, legumes, grain products.	Coenzyme used in DNA and RNA metabolism.	Megaloblastic anemia (large, immature red blood cells); digestive disturbances.	Masks vitamin B$_{12}$ deficiency; interferes with drugs used to control epilepsy.
B$_{12}$	Animal products.	Coenzyme used in DNA and RNA metabolism.	Megaloblastic anemia; pernicious anemia when due to inadequate intrinsic factor; nervous system damage.	Thought to be nontoxic.
Pantothenic acid	Animal products and whole grains; widely distributed in foods.	Coenzyme used in energy metabolism.	Fatigue, numbness, and tingling of hands and feet.	Generally nontoxic; occasionally causes diarrhea.
Biotin	Widely distributed in foods.	Coenzyme used in energy metabolism.	Scaly dermatitis.	Thought to be nontoxic.
C (ascorbic acid)	Fruits and vegetables, especially broccoli, cabbage, cantaloupe, cauliflower, citrus fruits, green pepper, kiwi fruit, strawberries.	Functions in synthesis of collagen; is an antioxidant; aids in detoxification; improves iron absorption; still under intense study.	Scurvy; petechiae (minute hemorrhages); weakness, delayed wound healing; impaired immune response.	GI upsets; confounds certain lab tests.

continued

Key Information About Vitamins and Many Essential Minerals (continued)

Mineral	Major dietary sources	Major functions	Signs of severe, prolonged deficiency	Signs of extreme excess
Calcium	Milk, cheese, dark green vegetables, legumes.	Bone and tooth formation; blood clotting; nerve transmission.	Stunted growth; perhaps less bone mass.	Depressed absorption of some other minerals; possible kidney damage.
Phosphorus	Milk, cheese, meat, poultry, whole grains.	Bone and tooth formation; acid-base balance; component of coenzymes.	Weakness; demineralization of bone.	Depressed absorption of some minerals.
Magnesium	Whole grains, green leafy vegetables.	Component of enzymes.	Neurological disturbances.	Neurological disturbances.
Sulfur	Sulfur-containing amino acids in dietary proteins.	Component of cartilage, tendons, and proteins.	(Related to protein deficiency.)	Excess sulfur-containing amino acid intake leads to poor growth; liver damage.
Sodium	Salt, soy sauce, cured meats, pickles, canned soups, processed cheese.	Body water balance, nerve and muscle function.	Muscle cramps; reduced appetite.	High blood pressure in genetically predisposed individuals.
Potassium	Meats, milk, many fruits and vegetables, whole grains.	Body water balance; nerve and muscle function.	Muscular weakness; paralysis.	Muscular weakness; cardiac arrest.
Chloride	(Same as sodium)	Plays a role in acid-base balance; formation of gastric juice.	Muscle cramps; reduced appetite; poor growth.	High blood pressure in genetically predisposed individuals.
Iron	Meats, eggs, legumes, whole grains, green leafy vegetables.	Components of hemoglobin and enzymes.	Iron-deficiency anemia, weakness, impaired immune function.	Acute: shock, death. Chronic: liver damage, cardiac failure.
Iodine	Marine fish and shellfish; dairy products; iodized salt; some breads.	Component of thyroid hormones.	Goiter (enlarged thyroid).	Iodide goiter.
Fluoride	Drinking water, tea, seafood.	Maintenance of tooth (and maybe bone) structure.	Higher frequency of tooth decay.	Acute: Digestive distress. Chronic: mottling of teeth; skeletal deformation.
Zinc	Meats, seafood, whole grains.	Component of enzymes; growth factor for senses of taste and smell.	Growth failure; scaly dermatitis; reproductive failure; impaired immune function; taste and smell deficits.	Acute: nausea; vomiting; diarrhea. Chronic: Adversely affects copper metabolism, anemia, and immune function.
Selenium	Seafood, meats, whole grains.	Components of enzymes; functions in close association with vitamin E.	Muscle pain; maybe heart muscle deterioration.	Nausea and vomiting; hair and nail loss.
Copper	Seafood, nuts, legumes, organ meats.	Component of enzymes.	Anemia; bone and cardiovascular changes.	Nausea: liver damage.
Cobalt	Vitamin B_{12} (animal products).	Component of vitamin B_{12}.	Not reported except as vitamin B_{12} deficiency.	With alcohol: heart failure.
Chromium	Brewer's yeast, liver, seafood, meat, some vegetables.	Involved in glucose and energy metabolism.	Impaired glucose metabolism.	Lung and kidney damage (occupational exposures only).
Manganese	Nuts, whole grains, vegetables and fruits, tea.	Component of enzymes.	Abnormal bone and cartilage.	Central nervous system damage (occupational exposures).
Molybdenum	Legumes, cereals, some vegetables.	Component of enzymes.	Disorder in nitrogen excretion.	Inhibition of enzymes; adversely affects copper metabolism.

References: (1) Shils, M. E. Magnesium. In *Modern Nutrition in Health and Disease,* eds. M. E. Shils and V. R. Young. Philadelphia: Lea & Febiger, 1988. (2) Fairbanks, V. F., and E. Beutler. Iron (From same as [1].) (3) Solomons, N. W. Zinc and Copper. (From same as [1].) (4) Underwood, E. J. *Trace Elements in Human and Animal Nutrition.* New York: Academic Press, 1977.

A.D.A.M.® Correlations

This appendix is a handy reference tool to enhance your knowledge of anatomy and physiology when you have access to the *A.D.A.M. Standard®* software program. Organized by chapter, the table below lists where links occur between *A.D.A.M. Standard®* and selected figures from *Essentials of Human Anatomy & Physiology* and its companion text, the *Anatomy & Physiology Coloring Workbook.*

As you make your way through this textbook, you will notice that some illustrations include a "walking man" icon. This icon identifies figures for which using *A.D.A.M. Standard®* would help you to learn the associated anatomy. In this appendix, figures are identified by figure number and key structure and are listed with the appropriate view and layer in *A.D.A.M.®*

Although your instructor will, no doubt, help you to get started with *A.D.A.M. Standard,®* here are a couple of hints that will be important in helping you navigate through the program. When in the "Full Anatomy" option, the term "View" refers to the "Anterior," "Lateral," "Posterior," or "Medial" view in the primary window of the computer screen. "Layer" refers to one or more of the layers listed in the right corner of the screen. Using the indicated layer will bring the structure into full view.

Essentials Figure Number	*Coloring Workbook* Figure Number	Key Structures	A.D.A.M.® View and Layer
1.2b	none	Skeletal system	Using "View" on the Menu Bar, choose Skeletal system
1.2c	none	Muscular system	Using "View" on the Menu Bar, choose Muscular system
1.2d	1.3	Nervous system	Using "View" on the Menu Bar, choose Nervous system
1.2e	none	Endocrine system	Using "View" on the Menu Bar, choose Endocrine system
1.2f	1.1	Cardiovascular system	Using "View" on the Menu Bar, choose Cardiovascular system
1.2g	none	Lymphatic system	Using "View" on the Menu Bar, choose Lymphatic system
1.2h	1.2	Respiratory system	Using "View" on the Menu Bar, choose Respiratory system
1.2i	1.5	Digestive system	Using "View" on the Menu Bar, choose Digestive system
1.2j	1.4	Urinary system	Using "View" on the Menu Bar, choose Urinary system; using "Options," choose Female; using "Options," choose Male
1.2k,l	1.6	Reproductive system	Using "View" on the Menu Bar, choose Reproductive system; using "Options," choose Female; using "Options," choose Male

continued

A.D.A.M.® Correlations (continued)

Essentials Figure Number	Coloring Workbook Figure Number	Key Structures	A.D.A.M.® View and Layer
4.1c	4.1	Pericardium	Go to "Full Anatomy" View: Anterior Layer: Heart-cut section
4.1c	4.1	Peritoneums, serous membranes	Go to "Full Anatomy" View: Medial Layer: Peritoneum
5.1	5.14	Skeletal system	Using "View" on the Menu Bar, choose Skeletal system
5.6	5.14	Skeletal system	Using "View" on the Menu Bar, choose Skeletal system
5.7	5.3a	Skull bones	Go to "Full Anatomy" View: Lateral Layer: Skull
5.10	5.4	Paranasal sinuses	Go to "Full Anatomy" View: Anterior Layer: Skull-coronal section
5.10	5.4	Paranasal sinuses	Go to "Full Anatomy" View: Medial Layer: Nerves of the adductor canal
5.11	5.3c	Skull bones	Go to "Full Anatomy" View: Anterior Layer: Skull
5.14	5.6	Vertebral column	Go to "Full Anatomy" View: Medial Layer: Central nervous system and sacral plexus
5.18	none	Sacrum and lumbar vertebrae	Go to "Full Anatomy" View: Posterior Layer: Bones-coronal section
5.19	5.7	Thorax	Using "View" on the Menu Bar, choose Skeletal system
5.19	none	Bony thorax	Go to "Full Anatomy" View: Anterior Layer: Ribs
5.20	5.8	Scapula	Go to "Full Anatomy" View: Posterior Layer: Scapula
6.15 and 6.20	6.6 and 6.10	Superficial muscles of the anterior body	Using "View" on the Menu Bar, choose Muscular system

A.D.A.M.® Correlations (continued)

Essentials Figure Number	*Coloring Workbook* Figure Number	Key Structures	A.D.A.M.® View and Layer
6.16 and 6.21	6.7 and 6.11	Superficial muscles of the posterior body	Using "View" on the Menu Bar, choose Muscular system
7.12	none	Regions of the brain	Go to "Full Anatomy" View: Medial Layer: Central nervous system and sacral plexus
7.13	7.3	Surface of the brain	Go to "Full Anatomy" View: Lateral Layer: Brain
7.15	7.4	Midsagittal brain	Go to "Full Anatomy" View: Medial Layer: Nerves of the adductor canal
7.16	7.5	Meninges of the brain	Go to "Full Anatomy" View: Anterior Layer: Bones-coronal section
7.19	none	Spinal cord	Go to "Full Anatomy" View: Posterior Layer: Spinal cord
7.23c	none	Peripheral nerves of the sacral plexus in the lower limb	Go to "Full Anatomy" View: Posterior Layer: Sciatic nerve and branches
7.25	none	Autonomic nervous system	Go to "Full Anatomy" View: Medial Layer: Autonomic nerve plexuses
8.2	8.1	Extrinsic muscles of the eye	Go to "Full Anatomy" View: Lateral Layer: Eye muscles-lateral
9.3	9.2	Endocrine system (Female)	Using "Options," choose Female; using "View" on the Menu Bar, choose Endocrine system
9.3	9.2	Endocrine system (Male)	Using "Options," choose Male; using "View" on the Menu Bar, choose Endocrine system
11.1	none	Pericardium	Go to "Full Anatomy" View: Anterior Layer: Heart-cut section
11.2a	11.1	Heart, coronary vessels, and great vessels	Go to "Full Anatomy" View: Anterior Layer: Cardiac veins

continued

A.D.A.M.® Correlations (continued)

Essentials Figure Number	Coloring Workbook Figure Number	Key Structures	A.D.A.M.® View and Layer
11.2b	11.2	Interior of the heart	Go to "Full Anatomy" View: Anterior Layer: Hear-cut section
11.3	none	Pulmonary circulation	Go to "Full Anatomy" View: Anterior Layer: Pulmonary arteries
11.8	none	Lymphatic vessels and nodes	Using "View" on the Menu Bar, choose Lymphatic system
11.10	11.6	Arteries of the systemic circulation	Using "View" on the Menu Bar, choose Cardiovascular system
11.11	11.5	Veins of the systemic circulation	Using "View" on the Menu Bar, choose Cardiovascular system
11.13	11.7	Hepatic portal circulation	Go to "Full Anatomy" View: Anterior Layer: Portal vein and tributaries
13.2	13.1	Upper respiratory tract	Go to "Full Anatomy" View: Medial Layer: Nasal conchae
14.1	14.1	Digestive system	Using "View" on the Menu Bar, choose Digestive system
14.2a	14.2	Oral cavity	Go to "Full Anatomy" View: Anterior Layer: Oral mucosa
14.4	14.3	Stomach	Go to "Full Anatomy" View: Anterior Layer: Stomach-coronal section
14.4	14.6	Liver, gallbladder, and associated ducts	Go to "Full Anatomy" View: Anterior Layer(s): Liver, gallbladder
14.4	14.6	Pancreas and duodenum	Go to "Full Anatomy" View: Anterior Layer: Pancreas
14.5	none	Peritoneal attachments	Go to "Full Anatomy" View: Anterior Layer: Greater omentum
15.1	15.1	Female urinary system; Male urinary system	Using "View" on the Menu Bar, choose Urinary system; using "Options," choose Female; using "Options," choose Male

A.D.A.M.® Correlations (continued)

Essentials Figure Number	Coloring Workbook Figure Number	Key Structures	A.D.A.M.® View and Layer
15.2	15.2	Internal kidney	Go to "Full Anatomy" View: Anterior Layer: Ureter
16.2	16.1	Male reproductive system	Using "Options," choose Male Go to "Full Anatomy" View: Medial Layer: Urinary bladder and prostate gland
16.6	16.5	Female reproductive system	Using "Options," choose Female Go to "Full Anatomy" View: Medial Layer: Uterus
16.10	16.9	Breast	Using "Options," choose Female Go to "Full Anatomy" View: Anterior Layer: Breast

Photographic Credits

Art Credits

Glossary

Pronunciations in the text and this glossary use the following rules:

1. Accent marks follow stressed syllables. The primary stress is shown by ′ and the secondary stress by ″.

2. Unless otherwise noted, assume that vowels at the ends of syllables are long and vowels followed by consonants are short. Exceptions to this rule are indicated by a bar (̄) over the vowel, which indicates a long vowel, or a breve sign (̆) over the vowel, indicating that the vowel is short.

For example, the phonetic spelling of "thrombophlebitis" is *throm″bo-flĕ-bi′tis*. The next-to-last syllable *(bi′)* receives the greatest stress, and the first syllable *(throm″)* gets the secondary stress. The vowel in the second syllable comes at the end of the syllable and is long. The vowel that comes at the end of the third syllable is short because it has a breve sign.

Abdomen *ab′do-men* the portion of the body between the diaphragm and the pelvis.

Abduct *ab-dukt′* to move away from the midline of the body.

Abortion *ab-bor′shun* termination of a pregnancy before the embryo or fetus is viable outside the uterus.

Absorption *ab-sorp′shun* passage of a substance into or across a blood vessel or membrane.

Accommodation (1) adaptation in response to differences or changing needs; (2) adjustment of the eye for seeing objects at close range.

Acetabulum *as″ĕ-tab′u-lum* the cuplike cavity on the lateral surface of the hip bone that receives the femur.

Acetylcholine *as″ĕ-til-ko′lēn* a chemical transmitter substance released by certain nerve endings.

Achilles tendon *ah-kil′ēz ten′don* the tendon that attaches the calf muscles to the heel bone; also called the calcaneal tendon.

Acid a substance that liberates hydrogen ions when in an aqueous solution (compare with *base*).

Acidosis *as″ĭ-do′sis* a condition in which the blood has an excess hydrogen ion concentration and a decreased pH.

Acne inflammatory disease of the skin; infection of the sebaceous glands.

Acromegaly *ak″ro-meg′a-le* an abnormal pattern of bone and connective-tissue growth characterized by enlarged hands, face, and feet and associated with excessive secretion of pituitary growth hormone in the adult.

Acromion *ah-kro′me-on* the outer projection of the spine of the scapula; the highest point of the shoulder.

Acrosome *ak′ro-sōm* an enzyme-containing structure covering the nucleus of the sperm.

Actin *ak′tin* a contractile protein of muscle.

Action potential an electrical event occurring when a stimulus of sufficient intensity is applied to a neuron or muscle cell, allowing sodium ions to move into the cell and reverse the polarity.

Active immunity immunity produced by an encounter with an antigen; provides immunologic memory.

Active transport net movement of a substance across a membrane against a concentration or electrical gradient; requires release and use of cellular energy.

Adaptation (1) any change in structure or response to suit a new environment; (2) decline in the transmission of a sensory nerve when a receptor is stimulated continuously and without change in stimulus.

Addison's disease condition resulting from deficient secretion of adrenal cortical hormones.

Adduct *ah-dukt′* to move toward the midline of the body.

Adenosine triphosphate (ATP) *ah-den″o-sin tri-fos′fāt* the compound that is the important intracellular energy source; cellular energy.

Adipose *ad′ĭ-pōs* fatty.

Adrenal glands *ah-dre′nal* hormone-producing glands located superior to the kidneys; each consists of a medulla and cortex areas.

Adrenaline *ah-dren″ah-lin* epinephrine.

Adrenergic fibers *ad″ren-er′jik* nerve fibers that release norepinephrine.

Adrenocorticotropic hormone (ACTH) *ad-rē′no-kor″tĭ-ko-trōp′ik* an anterior pituitary hormone that influences the activity of the adrenal cortex.

Aerobic *a-er-o′bik* requiring oxygen to live or grow.

Afferent *af′er-ent* carrying to or toward a center.

Afferent neurons *nu′ronz* nerve cells that carry impulses toward the central nervous system.

Agglutination *ah-gloo″tin-a′shun* clumping of (foreign) cells, induced by cross-linking of antigen-antibody complexes.

Agglutinins *ah-gloo′tĭ-ninz* antibodies in blood plasma that cause clumping of corpuscles or bacteria.

Agglutinogens *ag″loo-tin′o-jenz* (1) antigens that stimulate the formation of a specific agglutinin; (2) antigens found on red blood cells that are responsible for determining the ABO blood group classification.

Agonist *ag′o-nist* a muscle that bears the primary responsibility for causing a certain movement; a prime mover.

Albumin *al-bu′min* a protein found in virtually all animal tissue and fluid. The most abundant plasma protein.

Albuminuria *al″bu-mĭ-nu′re-ah* presence of albumin in the urine.

Aldosterone *al-dos-ter′ōn* a hormone produced by the adrenal cortex that is important in sodium retention and reabsorption by kidney tubules.

Alimentary *al″ĭ-men′tar-e* pertaining to the digestive organs.

Alkalosis *al″kah-lo′sis* a condition in which the blood has a lower hydrogen ion concentration than normal, and an increased pH.

Allergy *al′er-je* overzealous immune response to an otherwise harmless antigen; also called hypersensitivity.

Alopecia *al″o-pe′she-ah* baldness, condition of hair loss.

Alveolus *al-ve′o-lus* (1) a general term referring to a small cavity or depression; (2) an air sac in the lungs.

Amino acid *ah-me′no* an organic compound containing nitrogen, carbon, hydrogen, and oxygen; the building block of protein.

Amnion *am′ne-on* the fetal membrane that forms a fluid-filled sac around the embryo.

Amphiarthrosis *am″fe-ar-thro′sis* a slightly movable joint.

Anabolism *an-nab′o-lizm* the energy-requiring building phase of metabolism in which simpler substances are combined to form more complex substances.

Anaerobic *an-a′-er-ōb-ik* not requiring oxygen.

Anatomy the science of the structure of living organisms.

Androgen *an′dro-jen* a hormone that controls male secondary sex characteristics.

Anemia *ah-ne′me-ah* reduced oxygen-carrying capacity of the blood caused by a decreased number of erythrocytes or decreased percentage of hemoglobin in the blood.

Aneurysm *an′u-rizm* blood-filled sac in an artery wall caused by dilation or weakening of the wall.

Angina pectoris *an-ji′nah pek′tor-is* severe, suffocating chest pain caused by brief lack of oxygen supply to heart muscle.

Anorexia *an″o-rek′se-ah* loss of appetite or desire for food.

Anoxia *ah-nok′se-ah* a deficiency of oxygen.

Antagonists *an-tag′o-nists* muscles that act in opposition to an agonist or prime mover.

Anterior *an-ter′e-er* the front of an organism, organ, or part; the ventral surface.

Antibody *an″tĭ-bod′e* a specialized substance produced by the body that can provide immunity against a specific antigen.

Antidiuretic hormone (ADH) *an″ti-di″yer-eb′tik* a hormone released by the posterior pituitary that promotes the reabsorption of water by the kidney.

Antigen *an′tĭ-jen* any substance—including toxins, foreign proteins, or bacteria—that, when introduced to the body, is recognized as foreign and activates the immune system.

Anus *a′nus* the distal end of the digestive tract; the outlet of the rectum.

Aorta *a-or′tah* the major systemic artery; arises from the left ventricle of the heart.

Aortic body a receptor in the aortic arch sensitive to changing oxygen, carbon dioxide, and pH levels of the blood.

Apocrine gland *ap′o-krin* the less numerous type of sweat gland. Produces a secretion containing water, salts, and proteins.

Aponeurosis *ap′o-nu-ro′sis* fibrous or membranous sheet connecting a muscle and the part it moves.

Appendicular skeleton *ap″en-dik′u-lar* bones of the limbs and limb girdles that are attached to the axial skeleton.

Appendix *ah-pen′diks* a wormlike sac attached to the cecum of the large intestine.

Aqueous humor *a′kwe-us hu′mer* the watery fluid in the anterior chambers of the eye.

Arachnoid *ah-rak′noid* weblike; specifically, the weblike middle layer of the three meninges.

Areola *ah-re′o-lah* the circular, pigmented area surrounding the nipple.

Arrector pili *ah-rek′tor pi′li* tiny, smooth muscles attached to hair follicles, which cause the hair to stand upright when activated.

Arteriole *ar-tēr′e-ol* minute artery.

Arteriosclerosis *ar-tēr′e-o-skler-o′sis* any of a number of proliferative and degenerative changes in the arteries leading to their decreased elasticity.

Artery a vessel that carries blood away from the heart.

Arthritis *ar-thri′tis* inflammation of the joints.

Articulation joint; point where two bones meet.

Asthma *az′mah* disease or allergic response characterized by bronchial spasms and difficult breathing.

Astigmatism *ah-stig′mah-tizm* a visual defect resulting from irregularity in the lens or cornea of the eye causing the image to be out of focus.

Atherosclerosis *a"ther-o"skler-o'sis* changes in the walls of large arteries consisting of lipid deposits on the artery walls. The early stage of arteriosclerosis.

Atlas the first cervical vertebra; articulates with the occipital bone of the skull and the second cervical vertebra (axis).

Atom *at'um* the smallest part of an element; indivisible by ordinary chemical means.

Atomic mass number the sum of the number of protons and neutrons in the nucleus of an atom.

Atomic number the number of protons in an atom.

Atomic symbol a one- or two-letter symbol indicating a particular element.

Atomic weight average of the mass numbers of all of the isotopes of an element.

Atrioventricular node (AV node) *a"tre-o-ven-trik'u-lar* a specialized mass of conducting cells located at the atrioventricular junction in the heart.

Atrium *a'tre-um* a chamber of the heart receiving blood from the veins.

Atrophy *at'ro-fe* a reduction in size or wasting away of an organ or cell resulting from disease or lack of use.

Auditory *aw'dĭ-to"re* pertaining to the sense of hearing.

Auditory ossicles *aw'sih'kulz* the three tiny bones serving as transmitters of vibrations and located within the middle ear: the malleus, incus, and stapes.

Auditory tube tube connecting the middle ear and the pharynx.

Auscultation *aws"kul-ta'shun* the act of examination by listening to body sounds.

Autoimmune response the production of antibodies or effector T cells that attack a person's own tissue.

Automaticity *aw-to"mah-tis'ĭ-te* the ability of a structure, organ, or system to initiate its own activity.

Autonomic *aw"to-nom'ik* self-directed; self-regulating; independent.

Autonomic nervous system the division of the nervous system that functions involuntarily; innervates cardiac muscle, smooth muscle, and glands.

Axial skeleton *ak'se-al* the bones of the skull, vertebral column, thorax, and sternum.

Axilla *ak-sih'lah* armpit.

Axis (1) the second cervical vertebra; has a vertical projection called the dens around which the atlas rotates; (2) the imaginary line about which a joint or structure revolves.

Axon *ak'son* neuron process that carries impulses away from the nerve cell body; efferent process; the conducting portion of a nerve cell.

B cells lymphocytes that oversee humoral immunity; their descendants differentiate into antibody-producing plasma cells; also called B lymphocytes.

Bacteria any of a large group of microorganisms, generally one-celled; found in humans and other animals, plants, soil, air, and water; have a broad range of functions.

Basal metabolic rate *met"ah-bol'ik* the rate at which energy is expended (heat produced) by the body per unit time under controlled (basal) conditions: 12 hours after a meal, at rest.

Basal nuclei *nu'kle-i* gray matter areas deep within the white matter of the cerebral hemispheres.

Base a substance that accepts hydrogen ions; proton acceptor.

Basement membrane a thin layer of extracellular material to which epithelial cells are attached in mucosa surfaces.

Basophils *ba'zo-filz* white blood cells whose granules stain deep blue with basic dye; have a relatively pale nucleus and granular-appearing cytoplasm.

Benign *be-nīn'* not malignant.

Biceps *bi'seps* two-headed, especially applied to certain muscles.

Bicuspid *bi-kus'pid* having two points or cusps.

Bile a greenish-yellow or brownish fluid produced in and secreted by the liver, stored in the gallbladder, and released into the small intestine.

Biopsy *bi'ŏp-se* the removal and examination of live tissue; usually done to detect or rule out the presence of cancerous cells.

Blastocyst *blas'to-sist* a stage of early embryonic development.

Blood-brain barrier a mechanism that inhibits passage of materials from the blood into brain tissues.

Bolus *bo'lus* a rounded mass of food prepared by the mouth for swallowing.

Bony thorax *bōn'e tho'raks* bones of the thorax, including ribs, sternum, and thoracic vertebrae.

Bowman's capsule *bo'manz* the double-walled cup at the end of a nephron that encloses a glomerulus; also called a glomerular capsule.

Brachial *bra'ke-al* pertaining to the arm.

Bradycardia *brad"e-kar'de-ah* slow heart beat, usually defined as a rate under 60 beats per minute.

Brain stem the portion of the brain consisting of the medulla, pons, and midbrain.

Bronchitis *brong-ki'tis* inflammation of the mucous membranes of the bronchi.

Bronchus *brong'kus* one of the two large branches of the trachea leading to the lungs.

Buccal *buk'al* pertaining to the cheek.

Buffer a substance or substances that help to stabilize the pH of a solution.

Bundle branch block a blocking of heart action resulting from damage to one of the bundle branches; delayed contraction of one ventricle.

Bursa *ber'sah* a small sac filled with fluid and located at friction points, especially joints.

Calcitonin *kal"sĭ-to'nin* a hormone released by the thyroid that decreases calcium levels of the blood.

Calculus *kal'ku-lus* a stone formed within various body parts.

Calorie *kal'o-re* unit of heat; the large calorie is the amount of heat required to raise 1 kg of water 1° C; also used in metabolic and nutrition studies as the unit to measure the energy value of foods.

Calyx *ka'liks* a cuplike extension of the pelvis of the kidney.

Canal a duct or passageway; a tubular structure.

Canaliculus *kan"ah-lik'u-lus* extremely small tubular passage or channel.

Cancer a malignant, invasive cellular neoplasm that has the capability of spreading throughout the body or body parts.

Capillary *kap'ĭ-lar"e* a minute blood vessel connecting arterioles with venules.

Carbohydrate *kar"bo-hi'drāt* organic compound composed of carbon, hydrogen, and oxygen; includes starches, sugars, cellulose.

Carcinogen *kar-sin'o-jen* cancer-causing agent.

Carcinoma *kar"sĭ-no'mah* cancer; a malignant growth of epithelial cells.

Cardiac *kar'de-ak* pertaining to the heart.

Cardiac cycle sequence of events encompassing one complete contraction and relaxation of the atria and ventricles of the heart.

Cardiac muscle specialized muscle of the heart.

Cardiac output the blood volume (in liters) ejected per minute by the left ventricle.

Cardioesophageal sphincter *kar"de-o-ě-sof"ah-je'al sfingk'ter* valve between the stomach and esophagus.

Carotid *kah-rot'id* the main artery in the neck.

Carotid body a receptor in the common carotid artery sensitive to changing oxygen, carbon dioxide, and pH levels of the blood.

Carotid sinus *si'nus* a dilation of a common carotid artery; involved in regulation of systemic blood pressure.

Carpal *kar'pal* one of the eight bones of the wrist.

Cartilage *kar'tĭ-lij* white, semiopaque connective tissue.

Catabolism *kah-tab'o-lizm* the process in which living cells break down substances into simpler substances; destructive metabolism.

Cataract *kat'ah-rakt* partial or complete loss of transparency of the crystalline lens of the eye.

Catecholamines *kat"ě-kol'ah-menz* epinephrine and norepinephrine.

Caudal *kaw'dal* in humans, the inferior portion of the anatomy.

Cecum *se'kum* the blind-end pouch at the beginning of the large intestine.

Cell the basic biological unit of living organisms, containing a nucleus and a variety of organelles enclosed by a limiting membrane.

Cell membrane see *plasma membrane*.

Cellular immunity *sel'u-lar ĭ-mu'nĭ-te* immunity conferred by lymphocytes called T cells. Also referred to as cell-mediated immunity.

Cellular respiration metabolic processes in which ATP is produced.

Cementum *se-men'tum* the bony connective tissue that covers the root of a tooth.

Central nervous system (CNS) the brain and the spinal cord.

Centriole *sen'tre-ōl* a minute body found near the nucleus of the cell; active in cell division.

Cerebellum *ser"ě-bel'um* part of the hindbrain; involved in producing smoothly coordinated skeletal muscle activity.

Cerebral aqueduct *ser'ě-bral ak'we-dukt"* the slender cavity of the midbrain that connects the third and fourth ventricles; also called the aqueduct of Sylvius.

Cerebrospinal fluid the fluid produced in the cerebral ventricles; fills the ventricles and surrounds the central nervous system.

Cerebrum *ser'ě-brum* the largest part of the brain; consists of right and left cerebral hemispheres.

Cerumen *sě-roo'men* earwax.

Cervical *ser'vĭ-kal* refers to the neck or the necklike portion of an organ or structure.

Cervix *ser'viks* the inferior necklike portion of the uterus leading to the vagina.

Chemical bond an energy relationship holding atoms together; involves the interaction of electrons.

Chemical energy energy form stored in the bonds of chemicals.

Chemical reaction process in which molecules are formed, changed, or broken down.

Chemoreceptors *ke"mo-re-sep'torz* receptors sensitive to various chemicals in solution.

Chiasma *ki-as'mah* a crossing or intersection of two structures, such as the optic nerves.

Cholecystokinin *ko"le-sis"to-kin'in* an intestinal hormone that stimulates gallbladder contraction and pancreatic juice release.

Cholesterol *ko-les'ter-ol* a steroid found in animal fats as well as in most body tissues; made by the liver.

Cholinergic fibers *ko"lin-er'jik* nerve endings that, upon stimulation, release acetylcholine.

Chondrocyte *kon'dro-sīt* a mature cartilage cell.

Chorion *ko're-on* the outermost fetal membrane; helps form the placenta.

Choroid *ko'roid* the pigmented nutritive layer of the eye.

Chromatin *kro'mah-tin* the structures in the nucleus that carry the hereditary factors (genes).

Chromosome *kro'mo-sōm* barlike body of tightly coiled chromatin; visible during cell division.

Chyme *kīm* the semifluid stomach contents consisting of partially digested food and gastric secretions.

Cilia *sil'e-ah* tiny, hairlike projections on cell surfaces that move in a wavelike manner.

Circle of Willis a union of arteries at the base of the brain.

Circumduction *ser"kum-duk'shun* circular movement of a body part.

Cirrhosis *sĭ-ro'sis* a chronic disease of the liver, characterized by an overgrowth of connective tissue or fibrosis.

Cleavage *klēv'ij* an early embryonic phase consisting of rapid cell divisions without intervening growth periods.

Clitoris *kli'to-ris* a small, erectile structure in the female, homologous to the penis in the male.

Clonal selection *klo'nul* the process during which a B cell or T cell becomes sensitized through binding contact with an antigen.

Clone descendants of a single cell.

Coagulation clotting (of blood).

Cochlea *kok'le-ah* a cavity of the inner ear resembling a snail shell; houses the hearing receptor.

Coitus *ko'ĭ-tus* sexual intercourse.

Coma *ko'mah* unconsciousness from which the person cannot be aroused.

Complement a group of plasma proteins that normally circulate in inactive forms; when activated by complement fixation, causes lysis of foreign cells and enhances phagocytosis and inflammation.

Compound substance composed of two or more different elements, the atoms of which are chemically united.

Concave *kon'kāv* having a curved or depressed surface.

Conductivity *kon"duk-tiv'ĭ-te* ability to transmit an electrical impulse.

Condyle *kon'dīl* a rounded projection at the end of a bone that articulates with another bone.

Cones one of the two types of photoreceptor cells in the retina of the eye. Provides for color vision.

Congenital *kon-jen'ĭ-tal* existing at birth.

Conjunctiva *kon"junk-ti'vah* the thin, protective mucous membrane lining the eyelids and covering the anterior surface of the eyeball.

Conjunctivitis *kon"junk"tĭ-vi'tis* an inflammation of the conjunctiva of the eye.

Connective tissue a primary tissue; form and function vary extensively. Functions include support, storage, and protection.

Contraception *kon"trah-sep'shun* the prevention of conception; birth control.

Contraction *kon-trak'shun* to shorten or develop tension, an ability highly developed in muscle cells.

Contralateral *kon"trah-lat'er-al* opposite; acting in unison with a similar part on the opposite side of the body.

Convergence *kon-ver'jens* turning toward a common point from different directions.

Convoluted *kon'vo-lūted* rolled, coiled, or twisted.

Cornea *kor'ne-ah* the transparent anterior portion of the eyeball.

Coronal plane see *frontal plane.*

Corpus *kor'pus* body; the major portion of an organ.

Cortex *kor'teks* the outer surface layer of an organ.

Corticosteroids *kor"tĭ-ko-stĕ'roidz* the steroid hormones released by the adrenal cortex.

Cortisol *kor'tĭ-sol* a glucocorticoid produced by the adrenal cortex.

Costal *kos'tal* pertaining to the ribs.

Covalent bond *ko-va'lent* a bond involving the sharing of electrons between atoms.

Cramp painful, involuntary contraction of a muscle.

Cranial *kra'ne-al* pertaining to the skull.

Cranial nerves the 12 pairs of nerves that arise from the brain.

Crenation *kre-na'shun* the shriveling of a cell, for example an erythrocyte, resulting from loss of water.

Cretinism *kre'tin-izm* a severe thyroid deficiency in the young that leads to stunted physical and mental growth.

Cryptorchidism *krip-tor'kĭ-dizm* a developmental defect in which the testes fail (or one testis fails) to descend into the scrotum.

Cupula *ku'pu-lah* a domelike structure.

Cushing's syndrome *koosh'ingz sin'drōm* a disease produced by excess secretion of adrenocortical hormone; characterized by adipose tissue accumulation, weight gain, and osteoporosis.

Cutaneous *ku-ta'ne-us* pertaining to the skin.

Cutaneous membrane the skin; composed of epidermal and dermal layers.

Cyanosis *si"ah-no'sis* a bluish coloration of the mucous membranes and skin caused by deficient oxygenation of the blood.

Cystitis *sis-ti'tis* an inflammation of the urinary bladder.

Cytokinesis *si"to-kǐ-ne'sis* division of cytoplasm that occurs after the cell nucleus has divided.

Cytology *si-tol'o-je* the science concerned with the study of cells.

Cytoplasm *si'to-plazm"* the substance of a cell other than that of the nucleus.

Cytotoxic T cell see *killer T cell.*

Deciduous *de-sid'u-us* temporary.

Deciduous (milk) teeth the 20 temporary teeth replaced by permanent teeth; "baby" teeth.

Decomposition reaction a destructive chemical reaction in which complex substances are broken down into simpler ones.

Defecation *def"ih-ka'shun* the elimination of the contents of the bowels (feces).

Deglutition *deg"loo-tish'un* the act of swallowing.

Dehydration *de"hi-dra'shun* a condition resulting from excessive loss of water.

Dendrites *den'drīts* the branching extensions of neurons that carry electrical signals to the cell body; the receptive portion of a nerve cell.

Dentin *den'tin* the calcified tissue forming the major part of a tooth; deep to the enamel.

Deoxyribonucleic acid (DNA) *de-ok"se-ri"bo-nu"kla'ik* nucleic acid found in all living cells; carries the organism's hereditary information.

Depolarization *de-po"lar-i-za'shun* the loss of a state of polarity; the loss of a negative charge inside the plasma membrane.

Dermatitis *der"mah-ti'tis* an inflammation of the skin; nonspecific skin allergies.

Dermis *der'mis* the deep layer of the skin; composed of dense, irregular connective tissue.

Diabetes insipidus *di"ah-be'tēz in-sip'i-dus* a disease characterized by passage of large amounts of dilute urine plus intense thirst and dehydration; a hypothalamic disorder is the cause.

Diabetes mellitus *mel-li'tus* a disease caused by deficient insulin release or inadequate responsiveness to insulin, leading to inability of the body cells to use carbohydrates at a normal rate.

Dialysis *di-al'ǐ-sis* diffusion of solute(s) through a semipermeable membrane.

Diapedesis *di"ah-pě-de'sis* the passage of blood cells through intact vessel walls into the tissues.

Diaphragm *di'ah-fram* (1) any partition or wall separating one area from another; (2) a muscle that separates the thoracic cavity from the abdominopelvic cavity.

Diarthrosis *di"ar-thro'sis* a freely movable joint.

Diastole *di-as'to-le* a period (between contractions) of relaxation of the heart during which it fills with blood.

Diencephalon *di"en-sef'ah-lon* that part of the forebrain between the cerebral hemispheres and the midbrain including the thalamus, the third ventricle, and the hypothalamus.

Diffusion *dǐ-fu'zhun* the spreading of particles in a gas or solution with a movement toward uniform distribution of particles.

Digestion *di-jest'jun* the bodily process of breaking down foods chemically and mechanically.

Dilate *di'lāt* to stretch; to open; to expand.

Disaccharide *di-sak'ǐ-rīd* literally, double sugar; examples include sucrose and lactose.

Distal *dis'tal* farthest from the point of attachment of a limb or origin of a structure.

Diverticulum *di"ver-tik'u-lum* a pouch or sac in the walls of a hollow organ or structure.

Dorsal *dor'sal* pertaining to the back; posterior.

Duct *dukt* a canal or passageway.

Duodenum *du"o-de'num* the first part of the small intestine.

Dura mater *du'rah ma'ter* the outermost and toughest of the three membranes (meninges) covering the brain and spinal cord.

Dynamic equilibrium sense that reports on angular or rotatory movements of the head in space.

Dyspnea *disp'ne-ah* labored, difficult breathing.

Ectopic *ek-top'ik* not in the normal place; for example, in an ectopic pregnancy the egg is implanted at a place other than the uterus.

Edema *ě-de'mah* an abnormal accumulation of fluid in body parts or tissues; causes swelling.

Effector *ef-fek'tor* an organ, gland, or muscle capable of being activated by nerve endings.

Efferent *ef'er-ent* carrying away or away from.

Efferent neurons neurons that conduct impulses away from the central nervous system.

Ejaculation *e-jak"u-la'shun* the sudden ejection of semen from the penis.

Electrical energy energy form resulting from the movement of charged particles.

Electrocardiogram (ECG) *e-lek″tro-kar′de-o-gram″* a graphic record of the electrical activity of the heart.

Electroencephalogram (EEG) *e-lek″tro-en-sef′ah-lo-gram″* a graphic record of the electrical activity of nerve cells in the brain.

Electrolyte *e-lek′tro-līt* a substance that breaks down into ions when in solution and is capable of conducting an electric current.

Electron negatively charged subatomic particle; orbits the atomic nucleus.

Electron transport chain metabolic pathway within the mitochondria in which energy harvested from high-energy hydrogen atoms is used to make ATP. Final delivery of H to molecular oxygen produces water.

Element *el′ĕ-ment* any of the building blocks of matter; oxygen, hydrogen, carbon, for example.

Embolism *em′bo-lizm* the obstruction of a blood vessel by an embolus (blood clot, bubble of air, or fatty mass floating in the blood).

Embryo *em′bre-o* an organism in its early stages of development; in humans, the first 2 months after conception.

Emesis *em′ĕ-sis* vomiting.

Emphysema *em″fĭ-se′mah* a condition caused by overdistension of the pulmonary alveoli and fibrosis of lung tissue.

Enamel the hard, calcified substance that covers the crown of a tooth.

Endocarditis *en″do-kar-di′tis* an inflammation of the inner lining of the heart.

Endocardium *en″do-kar′de-um* the endothelial membrane lining the interior of the heart.

Endocrine glands *en′do-krin* ductless glands that empty their hormonal products directly into the blood.

Endocrine system body system that includes internal organs that secrete hormones.

Endometrium *en-do-me′tre-um* the mucous membrane lining of the uterus.

Endomysium *en″do-mis′e-um* the thin connective tissue surrounding each muscle cell.

Endoplasmic reticulum *en″do-plas′mik rĕ-tik′u-lum* a membranous network of tubular or saclike channels in the cytoplasm of a cell.

Endothelium *en″do-the′le-um* the single layer of simple squamous cells that line the walls of the heart and the vessels that carry blood and lymph.

Energy the ability to do work.

Enzyme *en′zīm* a substance formed by living cells that acts as a catalyst in bodily chemical reactions.

Eosinophils *e″o-sin′o-filz* granular white blood cells whose granules readily take up a stain called eosin.

Epidermis *ep″ĭ-der′mis* the outer layers of the skin; epithelium.

Epididymis *ep″ĭ-did′ĭ-mis* that portion of the male duct system in which sperm mature. Empties into the *ductus,* or *vas, deferens.*

Epiglottis *ep″ĭ-glot′is* the elastic cartilage at the back of the throat; covers the glottis during swallowing.

Epimysium *ep″ĭ-mis′e-um* the sheath of fibrous connective tissue surrounding a muscle.

Epinephrine *ep″ĭ-nef′rin* the chief hormone of the adrenal medulla.

Epiphysis *ĕ-pif′ĭ-sis* the end of a long bone.

Epithelium *ep″ĭ-the′le-um* one of the primary tissues; covers the surface of the body and lines the body cavities, ducts, and vessels.

Equilibrium *e″kwĭ-lib′re-um* balance; a state when opposite reactions or forces counteract each other exactly.

Erythrocytes *ĕ-rith′ro-sīts* red blood cells.

Erythropoiesis *ĕ-rith″ro-poi-e′sis* the process of erythrocyte formation.

Estrogens *es′tro-jenz* hormones that stimulate female secondary sex characteristics; female sex hormones.

Eupnea *ūp-ne′ah* easy, normal breathing.

Exchange reaction a chemical reaction in which bonds are both made and broken; atoms become combined with different atoms.

Excretion *ek-skre′shun* the elimination of waste products from the body.

Exocrine glands *ek′so-krin* glands that have ducts through which their secretions are carried to a particular site.

Expiration *eks″pĭ-ra′shun* the act of expelling air from the lungs; exhalation.

Extracellular *eks″trah-sel′u-lar* outside a cell.

Extracellular fluid fluid within the body but outside the cells.

Fallopian tube *fal-lo′pe-an* see *uterine tube.*

Fascia *fash′e-ah* layers of fibrous tissue covering and separating muscles.

Fascicle *fas′ĭ-kul* a bundle of nerve or muscle fibers bound together by connective tissue.

Fatty acid a building block of fats.

Feces *fe′sēz* material discharged from the bowel composed of food residue, secretions, and bacteria.

Fertilization *fer′tĭ-lĭ-za′shun* fusion of the nuclear material of an egg and a sperm.

Fetus *fe'tus* the unborn young; in humans the period from the third month of development until birth.

Fibrillation *fi-bri-la'shun* irregular, uncoordinated contraction of muscle cells, particularly of the heart musculature.

Fibrin *fi'brin* the fibrous insoluble protein formed during the clotting of blood.

Fibrinogen *fi-brin'o-jen* a blood protein that is converted to fibrin during blood clotting.

Filtration *fil-tra'shun* the passage of a solvent and dissolved substances through a membrane or filter.

Fissure *fis'zher* (1) a groove or cleft; (2) the deepest depressions or inward folds on the brain.

Fixators *fiks-a'torz* muscles acting to immobilize a joint or a bone; fixes the origin of a muscle so that muscle action can be exerted at the insertion.

Flaccid *flak'sid* soft; flabby; relaxed.

Flagella *flah-jel'ah* long, whiplike extensions of the cell membrane of some bacteria and sperm. Serve to propel the cell.

Flexion *flek'shun* bending; the movement that decreases the angle between bones.

Focus *fo'kus* creation of a sharp image by a lens.

Follicle *fol'li-kul* (1) structure in an ovary consisting of a developing egg surrounded by follicle cells; (2) colloid-containing structure in the thyroid gland.

Follicle-stimulating hormone (FSH) a hormone produced by the anterior pituitary that stimulates ovarian follicle production in females and sperm production in males.

Fontanels *fon"tah-nelz'* the fibrous membranes in the skull where bone has not yet formed; babies' "soft spots."

Foramen *fo-ra'men* a hole or opening in a bone or between body cavities.

Formed elements cellular portion of blood.

Fossa *fos'ah* a depression; often an articular surface.

Fovea *fo-ve'ah* a pit.

Frontal (coronal) plane a longitudinal plane that divides the body or an organ into anterior and posterior parts.

Fundus *fun'dus* the base of an organ; that part farthest from the opening of the organ.

Gallbladder the sac beneath the right lobe of the liver used for bile storage.

Gallstones particles of hardened cholesterol or calcium salts that are occasionally formed in gallbladder and bile ducts.

Gamete *gam'ēt* male or female sex cell (sperm/egg).

Gametogenesis *gam"ē-to-jen'ē-sis* the formation of gametes.

Ganglion *gang'gle-on* a group of nerve cell bodies located in the peripheral nervous system.

Gastrin *gas'trin* a hormone that stimulates gastric secretion, especially hydrochloric acid release.

Gene *jēn* one of the biological units of heredity located in chromatin; transmits hereditary information.

Genetics *je-net'iks* the science of heredity.

Genitalia *jen"i-ta'le-ah* the external sex organs.

Germ layers the initial or primary tissues formed in the embryo (ectoderm, mesoderm, endoderm) from which all body tissues arise.

Gingiva *jin-ji'vah* the gums.

Gland an organ specialized to secrete or excrete substances for further use in the body or for elimination.

Glaucoma *glaw-ko'mah* an abnormal increase of the pressure within the eye.

Glia *gli'ah* see *neuroglia.*

Glomerulus *glo-mer'u-lus* a knot of coiled capillaries in the kidney; forms filtrate.

Glottis *glot'is* the opening between the vocal cords in the larynx.

Glucagon *gloo'kah-gon* a hormone formed by the pancreatic islets of the pancreas; raises the glucose level of blood.

Glucocorticoids *gloo"ko-kor't i-koidz* the adrenal cortex hormones that increase blood glucose levels and aid the body in resisting long-term stress.

Glucose *gloo'kōs* the principal sugar in the blood; a monosaccharide.

Glycerol *glis'er-ol* a sugar alcohol; one of the building blocks of fats.

Glycogen *gli'ko-jen* the main carbohydrate stored in animal cells; a polysaccharide.

Glycogenesis *gli"ko-jen'ē-sis* formation of glycogen from glucose.

Glycogenolysis *gli"ko-jě-nol'i-sis* breakdown of glycogen to glucose.

Glycolysis *gli-kol'i-sis* breakdown of glucose to pyruvic acid; an anaerobic process.

Goblet cells individual cells (simple glands) that produce mucus.

Goiter *goi'ter* a benign enlargement of the thyroid gland.

Gonadotropins *gon"ah-do-tro'pinz* gonad-stimulating hormones produced by the anterior pituitary.

Gonads *go'nadz* organs producing gametes; ovaries or testes.

Graafian follicle *graf'e-an fol'li-kul* a mature ovarian follicle; also called a vesicular follicle.

Graded response a response that varies directly with the strength of the stimulus.

Gray matter the gray area of the central nervous system; contains unmyelinated nerve fibers and nerve cell bodies.

Groin the junction of the thigh and the trunk; the inguinal area.

Growth hormone a hormone that stimulates growth in general; produced in the anterior pituitary.

Gustation *gus-ta'shun* taste.

Gyrus *ji'rus* an outward fold of the surface of the cerebral cortex.

Hamstring muscles the posterior thigh muscles: the biceps femoris, semimembranosus, and semitendinosus.

Haversian system *ha-ver'shan* see *osteon*.

Heart block impaired transmission of impulses from atrium to ventricle.

Heart murmur an abnormal heart sound (usually resulting from valve problems).

Heat stroke the failure of the heat-regulating ability of an individual under heat stress.

Helper T cell the type of T lymphocyte that orchestrates cellular immunity by direct contact with other immune cells and by releasing chemicals called lymphokines; also helps to mediate the humoral response by interacting with B cells.

Hematocrit *he-mat'o-krit* the percentage of erythrocytes to total blood volume.

Hematopoiesis *he''mato-poi-e'sis* formation of blood cells.

Hemiplegia *hem''e-ple'je-ah* paralysis of one side of the body.

Hemocytoblasts *he''mo-si'to-blastz* stem cells that give rise to all the formed elements of the blood.

Hemoglobin *he''mo-glo'bin* the oxygen-transporting pigment of erythrocytes.

Hemolysis *he-mol'ĭ-sis* the rupture of erythrocytes.

Hemophilia *he''mo-fil'e-ah* an inherited clotting defect caused by absence of a blood-clotting factor.

Hemorrhage *hem'or-ij* the loss of blood from the vessels by flow through ruptured walls; bleeding.

Hepatic portal system *hĕ-pat'ik* the circulation in which the hepatic portal vein carries dissolved nutrients to the liver tissues for processing.

Hepatitis *hep''ah-ti'tis* inflammation of the liver.

Hilum, hilus *hi'lum, hi'lus* a depressed area where vessels enter and leave an organ.

Histamine *his'tah-mēn* a substance that causes vasodilation and increased vascular permeability.

Histology *his-tol'o-je* the branch of anatomy dealing with the microscopic structure of tissues.

Homeostasis *ho''me-o-sta'sis* a state of body equilibrium or stable internal environment of the body.

Homologous *ho-mol'o-gus* parts or organs corresponding in structure but not necessarily in function.

Hormones *hor'mōnz* chemical messengers secreted by endocrine glands; responsible for specific regulatory effects on certain parts or organs.

Humoral immunity *hu'mor-al* immunity provided by antibodies released by sensitized B cells and their plasma cell progeny. Also called antibody-mediated immunity.

Hyaline *hi'ah-lin* glassy; transparent.

Hydrochloric acid *hi''dro-klo'rik* HCl; aids protein digestion in the stomach; produced by parietal cells.

Hydrogen bond weak bond in which a hydrogen atom forms a bridge between two electron-hungry atoms. An important intramolecular bond.

Hydrolysis *hi-drol'ĭ-sis* the process in which water is used to split a substance into smaller particles.

Hyperopia *hi''per-o'pe-ah* farsightedness.

Hypertension *hi''per-ten'shun* an abnormally high blood pressure.

Hypertonic *hi''per-ton'ik* excessive, above normal, tone or tension.

Hypertrophy *hi-per'tro-fe* an increase in the size of a tissue or organ independent of the body's general growth.

Hypothalamus *hi''po-thal'ah-mus* the region of the diencephalon forming the floor of the third ventricle of the brain.

Hypothermia *hi''po-ther'me-ah* subnormal body temperature.

Hypotonic *hi-po-ton'ik* below normal tone or tension.

Hypoxia *hi-pok'se-ah* a condition in which inadequate oxygen is available to tissues.

Ileum *il'e-um* the terminal part of the small intestine; between the jejunum and the cecum of the large intestine.

Immune response antigen-specific defenses mounted by activated lymphocytes (T cells and B cells).

Immunity *ĭ-mu'nĭ-te* the ability of the body to resist many agents (both living and nonliving) that can cause disease; resistance to disease.

Immunocompetence *im''mu-no-kom'pĕ-tents* the ability of the body's immune cells to recognize (by binding) specific antigens; reflects the presence of plasma membrane–bound receptors.

Immunodeficiency disease *im''mu-no-de-fish'en-se* disease resulting from the deficient production or function of immune cells or certain molecules (complement, antibodies, and so on) required for normal immunity.

Immunoglobulin *im″mu-no-glob′u-lin* a protein molecule, released by plasma cells, that mediates humoral immunity; an antibody.

Infarct *in′farkt* a region of dead, deteriorating tissue resulting from a lack of blood supply.

Inferior (caudal) pertaining to a position near the tail end of the long axis of the body.

Inflammation *in″flah-ma′shun* a physiological response of the body to tissue injury; includes dilation of blood vessels and increased vessel permeability.

Inguinal *ing′gwĭ-nal* pertaining to the groin region.

Inner cell mass a cluster of cells in the blastocyst from which the embryo develops.

Innervation *in″er-va′shun* the supply of nerves to a body part.

Inorganic compound a compound that lacks carbon; for example, water.

Insertion *in-ser′shun* the movable attachment of a muscle as opposed to its origin.

Inspiration *in″spĭ-ra′shun* the drawing of air into the lungs; inhalation.

Insulin *in′su-lin* the hypoglycemic hormone produced in the pancreas affecting carbohydrate and fat metabolism, blood glucose levels, and other systemic processes.

Integumentary system *in-teg-u-men′tar-e* the skin and its accessory organs.

Intercellular *in″ter-sel′u-lar* between the body cells.

Intercellular matrix *ma′triks* the material between adjoining cells; especially important in connective tissue.

Internal respiration the use of oxygen by body cells; synonym: cellular respiration.

Interneurons also called association neurons; completes the pathway between afferent and efferent neurons.

Interstitial fluid *in″ter-stish′al* the fluid between the cells.

Intervertebral disks *in″ter-ver′tĕ-bral* the disks of fibrocartilage between the vertebrae.

Intracellular *in″trah-sel′u-lar* within a cell.

Intracellular fluid fluid within a cell.

Intrinsic factor *in-trin′sik* a substance produced by the stomach that is required for vitamin B_{12} absorption.

Invert to turn inward.

Ion *i′on* an atom with a positive or negative electric charge.

Ionic bond bond formed by the complete transfer of electron(s) from one atom to another (or others). The resulting charged atoms, or ions, are oppositely charged and attract each other.

Ipsilateral *ip″sĭ-lat′er-al* situated on the same side.

Iris *i′ris* the pigmented, involuntary muscle that acts as the diaphragm of the eye.

Irritability *ir″ĭ-tah-bil′ĭ-te* ability to respond to a stimulus.

Ischemia *is-ke′me-ah* a local decrease in blood supply.

Isometric *i″so-met′rik* of the same length.

Isotonic *i″so-ton′ik* having a uniform tension; of the same tone.

Isotope *i′sĭ-tōp* different atomic form of the same element. Isotopes vary only in the number of neutrons they contain.

Jaundice *jawn′dis* an accumulation of bile pigments in the blood producing a yellow color of the skin.

Jejunum *je-joo′num* the part of the small intestine between the duodenum and the ileum.

Joint the junction of two or more bones; an articulation.

Keratin *ker′ah-tin* a tough, insoluble protein found in tissues such as hair, nails, and epidermis of the skin.

Ketosis *ke-to′sis* an abnormal condition during which an excess of ketone bodies are produced.

Killer T cell effector T cell that directly kills foreign cells. Also called a cytotoxic T cell.

Kinetic energy energy of motion.

Kinins *ki′ninz* group of polypeptides that dilate arterioles, increase vascular permeability, and induce pain.

Krebs cycle the aerobic pathway occurring within the mitochondria, in which energy is liberated during metabolism of carbohydrates, fats, and amino acids and CO_2 is produced.

Labia *la′be-ah* lips.

Labyrinth bony cavities and membranes of the inner ear that house the hearing and equilibrium receptors.

Lacrimal *lak′rĭ-mal* pertaining to tears.

Lactation *lak-ta′shun* the production and secretion of milk.

Lacteal *lak′te-al* special lymphatic capillaries of the small intestine that take up lipids.

Lactic acid *lak′tik* the product of anaerobic metabolism, especially in muscle.

Lacuna *lah-ku′nah* a little depression or space; in bone or cartilage, lacunae are occupied by cells.

Lamina *lam′ĭ-nah* (1) a thin layer or flat plate; (2) the portion of a vertebra between the transverse process and the spinous process.

Laryngitis *lar"in-ji'tis* an inflammation of the larynx.

Larynx *lar'inks* the cartilaginous organ located between the trachea and the pharynx; voice box.

Lateral *lat'er-al* away from the midline of the body.

Lens the elastic, doubly convex structure in the eye that focuses the light entering the eye on the retina.

Lesion *le'zhun* a tissue injury or wound.

Leukemia *lu-ke'me-ah* a cancerous condition in which there is an excessive production of immature leukocytes.

Leukocyte *lu'ko-sīt* white blood cell.

Ligament *lig'ah-ment* a cord of fibrous tissue that connects bones.

Lipid *lip'id* organic compound formed of carbon, hydrogen, and oxygen; examples are fats and cholesterol.

Lumbar *lum'bar* the portion of the back between the thorax and the pelvis.

Lumen *lu'men* the space inside a tube, blood vessel, or hollow organ.

Luteinizing hormone *lu'te-in"īz-ing* an anterior pituitary hormone that aids maturation of cells in the ovary and triggers ovulation. Also causes the interstitial cells of the testis to produce testosterone.

Lymph *limf* the watery fluid in the lymph vessels collected from the tissue spaces.

Lymph node a mass of lymphatic tissue

Lymphatic system *lim-fat'ik* a system of lymphatic vessels, lymph nodes, and other lymphoid organs and tissues.

Lymphocytes *lim'fo-sītz* agranular white blood cells formed in the bone marrow that mature in the lymphoid tissue.

Lymphokines *lim'fo-kīnz* proteins involved in cells-mediated immune responses that enhance the immune and inflammatory responses.

Lysosomes *li'so-sōmz* organelles that originate from the Golgi apparatus and contain strong digestive enzymes.

Lysozyme *li'so-zīm* an enzyme found in sweat, saliva, and tears that is capable of destroying certain kinds of bacteria.

Macrophage *mak'ro-fāj"* a phagocytic cell particularly abundant in lymphatic and connective tissues; important as an antigen-presenter to T cells and B cells in the immune response.

Malignant *mah-lig'nant* life threatening; pertains to neoplasms that spread and lead to death, such as cancer.

Mammary glands *mam'mer-e* milk-producing glands of the breasts.

Mastication *mas"tĭ-ka'shun* the act of chewing.

Matter anything that occupies space and has mass.

Meatus *me-a'tus* the external opening of a canal.

Mechanical energy energy form directly involved in putting matter into motion.

Mechanoreceptors *mek"ah-no-rē-sep'terz* receptors sensitive to mechanical pressures such as touch, sound, or contractions.

Medial *me'de-al* toward the midline of the body.

Mediastinum *me"de-as-ti'num* the region of the thoracic cavity between the lungs.

Medulla *mĕ-dul'ah* the central portion of certain organs.

Meiosis *mi-o'sis* the two cell successive cell divisions in gamete formation producing nuclei with half the full number of chromosomes (haploid).

Melanin *mel'ah-nin* the dark pigment synthesized by melanocytes; responsible for skin color.

Melanocyte a cell that produces melanin.

Memory cell member of T cell and B cell clones that provides for immunologic memory.

Meninges *mĕ-nin'jēz* the membranes that cover the brain and spinal cord.

Meningitis *men"in-ji'tis* inflammation of the meninges of the brain or spinal cord.

Menopause *men'o-pawz* the physiological end of menstrual cycles.

Menses *men'sēz* monthly discharge of blood from the uterus. Menstruation.

Menstruation *men"stroo-a'shun* the periodic, cyclic discharge of blood, secretions, tissue, and mucus from the mature female uterus in the absence of pregnancy.

Mesentery *mes'en-ter"e* the double-layered membrane of the peritoneum that supports most organs in the abdominal cavity.

Metabolic rate *met"ah-bol'ik* the energy expended by the body per unit time.

Metabolism *mĕ-tab'o-lizm* the sum total of the chemical reactions that occur in the body.

Metabolize *mĕ-tab'o-līz* to transform substances into energy or materials the body can use or store by means of anabolism or catabolism.

Metacarpal *met"ah-kar'pal* one of the five bones of the palm of the hand.

Metastasis *mĕ-tas'tah-sis* the spread of cancer from one body part or organ into another not directly connected to it.

Metatarsal *met"ah-tar'sal* one of the five bones between the tarsus and the phalanges of the foot.

Microvilli *mi"kro-vil'i* the tiny projections on the free surfaces of some epithelial cells; increase surface area for absorption.

Micturition   *mik″tu-rish′un*   urination, or voiding; emptying the bladder.

Mineralocorticoid   *min″er-al-o-kor′tĭ-koid*   an adrenal cortical steroid hormone that regulates mineral metabolism and fluid balance.

Minerals   the inorganic chemical compounds found in nature; salts.

Mitochondria   *mi″to-kon′dre-ah*   the rodlike cytoplasmic organelles responsible for ATP generation for cellular activities.

Mitosis   *mi-to′sis*   the division of the cell nucleus; often followed by division of the cytoplasm of a cell.

Mixed nerves   nerves containing the processes of motor and sensory neurons; their impulses travel to and from the central nervous system.

Molecule   *mol′ĕ-kyool*   particle consisting of two or more atoms held together by chemical bonds.

Monoclonal antibodies   *mon″o-klōn′ul*   pure preparations of identical antibodies that exhibit specificity for a single antigen.

Monocyte   *mon′o-sīt*   large single-nucleus white blood cell; agranular leukocyte.

Monosaccharide   *mon″o-sak′ĭ-rīd*   literally, one sugar; the building block of carbohydrates; examples include glucose and fructose.

Mons pubis   *monz pu′bis*   the fatty eminence over the pubic symphysis in the female.

Motor unit   a motor neuron and all the muscle cells it supplies.

Mucous membrane (mucosa)   membrane that forms the linings of body cavities open to the exterior (digestive, respiratory, urinary, and reproductive tracts).

Mucus   *myoo′kus*   a sticky, thick fluid, secreted by mucous glands and mucous membranes, that keeps the free surface of membranes moist.

Multiple sclerosis   *skle-ro′sis*   a chronic condition characterized by destruction of the myelin sheaths of neurons in the spinal cord and the brain.

Muscle fibers   muscle cells.

Muscle spindle   encapsulated sensory receptor found in skeletal muscle that is sensitive to stretch.

Muscle tone   sustained partial contraction of a muscle in response to stretch receptor inputs; keeps the muscle healthy and ready to react.

Muscle twitch   a single rapid contraction of a muscle followed by relaxation.

Muscular dystrophy   *dis′tro-fe*   a progressive disorder marked by atrophy and stiffness of the muscles.

Muscular system   organ system consisting of skeletal muscles and their connective tissue attachments.

Myelin   *mi′ĕ-lin*   a white, fatty lipid substance.

Myelinated fibers   *mi′ĕ-lĭ-nāt″ed*   axons (projections of a nerve cell) covered with myelin.

Myocardial infarction   *mi″o-kar′de-al in-fark′shun*   a condition characterized by dead tissue areas in the myocardium caused by interruption of blood supply to the area.

Myocardium   *mi″o-kar′de-um*   the cardiac muscle layer of the heart wall.

Myofibrils   *mi″o-fi′brilz*   contractile organelles found in the cytoplasm of muscle cells.

Myofilament   *mi″o-fil′ah-ment*   filaments composing the myofibrils. Of two types: actin and myosin.

Myometrium   *mi″o-me′tri-um*   the thick uterine musculature.

Myopia   *mi-ō′pe-ah*   nearsightedness.

Myosin   *mi′o-sin*   one of the principal contractile proteins found in muscle.

Nares   *na′rēz*   the nostrils.

Necrosis   *nĕ-kro′sis*   the death or disintegration of a cell or tissues caused by disease or injury.

Negative feedback   feedback that causes the stimulus to decline or end.

Neoplasm   *ne′o-plazm*   an abnormal growth of cells. Sometimes cancerous.

Nephrons   *nef′ronz*   structural and functional units of the kidney.

Nerve   bundle of neuron processes (axons) outside the central nervous system.

Nerve fiber   axon of a neuron.

Neuroglia   *nu-rog′le-ah*   the nonneuronal tissue of the central nervous system that performs supportive and other functions; also called glia.

Neuromuscular junction   *nu″ro-mus′ku-lar*   the region where a motor neuron comes into close contact with a skeletal muscle cell.

Neurons   *nu′ronz*   cells of the nervous system specialized to transmit messages throughout the body.

Neurotransmitter   chemical released by neurons that may, upon binding to receptors of neurons or effector cells, stimulate or inhibit them.

Neutralization   *nu″tral-i-za′shun*   (1) a chemical reaction that occurs between an acid and a base; (2) blockage of the harmful effects of bacterial exotoxins or viruses by the binding of antibodies to their functional sites.

Neutron   *nu′tron*   uncharged subatomic particle; found in the atomic nucleus.

Neutrophils   *nu′tro-filz*   the most abundant of the white blood cells.

Nucleic acid   *nu-kle′ic*   class of organic molecules that includes DNA and RNA.

Nucleoli *nu-kle'o-li* small spherical bodies in the cell nucleus; function in ribosome synthesis.

Nucleotide *nu'kle-o-tīd* building block of nucleic acids.

Nucleus *nu'kle-us* a dense central body in most cells containing the genetic material of the cell.

Occipital *ok-sip'ĭ-tal* pertaining to area at the back of the head.

Occlusion *ŏ-kloo'zhun* closure or obstruction.

Olfaction *ol-fak'shun* smell.

Oogenesis *o"o-jen'ĕ-sis* the process of formation of the ova.

Ophthalmic *of-thal'mik* pertaining to the eye.

Optic *op'tik* pertaining to the eye.

Optic chiasma *op'tik ki-as'mah* the partial crossover of fibers of the optic nerves.

Oral relating to the mouth.

Organ a part of the body formed of two or more tissues that performs a specialized function.

Organ system a group of organs that work together to perform a vital body function; e.g., nervous system.

Organelles *or"gan-elz'* specialized structures in a cell that perform specific metabolic functions.

Organic compound a compound containing carbon; examples include proteins, carbohydrates, and fats.

Organism an individual living thing.

Origin attachment of a muscle that remains relatively fixed during muscular contraction.

Osmoreceptor *oz"mo-re-cep'tor* a structure sensitive to osmotic pressure or concentration of a solution.

Osmosis *oz-mo'sis* the diffusion of a solvent through a membrane from a dilute solution into a more concentrated one.

Ossicles *os'sĭ-kulz* the three bones of the middle ear: hammer, anvil, and stirrup.

Osteoblasts *os'te-o-blasts"* bone-forming cells.

Osteoclasts *os'te-o-klasts"* large cells that resorb or break down bone matrix.

Osteocyte *os'te-o-sīt"* a mature bone cell.

Osteon *os'te-on* a system of interconnecting canals in the microscopic structure of adult compact bone; unit of bone.

Osteoporosis *os"te-o-po-ro'sis* an increased softening of the bone resulting from a gradual decrease in rate of bone formation; a common condition in older people.

Otic *o'tik* pertaining to the ear.

Otitis media *o-ti'tis me'de-ah* middle-ear infection.

Otolith *o'to-lith* one of the small calcified masses in the utricle and saccule of the inner ear.

Ovarian cycle *o-va're-an* the monthly cycle of follicle development, ovulation, and corpus luteum formation in an ovary.

Ovary *o'var-e* the female sex organ in which ova (eggs) are produced.

Ovulation *ov"u-la'shun* the release of an ovum (or oocyte) from the ovary.

Ovum *o'vum* the female gamete (germ cell); an egg.

Oxidation *ok"sĭ-da'shun* the process of substances combining with oxygen or the removal of hydrogen.

Oxygen debt *ok'sĭ-jen* the volume of oxygen required after exercise to oxidize the lactic acid formed during exercise.

Oxyhemoglobin *ok"sĭ-he"mo-glo'bin* hemoglobin combined with oxygen.

Oxytocin *ok"sĭ-to'sin* hormone released by the posterior pituitary that stimulates contraction of the uterus during childbirth and ejection of milk during nursing.

Palate *pal'at* roof of the mouth.

Palpation *pal-pa'shun* examination by touch.

Pancreas *pan'kre-as* gland posterior to the stomach, between the spleen and the duodenum, produces both endocrine and exocrine secretions.

Pancreatic juice *pan"kre-at'ik* a secretion of the pancreas containing enzymes for digestion of all food categories.

Papilla *pah-pil'ah* small nipplelike projection.

Papillary muscles *pap'ĭ-ler"e* cone-shaped muscles found in the heart ventricles.

Paralysis *pah-ral'ĭ-sis* the loss of muscle function.

Paraplegia *par"ah-ple'je-ah* paralysis of the lower limbs.

Parasympathetic division *par"ah-sim"pah-thet'ik* a division of the autonomic nervous system; also referred to as the craniosacral division.

Parathyroid glands *par"ah-thi'roid* small endocrine glands located on the posterior aspect of the thyroid gland.

Parathyroid hormone (PTH) hormone released by the parathyroid glands that regulates blood calcium level.

Parietal *pah-ri'ĕ-tal* pertaining to the walls of a cavity.

Parotid *pah-rot'id* located near the ear.

Passive immunity short-lived immunity resulting from the introduction of "borrowed antibodies" obtained from an immune animal or human donor; immunological memory is not established.

Passive transport membrane transport processes that do not require cellular energy (ATP); e.g., diffusion, which is driven by kinetic energy.

Patella *pah-tel'ah* the kneecap.

Pathogen disease-causing microorganism (e.g., some bacteria, fungi, viruses, etc).

Pathogenesis *path"o-jen'ĕ-sis* the development of a disease.

Pectoral *pek'to-ral* pertaining to the chest.

Pectoral girdle see *shoulder girdle.*

Peduncle *pe-dung'kul* a stalk of fibers, especially that connecting the cerebellum to the pons, midbrain, and medulla oblongata.

Pelvic girdle incomplete bony basin formed by the two coxal bones that secures the lower limbs to the sacrum of the axial skeleton.

Pelvis *pel'vis* a basin-shaped structure; lower portion of the skeleton of the body trunk.

Penis *pe'nis* the male organ of copulation and urination.

Pepsin an enzyme capable of digesting proteins in an acid pH.

Pericardium *per"ĭ-kar'de-um* the membranous sac enveloping the heart.

Perimysium *per"ĭ-mis'e-um* the connective tissue enveloping bundles of muscle fibers.

Perineum *per"i-ne'um* that region of the body extending from the anus to the scrotum in males and from the anus to the vulva in females.

Periosteum *per"e-os'te-um* double-layered connective tissue that covers and nourishes the bone.

Peripheral nervous system (PNS) *pĕ-rif'er-al* a system of nerves that connects the outlying parts of the body with the central nervous system.

Peripheral resistance the resistance to blood flow offered by the systemic blood vessels; a measure of the amount of friction encountered by blood.

Peristalsis *per"ĭ-stal'sis* the waves of contraction seen in tubelike organs; propels substances along the tract.

Peritoneum *per"ĭ-to-ne'um* the serous membrane lining the interior of the abdominal cavity and covering the surfaces of the abdominal organs.

Peritonitis *per"ĭ-to-ni'tis* an inflammation of the peritoneum.

Permeability *per"me-ah-bil'ĭ-te* that property of membranes that permits passage of molecules and ions.

Peroneal *per"o-ne'al* pertaining to the lateral aspect of the leg.

pH the symbol for hydrogen ion concentration; a measure of the relative acidity or alkalinity of a solution.

Phagocyte *fag'o-sīt* a cell capable of engulfing and digesting particles or cells harmful to the body.

Phagocytosis *fag"o-si-to'sis* the ingestion of solid particles by cells.

Phalanges *fah-lan'jēz* the bones of the finger or toe.

Pharynx *far'inks* the muscular tube extending from the posterior of the nasal cavities to the esophagus.

Phospholipid *fos'fo-lip"id* a modified lipid containing phosphorus.

Photoreceptors *fo"to-re-sep'torz* specialized receptor cells that respond to light energy.

Physiology *fiz"e-ol'o-je* the science of the functioning of living organisms.

Pinocytosis *pi"no-si-to'sis* the engulfing of extracellular fluid by cells.

Pituitary gland *pĭ-tu'ĭ-tār"e* the neuroendocrine gland located beneath the brain that serves a variety of functions including regulation of the gonads, thyroid, adrenal cortex, water balance, and lactation.

Placenta *plah-sen'tah* the temporary organ that provides nutrients and oxygen to the developing fetus, carries away wastes, and produces the hormones of pregnancy.

Plantar *plan'tar* pertaining to the sole of the foot.

Plasma *plaz'mah* the fluid portion of the blood.

Plasma cell member of a B cell clone; specialized to produce and release antibodies.

Plasma membrane membrane that encloses cell contents; outer limiting membrane.

Platelet *plāt'let* one of the irregular cell fragments of blood; involved in clotting.

Pleura *ploor'ah* the serous membrane covering the lungs and lining the thoracic cavity.

Pleurisy *ploor'ĭ-se* inflammation of the pleurae, making breathing painful.

Plexus *plek'sus* a network of interlacing nerves, blood vessels, or lymphatics.

Plica *pli'kah* a fold.

Pneumothorax *nu"mo-tho'raks* the presence of air or gas in a pleural cavity.

Polar body a minute cell produced during meiosis in the ovary.

Polar molecules nonsymmetrical molecules that contain electrically unbalanced atoms.

Polarized *po'lar-īzd* the state of an unstimulated neuron or muscle cell in which the inside of the cell is relatively negative in comparison to the outside; the resting state.

Polycythemia *pol"e-si-the'me-ah* presence of an abnormally large number of erythrocytes in the blood.

Polypeptide *pol"e-pep'tīd* a chain of amino acids.

Polysaccharide *pol″e-sak′ĭ-rīd* literally, many sugars; a polymer of linked monosaccharides; examples include starch and glycogen.

Pons (1) any bridgelike structure or part; (2) the brain area connecting the medulla with the midbrain, providing linkage between upper and lower levels of the central nervous system.

Positive feedback feedback that tends to cause a variable to change in the same direction as the initial change; enhances the stimulus.

Postganglionic (postsynaptic) neuron *pōst″gang-gle-on′ik* a neuron of the autonomic nervous system having its cell body in a ganglion and its axon extending to an organ or tissue.

Precipitation formation of insoluble complexes that settle out of solution.

Preganglionic (presynaptic) neuron *pre″gang-gle-on′ik* a neuron of the autonomic nervous system having its cell body in the brain or spinal cord and its axon terminating in a ganglion.

Pressoreceptor *pres″o-re-sep′tor* a nerve ending in the wall of the carotid sinus and aortic arch sensitive to vessel stretching.

Pressure gradient difference in hydrostatic (fluid) pressure that drives filtration.

Primary (immune) response the initial response of the immune system to an antigen; involves clonal selection and establishes immunological memory.

Prime mover muscle whose contractions are primarily responsible for a particular movement; agonist.

Process (1) a prominence or projection; (2) a series of actions for a specific purpose.

Progesterone *pro-jes′tĕ-rōn* a hormone partly responsible for preparing the uterus for the fertilized ovum.

Pronation *pro-na′shun* the inward rotation of the forearm causing the radius to cross diagonally over the ulna—palms face posteriorly.

Prone refers to a body lying horizontally with the face downward.

Proprioceptor *pro″pre-oh-sep′tor* a receptor located in a muscle or tendon; concerned with locomotion, posture, and muscle tone.

Protein *pro′tēn* a complex nitrogenous substance; the main building material of cells.

Proteinuria *pro″te-in-u′re-ah* the passage of proteins in the urine.

Proton *pro′ton* subatomic particle that bears a positive charge; located in the atomic nucleus.

Proximal *prok′sĭ-mal* toward the attached end of a limb or the origin of a structure.

Puberty *pu′ber-te* the period at which reproductive organs become functional.

Pulmonary *pul′mo-ner″e* pertaining to the lungs.

Pulmonary circulation system of blood vessels that carry blood to and from the lungs for gas exchange.

Pulmonary edema *ĕ-de′mah* a leakage of fluid into the air sacs and tissue of the lungs.

Pulmonary ventilation breathing; consists of inspiration and expiration.

Pulse the rhythmic expansion and recoil of arteries resulting from heart contraction; can be felt from the outside of the body.

Pupil an opening in the center of the iris through which light enters the eye.

Purkinje fibers *pur-kin′je* the modified cardiac muscle fibers of the conduction system of the heart.

Pus the fluid product of inflammation composed of white blood cells, the debris of dead cells, and a thin fluid.

Pyelonephritis *pi″ĕ-lo-nĕ-fri′tis* an inflammation of the kidney pelvis and surrounding kidney tissues.

Pyloric region *pi-lor′ik* the final portion of the stomach; joins with the duodenum.

Pyramid any cone-shaped structure of an organ.

Pyrogen *pi′ro-jen* an agent or chemical substance that induces fever.

Quadriplegia *kwod″rĭ-ple′je-ah* the paralysis of all four limbs.

Radiant energy energy of the electromagnetic spectrum, which includes heat, light, ultraviolet waves, infrared waves, and other forms.

Radiate *ra′de-āt* diverge from a central point.

Radioactivity the process of spontaneous decay seen in some of the heavier isotopes, during which particles or energy is emitted from the atomic nucleus; results in the atom becoming more stable.

Radioisotope *ra″de-o-i′sĭ-tōp* isotope that exhibits radioactive behavior.

Ramus *ra′mus* a branch of a nerve, artery, vein, or bone.

Receptor *re-sep′tor* (1) a peripheral nerve ending specialized for response to particular types of stimuli; (2) molecule that binds specifically with other molecules, e.g., hormones and neurotransmitters.

Reduction restoring broken bone ends (or a dislocated bone) to its original position.

Reflex automatic reaction to a stimulus.

Refract bend; usually refers to light.

Refractory period *re-frak′to-re* the period of unresponsiveness to threshold stimulation.

Renal *re′nal* pertaining to the kidney.

Renal calculus *re′nal kal′ku-lus* a kidney stone.

Renin *re'nin* a substance released by the kidneys that is involved with raising blood pressure.

Reticulum *rĕ-tik'u-lum* a fine network.

Retina *ret'ĭ-nah* light sensitive layer of the eye; contains rods and cones.

Ribonucleic acid (RNA) *ri"bo-nu-kle'ic* the nucleic acid that contains ribose; acts in protein synthesis.

Ribosomes *ri'bo-sōmz* cytoplasmic organelles at which proteins are synthesized.

Rods one of the two types of photosensitive cells in the retina.

Rotate to turn about an axis.

Rugae *roo'je* elevations or ridges, as in the mucosa of the stomach.

Sacral *sa'kral* the lower portion of the back, just superior to the buttocks.

Sagittal plane *saj'ĭ-tal* a longitudinal plane that divides the body or any of its parts into right and left portions.

Saliva *sah-li'vah* the secretion of salivary glands ducted into the mouth.

Salt ionic compound that dissociates into charged particles (other than hydrogen or hydroxyl ions) when dissolved in water.

Sclera *skle'rah* the firm white fibrous outer layer of the eyeball; protects and maintains eyeball shape.

Scrotum *skro'tum* the external sac enclosing the testes.

Sebaceous glands *se-ba'shus* glands that empty their sebum secretion into hair follicles.

Sebum *se'bum* the oily secretion of sebaceous glands.

Second messenger intracellular molecule generated by binding of a chemical to a membrane receptor; mediates intracellular responses.

Secondary (immune) response second and subsequent responses of the immune system to a previously met antigen; more rapid and more vigorous than the primary response.

Secondary sex characteristics anatomical features that develop under influence of sex hormones (male or female pattern of muscle development, bone growth, body hair, etc.) that are not directly involved in the reproductive process.

Secretion *se-kre'shun* (1) the passage of material formed by a cell to its exterior; (2) cell product that is transported to the cell exterior.

Semen *se'men* fluid mixture produced by male reproductive structures; contains sperm, nutrients, and mucus.

Semilunar valves *sem"e-lu'nar* valves that prevent blood return to the ventricles after contraction.

Seminiferous tubules *se"mĭ-nif'er-us* highly convoluted tubes within the testes that form sperm.

Sensory nerve a nerve that contains processes of sensory neurons and carries nerve impulses to the central nervous system.

Sensory nerve cell an initiator of nerve impulses following receptor stimulation.

Serous fluid *ser'us* a clear, watery fluid secreted by the cells of a serous membrane.

Serous membrane membrane that lines a cavity without an opening to the outside of the body (except for joint cavities); serosa.

Sex chromosomes chromosomes that determine genetic sex; the X and Y chromosomes.

Shoulder girdle composite of two bones, scapula and clavicle, that attach the upper limb to the axial skeleton; also called the pectoral girdle.

Sinoatrial node *si"no-a'tre-al* the mass of specialized myocardial cells in the wall of the right atrium; pacemaker of the heart.

Sinus *si'nus* (1) a mucous membrane–lined, air-filled cavity in certain cranial bones; (2) a dilated channel for passage of blood or lymph.

Skull bony enclosure for the brain.

Smooth muscle muscle consisting of spindle-shaped, unstriped (nonstriated) muscle cells; involuntary muscle.

Solute *sol'yoot* the dissolved substance in a solution.

Solution a homogenous mixture of two or more components.

Somatic nervous system *so-mat'ik* a division of the peripheral nervous system; also called the voluntary nervous system.

Sperm mature male germ cell.

Spermatogenesis *sper"mah-to-jen'ĕ-sis* the process of sperm production in the male; involves meiosis.

Sphincter *sfink'ter* a circular muscle surrounding an opening; acts as a valve.

Spinal nerves the 31 pairs of nerves that arise from the spinal cord.

Squamous *skwa'mus* (1) flat, scalelike; (2) pertaining to flat, thin cells that form the free surface of some epithelial tissues.

Stasis *sta'sis* (1) a decrease or stoppage of flow; (2) a state of nonchange.

Static equilibrium *stat'ik e"kwĭ-lib're-um* balance concerned with changes in the position of the head.

Stenosis *stĕ-no'sis* abnormal constriction or narrowing.

Steroids *stĕ'roidz* a specific group of chemical substances including certain hormones and cholesterol.

Stimulus *stim'u-lus* an excitant or irritant; a change in the environment producing a response.

Stratum *stra'tum* a layer.

Stressor any stimulus that directly or indirectly causes the hypothalamus to initiate stress-reducing responses, such as the fight-or-flight response.

Striated muscle *stri'āt-ed* muscle consisting of cross-striated (cross-striped) muscle fibers; includes cardiac and skeletal muscle.

Stroke a condition in which brain tissue is deprived of a blood supply, as in blockage of a cerebral blood vessel.

Stroke volume a volume of blood ejected by a ventricle during systole.

Subcutaneous *sub″ku-ta'ne-us* beneath the skin.

Sublingual *sub-ling'gwal* beneath the tongue.

Sudoriferous glands *su″do-rif'er-us* see *sweat glands.*

Sulcus *sul'kus* a furrow on the brain, less deep than a fissure.

Summation *sum-ma'shun* the accumulation of effects, especially those of muscular, sensory, or mental stimuli.

Superficial (external) located close to or on the body surface.

Superior refers to the head or upper body regions.

Supination *su″pĭ-na'shun* the outward rotation of the forearm causing palms to face anteriorly.

Supine refers to a body lying with the face upward.

Suppressor T cells regulatory T lymphocytes that suppress the immune response.

Surfactant *sur-fak'tant* a chemical substance coating the pulmonary alveoli walls that reduces surface tension, thus preventing collapse of the alveoli after expiration.

Suspensory ligament of an eye *sus-pen'so-re* fibrous ligament that holds the lens in place in the eye.

Sutures *soo'churz* immovable fibrous joints that connect the bones of the adult skull.

Sweat glands the glands that produce a saline solution called sweat; also called sudoriferous glands.

Sympathetic division a division of the autonomic nervous system; opposes parasympathetic functions; called the fight-or-flight division.

Synapse *sin'aps* the region of communication between neurons.

Synaptic cleft *sĭ-nap'tik* the fluid-filled space at a synapse between neurons.

Synarthrosis *sin″ar-thro'sis* an immovable joint.

Synergists *sin'er-jists* muscles cooperating with another to produce a desired movement.

Synovial fluid *sĭ-no've-al* a fluid secreted by the synovial membrane; lubricates joint surfaces and nourishes articular cartilages.

Synovial joint freely movable joint exhibiting a joint cavity enclosed by an articular (fibrous) capsule lined with synovial membrane.

Synovial membrane membrane that lines the capsule of a synovial joint.

Synthesis reaction chemical reaction in which larger molecules are formed from simpler ones.

System a group of organs that function cooperatively to accomplish a common purpose; there are eleven major systems in the human body.

Systemic *sis-tem'ik* general; pertaining to the whole body.

Systemic circulation system of blood vessels that carries nutrient and oxygen-rich blood to all body organs.

Systemic edema *ĕ-de'mah* an accumulation of fluid in body organs or tissues.

Systole *sis'to-le* the contraction phase of heart activity.

Systolic pressure *sis-tol'ik* the pressure generated by the left ventricle during systole.

T cells lymphocytes that mediate cellular immunity; include helper, killer, suppressor, and memory cells. Also called T lymphocytes.

Tachycardia *tak″e-kar'de-ah* an abnormal, excessively rapid heart rate; over 100 beats per minute.

Tarsal *tahr'sal* one of the seven bones that form the ankle and heel.

Taste buds receptors for taste on the tongue, roof of mouth, pharynx, and larynx.

Tendons *ten'dunz* cords of dense fibrous tissue attaching a muscle to a bone.

Testis *tes'tis* the male primary sex organ that produces sperm.

Tetanus *tet'ah-nus* (1) the tense, contracted state of a muscle; (2) an infectious disease.

Thalamus *tha'luh-mis* a mass of gray matter in the diencephalon of the brain.

Thermoreceptor *ther″mo-re-sep'ter* a receptor sensitive to temperature changes.

Thoracic *tho-ras'ik* refers to the chest.

Thorax *tho'raks* that portion of the body trunk above the diaphragm and below the neck.

Threshold stimulus the weakest stimulus capable of producing a response in an irritable tissue.

Thrombin *throm'bin* an enzyme that induces clotting by converting fibrinogen to fibrin.

Thrombophlebitis *throm″bo-fle-bi'tis* an inflammation of a vein associated with blood clot formation.

Thrombus *throm'bus* a fixed clot that develops and persists in an unbroken blood vessel.

Thymus gland *thi'mus* an endocrine gland active in the immune response.

Thyroid gland *thi'roid* one of the largest of the body's endocrine glands; straddles the anterior trachea.

Tidal volume amount of air inhaled or exhaled with a normal breath.

Tissue a group of similar cells specialized to perform a specific function; primary tissue types are epithelial, connective, muscle, and nervous tissues.

Trachea *tra'ke-ah* The windpipe; the respiratory tube extending from larynx to bronchi.

Tract a collection of nerve fibers in the CNS having the same origin, termination, and function.

Transverse plane plane that divides the body or its parts into superior and inferior portions; also called a cross section.

Trauma *traw'mah* an injury, wound, or shock; usually produced by external forces.

Triglycerides *tri-glis'er-īdz* compounds composed of fatty acids and glycerol; fats and oils; also called neutral fats.

Trochanter *tro-kan'ter* a large, somewhat blunt process.

Tropic hormone *trōp'ik* a hormone that regulates the function of another endocrine organ.

Tubercle *tu'ber-kul* a nodule or small rounded process.

Tuberosity *tu"bĕ-ros'ĭ-te* a broad process, larger than a tubercle.

Tunica *too'nĭ-kah* a covering or tissue coat; layer.

Tympanic membrane *tim-pan'ik* the eardrum.

Ulcer *ul'ser* a lesion or erosion of the mucous membrane, such as gastric ulcer of stomach.

Umbilical cord *um-bĭ'lĭ-kul* a structure bearing arteries and veins connecting the placenta and the fetus.

Umbilicus *um-bĭ'lĭ-kus* navel; marks site of the umbilical cord in the fetal stage.

Urea *u-re'ah* the main nitrogen-containing waste excreted in the urine.

Ureters *u-re'terz* tubes that carry urine from kidney to bladder.

Urethra *u-re'thrah* the canal through which urine passes from the bladder to the outside of the body.

Uterine tube the oviduct. The tube through which the ovum is transported to the uterus; also called fallopian tube.

Uvula *u'vu-lah* tissue tag hanging from soft palate.

Valence shell *va'lens* the outermost energy level of an atom that contains electrons; the electrons in the valence shell determine the bonding behavior of the atom.

Vas *vas* a duct; vessel.

Vascular *vas'ku-lar* pertaining to blood channels or vessels.

Vasoconstriction *vas"o-kon-strik'shun* narrowing of blood vessels.

Vasodilation *vas"o-di-la'shun* relaxation of the smooth muscles of blood vessels producing dilation.

Vasomotor nerve fibers *vas-o-mo'tor* the nerve fibers that regulate the constriction or dilation of blood vessels.

Vein *vān* a vessel carrying blood away from the tissues toward the heart.

Ventral anterior or front.

Ventricles *ven'trĭ-kulz* (1) discharging chambers of the heart; (2) cavities within the brain.

Venule *ven'ūl* a small vein.

Vertebral column the spine, formed of a number of individual bones called vertebrae and two composite bones (sacrum and coccyx).

Vertigo *ver'tĭ-go* dizziness; the feeling of movement such as a sensation that the external environment is revolving.

Vesicular follicle a mature ovarian follicle.

Villi *vil'i* fingerlike projections of the small intestinal mucosa that tremendously increase its surface area for absorption.

Viscera *vis'era* the internal organs.

Visceral *vis'er-al* pertaining to the internal part of a structure or the internal organs.

Viscosity *vis-kos'ĭ-te* the state of being sticky or thick.

Visual acuity *ah-ku'ĭ-te* the ability of the eye to distinguish detail.

Vital capacity the volume of air that can be expelled from the lungs by forcible expiration after the deepest inspiration; total exchangeable air.

Vitamins organic compounds required by the body in minute amounts for physiological maintenance and growth.

Voluntary muscle muscle under control of the will; skeletal muscle.

Vulva *vul'va* female external genitalia.

White matter white substance of the central nervous system; the myelinated nerve fibers.

Zygote *zi'gōt* the fertilized ovum; produced by union of two gametes.

Index